AF801633

International Association of Geodesy Symposia

Fernando Sansò, Series Editor

International Association of Geodesy Symposia

Fernando Sansò, Series Editor

Symposium 101: Global and Regional Geodynamics
Symposium 102: Global Positioning System: An Overview
Symposium 103: Gravity, Gradiometry, and Gravimetry
Symposium 104: Sea SurfaceTopography and the Geoid
Symposium 105: Earth Rotation and Coordinate Reference Frames
Symposium 106: Determination of the Geoid: Present and Future
Symposium 107: Kinematic Systems in Geodesy, Surveying, and Remote Sensing
Symposium 108: Application of Geodesy to Engineering
Symposium 109: Permanent Satellite Tracking Networks for Geodesy and Geodynamics
Symposium 110: From Mars to Greenland: Charting Gravity with Space and Airborne Instruments
Symposium 111: Recent Geodetic and Gravimetric Research in Latin America
Symposium 112: Geodesy and Physics of the Earth: Geodetic Contributions to Geodynamics
Symposium 113: Gravity and Geoid
Symposium 114: Geodetic Theory Today
Symposium 115: GPS Trends in Precise Terrestrial, Airborne, and Spaceborne Applications
Symposium 116: Global Gravity Field and Ist Temporal Variations
Symposium 117: Gravity, Geoid and Marine Geodesy
Symposium 118: Advances in Positioning and Reference Frames
Symposium 119: Geodesy on the Move
Symposium 120: Towards an Integrated Global Geodetic Observation System (IGGOS)
Symposium 121: Geodesy Beyond 2000: The Challenges of the First Decade
Symposium 122: IV Hotine-Marussi Symposium on Mathematical Geodesy
Symposium 123: Gravity, Geoid and Geodynamics 2000
Symposium 124: Vertical Reference Systems
Symposium 125: Vistas for Geodesy in the New Millennium
Symposium 126: Satellite Altimetry for Geodesy, Geophysics and Oceanography
Symposium 127: V Hotine Marussi Symposium on Mathematical Geodesy
Symposium 128: A Window on the Future of Geodesy
Symposium 129: Gravity, Geoid and Space Missions
Symposium 131: Geodetic Deformation Monitoring: From Geophysical to Engineering Roles

Geodetic Deformation Monitoring: From Geophysical to Engineering Roles

IAG Symposium
Jaén, Spain
March 17-19, 2005

Edited by
Fernando Sansò and Antonio J. Gil

Springer

Volume Editors

Professor Fernando Sansò
Polytechnic of Milan
D.I.I.A.R. – Surveying Section
Piazza Leonardo da Vinci, 32
20133 Milan
Italy

Professor Antonio J. Gil
University of Jaén
Department of Cartografic, Geodetic
and Photogrammetric Engineering
Campus Las Lagunillas, Edificio A3
23071 Jaén
Spain

Series Editor

Professor Fernando Sansò
Polytechnic of Milan
D.I.I.A.R. – Surveying Section
Piazza Leonardo da Vinci, 32
20133 Milan
Italy

Library of Congress Control Number: 2006931677

ISSN 0939-9585
ISBN-10 3-540-38595-9 Springer Berlin Heidelberg New York
ISBN-13 3-540-38595-0 Springer Berlin Heidelberg New York

This work is subject to copyright. All rights are reserved, whether the whole or part of the material is concerned, specifically the rights of translation, reprinting, reuse of illustrations, recitation, broadcasting, reproduction on microfilm or in any other way, and storage in data banks. Duplication of this publication or parts thereof is permitted only under the provisions of the German Copyright Law of September 9, 1965, in its current version, and permission for use must always be obtained from Springer-Verlag. Violations are liable to prosecution under the German Copyright Law.

Springer is a part of Springer Science+Business Media
springer.com
© Springer-Verlag Berlin Heidelberg 2006
Printed in Germany

The use of general descriptive names, registered names, trademarks, etc. in this publication does not imply, even in the absence of a specific statement, that such names are exempt from the relevant protective laws and regulations and therefore free for general use.

Cover design: WMXDesign GmbH, Heidelberg
Production: Almas Schimmel
Typesetting: Camera ready by the authors

Printed on acid-free paper 32/3141/as 5 4 3 2 1 0

Foreword

According to a widely accepted definition, Geodesy is the science dealing with the determination of the position of points in space, the shape and gravity filed of the earth and with their time variations.

A consequence is that geodesists feel as a permanent subject of research, the detection, analysis and interpretation of spatial deformation as well as gravity field variation. Deformation in fact is what remains of time variation when reference systems effects are eliminated; as such, deformation is a concept much closer to the dynamics of bodies, included the body of the earth, than simple change of vector components, either for positions or for the gravity field. In addition all problems of temporal variations can be put on quite different spatial scales so that one and the same concept can be applied to a number of different physical contexts.

Also time scales of these phenomena can be very different from one another ranging from centuries down to hours; here in any case we shall consider only slow phenomena such that they could be described as a transition from one quasi-static state to another, very much as it is done in equilibrium thermodynamics. For instance the determination of the time history of the polyhedron of permanent satellite stations, spread all over the world, the same used to define the global terrestrial reference frame, has as geophysical background the plate motion of the crust taking place on a global scale (for the earth) and on time scale of the order of decades.

To the opposite side we can put the monitoring of structural deformation, e.g. of a dam or of a bridge, where the spatial scale is 0,1 to 1 kilometre and the time scale ranges between hours and days.

Also the gravity field variations can have a wide spectrum of spatial and temporal scales ranging from global changes due to a very large and intense displacement of masses in the interior of the earth, down to a continental scale affected, for instance, by a post glacial rebound, to a very local level where even a change in the depth of the water table is observed in surface gravity measurements. All these phenomena and the observations performed to detect them are object of nowadays research and results of recent advances for a large number of instances have been discussed during the IAG Symposium on March 2005 in Jaen (Spain).

One cannot say that final "conclusions" have been drawn, because this is probably not possible but a good overview of theoretical matters, models and results stemming from specific applications has been performed and presented in these proceedings thus allowing to define our today level of understanding such a field in the geodetic community.

Fernando Sansò

Report on the Symposium

The International Symposium on Geodetic Deformation Monitoring has been organized by the Geodesy Research Group of the Jaén University (Spain).

The International Scientific Committee has been composed by Fernando Sansò (Italy), Athanasios Dermanis (Greece), Ewald Brückl (Austria), Heinz G. Henneberg (Venezuela) and Antonio J. Gil (Spain).

The members of the Local Organizing Committee have been: Antonio J. Gil, Mª Jesús Borque, Mª Selmira Garrido, Clara de Lacy, Mª Isabel Ramos, Antonio M. Ruiz and Gracia Rodríguez-Caderot.

The main topics covered in the symposium have been: mathematical and statistical models for crustal deformation analysis, deformation monitoring from GPS and InSAR data: analysis and geophysical interpretation, geodetic monitoring of movements in civil engineering, integration of spatial and terrestrial techniques in deformation studies, geodynamical applications of gravimetric observations and present-day geodetic instrumentation for deformation monitoring, which were got into four sessions: SESSION A: General Methods for Deformation Analysis, SESSION B: Special Measurements Tools, Data Integration and Specific Models, SESSION C: Geodynamical Applications, Crustal Deformation and SESSION D: Local and Engineering Applications.

About eighty people from more than twenty countries have attended the symposium. Fifty-five papers have been presented in the oral and poster sessions. The most interesting papers have been accepted for publication in these ISGDM Proceedings after a revision process by the following reviewers: Z. Altamimi, D. Arabelos, R. Barzaghi, L. Bastos, B. Benciolini, B. Betti, L. Biagi, K. Borre, C. Braitenberg, M. Brovelli, E. Brückl, F. Brunner, J. Catalao, M. Crespi, M. Crosetto, A. Dermanis, M. Doufexopoulou, R. Eyers, A. Fotiou, J. Gárate, A.J. Gil, E.W. Grafarend, R. Hanssen, B. Hofmann-Wellenhof, U. Hugentobler, K. Jacobsen, B. Kearsley, C. Kotsakis, C. Lacy, M. Llubes, A. Manzino, G. Manzoni, G.F. Panza, J.A. Peláez, F. Riguzzi, G. Rodríguez-Caderot, A.M. Ruiz, F. Sacerdote, F. Sansò, I.N. Tziavos, G. Vassena, M. Vermeer and A. Wieser.

The symposium has been sponsored by the International Association of Geodesy, Ministerio de Educación y Ciencia (Spain), Leica Geosystems, Trimble, Department of Cartographic, Geodetic and Photogrammetric Engineering, Jaén University, Eurolingua Foundation, Provincial Council of Jaén, Caja de Granada and Colegio Oficial de Ingenieros Técnicos en Topografía (Andalusia, Spain).

Finally, as President of the Local Organizing Committee I would like to express my gratitude to the authorities of the Jaén University for all institutional support to organize this event; to the members of the committees and reviewers for their work really well done and to the attendees for coming to Jaén. I hope they all have found this symposium fascinating.

Antonio J. Gil

Contents

Invited Opening Address

H. G. Henneberg

Recent Crustal Movements, Geodetic Tasks, History, Present and Future

H.G. Henneberg
University of Zulia. Venezuela.

Distinguished Guests, Dear Colleagues, Ladies and Gentlemen,

In 1956 Prof. Angelos Galanopoulos, from Athens, informed about his visit to the remains of Santorin. Santorin Island exploded 1400 years before our time scale; it was the biggest explosion, destruction and movement reported of the Earth's crust. Galanopoulos estimated that this event was equivalent to the explosion of hundreds of hydrogen bombs together and it could be that the Diluvian Epoch written in the Bible references to that event.

The explosion of Krakatoa 1883 formed a cloud of dust into the sky of app. 50 km height above sea level, and rock parts of the crust were thrown as far as 80 km. The tsunamis destroyed 295 villages with a total of 36000 casualties. Ships were thrown up to 3 km land inwards from the coast and the sound was heard up to 3200 km. The cloud could be seen for several years around the globe.

It was during the General Assembly of the IAG in Berkeley 1963 when the Crustal Movement problem entered IAG in form of a Special Commission. It was Charles Whitten (USA) who reported the destructive event of the large Californian Earthquake with offsets of the San Andreas Fault of up to several meters. It was Prof. Kukkameki from Finland who described the slow but steady uplift of the Northern part of Europe. These were demonstrations of two types of changes of the Earth's crust: a sudden movement with great effect and destruction and another very slow and smooth deformation with also small rotations present.

Very recently, just some months ago, we had also one of the biggest destructions caused by a Tsunami which was originated by a sudden fracture of the crust (Marmot – Sumatra – Indonesia December 27th, 2004). In addition we have also an independent report that the Maldives Islands – South of India – are sinking very slowly into the sea, with estimates that in about 50 years these Islands will become inhabitable. This last mentioned movement should alert the NN.UU. to look for a timely adequate solution to relocate the inhabitants involved. Again we see a very sudden destructive event and a very slow movement in quite nearby regions.

Colleagues, let us have present also the real extended earthslides of Huascasan in Peru 1962 and 1970. In 1970 the volume of the earth and rock masses came up to 100 Millions of m^3 of rocks and earth and the mass velocity went up to 120 km/h, causing 18000 fatalities. Of this devastated area a very minute photogrammetric document was realized, which includes also all measures to rebuild on secure places the city for the people who were saved. And here I have to mention also another very big disaster, that of Vaiont in Italy with the disappearance of Longarone village where former inhabitants could never be found. This is another example of devastating Earth surface movements in recent time and there were others.

In 1999 the IAG reformed the Organization and did not follow with the RCM Commission. In spite of that fact, our Finnish Colleagues organized a Crustal Movement Symposium in 2001, which took place in Helsinki and turned out to be a real big success which encouraged to go on with our research on to the basic scientific level of that field and also to extend it.

We see today that other Institutions are following this important research area. It was demonstrated that our C.M. studies are a permanent process, pushed ahead by our International Geodetic Community. We have seen lately, that we have to deal in our geodetic field with always new facts to consider.

When we compare the real big events which have taken place in changing the Earth's crust by large, slow and sudden processes we have to consider also very seriously the two thirds of the Earth's Crust which lay UNDER the Oceans. There is a real big future field of geodetic applications to crustal changes. Submarine vessels as Alwyn, Mir I, Mir II and others already operating, can be working down to 6 km. The sound channel in 1200 m depth allows high precision horizontal geodetic measurements.

A geodetic network at that depth would be a permanent frame of reference which guarantees the precision of any changes of the crust in a large extension. At any rate it would be a big step forward to come to real scientific research due to the accuracy guaranteed by repetition measurements at any time and always under the same circumstances and conditions.

I would like to mention here that this proposition to establish stations on the seafloor with forced centering devices was already presented in the past during the "Marine Geodesy Symposia".

Now, let us come to the Calypso project. It is a prototype of a Plate Boundary Observatory that was installed after the "Lava Dome Collapse" on Montserrat Volcano, Caribbean Sea. This Seismic-Geodetic Observatory make permanent GPS measurements, deformation research, sea level control and others.

A real new challenge of determination of crustal changes in the deep sea was shown recently by the results of a report of the first exploration on Gakkle Ridge. G.R. is an undersea mountain chain which is located in the North Pole Region in direction to the coast of Siberia.

The research ship "Nordstern", Germany, and another one discovered in about 5 – 6 km depth a chain of active volcanoes. That report informs that there are also locations on the sea floor where the Earth crust does not exist. The water column lies directly over the Earth mantle and that is really something new to consider.

May I still mention the actually changing of the Ice Household of the Earth? Mean sea level is rising, glaciers are melting, the changing of the Earth crust arte quite everywhere; the gravity field is affected and many other things can still be discussed. And this we will do here during these days and on other occasions in the near future.

Looking at the presentations of this Symposium, we find a real wide diversity which shows that the movements, the deformations and the changes of the crust are a permanent process. It is to be mentioned that the human activity is also influencing this process by construction activities, big dams, water accumulation by artificial lakes, permanent amplifying of cities, augmenting their load onto the crust, producing so-called settlements, which influences also the gravity field. The inclusion of Sea Floor Areas, referred to before, by means of precision geodetic research, will be essential for future interpretation of crustal changes.

Thank you for your attention.

Session A

General Methods
for Deformation Analysis

Chairmen: F. Sansò and A. Dermanis

Estimating Crustal Deformation Parameters from Geodetic Data: Review of Existing Methodologies, Open Problems and New Challenges

Athanasios Dermanis and Christopher Kotsakis
Department of Geodesy and Surveying
The Aristotle University of Thessaloniki
University Box 390, GR-54124, Thessaloniki, Greece

Abstract. The study of crustal deformation using various types of geodetic data is a research topic whose practical importance needs hardly to be stressed, and its theoretical richness encompasses several scientific disciplines, including estimation theory, differential geometry, elasticity theory, geodynamics and physics. In this paper, an attempt is made to summarize the existing methodologies that are commonly applied in the geodetic practice for crustal deformation studies. Special emphasis is given on issues such as: (i) the definition and the estimability of frame-invariant quantities in time-dependent geodetic networks, (ii) the separation of rigid motion effects from actual body deformation changes, (iii) the problem of spatial and/or temporal interpolation of the crustal deformation field, and (iv) the separation of the total deformation field into a "horizontal" part and a "vertical" part. An assessment of the remaining open problems that exist within the currently used geodetic methodologies for crustal deformation analysis is also given, and finally a number of new challenges that are imposed by the availability of data types which are essentially continuous in space and/or time, is listed.

Keywords. Crustal deformation, estimation, interpolation, collocation, GPS, SAR

1 Introduction

The assessment of reliable information concerning crustal motion is based on the fulfillment of criteria concerning a satisfactory signal-to-noise ratio and sufficient spatial and temporal resolution.

Resolution is directly related to the nature of the studied deformation, which may range from the slow and spatially smooth plate motion (e.g. Soler 1977) to the spatially more complex local deformation patterns (e.g. Dermanis et al. 1981, Kogan et al. 2000), or to temporally more abrupt landslide-related deformations (e.g. Prescott 1981).

Signal-to-noise ratio, after a long period of unsatisfactory performance which required long monitoring periods to detect persistent secular deformations, has finally met the criteria for the detection of deformations of smaller magnitude within shorter time periods. Since this growth in precision comes essentially from GPS observations, some quality assessment questions remain open, concerning the difference between realistic and nominal accuracy measure and the danger of interpreting as deformation other systematic influences on the relevant data (Davis et al. 1989).

The main breakthrough has taken place in the increase of temporal resolution to at least daily values, which exceeds far beyond the resolution required for the detection of steady crustal deformations. Apart from its role in detecting abrupt deformations, related e.g. to seismic events, the high temporal resolution provides a tool for hopefully resolving the quality assessment problem.

With respect to spatial resolution, the cost of maintaining a dense network of permanent GPS stations or of often repeated GPS surveys, turns our interest to different sources of densification data, such as laser scanning and SAR interferometry (Crippa et al. 2002, Hanssen 2005, Wright et al. 2004, Lohman and Simons 2005).

The advancement in observational performance, typically resolves some theoretical problems and on the other hand poses some new challenges, as more elaborate models and data analysis techniques become necessary.

A critical problem that remains beyond the reach of observational advancement is the obligatory restriction of the data on the two-dimensional physical surface of the earth while crustal deformation is by nature a three-dimensional physical process. Only a parallel separation of geophysical deformation models into a horizontal and vertical part may overcome this problem to a certain extend.

While spatial or temporal resolution remains beyond the point when it may be operationally considered as a continuous observational process, interpolation is the most crucial issue in data analysis methodologies. Such interpolation may be either direct or "masked" under a stochastic prediction model, where smooth deformation is modeled by a stochastic process (Dermanis 1976, 1988). In this case, correlation is a measure of similarity between nearby displacements which secure a smooth deformation field, or in fact a piecewise smooth field between discontinuities which are either spatial (faults) or temporal (seismic events).

A usual prerequisite for a successful interpolation using the stochastic tool of prediction, is trend removal, so that the reduced field can be indeed modeled by a zero-mean stochastic process. This trend removal is based partly on a redefinition of the dynamic behavior of the reference system. The original system must be replaced by an optimal one, which absorbs, as much as possible, from the apparent displacements reflecting an inappropriate reference system definition and not crustal deformation itself. One may argue that proper deformation parameters are invariant under changes of the reference system (Dermanis 1985, Xu et al. 2000), but this is true once continuous information is available. The interpolation that provides such continuous information may well be system dependent, thus leading to different values for the theoretically invariant deformation parameters.

We shall present here a whole armory of possible interpolation approaches for various possible data availability situations with the warning that the proper choice of data analysis methodology is a critical issue, which calls for a deep understanding of the "geophysical" part of the studied phenomena and necessitates a dialectic relation between geodesy and geophysics. A crucial part in this dialogue is that the final product resulting from the analysis of geodetic data, is accompanied by an, as realistic as possible, description of its overall quality and the assumptions involved in its derivation.

2 Deformation parameters in relation to the type of available data

Deformation parameters are one of the three elements involved in models of the "mechanics of continuous media", the other two being the acting forces and the parameters describing the behavior of the deforming material (Sokolnikoff 1956, Sansò 1982). As is the usual case in physics, the relevant models are differential equations describing the local aspects of the problem at each point of the deforming body. These are the well known "constitutional equations", also known with the less formal term "stress-strain relations". The geodetic input in this type of analysis is the deformation function, relating the coordinates of any material point at any epoch with its identifying coordinates, which may be its coordinates at a particular reference epoch (Lagrangean approach) or its present coordinates (Eulerian approach), or even its coordinates in a reference "undeformed" state that is not realized at any particular epoch.

If

$$\mathbf{x} = \psi(P,t) = \psi_t(P) \tag{1}$$

denotes the Cartesian coordinates of point P at epoch t, we may use the coordinates $\mathbf{x}_0 = \psi(P,t_0) = \psi_{t_0}(P)$ as independent parameters and use the point identifying relation $P = \psi_{t_0}^{-1}(\mathbf{x}_0)$ in order to describe the coordinate at any point and epoch by a deformation function

$$\mathbf{x} = \mathbf{f}(\mathbf{x}_0,t) \equiv \psi(\psi_{t_0}^{-1}(\mathbf{x}_0),t) \tag{2}$$

In the constitutional relations enters not the deformation function itself but its local linear approximation by the corresponding "tangent mapping" described by the "deformation gradient" matrix

$$\mathbf{F}(\mathbf{x}_0,t) = \frac{\partial \mathbf{x}}{\partial \mathbf{x}_0}(\mathbf{x}_0,t) \tag{3}$$

Of course the particular value of the gradient matrix depends on the chosen instantaneous reference system at the epochs t_0 and t, and since t is an independent parameter, on the choice of the "dynamic" reference system $(O(t),\vec{\mathbf{e}}(t))$ which is a temporally smooth choice of reference systems for each particular epoch having $O(t)$ as the point of origin and $\vec{\mathbf{e}}(t) = [\vec{e}_1(t)\vec{e}_2(t)\vec{e}_3(t)]$ as the orthonormal vector basis.

A frame invariant description involves either numerical or physical invariants which are functions of the elements of the deformation gradient $\mathbf{F}$ (Boucher 1980, Sansò 1982, Dermanis 1981, 1985). The numerical invariants are eventually functions of

the singular values λ_1, λ_2, λ_3 of $\mathbf{F}$ which are roots of the common eigenvalues of either the right Cauchy strain tensor matrix $\mathbf{C}=\mathbf{F}^T\mathbf{F}=\mathbf{U}^2$ or the left Cauchy strain tensor matrix $\mathbf{B}=\mathbf{F}\mathbf{F}^T=\mathbf{V}^2$. Here $\mathbf{U}=\mathbf{C}^{1/2}$ is the right stretch tensor matrix and $\mathbf{V}=\mathbf{B}^{1/2}$ is the left stretch tensor matrix. They are both symmetric matrices appearing in the polar decomposition of the deformation gradient $\mathbf{F}=\mathbf{R}\mathbf{U}=\mathbf{V}\mathbf{R}$, where $\mathbf{R}$ is the orthogonal rotation matrix. All the above matrices represent in the chosen frame tensors bearing corresponding names.
More popular in classical deformation analysis is the strain tensor (Sansò 1982, Dermanis 1981)

$$\mathbf{E}=\frac{1}{2}(\mathbf{C}-\mathbf{I})=\frac{1}{2}(\mathbf{F}^T\mathbf{F}-\mathbf{I}) \tag{4}$$

or even its "infinitesimal approximation"

$$\mathbf{E}_{\text{inf}}=\frac{1}{2}(\mathbf{J}+\mathbf{J}^T) \tag{5}$$

based on neglecting 2nd order terms in the displacement gradient $\mathbf{J}=\frac{\partial \mathbf{u}}{\partial \mathbf{x}_0}=\frac{\partial(\mathbf{x}-\mathbf{x}_0)}{\partial \mathbf{x}_0}=\mathbf{F}-\mathbf{I}$.
The situation becomes more transparent with the use of the singular value decomposition (SVD)

$$\mathbf{F}=\mathbf{Q}^T\mathbf{L}\mathbf{P} \tag{6}$$

for which it can be easily verified that $\mathbf{C}=\mathbf{P}^T\mathbf{L}^2\mathbf{P}$, $\mathbf{B}=\mathbf{Q}^T\mathbf{L}^2\mathbf{Q}$, $\mathbf{U}=\mathbf{P}^T\mathbf{L}\mathbf{P}$, $\mathbf{V}=\mathbf{Q}^T\mathbf{L}\mathbf{Q}$ and $\mathbf{R}=\mathbf{Q}^T\mathbf{P}$. From the diagonalizations $\mathbf{C}=\mathbf{P}^T\mathbf{L}^2\mathbf{P}$ and $\mathbf{B}=\mathbf{Q}^T\mathbf{L}^2\mathbf{Q}$ it follows that the diagonal elements $(\mathbf{L}^2)_{kk}=\lambda_k^2$ are the common eigenvalues of the symmetric matrices $\mathbf{C}$ and $\mathbf{B}$, $\mathbf{P}^T=[\mathbf{p}_1\mathbf{p}_2\mathbf{p}_3]$ is the matrix with columns the eigenvectors $\mathbf{p}_k$ of $\mathbf{C}$ ($\mathbf{C}\mathbf{p}_k=\lambda_k^2\mathbf{p}_k$) and $\mathbf{Q}^T=[\mathbf{q}_1\mathbf{q}_2\mathbf{q}_3]$ is the matrix with columns the eigenvectors $\mathbf{q}_k$ of $\mathbf{B}$ ($\mathbf{B}\mathbf{q}_k=\lambda_k^2\mathbf{q}_k$). For the corresponding physical interpretation we note that a change in the instantaneous reference system bases from $\vec{\mathbf{e}}(t_0)$ and $\vec{\mathbf{e}}(t)$ to $\vec{\mathbf{e}}'(t_0)=\vec{\mathbf{e}}(t_0)\mathbf{P}^T$ and $\vec{\mathbf{e}}'(t)=\vec{\mathbf{e}}(t)\mathbf{Q}^T$ is accompanied by coordinate transformations $\mathbf{x}_0'=\mathbf{P}\mathbf{x}_0$ and $\mathbf{x}'=\mathbf{Q}\mathbf{x}$, so that the deformation gradient in the new systems becomes

$$\mathbf{F}'=\frac{\partial \mathbf{x}'}{\partial \mathbf{x}_0'}=\frac{\partial \mathbf{x}'}{\partial \mathbf{x}}\frac{\partial \mathbf{x}}{\partial \mathbf{x}_0}\frac{\partial \mathbf{x}_0}{\partial \mathbf{x}_0'}=\mathbf{Q}\mathbf{F}\mathbf{P}^T=\mathbf{L} \tag{7}$$

This means that, in the new systems, deformation does not change locally the directions of lines along the base vectors but only their lengths by a factor of λ_k. These directions are the principal directions which are physical invariants. The angular parameters defining the directions of the principal axes are not numerical invariants since they relate also to the axes of the reference system; see Dermanis (1981, 1985), Xu et al. (2000). The matrix $\mathbf{P}$ rotates the axes of the reference system at the reference epoch t_0 to the directions of the principal axes and $\mathbf{Q}$ does the same to the axes of the reference system at epoch t. The angular parameters defining the matrices $\mathbf{P}$ or $\mathbf{Q}$ are not numerical invariants. When the reference systems are changed their values are adjusted accordingly in order to define the same directions of the principal axes. Any other function of $\mathbf{F}$, which is a numerical invariant, is necessarily a function of the singular values of $\mathbf{F}$.

A different type of deformation analysis concerns the "rate of deformation" which is related to the time derivative of the deformation gradient $\dot{\mathbf{F}}=\frac{d}{dt}\mathbf{F}$, or the time derivatives $\dot{\varphi}=\frac{d}{dt}\varphi$ of numerical invariant functions $\varphi(\mathbf{F})=\varphi(\lambda_1.\lambda_2,\lambda_3)$.

The available data are typically coordinates of discrete points at discrete epochs $\mathbf{x}_{i,t_k}=\mathbf{x}(\mathbf{x}_{0i},t_k)$, for points P_i identified by their coordinates $\mathbf{x}_{0i}$ at the reference epoch t_0. In the case that both epochs t_0 and t_k are observation epochs, the calculation of the deformation gradient $\mathbf{F}$ at any point for the epoch t_k requires spatial interpolation which will provide spatial coordinates $\mathbf{x}_{t_k}=\mathbf{x}(\mathbf{x}_0,t_k)$ for "points" $\mathbf{x}_0$ other than the data points $\mathbf{x}_{0i}$. Temporal interpolation is required whenever deformation rates are required and the time derivative $\dot{\mathbf{F}}$ must be calculated. In addition it is required whenever observations at different points $\mathbf{x}_{0i}$ are performed not simultaneously (i.e. in a very short time period) but rather at different epochs t_j, in which case the observed coordinates must be reduced to a common epoch t_k.

The performance of the interpolation procedure is strongly related to the spatial and temporal resolution of the observations in relation to the spectral content of the relevant physical processes. In addition interpolation must be performed independently within separate regions and time intervals which are separated by discontinuities, i.e. faults and seismic events; for a finite element approach to the geodetic computation of two- and three-dimensional deformation parameters, see Dermanis and Grafarend (1992).

The establishment of permanent GPS networks has practically removed the need for temporal interpolation, since daily coordinate solutions provide a very satisfactory resolution for crustal deformations that evolves very slowly in time (Wernicke et al. 2004, Pietrantonio and Riguzzi 2004). However the spatial interpolation remains necessary, since densification of the observing networks raises considerably the relevant cost, while coordinates of nearby stations are affected by similar systematic errors which tend to be ultimately interpreted erroneously as additional contributions to deformation.

3 Definition of the reference frames as a tool of trend removal before interpolation.

Instead of spatially interpolating discrete coordinates $\mathbf{x}_i$ it is more convenient and effective to interpolate the corresponding displacements $\mathbf{u}_i = \mathbf{x}_i - \mathbf{x}_{0i}$ and to compute the deformation gradient $\mathbf{F} = \mathbf{I} + \mathbf{J}$ from the displacement gradient $\mathbf{J} = \dfrac{\partial \mathbf{u}}{\partial \mathbf{x}_0}$. The available displacements refer to an externally defined reference system (e.g. ITRFxx) which may be sufficient for global studies of plate motions, but not so for studying local deformation. Note that, in order to avoid the impact of reference frame definition on the computed displacement vectors, any redundant a-priori constraints should be first removed from the individually adjusted coordinate sets, before applying any trend removal and deformation analysis procedure. The displacements with respect to such an external reference frame reflect at the same time the intrinsic deformation of the observing network as well as its total rigid motion (displacement and rotation) with respect to the adopted reference system.

Although it is possible to perform trend removal (absorbing mainly the "rigid-body motion" effects such as the Eulerian motion of tectonic plates) and subsequent spatial interpolation to any local displacement field that is expressed with respect to a common global frame, we shall here discuss a particular strategy that is capable of separating the actual deformation from the rigid-body motion component by introducing an intrinsic to the network reference system. This separation will on one hand provide estimates of the motion of the network as a whole and on the other hand provide intrinsic displacements which are smaller and thus more convenient for interpolation, in particular when this is carried out by statistical tools as prediction of functionals related to a zero-mean stochastic process. Of course trend removal can be achieved also by other means in order to provide a known mean displacement function. The definition of a network-intrinsic reference system simply removes a significant part of the trends which has nothing to do with local physical processes but rather has to do with the arbitrary use of a particular reference system serving a different purpose.

When the available data are discrete in time, the definition of the reference system in each epoch is based on constraints which are introduced for the coordinate differences $\Delta\mathbf{x}_i(t_k) = \mathbf{x}_i(t_k) - \mathbf{x}_i(t_{k-1})$. In the case of classical geodetic observations the constraints were incorporated into the network adjustment as "inner constraints" on the unknown parameters, by using the coordinates of the previous epoch as approximate coordinates for the linearization of the observation equations at the current epoch. Nowadays the system defining conditions may be introduced a posteriori by requiring that *(a)* the "barycenter" of the network is preserved

$$\mathbf{m}(t_k) \equiv \frac{1}{N}\sum_{i=1}^{N} \mathbf{x}_i(t_k) = \mathbf{m}(t_{k-1}) \equiv \frac{1}{N}\sum_{i=1}^{N} \mathbf{x}_i(t_{k-1}) \tag{8}$$

and *(b)* that the vector of "relative angular momentum" of the network vanishes

$$\mathbf{h}_R \equiv \sum_{i=1}^{N} [\mathbf{x}_i(t_{k-1})\times]\,\mathbf{v}_i(t_{k-1}) = \tag{9}$$

$$= \sum_{i=1}^{N} [\mathbf{x}_i(t_{k-1})\times] \frac{\mathbf{x}_i(t_k) - \mathbf{x}_i(t_{k-1})}{t_k - t_{k-1}} = \mathbf{0} \tag{10}$$

These two conditions define respectively the position and orientation of the intrinsic to the network reference system, which is uniquely defined by the two equivalent conditions

$$\sum_{i=1}^{N} \Delta\mathbf{x}_i(t_k) = \mathbf{0} \tag{11}$$

and

$$\sum_{i=1}^{N} [\mathbf{x}_i(t_{k-1})\times]\,\Delta\mathbf{x}_i(t_k) = \mathbf{0} \tag{12}$$

If we remove the epoch t_k and set $\mathbf{x}_i(t_{k-1}) = \mathbf{x}_i^0$, we recognize 6 of the 7 inner constraints introduced by Meissl for the adjustment of so called "free networks" (Dermanis 2002). The 7th condition relates to the definition of scale in the network, by requiring that the quadratic mean distance of the network points from their barycenter remains constant

$$L(t_k) \equiv \sqrt{\frac{1}{N}\sum_{i=1}^{N} \| \mathbf{x}_i(t_k) - \mathbf{m}(t_k) \|^2} = L(t_{k-1}) \tag{13}$$

which under the barycenter preservation condition and neglecting second order terms in the small displacements $\Delta\mathbf{x}_i(t_k)$ simplifies to

$$\sum_{i=1}^{N} \mathbf{x}_i(t_{k-1})^T \Delta\mathbf{x}_i(t_k) = 0 \tag{14}$$

However, even if the maintenance of a constant unit of length in the observations is questionable, the incorporation of the last condition into the analysis leads to the interpretation of a common increase or decrease of the network size as a whole as simply variation of the unit of scale. The opposite possibility has the danger of interpreting variations in the scale of measuring unit of length (in fact unit of time) as a change in the overall size of the network. In our opinion one should use both approaches and interpret scale changes as deformation only when their magnitude is well above the influence of unit instability in the system of reference clocks used in the observations.

Remark. The adoption of a network-intrinsic reference frame for the determination of local displacement vectors $\Delta\mathbf{x}_i(t_k) = \mathbf{x}_i(t_k) - \mathbf{x}_i(t_{k-1})$ between different epochs in a common spatial reference system will provide "snapshots" of the deformation field, as viewed from an observer who is continuously situated at the "center of mass" of the deforming network. Note that the center of mass of a terrestrial deformable network is physically varying in space (with respect to other stable points outside the network), due to the alteration of its physical shape and/ size caused by its underlying dynamical behavior. In this way, the continuous fixation of the network barycenter, though the use of an inner constraint adjustment framework via a common set of initial approximate coordinates at some reference epoch, still gives a "relative" picture of the actual deformation field, since it will mask the part that causes variations in the geometrical position of the network barycenter.

Since the definition of the reference system epoch by epoch using an analog of the Meissl constraints we may characterize the relevant procedure as a "Meissl ladder", which is a discrete approximation to the definition of the reference system under time-continuous data (Dermanis 2002).
In the case of time-continuous coordinates $\mathbf{x}_i(t)$ a transformation to an optimal frame is realized by a time dependent change of reference system

$$\tilde{\mathbf{x}}_i(t) = \mathbf{R}(t)\mathbf{x}_i(t) + \mathbf{c}(t) \tag{15}$$

which satisfies two optimality conditions. The preservation of the "barycenter" of the network

$$\tilde{\mathbf{m}} \equiv \frac{1}{N}\sum_{i=1}^{N} \tilde{\mathbf{x}}_i = \text{const.} \tag{16}$$

determines the translation parameters, which in view of $\tilde{\mathbf{m}} = \mathbf{Rm} + \mathbf{c}$, become $\mathbf{c} = \tilde{\mathbf{m}} - \mathbf{Rm}$ and the system transformation reduces to

$$\tilde{\mathbf{x}}_i = \mathbf{R}(\mathbf{x}_i - \mathbf{m}) + \tilde{\mathbf{m}} \tag{17}$$

where $\mathbf{m}$ are the coordinates of the center of mass of the network and $\tilde{\mathbf{m}}$ an arbitrary constant vector, e.g. $\tilde{\mathbf{m}} = \mathbf{m}(t_0)$. The parameters of the rotation matrix $\mathbf{R}$ are determined by the "discrete Tisserand condition" that the vector of "relative angular momentum" of the network vanishes (Dermanis 2000, 2001)

$$\tilde{\mathbf{h}}_R \equiv \sum_{i=1}^{N} [(\tilde{\mathbf{x}}_i - \tilde{\mathbf{m}})\times]\frac{d(\tilde{\mathbf{x}}_i - \tilde{\mathbf{m}})}{dt} = \mathbf{0} \tag{18}$$

This is a discrete version of the condition for Tisserand axes for the earth, where the integral over the whole earth masses is replaced by a sum over the network points considered as point masses with the mass each. Introducing the relative rotation vector $\boldsymbol{\omega}$ as the axial vector of the antisymmetric matrix $[\boldsymbol{\omega}\times] = \frac{d}{dt}(\mathbf{R}^T)\mathbf{R}$, it can be shown that

$$\tilde{\mathbf{h}}_R = \mathbf{R}(-\mathbf{C}\boldsymbol{\omega} + \mathbf{h}_R) = \mathbf{0} \tag{19}$$

where

$$\mathbf{C} = -\sum_{i=1}^{N} [(\mathbf{x}_i - \mathbf{m})\times]^2 \tag{20}$$

is the inertia matrix and

$$\mathbf{h}_R = \sum_{i=1}^{N} [(\mathbf{x}_i - \mathbf{m})\times] \frac{d(\mathbf{x}_i - \mathbf{m})}{dt} \tag{21}$$

is the relative angular momentum vector of the network, with respect to original reference system. Thus

$$\boldsymbol{\omega} = \mathbf{C}^{-1}\mathbf{h}_R \tag{22}$$

and the parameters of the rotation matrix $\mathbf{R}$ are determined from the solution of the geometric Euler differential equations which are the axial part of the antisymmetric matrix relation $[\boldsymbol{\omega}\times] = \frac{d}{dt}(\mathbf{R}^T)\mathbf{R}$. To give a specific example consider the parameterization $\mathbf{R} = \mathbf{R}_m(\theta_m)\mathbf{R}_k(\theta_k)\mathbf{R}_n(\theta_n)$ by means of consecutive rotations around the n, k and m axis, in which case the angles $\boldsymbol{\theta} = [\theta_n \theta_k \theta_m]^T$ are determined by the solution of the differential equations

$$\frac{d\boldsymbol{\theta}}{dt} = \begin{bmatrix} \mathbf{i}_n^T \\ \mathbf{i}_k^T \mathbf{R}_n(\theta_n) \\ \mathbf{i}_m^T \mathbf{R}_k(\theta_k)\mathbf{R}_n(\theta_n) \end{bmatrix} \boldsymbol{\omega} \tag{23}$$

where $\mathbf{i}_p$ denotes the pth row of the identity matrix $\mathbf{I} = [\mathbf{i}_1 \mathbf{i}_2 \mathbf{i}_3]$. We may choose $\boldsymbol{\theta}(t_0) = \mathbf{0}$ which, in combination with $\breve{\mathbf{m}} = \mathbf{m}(t_0)$, makes the new and the original frames identical at the reference epoch t_0. The resulting angles $\boldsymbol{\theta}(t)$ describe the rotation of the network as a whole while its translational motion $\mathbf{c}(t)$ is given by

$$\mathbf{c} = \breve{\mathbf{m}} - \mathbf{R}(\boldsymbol{\theta})\mathbf{m}\,. \tag{24}$$

Further analysis will make use of the transformed coordinate functions $\breve{\mathbf{x}}(t)$ and displacements $\breve{\mathbf{u}}(t) = \breve{\mathbf{x}}(t) - \breve{\mathbf{x}}(t_0) = \breve{\mathbf{x}}(t) - \breve{\mathbf{x}}_0$, which we will further on denote by $\mathbf{x}(t)$ and $\mathbf{u}(t)$ for the sake of simplicity.

Let us also note that, apart from the trend removal related to linear (global) trends in the spatial displacement field, other seasonal (annual, semi-annual) station-dependent signals can also be removed from the available coordinate values, before applying any further interpolation procedure for local deformation analyses. Such information is nowadays often available and accessible through the analysis of GPS time series in various types of regional, continental and global networks.

4 Three-dimensional or two-dimensional deformation?

The physical process of crustal deformation is unquestionably three-dimensional (3D) and any deformation parameters entering the constitutional equations should naturally refer to the 3D case (Voosoghi 2000). On the other hand, there is a long tradition of separating vertical from horizontal information, leading to the study of a two-dimensional (2D) deformation of an abstract and somewhat unnatural entity, consisting of the projections of the material points on a horizontal plane. The fact is that available information is restricted on the 2D surface of the earth. In order to derive 3D deformation parameters, we need not only to spatially interpolate in the horizontal sense, but also to *extrapolate* in the vertical sense. Interpolation is always a less subject to errors process compared to extrapolation and it can be virtually overcome by further densification.

There have been attempts to derive 3D deformation from surface data by forming tetrahedrons and using a finite element approach where homogeneous deformation is assumed within each tetrahedron, which is an extension of the similar 2D approach where triangles are formed (Dermanis and Grafarend 1992, Voosoghi 2000). This approach may have some relevance when the central point of the tetrahedron is located higher than the external points, so that deforming earth masses are located inside the tetrahedron. In the opposite case one simply derives deformation for tetrahedrons consisting of "thin air". In any case, the low variation of surface altitude with respect to the usual mean distance between neighboring network stations leads to tetrahedrons with vertical dimension considerably smaller than the horizontal one and the homogeneity assumption hardly provides meaningful information in the vertical sense.

A proper unified approach which does not neglect, or treats separately, temporal height variation, is to study the deformation of the 2D earth surface as embedded in 3D space and to relate it to the 3D deformation of the crust. The extension of surface deformation to the only physically meaningful 3D crustal deformation is an improperly posed problem and can be solved only with of additional assumptions of geophysical nature. Geodesy can only provide boundary conditions on the surface of the earth for the elements of the deformation gradient of the actual crustal deformation.

The study of the deformation of the surface of the earth as embedded in 3D space is very complicated because the earth surface is not an Euclidean but rather a Riemannian manifold. In the literature for the mechanics of continuous media the use of deformation within Riemannian spaces can be found in the treatment of relativistic elasticity, which has its own particularities. We shall present the treatment of the surface deformation exploiting 3D coordinate information elsewhere. More details on the classical approach where 2D deformation is studied separately from height variation information can be found in Biagi and Dermanis (2005).

For the sake of generality, the analysis of the available data will be presented hereon, in a general way that it can incorporate both the 2D and the 3D case.

5 Interpolation methods: Interpretation and comparison

There are two basic approaches to interpolation, not completely independent from each other (Dermanis 1976, 1988; Dermanis and Rossikopoulos 1988). Analytic interpolation models the interpolating function by means of a number of unknown parameters, with most popular choice linear models of the form

$$f(P)=\sum_i a_i\, f_i(P) \tag{25}$$

where $f_i(P)$ are known base functions (e.g. algebraic polynomials, trigonometric polynomials, spline-type functions, etc.), and a_i are the unknown coefficients. For temporal interpolation the spatial variable P will be replaced by time t, while also simultaneous spatial-temporal interpolation is also possible. Discrete data are available of the form

$$b_k = f(P_k) + v_k = \sum_i a_i\, f_i(P_k) + v_k \tag{26}$$

polluted by the observation errors v_k. The objective is to determine parameter estimates $\hat{a}_i$ so that estimates of function values can be obtained at any point P by $\hat{f}(P)=\sum_i \hat{a}_i f_i(P)=\mathbf{f}_P^T\hat{\mathbf{a}}$.

There are two basic optimality criteria that can be also combined into a third hybrid approach. When the number of base functions is less than the number of data then the least-squares criterion $\mathbf{v}^T\mathbf{W}\mathbf{v}=$ min leads to the smoothing-interpolation solution

$$\hat{\mathbf{a}}=(\mathbf{G}^T\mathbf{W}\mathbf{G})^{-1}\mathbf{G}^T\mathbf{W}\mathbf{b} \tag{27}$$

where $\mathbf{G}$ is the matrix with elements $G_{ki}=f_i(P_k)$.

When the number of base functions is more than the number of data, then a minimum norm criterion $\mathbf{a}^T\breve{\mathbf{W}}\mathbf{a}=\min$ leads to the exact-interpolation solution

$$\hat{\mathbf{a}}=\breve{\mathbf{W}}^{-1}\mathbf{G}^T(\mathbf{G}\breve{\mathbf{W}}^{-1}\mathbf{G}^T)^{-1}\mathbf{b} \tag{28}$$

The interpolated function values in the last case are

$$\hat{f}(P)=\mathbf{f}_P^T\hat{\mathbf{a}}=\mathbf{f}_P^T\breve{\mathbf{W}}^{-1}\mathbf{G}^T(\mathbf{G}\breve{\mathbf{W}}^{-1}\mathbf{G}^T)^{-1}\mathbf{b} \tag{29}$$

The hybrid alternative refers to the criterion $\mathbf{v}^T\mathbf{W}\mathbf{v}+\mathbf{a}^T\breve{\mathbf{W}}\mathbf{a}=\min$, with solution

$$\begin{aligned}\hat{\mathbf{a}}&=(\mathbf{G}^T\mathbf{W}\mathbf{G}+\breve{\mathbf{W}})^{-1}\mathbf{G}^T\mathbf{W}\mathbf{b}\\&=\breve{\mathbf{W}}^{-1}\mathbf{G}^T(\mathbf{G}\breve{\mathbf{W}}^{-1}\mathbf{G}^T+\mathbf{W}^{-1})^{-1}\mathbf{b}\end{aligned} \tag{30}$$

$$\mathbf{C}_{\hat{\mathbf{a}}}=(\mathbf{G}^T\mathbf{W}\mathbf{G}+\breve{\mathbf{W}})^{-1}(\mathbf{G}^T\mathbf{W}\mathbf{G})(\mathbf{G}^T\mathbf{W}\mathbf{G}+\breve{\mathbf{W}})^{-1} \tag{31}$$

The interpolated values take in this case the form

$$\begin{aligned}\hat{f}(P)&=\mathbf{f}_P^T\hat{\mathbf{a}}=\mathbf{f}_P^T\breve{\mathbf{W}}^{-1}\mathbf{G}^T(\mathbf{G}\breve{\mathbf{W}}^{-1}\mathbf{G}^T+\mathbf{W}^{-1})^{-1}\mathbf{b}=\\&=\mathbf{k}_P^T(\mathbf{K}+\mathbf{W}^{-1})^{-1}\mathbf{b}\end{aligned} \tag{32}$$

It is possible to separate a factor α from the matrix $\breve{\mathbf{W}}=\alpha\breve{\mathbf{W}}_0$ in which case the criterion $\mathbf{v}^T\mathbf{W}\mathbf{v}+\alpha\mathbf{a}^T\breve{\mathbf{W}}_0\mathbf{a}=\min$ contains an additional regularization parameter α and the above solution is known as *Tikhonov regularization.*

At the antipodes of the deterministic analytical interpolation lies the interpolation by means of stochastic prediction, where the unknown function is assumed to be a zero-mean random field (stochastic process). The interpolating value $f(P)$ is a random variable which can be predicted from the sample values of the random variables $b_k = f(P_k) + v_k$ or $b_k = f(P_k)$. The minimum mean square prediction error leads to the well known (collocation) solution

$$\hat{f}(P) = \mathbf{c}_P^T \mathbf{C}_{\mathbf{b}}^{-1} \mathbf{b} = \mathbf{c}_P^T (\mathbf{C}_{\mathbf{s}} + \mathbf{C}_{\mathbf{v}})^{-1} \mathbf{b} \qquad (33)$$

where $\mathbf{C}_{\mathbf{b}} = \mathbf{C}_{\mathbf{s}} + \mathbf{C}_{\mathbf{v}}$ is the covariance matrix of the observations, $\mathbf{C}_{\mathbf{v}}$ is the noise covariance matrix, $(\mathbf{C}_{\mathbf{s}})_{jk} = \sigma(f(P_j), f(P_k)) = C(P_j, P_k)$ and $(\mathbf{c}_P)_k = \sigma(f(P), f(P_k)) = C(P, P_k)$,
where $C(P,Q)$ is the covariance function of the random field.

The equations (32) and (33) have a striking similarity, which is by no means accidental. Assuming a diagonal matrix $\tilde{\mathbf{W}}$ for the sake of simplicity, the matrix appearing in (32) have elements

$$K_{jk} = \sum_i \frac{1}{\tilde{W}_{ii}} f_i(P_j) f_i(P_k) = k(P_j, P_k) \qquad (34)$$

$$(\mathbf{k}_P)_k = \sum_i \frac{1}{\tilde{W}_{ii}} f_i(P) f_i(P_k) = k(P, P_k) \qquad (35)$$

where $k(P,Q) = \sum_i \frac{1}{W_{ii}} f_i(P) f_i(Q)$ is a "reproducing kernel" in the set of all linear combinations of the base functions, whose number may even be infinite. It suffices to select $1/\tilde{W}_{ii} = \sigma^2(a_i)$, interpreting the coefficients as random variables (weight=reciprocal variance!), so that the reproducing kernel becomes in the light of covariance propagation identical with the covariance function

$$C(P,Q) = \sigma(f(P), f(Q)) = \sum_i \frac{\partial f(P)}{\partial a_i} \frac{\partial f(Q)}{\partial a_i} \sigma^2(a_i) =$$

$$= \sum_i \sigma^2(a_i) f_i(P) f_i(Q) = k(P,Q) \qquad (36)$$

Therefore, minimum-norm/least-squares (hybrid) becomes numerically equivalent to minimum mean square error stochastic prediction (Dermanis 1976).

Although the estimates in both cases are numerically the same, their corresponding error assessment differs. In the deterministic case of hybrid interpolation, only the influence of the measurement errors can be taken into account to obtain the covariance of interpolated values

$$\sigma(\hat{f}(P), \hat{f}(Q)) = \mathbf{f}_P^T \mathbf{C}_{\hat{\mathbf{a}}} \mathbf{f}_Q =$$

$$= \mathbf{k}_P^T (\mathbf{K} + \mathbf{C}_{\mathbf{v}})^{-1} \mathbf{C}_{\mathbf{v}} (\mathbf{K} + \mathbf{C}_{\mathbf{v}})^{-1} \mathbf{k}_Q \qquad (37)$$

In the case of stochastic prediction the covariance of the observed function values must be taken also into account. Assuming that the observation noise is uncorrelated with the random field, the covariance of the interpolated values is

$$\sigma(\hat{f}(P), \hat{f}(Q)) = \mathbf{c}_P^T \mathbf{C}_{\mathbf{b}}^{-1} \mathbf{c}_Q = \mathbf{c}_P^T (\mathbf{C}_{\mathbf{s}} + \mathbf{C}_{\mathbf{v}})^{-1} \mathbf{c}_Q =$$

$$= \mathbf{c}_P^T (\mathbf{C}_{\mathbf{s}} + \mathbf{C}_{\mathbf{v}})^{-1} \mathbf{C}_{\mathbf{v}} (\mathbf{C}_{\mathbf{s}} + \mathbf{C}_{\mathbf{v}})^{-1} \mathbf{c}_Q +$$

$$+ \mathbf{c}_P^T (\mathbf{C}_{\mathbf{s}} + \mathbf{C}_{\mathbf{v}})^{-1} \mathbf{C}_{\mathbf{s}} (\mathbf{C}_{\mathbf{s}} + \mathbf{C}_{\mathbf{v}})^{-1} \mathbf{c}_Q \qquad (38)$$

Taking into account $\mathbf{C}_{\mathbf{s}} = \mathbf{K}$, $\mathbf{c}_P = \mathbf{k}_P$ and $\mathbf{c}_Q = \mathbf{k}_Q$, we can see that the propagated covariance in the stochastic approach contains an additional term (last line of Eq. (38)) expressing the uncertainty in the interpolation itself. Both approaches have their disadvantages. In the deterministic approach we assume that the chosen model (choice of type and number of base functions) describes perfectly the physical reality (no interpolation error!), which is not always correct. Hypothesis testing may be a helpful tool for the elimination of non-significant model parameters in this case, but the problem of choosing the proper type of base functions (e.g. polynomials, trigonometric functions, etc.) remains open. Even the same type of base functions expressed with the help of, e.g., plane horizontal coordinates $f_i(P) = f_i(x, y)$, gives different base functions for different choices of the coordinate system involved. As a consequence, the resulting interpolating function is not invariant under changes of the used coordinate system.

In the stochastic approach we assume that the chosen covariance function describes perfectly the stochastic behavior (e.g. smoothness) of the random field. However, the choice of the covariance function is to a certain degree arbitrary, since we do not have a sufficiently large data sample to secure a reliable empirical estimate. An advantage of the stochastic interpolation approach is its invariance under reference system transformations, when the covariance function depends only on the distance between the two relevant points

$$C(P,Q) = C(|PQ|) = C(r_{PQ}) .$$

In addition to the analytical exact least-squares interpolation and to the hybrid interpolation (with a deterministic or a stochastic interpretation), it is possible to combine the two in a single approach using the model

$$f(P) = \sum_i a_i \, f_i(P) + s(P) \tag{39}$$

where the function is decomposed into an analytic deterministic trend and an additional stochastic part. The model parameters $\{a_i\}$ are free to take any values, since they do not participate in the optimization principle. The assumption of a zero-mean random field for the validity of the minimum mean square error prediction requires the removal of the basic trend from the initial observed values, either simultaneously as seen in the above model, or a priori by a preceding analytical least-squares interpolation with a small number of base functions.

6 Implementation strategies for the interpolation of displacement fields

In general, spatial and/or temporal interpolation is required when the available data are not continuous. By "continuous" we mean data with a sufficiently high sampling resolution which can fully recover the full spectrum of the underlying function. In such cases, there is no need for interpolation per se, apart from the removal of the measurement noise with a filtering technique which indirectly implies a type of smoothing interpolation. The resulting filtered data are again discrete (with the same high sampling resolution) and not given in a continuous analytic form which allows the straightforward analytical computation of the required derivatives. Therefore, the elements of the deformation gradient or displacement gradient or velocity gradient matrix must be obtained by numerical differentiation techniques.

When the available data are not continuous, either in the time or the space domain, then the choice of the interpolation method depends on the characteristics of the underlying physical process. Experience from the daily solutions of permanent GPS stations has shown a linear temporal displacement with additive almost-white noise, while periodic terms of no relevance to crustal deformation have been properly computed and removed (e.g. Pietrantonio and Riguzzi 2004, Dong et al. 1998). This is the reason why ITRF models so far consist of initial coordinates and velocities only. Therefore, in the time domain one should use least-squares interpolation with a linear, or at most quadratic, analytic model.

Due to the lack of similar experiences with spatial data, one needs to experiment with either purely analytic models, or to use a combination of an analytic trend with an additive stochastic component in a hybrid interpolation scheme. The choice depends on the spectral behavior of the crustal deformation field in the particular application area. The analytic trend takes care of the lower frequencies of the field, whereas the stochastic signal represents, in a smooth way, the somewhat higher frequencies that may be present.

Another issue is whether the primary observations should be a-priori adjusted to compute coordinates and thus displacements, or interpolation and adjustment of the raw observations should be implemented simultaneously. The two-step procedure (adjustment, followed by interpolation) is usually dictated by the lack of access to the original data. From a strictly formal point of view, a simultaneous treatment is preferable provided the functional and stochastic models are correct. Since this is not the case, particularly for GPS data, a two-step procedure avoids interpreting observational noise as a physical crustal deformation process.

Apart from the above general guidelines, the data analysis strategy to be implemented depends on how observations are distributed in space and time, in relation to the overall design of the measurement procedure. In the sequel we discuss some specific data collection schemes that can appear in the geodetic practice.

Case 1: GPS repeated surveys

A network of stations is observed in repeated campaigns, within short-time interval (e.g. a week), so that they refer to discrete epochs. In this case, deformation analysis refers to the comparison of any two epochs which coincide with the actual observation epochs. Spatial-only interpolation is needed in order to compute the deformation gradient matrix at any point within the study area. However, since temporal coordinate variations are known to be of a linear nature, it may be advantageous to adjust all observation epochs simultaneously with a analytical linear interpolation $\mathbf{x}(P,t) = \mathbf{x}_0(P) + (t - t_0)\mathbf{v}(P)$, with respect to time. In this case, the station coordinates $\mathbf{x}(P_k, t_j)$ at the various observation epochs t_j are replaced by a much smaller number of unknowns, namely the coordinates $\mathbf{x}_0(P_k) = \mathbf{x}(P_k, t_0)$ at the reference epoch t_0 and the station velocities $\mathbf{v}(P_k)$. It remains to perform a spatial interpolation for the station velocities to obtain a velocity field from which displacement can be determined for any desired pair of epochs. Deformation analysis can be performed with the strain rate approach by using the velocity gradient instead of the displacement gradient.

Case 2: GPS "network scanning" surveys

Such situations arise when the number of available GPS receivers is not sufficient to cover with "simultaneous" observations the studied area. In order to complete the survey at a specific epoch, different "patches" of the monitoring network are sequentially observed until the entire area is properly covered. Repetitions of the above procedure at different time periods provide us with the necessary kinematic information for the estimation of the crustal deformation field.

In this case, temporal interpolation is required and it must be simultaneously performed with the least-squares data adjustment in order to obtain reference epoch coordinates and velocities values. The rest of the analysis follows the same guidelines as in the previous case.

Case 3: Monitoring by GPS permanent stations

In this case, the enormous amount of the available data makes the simultaneous adjustment for the determination of station coordinates and velocities impossible from a practical point of view. In practice daily adjustments are performed producing noisy coordinate estimates. A separate linear regression for every coordinate at every station provides the required parameters. In addition, it provides a great amount of residuals that can be used to estimate the variances and covariances for any pair of coordinates and velocity components of the same or different stations. Moreover, it is possible to detect temporal correlations by empirically estimating auto-covariance and cross-covariance functions. From limited experience, we may report that some small correlation exists for time intervals of one day (most probably due to unmodeled ionospheric effects), which becomes already negligible after two or more days. Thus, the quality assessment problem can be solved in a way completely independent of the formal covariance propagation which is known to be overly optimistic in the case of GPS data. As in the previous two cases, spatial interpolation of velocities must follow before performing the deformation analysis for a pair of epochs or a strain rate analysis.

Case 4: GPS permanent stations and repeated SAR surveys

There is little experience regarding the optimal merging of GPS and SAR data for deformation analysis (e.g. Bos et al. 2004, Lohman and Simons 2005). The simplest mind approach is to perform the analysis for the permanent GPS stations as in the previous case, and introduce the computed coordinates at the SAR campaign epoch as constraint values. Indeed, SAR is known to provide very good results in high spatial resolution while lacking a similar accuracy level in the longer wavelength parts. In this case the GPS information complements and enhances the information provided by SAR. When the deformation analysis involves comparison between time epochs that are covered by SAR campaigns, there is absolutely no need for interpolation in the spatial domain. For an arbitrary pair of epochs and/or strain rate deformation analysis, a temporal interpolation can be performed for all SAR points. The coordinates of each SAR densification point may not perfectly fit into a linear model (as in the case of GPS stations). The question that arises is whether a linear behavior should be imposed a-posteriori, or implemented in the simultaneous analysis of all SAR campaigns. The answer requires a deeper understanding of the data analysis methods used in the SAR data processing, which differs from the typical least-squares geodetic adjustment techniques (Hanssen 2001, 2005; Lohman and Simons 2005).

Case 5: GPS permanent stations and "permanent" SAR surveys

With a futuristic outlook, we envisage the case where SAR data sets are obtained with a very high temporal resolution (e.g. by involving a number of satellites). In this case, neither temporal or spatial interpolation is required, and one has to fully switch from the classical geodetic data analysis techniques to spectral methods which are continuous in principle but they are numerically implemented on dense discrete data. Instead of the problem of choosing the base functions or the covariance function, we face the problem of choosing the appropriate filter(s) which separate the noise attributed to observation errors from the signal attributed to crustal deformation.

7 Accuracy assessment of the results

The accuracy assessment of the results faces three fundamental problems:

(a) Quality assessment of the input data. This is a most critical problem for GPS derived coordinates which are accompanied by formal variance and covariance values that are commonly recognized to be too optimistic. The reason is that the raw data are supposed to be affected by uncorrelated noise while as in reality there are significant correlations due to physical processes (e.g. non-modeled atmospheric effects) or instrumental behavior (Williams 2003, Langbein and Johnson 1997, Mao et al. 1999). The hope for independent quality assessment lies in variance component estimation techniques, which take rather simplified forms when a linear trend is a realistic model for temporal evolution. The residuals after removal of the linear trend can be used to detect time-independent coordinate error variances and covariances or temporal correlations which however seem to be negligible for a two or more days interval.

(b) The independently assessed realistic covariance matrix of the input data must be propagated to the entries of the deformation gradient matrix. In this step the uncertainties due to the more-or-less arbitrary interpolation in between the discrete data points must be taken into account. This can be done only when a stochastic interpolation-prediction is employed, while in an analytic trend determination only one has to accept the interpolating smooth function as error-free. Both approaches (and therefore their combination) have their shortcomings. The trend removal depends on the particular model for the parameterization of the trend function. The stochastic prediction suffers from the arbitrariness of the chosen covariance function, summarized by the common variance of the function value at each point (stationarity assumption) and the correlation length under the isotropic hypothesis (distance where covariance drops at half the variance value).
(c) Covariance propagation is straightforward in the previous step where linear prediction algorithms are used and the predicted elements of the deformation gradient matrix **F** are linear functions of the input coordinates or displacements. This is not the case though for the computed deformation invariant parameters which, as functions of the singular values of the deformation gradient matrix, are highly nonlinear functions of the elements of **F** (Soler and van Gelder 1991). More realistic covariance propagation can be assessed by Monte-Carlo simulation methods, or extension of the linear covariance propagation scheme where at least second order partial derivatives of the non-linear expressions are implemented.

8 Conclusions

The scope of this work was to give an overview of some theoretical and practical aspects that are critical in the geodetic-based determination of crustal deformation fields. The importance of studying the deformation of the Earth's crust needs hardly to be stressed, especially in those parts of the world where the tectonic activity can cause disastrous effects that affect the lives of millions of people. Geodesy, being the science of measuring and mapping the surface of the Earth, plays a key role in crustal deformation studies by determining the temporal variations of its shape/size at various spatial and time scales (Dermanis and Livieratos 1983). Our focus in this paper was placed on the description of the basic principles and the mathematical techniques for the estimation of crustal deformation parameters from geodetic data, as well as on some special types of applications with particular interest within the field of modern four-dimensional geodesy.
For reasons of economy in our presentation, we have deliberately avoided a detailed treatment of SAR data processing techniques for crustal deformation analysis. Some details for this interesting and relatively unexplored topic in the geodetic literature can be found in Hanssen (2001) and the references given therein. Also, the contribution of the temporal variations of the gravity field, in terms of corresponding mass changes and redistributions and/or changes in the Earth's flattening (i.e. dynamic form factor J_2) and rotation axis as determined, for example, by the CHAMP and GRACE missions, plays an important role for crustal deformation studies in large spatial scales. Such an issue also has not been discussed here.

We should finally note that a topic of special interest is the development of optimal methods and algorithms for the estimation of crustal deformation parameters using a combination of geodetic and non-geodetic (e.g. seismic soundings, strain-meter and tiltmeter readings) observations, thus leading to an integrated inter-disciplinary approach for crustal deformation analysis.

References

Biagi L, Dermanis A (2005) The treatment of time-continuous GPS observations for the determi-nation of regional deformation parameters. This volume.

Bos AG, Usai S, Spakman W (2004) A joint analysis of GPS motions and InSAR to infer the coseismic surface deformation of the Izmit, Turkey earthquake. Geoph J Int, 158: 849-863.

Boucher C (1980) The general theory of deformations and its applications in geodesy. Boll Geod Sci Affi, 22(1): 13-35.

Crippa B, Crosetto M, Blazquez M, Barzaghi R (2002) SAR interferometry for deformation control. IAG Symposia, vol. 126, pp. 333-338, Springer-Verlag, Berlin Heidelberg New York.

Davis JL, Prescott WH, Svarc JL, Wendt KJ (1989) Assessment of Global Positioning System Measurements for Studies of Crustal Deformation. J Geoph Res, 94(B10): 13635-13650.

Dermanis A (1976) Probabilistic and deterministic aspects of linear estimation in geodesy. Report No. 244, Dept. of Geodetic Science, Ohio State University, Columbus, Ohio.

Dermanis A (1981) Geodetic estimability of crustal deformation parameters, Quarteniones Geodasiae, 2(2): 159-169.

Dermanis A (1985) The role of frame definitions in the geodetic determination of crustal deformation parameters. Bull Geod, 59: 247-274.

Dermanis A (1988) Geodetic applications of interpolation and prediction. Eratosthenes, 22: 229-262.

Dermanis A (2000) Establishing Global Reference Frames. Nonlinear, Temporal, Geophysical and Stochastic Aspects. IAG Symposia, vol. 123, pp. 35-42, Springer-Verlag, Berlin Heidelberg New York.

Dermanis A (2001) Global Reference Frames: Connecting Observation to Theory and Geodesy to Geophysics. Presented at the IAG Scientific Assembly "*Vistas for Geodesy in the New Millenium*", September 2-8, Budapest, Hungary.

Dermanis A (2002) The rank deficiency in estimation theory and the definition of reference frames. Presented at the V Hotine-Marussi Symposium on Mathematical Geodesy, June 17-21, Matera, Italy.

Dermanis A, Livieratos E, Rossikopoulos D, Vlachos D (1981) Geodetic prediction of crustal deformations at the seismic area of Volvi. Proceedings of the International Symposium on "*Geodetic Networks and Computations*", Munich, Germany, DGK, Reihe B, Nr. 258/V, pp. 234-248.

Dermanis A, Livieratos E (1983) Applications of deformation analysis in geodesy and geodynamics. Rev Geoph Spac Phys, 21(1): 41-50.

Dermanis A, Rossikopoulos D (1988) Modeling alternatives in four-dimensional geodesy. Proc. IAG International Symposium on "*Instrumentation, theory and analysis for integrated geodesy*", Sopron, Hungary, May 16-20, vol. 2, 115-145.

Dermanis A, Grafarend EW (1992) The finite element approach to the geodetic computation of two- and three-dimensional deformation parameters: a study of frame invariance and parameter estimability. Presented at the International Conference "*Cartography-Geodesy*", Maracaibo, Venezuela, November 24 – December 3, 1992.

Dong D, Herring TA, King RW (1998) Estimating regional deformation from a combination of space and terrestrial geodetic data. J Geod, 72:200-214.

Hanssen RF (2001) Radar interferometry: data interpretation and error analysis. Kluwer Academic Publishers, Dordrecht, The Netherlands.

Hanssen RF (2005) Satellite radar interferometry for deformation monitoring: a priori assessment of feasibility and accuracy. Inter J of Appl Earth Obs and Geoinf, 6: 253-260.

Kogan MG, Steblov GM, King RW, Herring TA, Frolov DI, Egorov SG, Levin VY, Lerner-Lam A, Jones A (2000) Geodetic constraints on the rigidity and relative motion of Eurasia and North America. Geoph Res Lett, 27(14): 2041-2044.

Langbein J, Johnson H (1997) Correlated errors in geodetic time series: implications for time-dependent deformation. J Geoph Res, 102(B1): 591-603.

Lohman RB, Simons M (2005) Some thoughts on the use of InSAR data to constrain models of surface deformation: noise structure and data downsampling, Geochem Geophys Geosyst, 6, Q01007, doi:10.1029/2004GC000841.

Mao A, Harrison CGA, Dixon TH (1999) Noise in GPS coordinate time series. J Geoph Res, 104(B2): 2797-2816.

Pietrantonio G, Riguzzi F (2004) Three-dimensional strain tensor estimation by GPS observations: methodological aspects and geophysical applications. J Geodyn, 38: 1-18.

Prescott WH (1981) The determination of displacement fields from geodetic data along a strike slip fault. J Geoph Res, 86(B7): 6067-6072.

Sansò F (1982) Three lectures on mathematical theory of elasticity. In Moritz H and Sunkel H (eds.) "Geodesy and Geodynamics", Lecture Notes, Third International Summer School in the Mountains on Geodesy and Global Geodynamics, Frauenberg, Austria, August 30 – September 10, 1982, pp. 461-530.

Sokolnikoff IS (1956) Mathematical theory of elasticity. McGraw Hill, New York.

Soler T (1977) Global plate tectonics and the secular motion of the pole. Report No. 252, Dept. of Geodetic Science, Ohio State University, Columbus, Ohio.

Soler T, van Gelder B (1991) On covariances of eigenvalues and eigenvectors of second rank symmetric tensors. Geoph J Int, 105: 537-546.

Voosoghi B (2000) Intrinsic deformation analysis of the earth surface based on 3-dimensional displacement fields derived from space geodetic measurements. Report Nr. 2003, Institut für Photogrammetrie, Universität Stuttgart.

Wernicke B, Davis JL, Bennett RA, Normandeau JE, Friedrich AM, Niemi NA (2004) Tectonic implications of a dense continuous GPS velocity field at Yucca Mountain, Nevada. J Geoph Res, 109, B12404, doi: 10.1029/2003JB002832.

Williams SDP (2003) The effect of coloured noise on the uncertainties of rates estimated from geodetic time series. J Geod, 76: 483-494.

Wright TJ, Parsons BE, Lu Z (2004) Toward mapping surface deformation in three dimensions using InSAR. Geoph Res Lett, 31, L01607, doi:10.1029/2003GL018827.

Xu P, Shimada S, Fujii Y, Tanaka T (2000) Invariant geodynamical information in geometric geodetic measurements. Geoph J Int, 142: 586-602.

The Bayesian Approach Applied to Significant Deformation Identification

F. Sansò
Polytechnic of Milan, D.I.I.A.R.- Surveying Section. Piazza Leonardo da Vinci, 32. 20133 Milan, Italy. e-mail: fernando.sanso@polimi.it

M. Clara de Lacy
Dept. Ingeniería Cartográfica, Geodésica y Fotogrametría. Escuela Politécnica Superior. Universidad de Jaén. Campus Las Lagunillas. 23071 Jaén. Spain. e-mail:mclacy@ujaen.es

Abstract. Geodetic monitoring is a complex of measurement procedures and data analysis which have primarily the purpose of answering the question: is the surveyed object undergoing a deformation agreeing with a same theoretical model? The simplest of such models is indeed that no deformation at all is verified and in this case we just ask the question whether the body is deforming or not. It is classical to answer to such problems by a significance analysis based on statistical testing theory. It is also a common feeling that many times testing procedures give a result of non significance (there is no evidence that the basic hypothesis H_0 can be refused) where the human eye and brain system tend to see a significant deformation. This "feeling" can be formalized by the Bayesian theory. This introduces some prior probability distribution in the model space and more acceptable results are obtained. In particular this seems to be effective when correlations are introduced which embody into the prior distribution very general behaviours of physical objects. This attenuates the pure geodetic point of view which often sees the monitoring as the determination of coordinates variations at the knots of a network where each point is physically isolated from the others.

1 Introduction

The problem we are facing is to find the deformation vector that satisfies equations established by the elasticity theory under hypothesis of linear approximation, small deformations, homogeneity and isotropy.

More precisely our purpose is to find the field of deformations from discrete values derived by monitoring deformations by geodetic methods. The standard approach is based on the establishment of a geodetic network. Measurements are repeated at different epochs and the coordinates of the network points are obtained at different epochs. The following step is to test the difference of coordinates. By hypothesis testing a decision is reached whether a coordinate difference is significantly different from zero so that deformations can be assumed to be active. Nevertheless, this approach has two main disadvantages: the deformations to be estimated are sometimes small with respect to the measurements precision and mathematical models used are often very simple. Therefore the error of the observations can collect a part of the parametric model error and we cope with a situation in which significant deformations are not detected by classical tests while the "human eye" tends to see them. The proposed solution in this paper is considering the problem in a Bayesian framework. In this case some prior information can be introduced. The deformation vector has by definition a prior information. It's here where the geophysical information can be included. For example, from the experience we can know if our problem corresponds to a compressive fault or if the study area is moving $5mm/year$ at least as an order of magnitude. If a classical test is carried out, we close the eyes to this information and we only compare coordinates. If a Bayesian test is considered, we use our prior information and we introduce it into the mathematical model. This procedure will allow us to put in evidence situations in which a classical test does not detect significant deformations while a Bayesian test does. This paradox has been studied in the last years by Koch (1984 and 1988) and Betti et al, (2001). In this paper we step forward in the study of the problem. In particular, in Section 2 a short review of the Bayesian approach is presented. In Section 3 a theorem is introduced. It will provide us with a sufficient condition to have situations in which a classical test accepts the null hypothesis (H_0) at a significance level α while a Bayesian test rejects H_0 at the same significance level. Two elementary examples illustrating this situation are shown. In Section 4 the Bayesian prediction at an arbitrary point, generally different from the observation sites, is presented along with two simple examples regarding non parametric and parametric approaches, respectively. The former pretends to be an engineering example and the latter pretends to represent a geophysical case.

2 Recalls of Bayesian estimation theory

2.1 Bayes' theorem

Bayes' theorem of elementary probability theory is as follows (Sansò and Venuti, 1997):

Theorem 1 *Given a set Ω with a probability distribution P, given a partition of Ω in non-overlapping measurable sets D_i,*

$$\bigcup_{i=1}^{n} D_i = \Omega, \quad D_i \cap D_j = \emptyset \quad i \neq j$$

and given an event A, the following holds:

$$P(D_i|A) = \frac{P(A|D_i)P(D_i)}{\sum_k P(A|D_k)P(D_k)} \tag{1}$$

where $P(D_i)$ is the prior information on D_i, $P(A|D_i)$ is the probability of sampling in A given D_i and $P(D_i|A)$ is the posterior probability.

When D_i is just a partition associated with a discrete random variable

$$D_i \equiv \{\omega; x(\omega) = x_i, i = 1, 2, \ldots, n\} \tag{2}$$

and A is also an event associated with an "observable" discrete random variable Y

$$A \equiv \{\omega; Y(\omega) = y_k\} \tag{3}$$

we write (1) as

$$P(X = x_i|Y = y_k) = \\ = \frac{P(Y = y_k|X = x_i)P(X = x_i)}{\sum_j P(Y = y_k|X = x_j)P(X = x_j)} \tag{4}$$

The expression (4) says that observing the value y_k for Y modifies the prior distribution of X, $P(X = x_i)$ into the posterior distribution $P(X = x_i|Y = y_k)$.

Let us notice that if $X = x_i$ is thought of as a fixed value of the parameter x, $P(Y = y_k|X = x_i)$ can be considered as the probabilistic model of the observation of Y, given $X = x_i$, i. e., the likelihood of the observations given the value of the parameter; for this reason we will also call

$$P(Y = y_k|X = x_i) = L(y_k|x_i). \tag{5}$$

If (X, Y) are continuous variables, the expression (5) is written in terms of probability densities as

$$P(x|y) = \frac{L(y|x)\tilde{P}(x)}{\int L(y|x)\tilde{P}(x)dx} \tag{6}$$

where X and Y can be scalar as well as vector variates and the integral ranges over the whole parameter space.

In (6), $\tilde{P}(x)$ is the prior distribution of the parameter X, while $P(x|y)$ is the posterior distribution, i. e. $P(x|y) = f_{X|Y}(x|y)$, $f_{X|Y}(x|y)$ indicating the conditional probability density.

All the information on X after observing Y is now contained in $P(x|y)$ and of course if we want to determine representative values we could for instance compute $\hat{x}_B = E_X\{X|Y = y\}$ interpreted as a Bayesian estimator, while the variance associated with the distribution $P(x|y)$ can be interpreted as the estimation error of $\hat{x}_B$.

2.2 Highest Posterior Density Region (HPDR)

If the posterior distribution of a parameter has been determined by Bayes' theorem (1), we may compute the probability that the parameter lies in a subset of the parameter space. Furthermore, often we are interested in finding the subset, where most, for instance 95%, of the probability is concentrated. Obviously there are an infinite number of ways to specify such a region. To obtain a unique solution, provided the density function has only one modal value, i. e. one maximum, the region should be defined such that the density of every point inside is at least as large as for any point outside of it. A region with such a property is called a region of highest posterior density (HPDR) or Bayesian confidence region (Koch 1990). In addition it has the property that for a given probability level it occupies the smallest possible volume in the parameter space.

Definition 1 *Let $\underline{x}$ be distributed as a gaussian with mean $\underline{\mu}$ and covariance C, namely*

$$\underline{x} \sim g(\underline{x}|\underline{\mu}, C) \tag{7}$$

with dim $\underline{x} = p$, the HPDR at significance level α is given by

$$\begin{aligned} &HPDR(\underline{\mu}, C, 1-\alpha) = \\ &= \{\underline{x}; (\underline{x}-\underline{\mu})^+ C^{-1}(\underline{x}-\underline{\mu}) \leq \chi^2_{p,\alpha}\} \end{aligned} \tag{8}$$

2.3 g-Lemma

The result of this well known algebraic lemma can be seen in Box and Tiao (1992). Its proof is shown in Appendix.

Lemma 1 *If $g(\underline{x}|\underline{\xi}, K)$ and $g(\underline{x}|\underline{\eta}, H)$ are given gaussian distributions, then*

$$g(\underline{x}|\underline{\xi}, K) g(\underline{x}|\underline{\eta}, H) = g(\underline{x}|\bar{\underline{x}}, W^{-1}) g(\underline{\xi}-\underline{\eta}|\underline{0}, C) \tag{9}$$

with

$$\begin{aligned} \bar{\underline{x}} &= W^{-1}(K^{-1}\underline{\xi} + H^{-1}\underline{\eta}) \\ W &= K^{-1} + H^{-1} \\ C &= K + H \end{aligned}$$

3 Classical testing and Bayesian testing

3.1 Theorem

Theorem 2 *Let $\underline{u}_T = (\ldots \underline{u}(\underline{x}_i) \ldots)^+$ be the deformation vector at N points $\underline{x}_i$ $(i = 1, \ldots N)$. Let us consider the mathematical model*

$$\underline{u}_0(\underline{x}_i) = \underline{u}_T(\underline{x}_i) + \underline{\epsilon}(\underline{x}_i) \tag{10}$$

where $\underline{u}_0$ represents the observation vector of $\underline{u}_T$ and $\underline{\epsilon}$ is the observation noise; we call C_0 the covariance matrix coming from geodetic measurements.

Let us also assume the prior distribution of $\underline{u}_T$ to be given by a normal distribution, namely

$$\underline{u}_T \sim g(\underline{u}_T|\tilde{\underline{u}}, \tilde{C}) \tag{11}$$

Then:

1. *The posterior distribution of $\underline{u}_T$ is given by*

$$\underline{u}_T \sim g(\underline{u}_T|\bar{\underline{u}}, W^{-1}) \tag{12}$$

with

$$\begin{aligned} W &= \tilde{C}^{-1} + C_0^{-1} \\ \bar{\underline{u}} &= W^{-1}(\tilde{C}^{-1}\tilde{\underline{u}} + C_0^{-1}\underline{u}_0) \end{aligned}$$

2. *There are situations determined by the values of $\underline{u}_0$, $\tilde{\underline{u}}$, $\tilde{C}$, C_0 and α in which it holds simultaneously*

$$\begin{aligned} \underline{u}_0^+ C_0^{-1}\underline{u}_0 &\leq \chi^2_{p,\alpha} \\ \bar{\underline{u}}^+ W^{-1}\bar{\underline{u}} &> \chi^2_{p,\alpha} \end{aligned} \tag{13}$$

Proof:

1. From (1), the posterior distribution of $\underline{u}_T$ is given by

$$\begin{aligned} P(\underline{u}_T|\underline{u}_0) &= \frac{P(\underline{u}_T, \underline{u}_0)}{\int P(\underline{u}_T, \underline{u}_0) d\underline{u}_T} = \\ &= \frac{g(\underline{u}_0|\underline{u}_T, C_0) g(\underline{u}_T|\tilde{\underline{u}}, \tilde{C})}{\int P(\underline{u}_T, \underline{u}_0) d\underline{u}_T} = \\ &= \frac{g(\underline{u}_T|\underline{u}_0, C_0) g(\underline{u}_T|\tilde{\underline{u}}, \tilde{C})}{\int P(\underline{u}_T, \underline{u}_0) d\underline{u}_T} \end{aligned} \tag{14}$$

Taking the Lemma 1 into account

$$\begin{aligned} P(\underline{u}_T|\underline{u}_0) &= \\ &= \frac{g(\underline{u}_T|\bar{\underline{u}}, W^{-1}) g(\underline{u}_0 - \tilde{\underline{u}}|\underline{0}, C)}{g(\underline{u}_0 - \tilde{\underline{u}}|\underline{0}, C)} = \\ &= g(\underline{u}_T|\bar{\underline{u}}, W^{-1}) \end{aligned} \tag{15}$$

2. To prove (13) it is enough to demonstrate that, for α suitably chosen, there are cases in which

$$\bar{\underline{u}}^+ W^{-1}\bar{\underline{u}} > \underline{u}_0^+ C_0^{-1}\underline{u}_0 \tag{16}$$

Let us suppose

$$\begin{aligned} P_0 &= C_0^{-1} \\ \tilde{P}(\lambda) &= \lambda\tilde{P}; \text{ with } \tilde{P} = \tilde{C}^{-1} \\ V(\lambda) &= \lambda\tilde{P} + P_0 \\ \bar{\underline{u}}(\lambda) &= (\lambda\tilde{P} + P_0)^{-1}(\lambda\tilde{P}\tilde{\underline{u}} + P_0\underline{u}_0) \end{aligned}$$

where λ is the inverse of the prior variance. It is important to note that if $\lambda \to 0$, the prior is actually a non informative prior. In fact, when $\lambda = 0$ the procedure goes to a classical test situation, nevertheless if $\lambda \neq 0$ this goes to the Bayesian approach.

We consider

$$\begin{aligned}\Phi(\lambda) &= \underline{\bar{u}}^+ V^{-1} \underline{\bar{u}} = \\ &= (\lambda \tilde{P} \underline{\tilde{u}} + P_0 \underline{u}_0)^+ (\lambda \tilde{P} + P_0)^{-1} \\ &\quad (\lambda \tilde{P} \underline{\tilde{u}} + P_0 \underline{u}_0) \qquad (17)\end{aligned}$$

and we are going to study if among $\Phi(\lambda)$ there is a quadratic form that verifies (16).

It is clear that

$$\Phi(0) = \underline{u}_0^+ P_0 \underline{u}_0 = \underline{u}_0^+ C_0^{-1} \underline{u}_0 \qquad (18)$$

coinciding with the right hand of (16); thus (16) is transformed into the equivalent condition

$$\Phi(\lambda) > \Phi(0) \qquad (19)$$

which is true in a neighborhood of zero if

$$\Phi'(0) > 0 \qquad (20)$$

The derivative of Φ at the origin is

$$\begin{aligned}\Phi'(0) &= 2(\tilde{P}\underline{\tilde{u}})^+ P_0^{-1} P_0 \underline{u}_0 - \\ &- (P_0 \underline{u}_0)^+ P_0^{-1} \tilde{P} P_0^{-1} P_0 \underline{u}_0 = \\ &= 2\underline{\tilde{u}}^+ \tilde{P} \underline{u}_0 - \underline{u}_0^+ \tilde{P} \underline{u}_0 = \\ &= \underline{\tilde{u}}^+ \tilde{P} \underline{\tilde{u}} - (\underline{u}_0 - \underline{\tilde{u}})^+ \tilde{P} (\underline{u}_0 - \underline{\tilde{u}}) \\ &\qquad (21)\end{aligned}$$

It is obvious that there are particular choices of $\underline{\tilde{u}}$ and $\underline{u}_0$ for which $\Phi'(0) > 0$. For example, if we take $\underline{u}_0 = \underline{\tilde{u}}$ we obtain

$$\Phi'(0) = \underline{\tilde{u}}^+ \tilde{P} \underline{\tilde{u}} > 0. \qquad (22)$$

In addition, we can say that if

$$\underline{u}_0 = \underline{\tilde{u}} + \underline{v} \qquad (23)$$

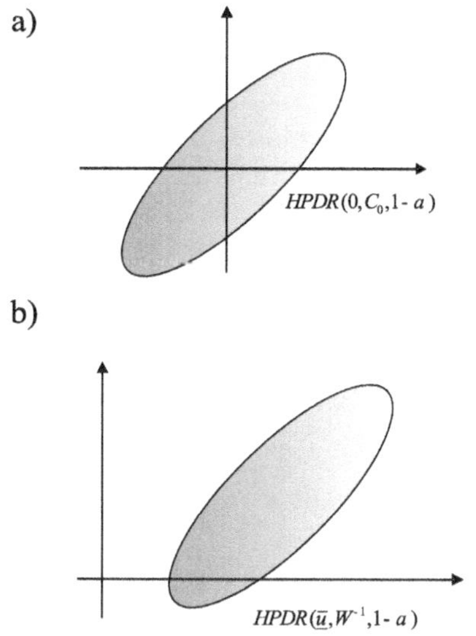

Fig. 1 a) HPDR in the classical hypotesis testing. b) HPDR in the Bayesian approach.

the condition (20) is verified if

$$\underline{\tilde{u}}^+ \tilde{P} \underline{\tilde{u}} > \underline{v}^+ \tilde{P} \underline{v} \qquad (24)$$

which becomes a sufficient condition for (16) to be true.

Remark 1 *The above theorem shows evidences of situations in which a classical test accepts the null hypothesis H_0 at significance level α while a Bayesian test rejects H_0 at the same significance level. From a geometrical point of view, we can say that (19) constitutes a sufficient condition for*

$$\underline{u}_0 \in HPDR(0, C_0, 1-\alpha) \qquad (25)$$

and

$$\underline{u}_T = 0 \notin HPDR(\underline{\bar{u}}, W^{-1}, 1-\alpha) \qquad (26)$$

simultaneously (Fig. 1). It is to say that there is a difference between observation and prediction, not so large, but it does exist.

In the deformation context, we can say that the above sufficient condition allows us to find particular cases in which a classical test would not detect significant deformations and a Bayesian test would.

3.2 Examples

In this section some particular examples in which (19) is verified are presented.

Let us assume

$$P_0 = C_0^{-1} = \frac{1}{\sigma_0^2} \cdot I$$
$$\tilde{P} = \tilde{C}^{-1} = \frac{\lambda}{\tilde{\sigma}^2} \cdot I$$

where σ_0^2 and $\tilde{\sigma}^2$ are the observation variance and prior variance, respectively and I represents the identity matrix.

We will consider

$$V(\lambda) = (\frac{\lambda}{\tilde{\sigma}^2} + \frac{1}{\sigma_0^2}) \cdot I = \frac{1}{\sigma_0^2}(\gamma\lambda + 1) \cdot I \quad (27)$$

where

$$\gamma = \frac{\sigma_0^2}{\tilde{\sigma}^2}$$

and

$$\underline{u}(\lambda) = \frac{\sigma_0^2}{(\gamma\lambda+1)}(\frac{\lambda}{\tilde{\sigma}^2}\underline{\tilde{u}} + \frac{1}{\sigma_0^2}\underline{u}_0) \quad (28)$$

$$\Phi(\lambda) = \frac{\sigma_0^2}{(\gamma\lambda+1)}(\frac{\lambda}{\tilde{\sigma}^2}\underline{\tilde{u}} + \frac{1}{\sigma_0^2}\underline{u}_0)^+ (\frac{\lambda}{\tilde{\sigma}^2}\underline{\tilde{u}} + \frac{1}{\sigma_0^2}\underline{u}_0) \quad (29)$$

Considering (23) we can write

$$\Phi(\lambda) = \frac{1}{\sigma_0^2}\frac{1}{(\gamma\lambda+1)}[(\gamma\lambda+1)\underline{\tilde{u}} + \underline{v}]^+ [(\gamma\lambda+1)\underline{\tilde{u}} + \underline{v}] \quad (30)$$

If we put

$$\tau = \gamma\lambda + 1$$

then

$$\Phi(\tau) = \frac{1}{\sigma_0^2}\frac{1}{\tau}(\tau\underline{\tilde{u}} + \underline{v})^+(\tau\underline{\tilde{u}} + \underline{v}) = \frac{1}{\sigma_0^2}(\frac{\tau^2|\underline{\tilde{u}}|^2 + 2\tau\underline{\tilde{u}}^+\underline{v} + |\underline{v}|^2}{\tau}) \quad (31)$$

the condition (19) is then equivalent to

$$\Phi(\tau) > \Phi(1)$$

for $\tau > 1$.

As

$$\Phi(1) = \frac{1}{\sigma_0^2}(|\underline{\tilde{u}}|^2 + 2\underline{\tilde{u}}^+\underline{v} + |\underline{v}|^2), \quad (32)$$

the condition we are looking for is

$$(\tau^2 - \tau)|\underline{\tilde{u}}|^2 + (1-\tau)|\underline{v}|^2 > 0 \quad (33)$$

Since $\tau > 1$, the above expression reads

$$\tau|\underline{\tilde{u}}|^2 - |\underline{v}|^2 > 0 \quad (34)$$

or equivalently

$$|\underline{v}| < \sqrt{\tau}|\underline{\tilde{u}}| \quad (35)$$

In the following we are giving a very elementary example to evidence a situation in which the sufficient condition (19) is verified. We consider vectors in R^2, a covariance matrix of the observations $C_0 = I$, a prior distribution $\tilde{P} = \lambda I$ and a significance level $\alpha = 5\%$. This last hypothesis implies that $\underline{u}_0^+\underline{u}_0 < 5.99$ is the region in R^2 where H_0 is accepted for an ordinary test.

Taking (23) and (35) into account,

$$|\underline{v}|^2 = |\underline{u}_0 - \underline{\tilde{u}}|^2 < \tau|\underline{\tilde{u}}|^2 \quad (36)$$

Let us suppose $\lambda = 1$, then $\tau = 2$ and the above expression reads

$$|\underline{u}_0|^2 - 2\underline{u}_0 \cdot \underline{\tilde{u}} + |\underline{\tilde{u}}|^2 < 2|\underline{\tilde{u}}|^2 \quad (37)$$

or equivalently

$$|\underline{\tilde{u}}|^2 + 2\underline{u}_0 \cdot \underline{\tilde{u}} + |\underline{u}_0|^2 > 2|\underline{u}_0|^2 \quad (38)$$

that is to say that

$$|\underline{\tilde{u}} + \underline{u}_0|^2 > 2|\underline{u}_0|^2 \quad (39)$$

is the condition we are looking for (Fig.2).

If we consider for example

$$\underline{\tilde{u}} = (1,1)^+$$
$$\underline{u}_0 = (3/2, 3/2)^+$$
$$\underline{v} = (1/2, 1/2)^+,$$

we have

$$\Phi(\lambda) = 6.25 > \chi^2_{\alpha=5\%} = 5.99 > \Phi(0) = \underline{u}_0^+\underline{u}_0 = 4.5$$

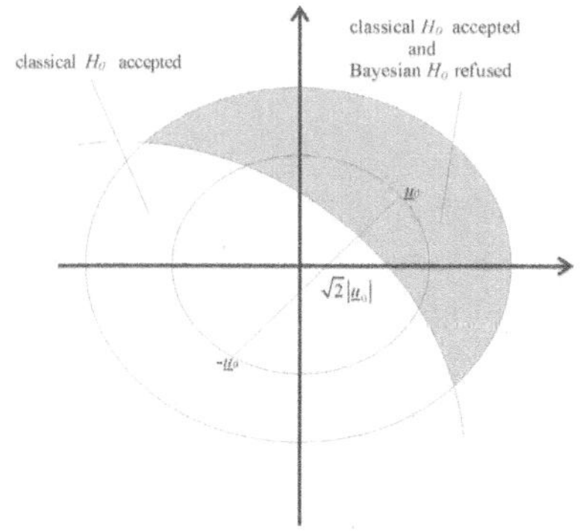

Fig. 2 Classical zone and Bayesian zone

4 Bayesian prediction at an arbitrary point

In this paragraph we try to look at a slightly more general problem where, contrary to Section 3, the deformation of a certain region R is represented by a field $\underline{u}(x)$ and not just by a value vector of $\underline{u}$ at a discrete number of points.

If we stay within the hypothesis, mentioned in the introduction, such a field will satisfy a classical system of partial differential equations, namely (Sansò, 1982)

$$\left\{ \begin{array}{l} \mu\Delta\underline{u} + (\lambda+\mu)\nabla(\nabla\cdot\underline{u}) = \underline{F} \\ +\ boundary\ conditions \end{array} \right. \tag{40}$$

where $\underline{u}$ is the deformation vector, Δ is the Laplace operator, ∇ is the Nabla operator, $\underline{F}$ is an elastic force and λ, μ are the Lamé constants. In the following examples we shall assume that $\underline{u}$ is zero in the boundary.

Now, we reduce the deformation field to zero mean, it is to say that we assume that by using some vague prior knowledge on the "average" $\underline{\tilde{u}}(\underline{x})$, we can define a residual field $\underline{S}(\underline{x})$ with the following characteristics

$$\begin{array}{rcll} \underline{S}(\underline{x}) & = & \underline{u}(\underline{x}) - \underline{\tilde{u}}(\underline{x}) & C_{\underline{SS}} = \Sigma(\underline{x}) \\ \underline{S}_0 & = & \{\underline{S}_0(\underline{x}_i)\} = & \\ & = & \{\underline{u}_0(\underline{x}_i) - \underline{\tilde{u}}(\underline{x}_i)\} & C_{\underline{S}_0\underline{S}_0} = \tilde{C} + C_0 \\ \underline{S}_T & = & \{\underline{S}(\underline{x}_i)\} = & \\ & = & \{\underline{u}(\underline{x}_i) - \underline{\tilde{u}}(\underline{x}_i)\} & C_{\underline{S}_T\underline{S}_T} = \tilde{C} \\ \underline{S}_0(\underline{x}_i) & = & \underline{S}(\underline{x}_i) + \underline{\epsilon}(\underline{x}_i) & C_{\epsilon\epsilon} = C_0 \end{array} \tag{41}$$

where $\underline{S}$ represents the unknown field to be predicted and $C_{\underline{SS}}$ is its matrix covariance function; $\underline{S}_0$ and $C_{\underline{S}_0\underline{S}_0}$ are the observation vector of the unknown field and its covariance matrix, respectively and $\underline{S}_T$ is a vector whose components are the subvectors $\underline{S}(\underline{x}_i)$. The dimension of $\underline{S}_T$ is $3 \times N$. The observation vector is composed by the signal and the noise $\underline{\epsilon}$.

Making the hypothesis that all random variables are normally distributed, the posterior distribution is given by

$$\underline{S}(\underline{x})|_{\underline{S}_0} \sim g(\underline{S}(\underline{x})|C_{\underline{SS}_0}C^{-1}_{\underline{S}_0\underline{S}_0}\underline{S}_0; \Delta) \tag{42}$$

where

$$\begin{array}{rcl} \Delta & = & \Sigma(\underline{x}) - C_{\underline{SS}_0}C^{-1}_{\underline{S}_0\underline{S}_0}C_{\underline{S}_0\underline{S}} \\ C_{\underline{SS}_0} & = & C_{\underline{SS}_T} = \underline{C}(\underline{x})^+ \end{array}$$

The Bayesian estimator and its error are given by the mean and covariance, respectively

$$E\{\underline{S}(\underline{x})|_{\underline{S}_0}\} = \underline{C}(\underline{x})^+(\tilde{C} + C_0)^{-1}\underline{S}_0 \tag{43}$$

$$C(\underline{S}(\underline{x})|_{\underline{S}_0}) = \Sigma(\underline{x}) - \underline{C}(\underline{x})^+(\tilde{C} + C_0)^{-1}\underline{C}(\underline{x}) \tag{44}$$

as in the classical collocation solution (Barzaghi and Sansò, 1983).

As these expressions show, for the actual computation of the Bayesian predicted field, the covariances are required. In the following we present two examples regarding the importance of the cavariance matrix in the Bayesian prediction and how we can draw information about them.

4.1 Non parametric approach

Let us assume the deformation field given by an equation system under hypothesis of small, elastic and linear deformation. That is equivalent to say that the deformation field is governed by (40) and the covariance function we are looking for will have to take into account this equation. We also consider that the region R, where (40) has to hold, is a given domain with boundary B.

Let us assume deformations generated by a concentrated random force

$$\underline{F}(\underline{x}, \underline{x}_0) = \underline{F}\delta(\underline{x}, \underline{x}_0) \tag{45}$$

where δ is the Dirac Delta function and $\underline{x}_0$ is a random point for example uniformly distributed in the region of the deforming body and $\underline{F}$ is distributed according to

$$\underline{F} \sim \mathcal{N}[0, F_0^2 \cdot I] \tag{46}$$

It is known that the reaction to this force is related to the Green function of the equation (40). Under elastic hypothesis it means that there is a 3×3 matrix $\{G(\underline{x}, \underline{x}_0)\}$ which overall satisfies the system of differential equations (Love, 1994)

$$\begin{aligned} \mu\Delta\{G(\underline{x}, \underline{x}_0)\} &+ (\lambda+\mu)\nabla\nabla \cdot \{G(\underline{x}, \underline{x}_0)\} = \\ &= \delta(\underline{x}, \underline{x}_0) \cdot I_3 \end{aligned} \tag{47}$$

together with suitable boundary conditions, of which the case of Dirichlet conditions is the most classical, namely $G(\underline{x}, \underline{x}_0)|_B = 0$.

The existence and uniqueness of such a general Green functions is guaranteed by the mathematical theory of elasticity (Landau and Lifsits, 1979; Courant and Hilbert, 1953). In particular when the differential operator is selfadjoint, like in (47), one can prove that $\{G(\underline{x}, \underline{x}_0)\} = \{G^+(\underline{x}_0, \underline{x})\}$ which is also known as reciprocity theorem (Courant and Hilbert, 1953). With the aid of the matrix $\{G(\underline{x}, \underline{x}_0)\}$, the general solution of (40) with zero boundary conditions can be expressed by

$$\underline{u}(\underline{x}) = \int_R \{G(\underline{x}, \underline{y})\}\underline{F}(\underline{y})d\underline{y} \tag{48}$$

Of course when $F(\underline{y})$ has the form (45), the equation (48) becomes

$$\underline{u}(\underline{x}) = \{G(\underline{x}, \underline{x}_0)\}\underline{F}. \tag{49}$$

At this point we can compute the matrix covariance function of the field $\underline{u}(\underline{x})$ by the standard formula

$$\begin{aligned} C(\underline{x}, \underline{y}) &= E_{\underline{x}_0, \underline{F}}\{\underline{u}(\underline{x})\underline{u}(\underline{y})^+\} = \\ &= \frac{1}{|R|}\int_R dx_0\{G(\underline{x}, \underline{x}_0)\}E\{\underline{F}\underline{F}^+\}\{G^+(\underline{y}, \underline{x}_0)\} = \\ &= \frac{F_0^2}{|R|}\int_R dx_0\{G(\underline{x}, \underline{x}_0)\}\{G^+(\underline{y}, \underline{x}_0)\} = \\ &= \frac{F_0^2}{|R|} \cdot H(\underline{x}, \underline{y}) \end{aligned} \tag{50}$$

where $|R|$ is just the measure of the region R. It is clear that the covariance function (50) inherits some properties of the typical solutions of the equation of elasticity through the presence of the Green matrix-function convolved with itself.

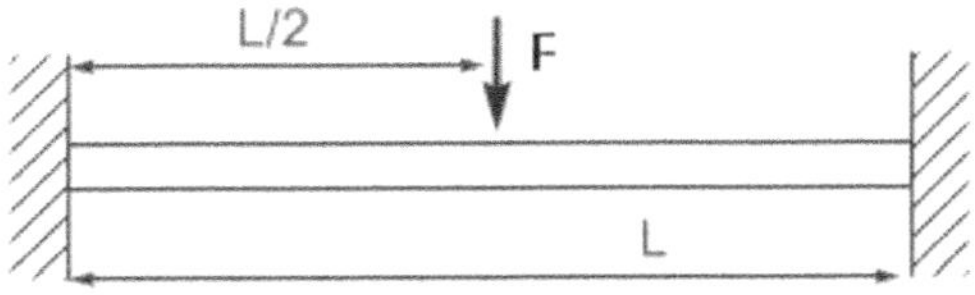

Fig. 3 Non parametric example: a clamped beam with $L=1$.

4.1.1 Example

This following simple case pretends to present an engineering example. Let us assume the equation of a clamped beam with length equal to 1 (Fig. 3), so that our domain R is just the interval $[0, 1]$. In this case, assuming that the beam charged by a force F orthogonal to the beam itself and that most of its deformation is verified in this orthogonal direction (axis z), it can be shown (Landau and Lifsits, 1979) that the deformation pattern $z(x)$ (with $0 \le z \le 1$) is governed by a simplified form of equation (40), namely

$$\begin{aligned} &z^{iv} \equiv F \\ &\begin{cases} z(0) = z(1) = 0 \\ z'(0) = z'(1) = 0 \end{cases} \end{aligned} \tag{51}$$

Since we are in dimension one, the Green matrix function becomes here just a scalar function $G(x, x_0)$, satisfying the simple equation

$$G^{iv}(x, x_0) = \delta(x - x_0) \tag{52}$$

with the same boundary conditions as in (51). Therefore the "random" deformation field satisfies

$$z^{iv} = F\delta(x - x_0) \tag{53}$$

Thus, once $G(x, x_0)$ is found, the deformation field is given by

$$z = FG(x, x_0) \tag{54}$$

and the corresponding covariance function is

$$C(x, y) = F_0^2 H(x, y) \tag{55}$$

We turn now to the exercise of finding $G(x, x_0)$. We try with a function of the form

$$G(x,x_0) = \frac{1}{6}\chi(x-x_0)(x-x_0)^3 + \\ + x^2(Ax+B) \quad (56)$$

where χ is the Heaviside function

$$\chi(x-x_0) = \begin{cases} 0 & x < x_0 \\ 1 & x > x_0 \end{cases} \quad (57)$$

A direct computation will show that $G^{iv}(x,x_0) = \delta(x-x_0)$. In fact

$$\begin{aligned} D^4x^2(Ax+B) &= 0 \\ D\chi(x-x_0) &= \delta(x-x_0) \\ \delta(x-x_0)(x-x_0)^\alpha &= 0 \ \forall\alpha \geq 1 \end{aligned} \quad (58)$$

where D represents the derivative. Since the conditions $z(0)=0$, $z'=0$ are automatically satisfied by the family (56), the two constants A and B can be determined from the boundary conditions only

$$\begin{cases} G(1,x_0) = \frac{1}{6}(1-x_0)^3 + A + B = 0 \\ G'(1,x_0) = \frac{1}{2}(1-x_0)^2 + 3A + 2B = 0 \end{cases} \quad (59)$$

Solving the equation system, we obtain

$$\begin{cases} A = -\frac{1}{2}(1-x_0)^2 + \frac{1}{3}(1-x_0)^3 \\ B = \frac{1}{2}(1-x_0)^2 - \frac{1}{2}(1-x_0)^3 \end{cases} \quad (60)$$

and then the Green function reads

$$\begin{aligned} G(x,x_0) &= \frac{1}{6}\chi(x-x_0)(x-x_0)^3 + \\ &+ x^2[\frac{1}{2}(1-x_0)^2(1-x) + \\ &+ (1-x_0)^3(\frac{1}{3}x - \frac{1}{2})] \end{aligned} \quad (61)$$

To obtain the Bayesian estimator we first need to calculate

$$H(x,y) = \int_0^1 G(x,x_0)G(y,x_0)dx_0 \quad (62)$$

Solving this integral, we obtain:

$$\begin{aligned} H(x,y) &= (\frac{1}{20160})\Big[(x^6 - 6x^5 + 15x^4y^2 - \\ &- 20x^3y^3 + 15x^2y^4 - 6xy^5 + \\ &+ y^6)|x-y| + (y^3(35x^3 - 21x^2y + \\ &+ 7xy^2 - y^3)|y| - x^4(x^3(4y^3 - 6y^2 + \\ &+ 1) - 7x^2y(2y^2 - 4y + 1) - 21xy^2 + \\ &+ 35y^3)) + (y-1)^3(x^3(4y^3 + 2y^2 - \\ &- 16y - 25) - x^2(6y^3 - 4y^2 - 31y - \\ &- 76) - 7x(y^2 + 4y + 10) + \\ &+ y^3 + 4y^2 + 10y + 20)|y-1| - \\ &- y^3(x^3(4y^3 - 14y^2 + 35) - \\ &- x^2y(6y^2 - 28y + 21) - 7xy^2 + \\ &+ y^3)|y| + x^3(68y^3 - 102y^2 + 84y - \\ &- 25) - x^2(102y^3 - 300y^2 + 273y - \\ &- 76) + 7x(12y^3 - 39y^2 + 36y - 10) - \\ &- 25y^3 + 76y^2 - 70y + 20\Big] \end{aligned} \quad (63)$$

In Fig. 4 the behaviour of the covariance function is studied from the behaviour of $H(x,y)$. In particular $H(x,1/2)$ and $H(x,x)$ are shown. As we can see the behaviour of the covariance $C(x,y) = F_0^2H(x,y)$ nicely reflects the zero boundary conditions as well as the expected symmetry properties. We can try to compute this expression numerically with H given by (63) and F_0 given by the elasticity theory as

$$F_0 = \frac{fL^3}{192EI} \quad (64)$$

where EI is the so called inflectional rigidity, while f is the pointwise force acting on the beam with length equal to L. Let us suppose a steel beam ($E = 200000\,MPa$), $4\,m$ long, with section equal to $0.2 \times 0.2\,m^2$ ($I = 1.333 \cdot 10^8\,mm^4$), charged at $x_0 = 2\,m$ with a force $f = 1\,MN$. From (64), $F_0 = 0.0125\,m$. If we consider (50) and (63) between 0 and 4 we obtain

$$\begin{aligned} C(1/2,1/2)^{1/2} &= F_0 \cdot H(1/2,1/2)^{1/2} \\ &\approx 0.0025m \end{aligned}$$

that seems to be a quite reasonable number.

4.2 Parametric approach

In this case the deformation field is not given by a system of equations. A situation like this is present when the material is unknown or when we face a non linear phenomena (for example a structure problem). In these cases an empirical and "ad hoc" model is used, it means that we write the deformation field as

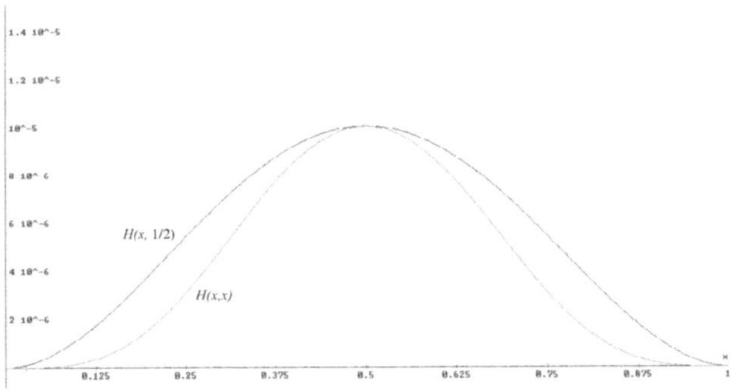

Fig. 4 $H(x, 1/2)$ and $H(x, x)$

$$\underline{u}(\underline{x}) = \underline{u}(\underline{x}|\underline{\theta}) \quad (65)$$

where all parameters can be treated in the Bayesian framework, then all of them are considered random variables with a prior distribution. If we call $\tilde{P}(\underline{\theta})$ the prior of the parameter $\underline{\theta}$, the covariance matrix is given by

$$\underline{C}(\underline{x}, \underline{y}) = E_{\theta}\{\underline{S}(\underline{x}|\underline{\theta})\underline{S}(\underline{y}|\underline{\theta})^+\} \quad (66)$$

4.2.1 Example

This simple case pretends to be a geophysical example. Let us consider a fault model with deformations only across the X axis, and suppose that the upper layer slides in the positive direction and the under layer slides in the opposite sense with approximately the same velocity, (Fig. 5).

The empiric model is given in terms of (scalar) deformation along the x axis only, according to the formulas

$$\begin{aligned} u(x,y) &\sim g(u|\tilde{u}, \tilde{\sigma}^2) \quad if \quad L > y > 0 \\ u(x,y) &\sim g(u|-\tilde{u}, \tilde{\sigma}^2) \quad if \quad -L < y < 0 \end{aligned}$$

$$\begin{aligned} E &\quad \{[u(x,y) - \tilde{u}]^+[u(x',y') - \tilde{u}]\} = \\ &= \tilde{\sigma}^2\rho \quad if \quad L > y, y' > 0 \\ E &\quad \{[u(x,y) - \tilde{u}]^+[u(x',y') - \tilde{u}]\} = \\ &= \tilde{\sigma}^2\rho \quad if \quad -L < y, y' < 0 \\ E &\quad \{[u(x,y) - \tilde{u}]^+[u(x',y') - \tilde{u}]\} = \\ &= -\tilde{\sigma}^2\rho \quad if \quad 0 < y < L, -L < y' < 0 \\ &\qquad or\ viceversa \end{aligned} \quad (67)$$

Then

$$\begin{aligned} \tilde{C} &= \tilde{\sigma}^2 \begin{pmatrix} 1 & \rho & -\rho & -\rho \\ \rho & 1 & -\rho & -\rho \\ -\rho & -\rho & 1 & \rho \\ -\rho & -\rho & \rho & 1 \end{pmatrix} = \\ &= \tilde{\sigma}^2\{(1-\rho)\cdot I + 4\rho\underline{a}\underline{a}^+\} \end{aligned} \quad (68)$$

with

$$\underline{a} = \frac{1}{2}(1, 1, -1, -1)^+$$

a vector of modulus 1. Since $\underline{a}\underline{a}^+$ is a projector, the inverse of (68) is given by

$$\tilde{P} = \tilde{C}^{-1} = \frac{1}{\tilde{\sigma}^2}\{\frac{1}{1-\rho}I - \frac{4\rho}{(1-\rho)(1+3\rho)}\underline{a}\underline{a}^+\} \quad (69)$$

Let us consider

$$\underline{\tilde{u}} = \tilde{u}(1, 1, -1, -1)^+$$

and

$$\underline{v} = (v_1, v_2, v_3, v_4)^+$$

If we substitute these vectors into (24) we obtain the displacement condition

$$\begin{aligned} \underline{v}^+\tilde{P}\underline{v} &= \frac{1}{\tilde{\sigma}^2}[\frac{1}{(1-\rho)}(v_1^2 + v_2^2 + v_3^2 + v_4^2) - \\ &- \frac{\rho}{(1-\rho)(1+3\rho)}(v_1 + v_2 - v_3 - v_4)^2] \\ &< \frac{\tilde{u}^2}{\tilde{\sigma}^2}\frac{4}{(1+3\rho)} \end{aligned} \quad (70)$$

Let us take the particular case $\underline{\tilde{u}} = (3, 3, -3, -3)^+$, $\underline{u}_0 = (2, 2, -2, -4)^+$, $\underline{v} = (-1, -1, 1, -1)$. From the displacement condition (70), ρ must be less than 0.9. Let us put $\rho = 0.7$, $\tilde{\sigma} = 5\,mm$, $C_0 = 4I\,mm^2$ and $\alpha = 1\%$; inserting them into (16), we obtain

$$\underline{\bar{u}}^+W^{-1}\underline{\bar{u}} = 100 > \chi^2_{1\%} = 13 > \underline{u}_0^+C_0^{-1}\underline{u}_0 = 7 \quad (71)$$

We have found an elementary example in which a classical test accepts H_0 and the correspondent Bayesian test rejects it, or equivalently

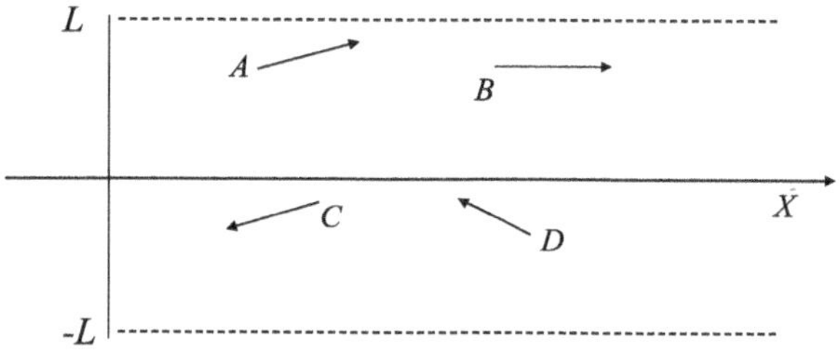

Fig. 5 Parametric example: fault model

a classical test does not detect significant deformations while the Bayesian test does. Readers can prove themselves that the particular choice of numbers is not unique but there are several combinations which give rise to the same conclussion.

5 Conclusions

In this paper we have stepped forward in the study of the Bayesian theory for detecting significant deformations by introducing a new theorem. This has allowed us to establish a sufficient condition to evidence situations in which a Bayesian test can detect deformations while a classical test fails. The benefits of the Bayesian approach versus classical tests have been illustrated by some elementary examples.

APPENDIX

Let us suppose that $g(\underline{x}|\underline{\xi}, K)$ and $g(\underline{x}|\underline{\eta}, H)$ are given gaussian distributions, then

$$g(\underline{x}|\underline{\xi}, K)g(\underline{x}|\underline{\eta}, H) = g(\underline{x}|\bar{\underline{x}}, W^{-1})g(\underline{\xi} - \underline{\eta}|\underline{0}, C) \tag{A.1}$$

with

$$\begin{aligned} \bar{\underline{x}} &= W^{-1}(K^{-1}\underline{\xi} + H^{-1}\underline{\eta}) \\ W &= K^{-1} + H^{-1} \\ C &= K + H \end{aligned}$$

Proof: by using the explicit formula of gaussian distribution, we have

$$g(\underline{x}|\underline{\xi}, K)g(\underline{x}|\underline{\eta}, H) = \frac{1}{(2\pi)^N\sqrt{D\Delta}}e^{\frac{-1}{2}Q} \tag{A.2}$$

where

$$\begin{aligned} D &= detK \\ \Delta &= detH \\ Q &= (\underline{x} - \underline{\xi})^+K^{-1}(\underline{x} - \underline{\xi}) + (\underline{x} - \underline{\eta})^+H^{-1}(\underline{x} - \underline{\eta}) \end{aligned}$$

The quadratic form Q can be written as

$$\begin{aligned} Q &= (\underline{x} - \underline{\xi})^+K^{-1}(\underline{x} - \underline{\xi}) + \\ &+ (\underline{x} - \underline{\eta})^+H^{-1}(\underline{x} - \underline{\eta}) = \\ &= \underline{x}^+K^{-1}\underline{x} - 2\underline{x}^+K^{-1}\underline{\xi} + \\ &+ \underline{\xi}^+K^{-1}\underline{\xi} + \underline{x}^+H^{-1}\underline{x} - \\ &- 2\underline{x}^+H^{-1}\underline{\eta} + \underline{\eta}^+H^{-1}\underline{\eta} \end{aligned} \tag{A.3}$$

Introducing

$$W^{-1} = H(H+K)^{-1}K = K(H+K)^{-1}H$$

Q reads

$$\begin{aligned} Q &= \underline{x}^+W\underline{x} - 2\underline{x}^+WW^{-1}(K^{-1}\underline{\xi} + H^{-1}\underline{\eta}) + \\ &+ \underline{\eta}^+H^{-1}\underline{\eta} + \underline{\xi}^+K^{-1}\underline{\xi} = \\ &= \underline{x}^+W\underline{x} - 2\underline{x}^+W\bar{\underline{x}} + \bar{\underline{x}}^+W\bar{\underline{x}} - \bar{\underline{x}}^+W\bar{\underline{x}} + \\ &+ \underline{\xi}^+K^{-1}\underline{\xi} + \underline{\eta}^+H^{-1}\underline{\eta} = \\ &= (\underline{x} - \bar{\underline{x}})^+W(\underline{x} - \bar{\underline{x}}) - L(\underline{\xi}, \underline{\eta}) \end{aligned} \tag{A.4}$$

and furthermore $L(\underline{\xi}, \underline{\eta})$ can be written as

$$\begin{aligned} L(\underline{\xi}, \underline{\eta}) &= \bar{\underline{x}}^+W\bar{\underline{x}} - \underline{\xi}^+K^{-1}\underline{\xi} - \underline{\eta}^+H^{-1}\underline{\eta} = \\ &= (K^{-1}\underline{\xi} + H^{-1}\underline{\eta})^+W^{-1}WW^{-1} \\ &\quad (K^{-1}\underline{\xi} + H^{-1}\underline{\eta}) - \\ &- \underline{\xi}^+K^{-1}\underline{\xi} - \underline{\eta}^+H^{-1}\underline{\eta} = \\ &= (K^{-1}\underline{\xi})^+H(H+K)^{-1}KK^{-1}\underline{\xi} + \\ &+ 2(H^{-1}\underline{\eta})^+H(H+K)^{-1}KK^{-1}\underline{\xi} + \\ &+ (H^{-1}\underline{\eta})^+H(H+K)^{-1}KH^{-1}\underline{\eta} - \\ &- \underline{\xi}^+K^{-1}\underline{\xi} - \underline{\eta}^+H^{-1}\underline{\eta} \end{aligned} \tag{A.5}$$

After some manipulations and putting

$$\begin{aligned} H &= H + K - K \\ K &= K + H - H \end{aligned}$$

we obtain

$$\begin{aligned} L(\underline{\xi}, \underline{\eta}) &= \underline{\xi}^+K^{-1}H(H+K)^{-1}\underline{\xi} + \\ &+ 2\underline{\eta}^+(H+K)^{-1}\underline{\xi} + \\ &+ \underline{\eta}^+(H+K)^{-1}KH^{-1}\underline{\eta} - \\ &- \underline{\xi}^+K^{-1}\underline{\xi} - \underline{\eta}^+H^{-1}\underline{\eta} = \\ &= \underline{\xi}^+K^{-1}\underline{\xi} - \underline{\xi}^+(H+K)^{-1}\underline{\xi} + \\ &+ 2\underline{\eta}^+(H+K)^{-1}\underline{\xi} + \underline{\eta}^+H^{-1}\underline{\eta} - \end{aligned}$$

$$
\begin{aligned}
- \quad & \underline{\eta}^{+}(H+K)^{-1}\underline{\eta} - \underline{\xi}^{+}K^{-1}\underline{\xi} - \\
- \quad & \underline{\eta}^{+}H^{-1}\underline{\eta} = \\
= \quad & -(\underline{\xi}-\underline{\eta})^{+}(H+K)^{-1}(\underline{\xi}-\underline{\eta})
\end{aligned} \tag{A.6}
$$

Therefore

$$
\begin{aligned}
Q \quad = \quad & (\underline{x}-\underline{\bar{x}})^{+}W(\underline{x}-\underline{\bar{x}}) + \\
+ \quad & (\underline{\xi}-\underline{\eta})^{+}(H+K)^{-1}(\underline{\xi}-\underline{\eta})
\end{aligned} \tag{A.7}
$$

Finally, the denominator of (A.2) is obtained from

$$
\begin{aligned}
& \sqrt{detW^{-1}det(H+K)} = \\
= \quad & \sqrt{detK det(H+K)^{-1} detH det(H+K)} = \\
= \quad & \sqrt{detK detH det(H+K)^{-1} det(H+K)} = \\
= \quad & \sqrt{D\Delta}
\end{aligned} \tag{A.8}
$$

References

Barzaghi, R. and Sansò, F. (1983). Sulla stima empirica della funzione di covarianza. *Anno XLII-Bollettino di Geodesia e Scienze Affini*,N.4, pp. 389-415.

Betti, B., Sansò, F. and Crespi, M. (2001). Deformation Detection According to a Bayesian Approach. In *IV Hotine-Marussi Symposium on Mathematical Geodesy*, Trento, Italy, September 14-17, Vol. 122, Springer Berlin Heidelberg New York, pp. 83-88.

Box, G.E.P. and Tiao, G.C. (1992). *Bayesian inference in statistical analysis.* John Wiley & sons, New York

Courant, R., Hilbert, D. (1953). *Methods of Mathematical physics.* Vol. I-II, Interscience Publ., Jdm Wiley Sons.

Koch, K. R. (1984). Statistical Tests for Detecting Crustal Movements Using Bayesian Inference. NOAA Technical Report NOS NGS 29, pp.1-6.

Koch, K. R. (1988).*Parameter estimation and hypothesis testing in linear models.* Springer Berlin Heidelberg New York Tokyo.

Koch, K. R. (1990).*Bayesian Inference With Geodetic Applications.* Number 31 in Lecture Notes in Earth Sciences. Springer, Berlin Heidelberg New York.

Landau, L., Lifsits, E. (1979). *Teoria dell'elasticit.* Course of Theoretical Physics, Vol. 7 (It. translation), Editor Riuniti.

Love, A. E. H. (1994). *A treatise on the mathematical theory of elasticity.* Dover Pub.

Sanso', F. (1982). Three lectures on mathematical theory of elasticity. Geodesy and Global Geodynamics, pp. 461-477. In Technischen Universitat Graz. Editors Helmut Moritz and Hans Snkel.

Sanso', F., Venuti G. (1997): Integer variables estimation problems: the Bayesian Approach. Annali di Geofisica Special Issue dedicated to Professor Michele Caputo, October 1997, Vol. XL, N.5, - pp. 1415-1431.

Deformations Detection by a Bayesian Approach: Prior Information Representation and Testing Criteria Definition

A. Albertella, N. Cazzaniga, F. Sansò
DIIAR, Dipartimento di Ingegneria Idraulica Ambientale, Sezione Rilevamento
Politecnico di Milano, p.za Leonardo da Vinci, 32 - 20133 Milan, Italy

F. Sacerdote
DIC, Dipartimento di Ingegneria Civile
Università di Firenze, via S.Marta, 3 - 50139 Florence, Italy

M. Crespi, L. Luzietti
DITS, Dipartimento di Idraulica Trasporti e Strade, Area di Geodesia e Geomatica
Università di Roma "La Sapienza", via Eudossiana, 18 - 00184 Rome, Italy

Abstract. It is well known that the classical testing procedures are not able to detect any significant deformation when the estimated displacements stemming from repeated surveys are small with respect to their precisions. This is true even if the displacements show some internal consistency (e.g. all the displacements have a common direction) or agree with prior hypotheses based on ancillary data (e.g. geological and geotechnical investigations about a landslide) or information derived by previous surveys.

In this paper we discuss an alternative (Bayesian) approach able to overcome this problem often rising in deformation monitoring and to establish a more powerful testing procedure by considering prior information about the displacements.

In details, we investigate the main problems related to the stochastic representation of the prior information by a suited covariance matrix and to the definition of a criterion for assessing the displacements significance.

Finally the approach is checked on synthetic examples.

Keywords. Deformation monitoring, Bayes' theorem, displacements significance

1 Introduction

The aim of this work is to continue and develop some ideas already presented in (Betti et al., 1998) related to the definition of a new testing procedure useful in deformations analysis when the estimated displacements are small with respect to the measurement precision.

In fact, in these cases, the classical testing procedure is not able to single out significant displacements even if they show an internal consistency, e.g. they have a common direction.

To overcome this limit we have already introduced a Bayesian approach, which allows to account for all the available prior information on the phenomenon under examination in order to increase the power of the test.

Usually, in fact, starting from preliminary investigations (e.g. coming from geology, geophysics, geotechnics, civil and structural engineering), both the boundaries of the deformation area and the approximate displacement directions and magnitudes may be inferred.

The Bayesian inference allows to evaluate the probability of an event in the light of the available prior evidences and of the collected observations.

We shortly recall Bayes' theorem. Let Y be a random vector of N observations $Y=(Y_1, Y_2, \ldots, Y_N)$; let the probability density function of the random vector Y be dependent on k unknown parameters. They are defined as random variables and collected in the random vector $x=(x_1, x_2, \ldots, x_k)$. The probability density function $f(x \mid Y)$ of the parameters given the observations is obtained from the density $f(x)$, which summarizes what is known about the parameters x before the data y are collected (prior distribution), and the density

$f(Y \mid x)$ of the observations given the parameters (likelihood function) by:

$$f(x \mid Y) = \frac{f\ (Y \mid x) \cdot f\ (x)}{\int f(Y \mid x) \cdot f\ (x)dx} \tag{1}$$

where:

$\int f(Y \mid x) \cdot f(x)dx$ is a normalization constant, which ensures that the density function $\mathbf{f(x \mid Y)}$ fulfills the conditions:

$$f(x \mid Y) \geq 0 \text{ and } \int f(x \mid Y)dx = 1 .$$

In equation (1) Bayes' theorem underlines that the observations modify through the likelihood function the prior knowledge of the parameters, thus leading to the posterior density function of the parameters themselves.

Particularly, in this work we investigate on a stochastic model to represent the prior information and a criterion to decide about displacements significance.

2 Bayesian model

The goal is to derive $\mathbf{f(x \mid Y)}$ by incorporating all the prior information in $\mathbf{f(x)}$; usually this information consists in hypothesized displacements with their covariance matrix, representing prior information precision and possibly a correlation pattern.

In this case, the (pseudo-)observations $\hat{D}$ are the estimated coordinate variations and the parameters are Δ, τ and p, where:

- Δ represents the (vector of) displacements
- $\tau = 1/\hat{\sigma}_0^2$ stands for the reciprocal estimated variance of unit weight
- p is the probability of non-zero (significant) displacements

The prior probability density function of the parameters is:

$$f(\Delta, \tau, p) = f(\Delta|p)f(\tau)f(p) \tag{2}$$

assuming that $\mathbf{\Delta}$ depends on p only.

We choose for $f(\Delta|p)$ a probability density function composed by a normal density function with probability p for non-zero displacements and by the Dirac delta function with probability (1-p) in case of no displacements:

$$f(\Delta|p) = (1-p)\delta(\Delta) + p\frac{1}{\left(2\pi\sigma_p^2\right)^{n/2}|C|^{1/2}} \cdot \exp\left[-\frac{1}{2\sigma_p^2}(\Delta-\overline{\Delta})^T C^{-1}(\Delta-\overline{\Delta})\right] \tag{3}$$

where C is the prior cofactor matrix accounting for correlation pattern and $\overline{\Delta}$ is the vector of prior displacements. It is well-known that the Dirac delta function represents a probability distribution concentrated in one point only:

$$\delta(\Delta) = \begin{cases} \infty & \Delta = 0 \\ 0 & \text{elsewhere} \end{cases}$$

so that the normalization rule holds:

$$\int_{-\infty}^{+\infty} \delta(\Delta)d\Delta = 1$$

The reciprocal variance of unit weight τ is assumed independent of (Δ,p) and gamma distributed with parameters a and b, that is:

$$f(\tau) = \frac{b^a \tau^{a-1} e^{-b\tau}}{\Gamma(a)} \qquad (\tau > 0) \tag{4}$$

as it is often done in literature on Bayesian methods. As regards (a,b) values, we consider $\hat{\sigma}_0^2 \sim \chi^2$, so that assuming $Var(\hat{\sigma}_0^2) = \overline{\sigma}_0^2$, we have $a = \overline{\sigma}_0^2 + 2$ and $b = (\overline{\sigma}_0^2 + 1)\overline{\sigma}_0^2$ (Koch, 1990, pag. 28), where $\overline{\sigma}_0^2$ is the expected value of the prior variance of unit weight.

At last we choose for p the uniform distribution in the interval [0,1]:

$$f(p) = c \quad (c = 1) \tag{5}$$

meaning that we assume a non informative prior situation for this parameter.

The likelihood function is a normal variate:

$$f(\hat{D}|\Delta, \tau, p) = \frac{\tau^{n/2}}{(2\pi)^{n/2}|Q|^{1/2}} \cdot \exp\left[-\frac{\tau}{2}(\hat{D}-\Delta)^T Q^{-1}(\hat{D}-\Delta)\right] \tag{6}$$

where Q is the cofactor matrix of the (pseudo-) observations at epoch j thus defined:

$$Q = Q_1 + Q_j \tag{7}$$

where Q_1 e Q_j are the cofactor matrix of the estimated coordinates at the first and generic epoch j, which are supposed to be independent.

The normalizing constant then reads:

$$f(\hat{D}) = \int_{\Delta}\int_{\tau}\int_{p} f(\hat{D}|\Delta,\tau,p)\, f(\Delta|p)\, f(\tau)\, f(p)\, d\Delta\, d\tau\, dp \qquad (8)$$

In the present framework, just to perform the analytical integrations of both the normalizing constant and the posterior marginal density functions, we suppose:

$$\sigma_p^2 = k\hat{\sigma}_0^2 = \frac{k}{\tau} \qquad (9)$$

This hypothesis simply relates in a useful way the prior information precision σ_p^2 and the (pseudo-) observations precision $\hat{\sigma}_0^2$ and does not introduce any theoretical limitation about the applicability of the new testing procedure.

Working on the exponent E of the equation (8):

$$E = (\hat{D}-\Delta)^T Q^{-1}(\hat{D}-\Delta) + \frac{1}{k}(\Delta-\overline{\Delta})^T C^{-1}(\Delta-\overline{\Delta})$$

introducing the following quantities:

$$\overline{x} = (Q^{-1}+H^{-1})^{-1}(Q^{-1}\hat{D}+H^{-1}\overline{\Delta})$$

$$W = (Q^{-1}+H^{-1})$$

with $H = kC$ it is possible to obtain:

$$E = (\hat{D}-\overline{\Delta})^T (Q+H)^{-1}(\hat{D}-\overline{\Delta}) + (\Delta-\overline{x})^T W(\Delta-\overline{x}) \qquad (10)$$

This expression is useful to perform the analytical integration in equation (8) because the exponent E is now divided into two terms: the first is independent from the parameters and the second one depends on Δ only.

The posterior density function we finally derive reads:

$$f(\Delta,\tau,p|\hat{D}) = \frac{1}{f(\hat{D})}\cdot\left[\frac{b^a\tau^{a-1}e^{-b\tau}}{\Gamma(a)}\right]\cdot[c]$$

$$\cdot\left\{\left(\frac{\tau}{2\pi}\right)^{n/2}\frac{1}{|Q|^{1/2}}\exp\left[-\frac{\tau}{2}(\hat{D}-\Delta)^T Q^{-1}(\hat{D}-\Delta)\right]\right\}$$

$$\cdot\left\{p\left(\frac{\tau}{2\pi k}\right)^{n/2}\frac{1}{|C|^{1/2}}\right.$$

$$\left.\exp\left[-\frac{\tau}{2k}(\Delta-\overline{\Delta})^T C^{-1}(\Delta-\overline{\Delta})\right] + (1-p)\delta(\Delta)\right\} \qquad (11)$$

therefore, we can compute the parameter (Δ,τ,p) posterior marginal density functions:

$$f(\Delta|\hat{D}) = \left[\frac{1}{T_1+T_2}\left(\frac{1}{2\pi}\right)^{n/2}\frac{1}{|Q|^{1/2}}\frac{b^a}{\Gamma(a)}c\right]$$

$$\cdot\left[\left(\frac{1}{2\pi k}\right)^{n/2}\frac{1}{|C|^{1/2}}\frac{\Gamma(n+a)}{2M^{(n+a)}} + \delta(\Delta)\frac{\Gamma(n/2+a)}{2N^{(n/2+a)}}\right]$$

$$f(\tau|\hat{D}) = \left[\frac{1}{T_1+T_2}\left(\frac{1}{2\pi}\right)^{n/2}\frac{1}{|Q|^{1/2}}\frac{b^a}{\Gamma(a)}c\right]$$

$$\cdot\left[\left(\frac{1}{2\pi k}\right)^{n/2}\frac{1}{|C|^{1/2}}\frac{F}{2}\exp(-G\tau) + \frac{\exp(-L\tau)}{2}\right]\tau^{n/2+a-1}$$

$$f(p|\hat{D}) = \left[\frac{1}{T_1+T_2}\left(\frac{1}{2\pi}\right)^{n/2}\frac{1}{|Q|^{1/2}}\frac{b^a}{\Gamma(a)}c\right]$$

$$\cdot\left[\left(\frac{1}{2\pi k}\right)^{n/2}\frac{1}{|C|^{1/2}}\frac{pF}{G^{(n/2+a)}} + \frac{(1-p)}{L^{(n/2+a)}}\right]\Gamma(n/2+a)$$

where T_1, T_2, F, G L, M and N are defined as follows:

$$T_1 = \frac{1}{(2\pi k)^{n/2}}\left(\frac{|W^{-1}|}{|Q||C|}\right)^{1/2}\frac{b^a c}{2G^{(n/2+a)}}\frac{\Gamma(n/2+a)}{\Gamma(a)}$$

$$T_2 = \frac{1}{(2\pi)^{n/2}}\frac{1}{(|Q|)^{1/2}}\frac{b^a c}{2L^{(n/2+a)}}\frac{\Gamma(n/2+a)}{\Gamma(a)}$$

$$F = (2\pi)^{n/2}|W^{-1}|^{1/2}$$

$$G = b + \frac{(\hat{D}-\overline{\Delta})^T(Q+H)^{-1}(\hat{D}-\overline{\Delta})}{2}$$

$$L = b + \frac{\hat{D}^T Q^{-1}\hat{D}}{2}$$

$$M = G + \frac{(\Delta-\overline{x})^T W(\Delta-\overline{x})}{2}$$

$$N = b + \frac{(\hat{D}-\Delta)^T Q^{-1}(\hat{D}-\Delta)}{2}$$

From these marginal distribution we can estimate the parameters, e.g. their expected values, to use in the testing procedure:

$$\Delta_B = E[\Delta] = \int_{-\infty}^{+\infty} \Delta f(\Delta|\hat{D}) d\Delta$$

$$\tau_B = E[\tau] = \int_{-\infty}^{+\infty} \tau f(\tau|\hat{D}) d\tau \qquad (12)$$

$$p_B = E[p] = \int_{0}^{1} p f(p|\hat{D}) dp$$

3 Bayesian testing procedure

The Bayesian model previously set up is used to establish the significance of the estimated displacements. In details, on the basis of the estimated p_B, we define the test statistic T_B as a probability contrast:

$$T_B = \frac{p_B}{1 - p_B} \qquad (13)$$

This statistic can be used to discriminate significant from non-significant displacements at the significance level α. We choose to perform a two-tails test, where the two hypotheses acceptance regions are:

H_0: non-significant displacements - $p_B \leq \frac{1}{2} - \frac{\alpha}{2}$

H_1: significant displacements - $p_B \geq \frac{1}{2} + \frac{\alpha}{2}$

and we derive the following test rule:

$$T_B < \frac{1-\alpha}{1+\alpha} \quad \text{accept} \quad H_0$$

$$T_B > \frac{1+\alpha}{1-\alpha} \quad \text{accept} \quad H_1 \qquad (14)$$

discriminant values of the test statistic T_B may be derived just inserting in these expressions the significance level α.

4 An elementary example

To show the effectiveness of the Bayesian testing procedure we analyse a simple synthetic example where new and classical approaches are compared.

We suppose to have independently surveyed 10 times a leveling line constituted by 10 benchmarks, 2 in the ends with known heights (1 and 10) and 8 in the middle to be controlled (typical set up for bridge monitoring). We have therefore m=9 height difference observations and n=8 unknown heights for each measurement epoch.

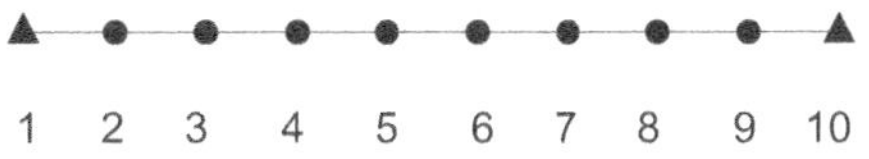

Figure 1. The leveling line

In this example we consider as parameter vector Δ the height variations with respect to epoch 1, so that the variance of unit weight $\hat{\sigma}_0^2$ at the generic epoch j is:

$$\hat{\sigma}_0^2 = \frac{r_1 \hat{s}_{01}^2 + r_j \hat{s}_{0j}^2}{r_1 + r_j} \qquad (15)$$

where $\hat{s}_{01}^2$ and $\hat{s}_{0j}^2$ are the estimated variances of unit weight coming from the least squares adjustments of the leveling measurements at epochs 1 and j and r_1, r_j are the corresponding global redundancies (always equal to 1 in our case).

The benchmark displacements (height variations) with respect to epoch 1 are obtained by synthetically corrupting the chosen displacements (Table 1) shown in Figure 2 with a noise normally distributed with standard deviation equal to 0.5 mm (the prior variance of unit weight in each leveling adjustment is just assumed equal to 0.25 mm^2). The synthetic series are displayed in Figure 3.

The classical procedure is based on few well-known simple assumptions:

- Y is the m-dimension (pseudo-)observations vector with normal distribution
- Y=A Δ is the linear model linking observations and n parameters
- $\hat{D}$ is the least squares estimate of Δ with $\hat{D} \sim N[\Delta, \hat{\sigma}_0^2 Q]$

which lead to the quadratic form distributed as a Fisher variate with (n,m-n) degrees of freedom suited for hypothesis testing on the parameters:

$$F_{sp} = \frac{\hat{D}^T Q^{-1} \hat{D}}{n\, \hat{\sigma}_0^2} \sim F_{n,m-n} \qquad (16)$$

To apply the Bayesian approach we introduce the prior information through the prior displacements $\overline{\Delta}$ with their prior cofactor matrix C, as explained in what follows. We considered three different situations:

1. prior information coming from previous surveys
2. prior information coming from structural hypotheses, i.e. displacements shown in Figure 2
3. wrong prior information, i.e. displacements shown in Figure 2 with opposite signs

Table 1. Displacements used to produce synthetic series of the benchmark (B) height variations at different epochs (Ep)

Ep/B	2	3	4	5	6	7	8	9
2	-0.19	-0.30	-0.33	-0.30	-0.19	0.00	0.07	-0.03
3	-0.37	-0.59	-0.67	-0.59	-0.37	0.00	0.13	-0.07
4	-0.56	-0.89	-1.00	-0.89	-0.56	0.00	0.20	-0.10
5	-0.74	-1.19	-1.33	-1.19	-0.74	0.00	0.27	-0.13
6	-0.93	-1.48	-1.67	-1.48	-0.93	0.00	0.33	-0.17
7	-1.11	-1.78	-2.00	-1.78	-1.11	0.00	0.40	-0.20
8	-1.30	-2.07	-2.33	-2.07	-1.30	0.00	0.47	-0.23
9	-1.48	-2.37	-2.67	-2.37	-1.48	0.00	0.53	-0.27
10	-1.67	-2.67	-3.00	-2.67	-1.67	0.00	0.60	-0.30

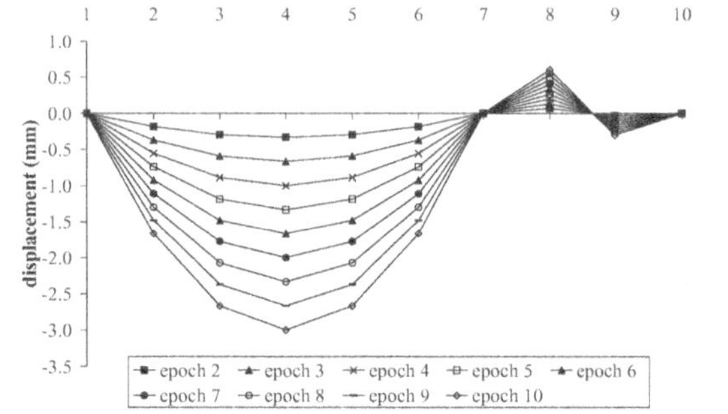

Figure 2. Sketch of the displacements used to produce the synthetic series of the benchmark height variations

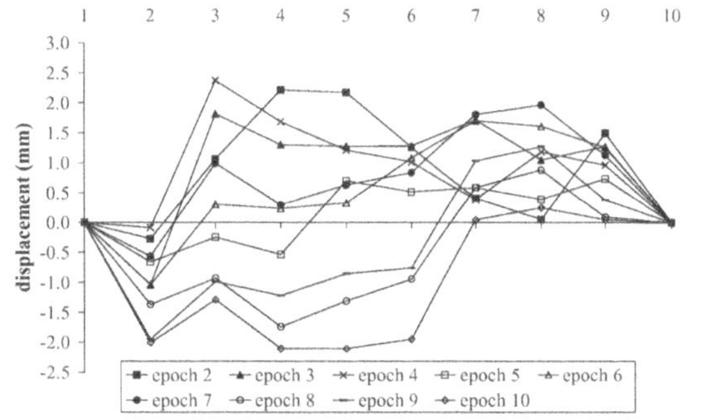

Figure 3. Synthetic series of the benchmark height variations

4.1 Prior information from previous surveys

We calculate $\overline{\Delta}$ for the generic epoch j using the measured displacements until the epoch j-1, so that the prior information can be improved epoch by epoch on the basis of previous measurements. Particularly, we hypothesize that displacements follow a linear trend in time for each benchmark, so that $\overline{\Delta}$ and its precision may be extrapolated by a least square fit of the linear trend itself (Figure 4).

Only for the epochs 2 and 3, when there are not enough actual measurements to estimate the trend, we are forced to assume acceptable prior displacements on the basis of more vague external information. In details, we choose:

- epoch 2 - $\overline{\Delta}=1$ mm, $\overline{\sigma}_0^2 = 2$ mm
- epoch 3 - $\overline{\Delta}=2\hat{D}_1$, $\overline{\sigma}_0^2 = 2$ mm

As regards prior cofactor matrix C, we underline that in this case we are providing a "direct" prior information on each benchmark, so that the possibility to introduce a correlation pattern is not used here and C is chosen equal to the unit matrix. At each epoch, the value of k requested to define the precision of the prior information according to (9) is computed as:

$$k = \frac{\overline{\sigma}^2}{\hat{\sigma}_0^2} \qquad (17)$$

where $\overline{\sigma}^2$ is the average of the extrapolated displacement precisions (Figure 4) for all the benchmarks and $\hat{\sigma}_0^2$ is the variance of unit weight at epoch j already defined in equation (15).

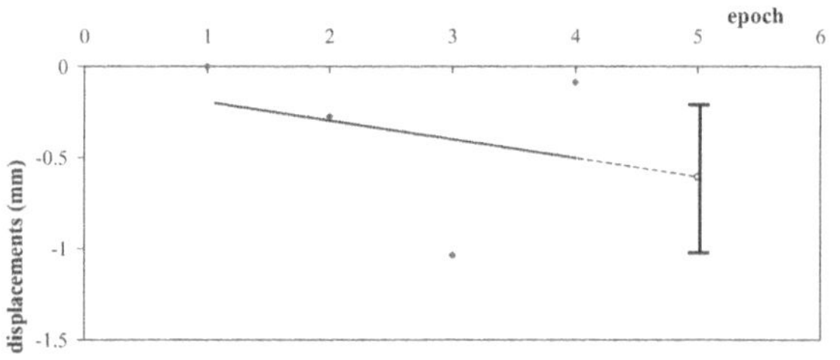

Figure 4. Example of the estimated linear trend for benchmark 2 with extrapolated height variation and precision (1 sigma) at epoch 5

The comparison between the two testing procedures is performed in terms of the ratios between the empirical test statistics F_{sp} and T_B (Table 3) and their corresponding critical values F_α and $T_{B,\alpha}$ at the α significance level (Table 2). In this respect it has to be noted that the meaning of α in the two tests is quite different: in the classical approach α obviously represents an upper bound of the H_0 acceptance region, whereas in the Bayesian approach α just bounds the region of indecision. Nevertheless, for both test statistics a ratio larger than 1 implies fundamental hypothesis H_0 rejection, therefore significant displacements at the α significance level.

The results show that the Bayesian approach is able to detect significant displacements since epoch 6 at both significance levels, whereas the classical approach empirical test statistics are by far below the critical value for every epoch. It is worth emphasizing that displacements are detected at epoch 6 even if 80% of them are within 1.0 mm, with mean and maximum values of 0.7 mm and 1.3 mm respectively, whereas the normally distributed noise has a standard deviation of 0.5 mm.

Table 2. Critical values of the classical and Bayesian test statistics

SIGNIFICANCE LEVEL α	CLASSICAL APPROACH F_α	BAYESIAN APPROACH $T_{B,\alpha}$
5%	5.59	1.11
1%	12.25	1.02

Table 3. Ratios between the empirical test statistics F_{sp} and T_B and their corresponding critical values F_α and $T_{B,\alpha}$ - prior information from previous surveys

EPOCH	CLASSICAL APPROACH F_{sp}/F_α		BAYESIAN APPROACH $T_B/T_{B,\alpha}$	
	α = 5%	α = 1%	α = 5%	α = 1%
2	0.41	0.19	0.46	0.50
3	0.52	0.24	0.52	0.57
4	0.39	0.18	1.03	1.12
5	0.13	0.06	0.45	0.49
6	0.25	0.11	1.57	1.71
7	0.28	0.13	1.79	1.95
8	0.27	0.12	1.18	1.28
9	0.41	0.19	1.80	1.96
10	0.41	0.19	1.79	1.95

Moreover values of k decrease epoch by epoch witnessing that prior information accuracy increases with time (Table 4).

Looking at the synthetic series it is possible to see that the leveling measurements of epoch 1 show some inconsistencies due to large noise, so they are not a good reference for assessing possible deformations. We therefore analyzed the case where the benchmark height variations are referred to epoch 2 too, excluding the measurements of epoch 1; the synthetic series are displayed in Figure 5. In Table 5 it is possible to see that the Bayesian testing procedure becomes more powerful and able to detect significant displacements one epoch before.

Table 4. Estimated parameters τ, p and k

EPOCH	k	τ	p
2	16.00	2.369	0.339
3	8.00	2.117	0.367
4	4.99	5.156	0.533
5	2.85	5.716	0.333
6	3.32	10.781	0.635
7	2.10	14.958	0.666
8	1.41	5.320	0.566
9	1.55	11.805	0.666
10	1.22	9.094	0.666

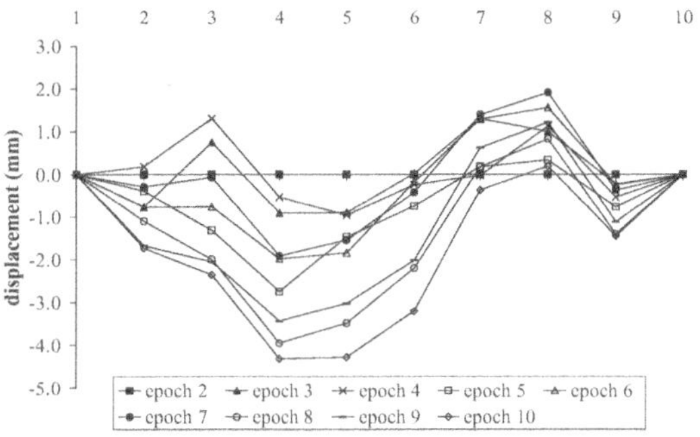

Figure 5. The synthetic series of the benchmark height variations after the epoch 1 measurements exclusion

Table 5. Ratios between the empirical test statistics F_{sp} and T_B and their corresponding critical values F_α and $T_{B,\alpha}$ after the epoch 1 measurements exclusion - prior information from previous surveys

EPOCH	CLASSICAL APPROACH F_{sp}/F_α		BAYESIAN APPROACH $T_B/T_{B,\alpha}$	
	α = 5%	α = 1%	α = 5%	α = 1%
3	0.44	0.20	0.49	0.53
4	0.44	0.20	0.66	0.72
5	0.35	0.16	0.49	0.53
6	0.47	0.21	1.78	1.94
7	0.62	0.28	1.79	1.95
8	0.93	0.42	1.79	1.95
9	0.90	0.41	1.80	1.96
10	0.97	0.44	1.80	1.96

4.2 Prior information from structural hypotheses

We consider the measurements collected in all the 10 epochs and compute $\overline{\Delta}$ for the generic epoch j on the basis of structural hypotheses, i.e. displacements reported in Table 1 and shown in Figure 2, with a fixed precision.

In order to set up the stochastic model at the generic epoch, we suppose that displacements of the benchmarks i and j may be written as:

$$\begin{aligned} z_i &= a_i\delta + \varepsilon_i \\ z_j &= a_j\delta + \varepsilon_j \end{aligned} \tag{18}$$

where δ is the maximum displacement (benchmark 4, Figure 2), a_i and a_j are coefficients derived from the structural hypotheses which relate the maximum displacement to those at benchmarks i and j and ε_i, ε_j are random noises. Therefore, from the stochastic point of view, we can split the generic displacement into two components: the first one (correlated) dependent on δ, which is assumed known (Table 1, Figure 2) with variance σ_δ^2; the second (uncorrelated) with $\varepsilon \sim N[0, \sigma_\varepsilon^2]$. The generic covariance matrix $\mathbf{C}_{z_iz_j}$ is so defined by the well known covariance propagation law:

$$C_{z_iz_j} = \begin{vmatrix} a_i^2\sigma_\delta^2 + \sigma_\varepsilon^2 & a_i a_j \sigma_\delta^2 \\ a_i a_j \sigma_\delta^2 & a_j^2\sigma_\delta^2 + \sigma_\varepsilon^2 \end{vmatrix} \tag{19}$$

and the correlation coefficient reads:

$$\rho_{z_iz_j} = \frac{a_i a_j \sigma_\delta^2}{\sqrt{\left(a_i^2\sigma_\delta^2 + \sigma_\varepsilon^2\right)\left(a_j^2\sigma_\delta^2 + \sigma_\varepsilon^2\right)}} \tag{20}$$

In our case, we choose $\sigma_\delta^2 = 1$ mm^2 and $\sigma_\varepsilon^2 = 1$ mm^2 and the prior variance of unit weight in each leveling adjustment is still assumed equal to 0.25 mm^2.

In the same manner as before, k is computed according to (17), where now $\overline{\sigma}^2$ is the average of the displacement precisions extrapolated by the covariance matrix (19).

The comparison between the two testing procedures is performed as before (Table 6).

The results show that the Bayesian approach is able to detect significant displacements when they become coherent with prior information since epoch 8 at both significance levels, whereas the classical approach empirical test statistics are by far below the critical value for every epoch.

It is remarkable to note that Bayesian test singles out significant displacements also in epochs 1 and 2, when the leveling measurements show some inconsistencies due to large noise, and drive the height difference estimates to some large positive values. These values are detected as significant even if they are not coherent (opposite) to the hypothesized prior displacements, since the prior information is quite vague in comparison with measurements precision. On the contrary, if we choose σ_δ^2=0.01 mm^2 and σ_ε^2=0.01 mm^2, with the same prior variance of unit weight equal to 0.25 mm^2 (prior information remarkably better than measurements precision, which is obviously a nonsense in practice), only estimated displacements coherent with prior information are recognized as significant (Table 7). In this respect a careful definition of prior information is crucial in order to avoid wrong conclusion: for example, in our case it would be useful to exclude positive displacements for benchmarks 2 to 6, since from structural hypotheses only negative ones are allowed.

4.3 Wrong prior information

In order to evaluate the effect of a wrong prior information, we consider again the measurements collected in all the 10 epochs, but assume $\overline{\Delta}$ for the generic epoch j with signs opposite than in Table 1.

Table 6. Ratios between the empirical test statistics F_{sp} and T_B and their corresponding critical values F_α and $T_{B,\alpha}$ - σ_δ^2=1 mm^2 and σ_ε^2=1 mm^2

EPOCH	CLASSICAL APPROACH F_{sp}/F_α		BAYESIAN APPROACH $T_B/T_{B,\alpha}$	
	$\alpha = 5\%$	$\alpha = 1\%$	$\alpha = 5\%$	$\alpha = 1\%$
2	0.41	0.19	1.00	1.09
3	0.53	0.24	1.18	1.28
4	0.39	0.18	0.65	0.71
5	0.13	0.06	0.52	0.57
6	0.25	0.11	0.56	0.61
7	0.28	0.13	0.51	0.56
8	0.27	0.12	1.67	1.82
9	0.41	0.19	1.64	1.79
10	0.41	0.19	1.74	1.89

Table 7. Ratios between the empirical test statistics F_{sp} and T_B and their corresponding critical values F_α and $T_{B,\alpha}$ - σ_δ^2 =0.01 mm^2 and σ_ε^2 =0.01 mm^2

EPOCH	CLASSICAL APPROACH F_{sp}/F_α		BAYESIAN APPROACH $T_B/T_{B,\alpha}$	
	α = 5%	α = 1%	α = 5%	α = 1%
2	0.41	0.19	0.73	0.79
3	0.53	0.24	0.80	0.87
4	0.39	0.18	0.57	0.62
5	0.13	0.06	0.57	0.62
6	0.25	0.11	0.80	0.87
7	0.28	0.13	0.66	0.72
8	0.27	0.12	1.78	1.94
9	0.41	0.19	1.71	1.86
10	0.41	0.19	1.72	1.87

As regards prior cofactor matrix C at each epoch, as before we derive it by computing the correlation coefficients among the 8 monitored benchmarks with σ_δ^2 =1 mm^2 and σ_ε^2 =1 mm^2 and the same prior variance of unit weight equal to 0.25 mm^2. The comparison between the two testing procedures is performed as before. The results show that the more prior information is in conflict with actual (pseudo-)observation (more precise prior displacements) the more Bayesian approach fails to detect significant displacements (Table 8).

Table 8. Ratios between the empirical test statistics F_{sp} and T_B and their corresponding critical values F_α and $T_{B,\alpha}$ - wrong prior information with σ_δ^2 =1 mm^2 and σ_ε^2 =1 mm^2

EPOCH	CLASSICAL APPROACH F_{sp}/F_α		BAYESIAN APPROACH $T_B/T_{B,\alpha}$	
	α = 5%	α = 1%	α = 5%	α = 1%
2	0.41	0.19	1.39	1.51
3	0.53	0.24	1.49	1.62
4	0.39	0.18	1.48	1.61
5	0.13	0.06	0.47	0.51
6	0.25	0.11	0.49	0.53
7	0.28	0.13	0.50	0.54
8	0.27	0.12	0.45	0.49
9	0.41	0.19	0.46	0.50
10	0.41	0.19	0.45	0.49

5 Conclusions and prospects

In this paper we developed a testing procedure suited for deformation monitoring based on a Bayesian approach, which allows to account for prior information about the displacements. In details, we investigated the representation of the prior information precision and correlation pattern by a suited covariance matrix C and the definition of a criterion for assessing the displacements significance, based on a probability ratio. This procedure was compared to the classical one on a synthetic example showing that testing power increases remarkably if actual (pseudo-) observation are coherent with prior information; on the other hand, when this does not happen, Bayesian approach fails.

Therefore the proposed approach is a powerful tool in deformation analysis but the careful choice of prior information is crucial: if it is not available with enough accuracy, a vague hypothesis with eventual constraints is probably the most safe choice, which may be improved by repeated survey. Future researches will be devoted to introduce constraints on prior information and to a deeper investigation on the definition of the test statistic.

Acknowledgements

This work was partially supported by the Project INGV-DPC 2005-2006 Vulcanology – Subproject V3_1 – Colli Albani.

Authors thank Barbara Betti and a second anonymous Reviewer very much for their extremely careful reading of the paper and suggestions which contributed to improve the manuscript.

References

Betti B., Crespi M., Sansò F. (1998). Deformation Detection According to a Bayesian Approach. *IV Hotine-Marussi Symposium on Mathematical Geodesy*, Trento, September 14-17, pp. 83-88.

Box G. E. P., Tiao G. C. (1992). *Bayesian Inference in Statistical Analysis*. Wiley Classics Library ed., New York Chichester Brisbane Toronto Singapore.

Koch K. R. (1990). *Bayesian Inference with Geodetic Applications*. Number 31 in Lecture Notes in Earth Sciences. Springer, Berlin Heidelberg New York.

Spectral Analysis of Geoidal Signals at Points of Geodynamical Interest Used in the Investigation of the Depth of Mass-Density Causal "Sources" of Ground Deformations

M.G. Doufexopoulou, B.A. Massinas
Department of Surveying Engineering, Higher Geodesy Lab & Dionysos Satellite Observatory
National Technical University of Athens, Zografou Campus, 9 Heroon Polytechniou, 15780 Athens, Greece

G. Bartha
Faculty of Earth Science and Engineering, Dept. of Geodesy & Mine Surveying
The University of Miskolc, Egyetemvaros 1, HU-3515 Miskolc, Hungary

Abstract. Geoid undulations produced as the difference between ellipsoidal heights h and leveled or trigonometric heights H are observed point signals that bear information about the interior of the Earth. Exploring this information *in a lateral sense* is possible by the use of statistical methods which do not involve geophysical inversion and are capable of pre-evaluating regions of geodynamical interest. Such investigations may be carried out by means of test computations of the depth of possible sources in the Earth's interior that might cause deformations on its surface.

In this paper, eight data series are built from point observed geoid undulations at eight different points. At these points, we gradually subtract from the observed undulations the variable-truncated contribution of CHAMP and GRACE based Gravity Field Models at lag $n = 2$. Each data series then represents a point signal gradually filtered numerically at varying frequencies as these are being revealed in the two models. The 16 series are then analyzed spectrally by using the Fast Fourier Transform (FFT) and the Maximum Entropy (MEM) methods with the objectives: (1) to investigate the performance of the CHAMP and GRACE models in an area with strong varying observed signals; (2) to detect whether there are dominant frequencies in terms of the degree of expansion n and, if so, to associate the dominating range of degrees with the depth of mass-density causal "source" which, in the case of the test points, is known to be in the vicinity of 40 km and less; and (3) to compare the observed undulation differences δN between test points with the differences of computed undulations with the aim of drawing conclusions on the use of these models in the process of height reference.

Keywords. Geoidal signals, frequency spectrum, mass-density discontinuities.

1 Introduction

Geodesy contributes to kinematics and deformation studies for all engineering or scientific purposes through measurements of distances and coordinates. These objectives are usually a part of extended scientific or engineering aims that include further tasks, such as the evaluation of the impact in natural hazards, the behavior of constructions artificial structures under loads or when subjected to forces, the investigation of the Earth's kinematical behavior (e. g. Abers, G.A., and R. McCaffrey 1994, Wallace et al. 2004). Usually the measuring campaign is designed independently through a control network or by selecting the reference and control points of simple configuration schemes in the region of interest. The determination of the kinematical/deformational parameters is thus based on the adoption of a relevant function of the motion that is grounded on theoretical or empirical assumptions. In the evaluation process, then, this formulation represents a "deterministic" model, so that the focuses on data evaluation. The topographic surface is however a dynamic interface between the lithosphere, hydrosphere and biosphere, the Earth being considered as a physical body. Under this model, topography represents the surface on which all interactions of most direct and indirect occurring impacts of solid-Earth science like landslides, floods, tsunamis, earthquakes, and volcanic eruptions, as well as human activities, take place. The challenge presented by monitoring the dynamics of this surface is three-fold: (1) to unravel the record of past interactions embedded in it (i. e.

the surface); (2) to determine the relative roles of natural and human-induced changes, and (3) to understand processes that act on this surface in order to predict or to study natural hazards. Modelling these unknown processes *requires the involvement of approaches of geophysical inverse problems.* In engineering practice the geophysical inverse processes are restricted to the adoption of a deterministic configuration of the main causal factors that produce a motion.

The duality of Geodesy is well illustrated in the previous objectives: positions are monitored in a geometric space that is subject to the Earth's physical activity. This duality exists also in most types of actual geodetic measurements that also depend on physical parameters of the Earth. As for the objective to determine from geodetic measurements the kinematic/deformation parameters, there are a number of questions to consider. These relate to the feasibility of all the modeling parts of a kinematical or deformation process. Sometimes the results of deformation studies may be more confusing than enlightening (Oezener et al, 2003). Associated questions can be: (1) How feasible is the rate and the scale of an expected motion or deformation? The rates are usually estimated on the basis of experience, so it may be difficult to reveal the real ones using few or an insufficient data coverage or even using over-determined evaluation models with large data sets. (2) How much can the inaccuracy of poorly known values of physical parameters (such as the troposphere, the atmosphere, the measuring method etc.) affect the measurements, upon which the final accuracy of the evaluation of a motion/deformation in using such data depend? And (3) *How long is the reasonable time lag* in a measuring campaign so that the assumption that discontinuity during the measurement campaign (i.e. changed physical conditions in the measuring signals, stable atmospheric or other effects) have not occurred, is valid?

In the design of a measuring campaign to monitor the kinematical behavior of a "site" (e.g. a region, a construction, etc.), the selection of reference and control points usually follows some previous knowledge or guess about the kind of expected motion; alternatively, it may be based on basic features of the Earth's upper crust known from geophysical studies. The present approach seeks to develop a new independent tool of design that can be used *without any physical knowledge* or previous results coming from geophysical inversion of the site. The approach follows the assumption that there is a relationship between the spectral content (or the information per wavelength) of the geoid and the depth of mass-density anomaly which generates the particular wavelength (e.g. Bowin 1983) and uses only observed geodetic data and coefficients of global gravity field models. This approach has the advantage of being independent of prior knowledge about the site, simple and of low cost. The aim of this study is to contribute to the means by which we can estimate the rate and the geographical scale of a possible motion/deformation assuming that the depth of the causing "sources" of a motion/deformation *is related to a topographical area of such size* that the motion is expected to be sensed through the measurements. The particular objectives are twofold:

A) The first objective is to associate the depth of causal "sources" with the degree *n* of expansion of recently released spherical harmonic models of the missions CHAMP and GRACE. This will be done by using existing data (GPS coordinates and trigonometric or leveled heights H at these points) collected for ordinary/ conventional mapping control. This task is accomplished by producing an FFT transform of numerical series constructed from local observed geoid undulation signals at particular points. We are using a quite small sample of data from the existing ones because the spectral analysis approach is first time applied in this area, so we depicted only such points for which the existence of causal "sources" was certified by independent investigations or indicated by real events. Then, from the computed value we subtract a theoretical value calculated with by use of the harmonic coefficients of the CHAMP and GRACE models truncated regularly with a lag of n=2. This de-trending approach is supported by previous work (e.g. Featherstone, W. E., 1997, Doufexopoulou M. et al., 1999) and is based on the geophysical work of Kahn (1977) and Bowin (1983), among others. As may be appreciated, the kinematical studies are based on methods that approach an inherently geophysical inverse problem (e.g. Kaipio et al. 1999).

B) The second objective is to present initial results about the adaptation of the CHAMP and GRACE geoid models with respect to raw observed geoid signals in Greece. This region has been previously investigated under this concept on a national scale as well as locally (e.g. Koumbis, 2001; Kakleas & Prodromou, 2000; Doufexopoulou, 2003). Indeed, this part of Europe is of quite high geodynamical interest. Earlier results using observed geoid signals and the global gravity field model expansions OSU91A and EGM96 have shown that this methodological approach could reveal the tectonic constellation without geophysical inversion (Doufexopoulou & Pagounis 1997; Pagounis 2000; Doufexopoulou 2002). Also, using the observed geoid in the sense of a local raw geo-signal (N_{signal}) and a computed value of this signal by the models OSU91A and EGM96 truncated at various wavelengths, several ignored geo-features within a small area were quite well depicted (Doufexopoulou 2002, 2003). Previous results and ongoing research in this topic provide the conviction that the GPS – leveled geoid is a useful geodetic measure which can support the two objectives of the study, especially after the release of gravity models by the CHAMP and GRACE satellite missions.

The data are GPS coordinates, trigonometric or leveled heights H at these points and coefficients of geopotential solutions based on the CHAMP and GRACE missions. From these ten points, 5 are distributed across the country and the other 5 are located in a small region 200X200 km nearby Attika, Athens (Fig. 1). The size of the country is about 6^oX $6^{o,}$ which is ideal for the evaluation of the performance of the CHAMP model at a resolution level of 5° X 5°, as the rms of the EIGEN (European Improved Gravity model of the Earth by New Techniques) is below 1cm in the geoid undulation (Reigber et al. 2002).

Both objectives aim to contribute to geo-kinematical investigations in a lateral and radial sense as follows. By knowing the adaptation performance of the CHAMP and GRACE missions in this particular area, several existing problems in the height reference (e.g. the strong variations of the slope of geoid over short distances, problems in interpolating the geoid at medium wavelengths) can be bypassed, at least in national scale geodynamical studies. Thus a national geoid and its relation to the GRS80 can be established. Also, in the hope that the spectral analysis of individual signal undulations in the vicinity of Newtonian effects of causal "sources" may reveal the dominating depth of the main "source" of a particular geoid anomaly with respect to the degree *n* of a global expansion, it can be possible to design the geometry of the kinematics after using feasible criteria. The wavelengths of observed geoid most affected from a local "source" may be used to configure the design. Though in the initial design of this research the MEM was configured in parallel to FFT, this was not applied due to the raw character of the signals that were spectrally analyzed which is a non relevant feature for the MEM method.

2 Problem Description, Data Arrangement, Past Investigation and Objectives

The observed geoid undulations as the difference:

$$N_{signal} = h_{GPS} - H_{(measured)} \quad (1)$$

can in fact be considered as raw Earth signals carrying information about Newtonian effects caused by the total mass-density distribution within the Earth.

The goal is therefore to *depict* from these signals information about the local Newtonian effects from upper mass layers. This can be done by using Global Gravity Field Models to eliminate numerically the effect of trends or global / regional Newtonian effects that exist in these raw signals. In doing so, different residual values δN_{signal} corresponding to different degrees *n* of the model's truncation can be constructed. Their content can be further analysed for information on local Newtonian effects by use through various methods. The existing information within these signals can be used in two ways: 1) *to confirm* existing a priori assumptions about the expected kinematics of the site that come usually from *geophysical inversion* (e.g. J P Kaipio *et al* 1999) or *to reveal* unknown local tectonic or geo-features; 2) *to evaluate* the performance of existing low cost geodetic data, like GPS points and leveled or trigonometric heights, with a view to detecting the possibility of traceable kinematics. Therefore, a control network for kinematics may be based on a feasible site selection of the points to use.

In the region of Greece numerical investigations have been carried out previously for similar purposes. Using observed signals, N_{signal} and the coefficients of OSU91A, EGM96, EGM98 as references, the main lateral mass discontinuities have been traced. The methodology was also successful in using raw data from Greece and Italy (Doufexopoulou et al. 1999). Previous results support the expectation that the depth variability of local Newtonian sources in the upper layers of the Earth may be sensed in respect of a global gravity field spectral form by analysing such point residuals. Especially with respect to CHAMP and GRACE models, this belief is stronger due to their good recovery at the wavelengths corresponding to the degree n=100 compared to earlier models (Koening et al. 2002). Throughout this study the observed signals in Eq. (1) are not considered strictly as geoid undulations but as raw Earth signals sensitive enough to be affected by local and regional Newtonian effects from the mass layers nearby the Earth's surface. The computed value of these signals by the coefficients of Global Gravity Field Models, N (φ, λ) may contain global scale mass signatures through the coefficients of the expansion (Blais 1996).

So, by subtracting expansions of the N (φ, λ) at certain degrees "n" of truncation equals to a low wavelength elimination of Newtonian effects:

$$N_{model}(\varphi,\lambda,n)=\frac{k\,M}{R_\gamma}\sum_{i=2}^{n}\left(\frac{a}{R}\right)^{i}\sum_{j=0}^{i}\left(C_{ij}\cos j\lambda + S_{ij}\sin j\lambda\right)P_{ij}(\sin\varphi) \qquad (2)$$

The subtraction from the local signals of the contribution of the model value at gradually different degrees "n" of truncation gives:

$$\delta N(\varphi,\lambda,n)=N_{signal}(\varphi,\lambda,n)-\frac{k\,M}{R_\gamma}\sum_{i=2}^{n}\left(\frac{a}{R}\right)^{i}\times\left[\sum_{j=0}^{i}\left(C_{ij}\cos j\lambda+S_{ij}\sin j\lambda\right)P_{ij}(\sin\varphi)\right] \qquad (3)$$

with n varying from low to high degrees, provides classes of residuals δN_n in an area or residual series that correspond only to a particular point (φ_o, λ_o):

$$\delta N_n=\delta N(\varphi_0,\lambda_0,n)= = N_{signal}(\varphi_0,\lambda_0)-N_{model}(\varphi_0,\lambda_0,n) \qquad (4)$$

Such residual sets δN_n may be formed at GPS points at which the orthometric height H is leveled or trigonometrically observed. In using residuals after the models OSU91A and EGM96 at a number of points it has been possible to reveal the locations of faults and focal depths of earthquakes in lateral sense, established independently from previous information (Fig. 2).

In order to associate the existing lateral variations revealed by previous studies to the depth of the "source" in more detail (under the inherent non-uniqueness of the geopotential inversion), in this study we analyze some series of residuals δN_n that correspond to particular points that are located in the vicinity of a local mass-density "source" (fault, focal depths of earthquakes) by FFT transform.

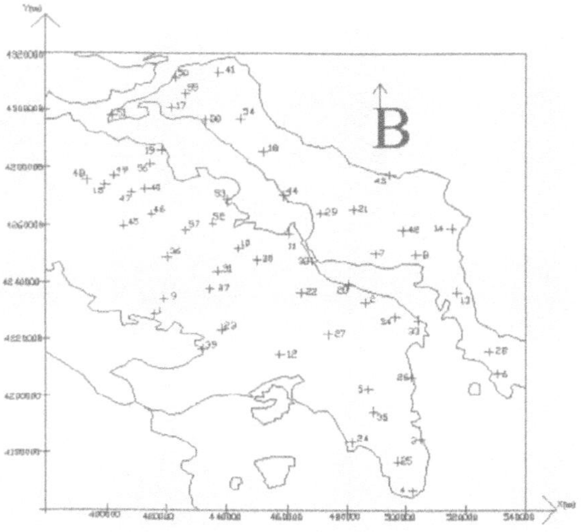

Fig. 1 The local region of the test study

A series of residual signals $N_{model}(n)$ with regular lag of truncation degree n of the CHAMP and GRACE expansions is expected to behave statistically as a stochastic signal. The stochastic character of the residuals can be tested by Gaussian measures of variance and the amplitude power spectrum of the geoid undulation components. In doing so, it is expected that the range of degree *n* of expansion of a particular model which is in excess or deficit of the expected stochastic behavior of residuals δN_n will be revealed.

Alternatively, it will be possible to monitor reasonably sized differences among amplitude and power distribution in the spectra corresponding to different points. In any of these cases there is evidence to refine the method further. The deviations of the signal series from stochastic expectation are assumed to come mainly from the effects of local Newtonian masses rather than from other error sources possibly existing in the data. Particularly strong deviations from the stochastic behavior can be due to mass discontinuities resulting from geological or tectonic faults. Such discontinuities are caused by various phenomena and have their origins at various depths.

Apart from the effect of local mass-density constellation, this type of residual is normally expected to contain: a) possible inconsistencies between the GPS datum and the local height datum b) measuring inaccuracies in both the GPS height h, and of the leveled or trigonometric height, H.

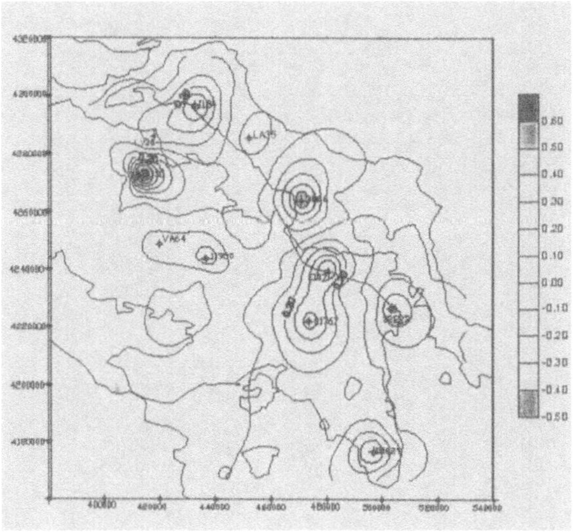

Fig. 2 Errors of signal interpolation with EGM96 model truncated at n=90

In this study, the observed signals N_{signal} reflect the variable geophysical and tectonic local background within a limited geographical region (Doufexopoulou 1984). Thus, it is reasonable to assume that local Newtonian effects dominate the effects coming from the error sources of the geodetic data.

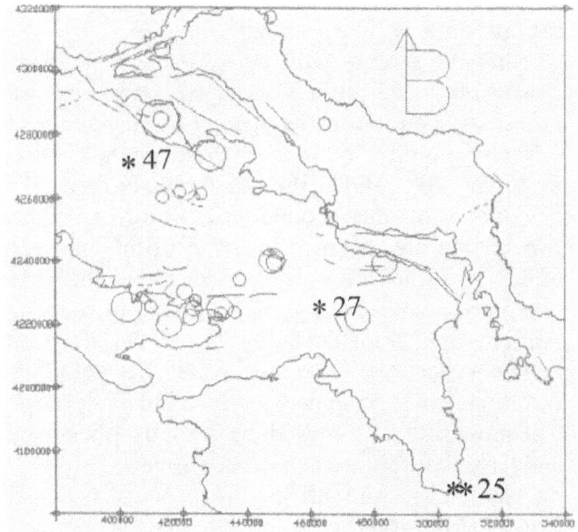

Fig. 3 Faults & focal places associated to geo-signal interpolation errors (numbered points with * refer to the FFT spectra)

The belief has been formed on the basis of (the data arising from) all previous investigations of this region and especially - reinforced by the results of the local study in which the locations of local tectonic faults and of focal depths of earthquakes below and above 60 km could be depicted (Fig. 3).

The present investigation is based on 10 signals distributed in two 5-point groups. By using the whole number of points we first tested how the CHAMP and GRACE models perform in the test region, knowing that both the depth local Newtonian "sources" (faults, foci or earthquakes) and the lateral mass-density variations are quite strong and varying within this area. The observed N signals at these points vary between 15.68 m and 44.67 m, indicating strong changes of the geoid with respect to the computed geoid of early gravity field models and after EGM96 and OSU91A coefficients related to the tectonic constellation have been omitted (see Doufexopoulou 1984, Doufexopoulou & Papafitsorou 1986, Pagounis 2000).

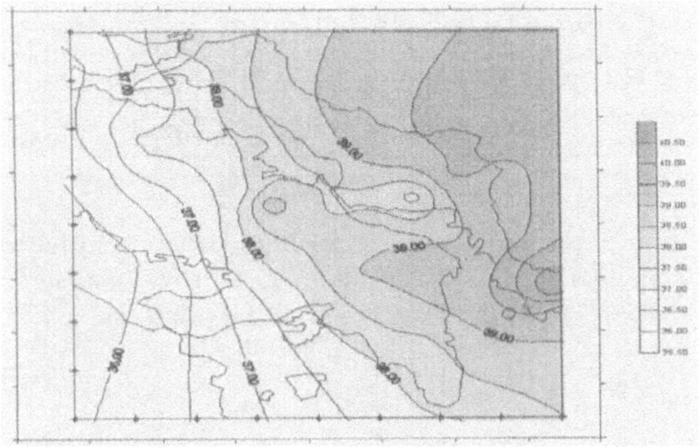

(a)

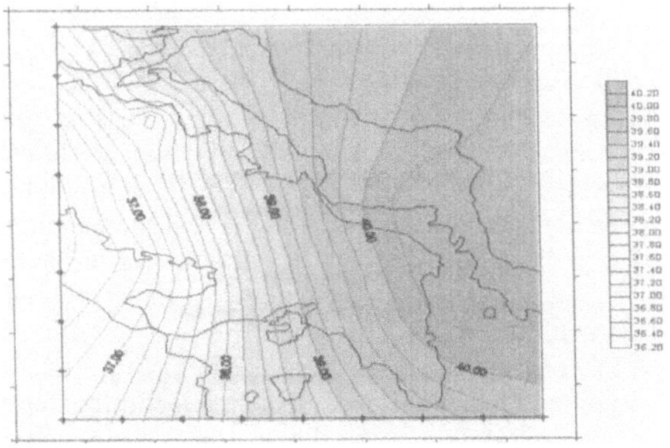

(b)

Fig. 4 The observed signal geoid (**a**) and the geoid after the OSU91A coefficients (**b**) in the investigated region

Another indication supporting this argument comes from recent results within the small region of 200 km X 200 km within which 5 points of the used sample are located. In this subset the observed signal undulations at 58 points have produced strong differences with respect to the computed geoid with the coefficients of EGM96 and OSU91A GGFM (reference) (Fig. 4).

A good adaptation of residuals $\delta N_n = N_{signal} - N_{model}(n)$ at the full model expansion of CHAMP and GRACE models may provide a result that contributes to existing practical needs in height control.

In particular, it may provide a first indication whether the new models CHAMP and GRACE can be used to replace the terrestrial methods for orthometric height control points, at least for engineering use.

The second objective aims to contribute to the site selection of a GPS control network to monitor geodynamical deformations. In fact, if the depth of a causal "source" is known approximately and expressed with respect to the harmonic degree of expansion of a Gravity Field Model, then the feasibility in the design of a geodynamical network can be improved in two ways: a) by selecting the places of reference and control points, or b) by interpreting results of geodynamical investigations with respect to feasibility according to global / local "causal" effects, introducing, for instance, two different time scales in the data evaluation. Last but not least, the second objective aims to present low cost alternative uses of the geoid undulation in the initial geo-prospecting investigation of an area that might be useful in engineering.

3 Description of Numerical Experiments with CHAMP and GRACE Models

The observed and computed values for the 10 signals at n=360 using both the CHAMP and GRACE models are shown in Table 1. Out of these points, 5 are scattered over the country between $\varphi = 35^o - 40^o.53$ and $\lambda = 21^o.50 - 27^o.13$, with spherical distances varying from 760 km to 256.3 km. The other 5 points belong to the smaller test area (Fig. 1) with distances between 78.61 km to 161. 4 km. This large range of distances in both groups and the de-trending of the observed signals for all wavelengths longer than 110 km after the empirical rule λ=360/n (λ is the wavelength expressed in km and n is the degree of the model's expansion), corresponds roughly to the possibility to detect causal "sources" at depths till about 18km (Kahn, 1977). Naturally, in this estimation, the non-uniqueness of the geophysical inverse problem is ignored. The points in the second group come from a dense sample of 58 points and are selected according to their proximity to areas where strong interpolation errors of the observed geoid have previously been discovered. These errors are associated with independently known causal "sources" (Figs. 3, 4). Based on this small sample a first evaluation of how the CHAMP and GRACE GGFM models adapt into this area can be carried out. The evaluation is based on (i) the standard deviation of the signal differences (Table 1) between observed and computed values with both models, on the largest difference found between observed and computed signal, (ii) on the rates of these differences between the wide area group and the small area group and - finally – (iii) on differences of geoid slopes in both groups. The comparison is based on a rms accuracy of the computed values of 0.27 cm which is given for resolution $5^o X 5^o$ of the Eigen-CH1-52514 coefficients (http://op.gfzpotsdam. de/CHAMP/results). The results of comparisons are illustrated in Table 2.

The differences between observed raw signals and their computed value among individual points are between zero and 0.35 m. As these points were identified by numbers belonging to the vicinity of previously revealed tectonic traces in using the same observed signals and the OSU91A and GRM96 models, in Table 3 the type of mass-"source" is indicated next to the point number. Without further comment, it is worth mentioning that Point 27 with the largest difference is in the vicinity of a geological fault that initiated the earthquake in Athens on 7th September 1999 with a focal depth estimated 9-14 km (Athens Metropolitan Earthquake - field report. Pavlides et al, 1999). The slope variation δN between the observed and computed values, along pairs of points at various distances (Table 3) illustrates how the two models compare with the observed value δN in the particular area.

The signals $\delta N(n)$ were calculated for values n=1, 2, ... 360. The power spectra of these series were calculated by FFT for n/2=180 points. The calculations were performed by using the free software **TimeStat** in a Windows environment. Figs. 5, 6, 7 depict the FFT power spectra of three individual geo-signals after frequency dependent de-trending with coefficients of geopotential models based on the CHAMP and GRACE missions. The type of "source" and other existing physical information is shown in brackets.

Table 1. Observed and computed undulation signals at the 10 sample points. The first 5 belong to the wide area and the second 5-point group (numbered) belongs to the small test area. The standard deviation of differences with respect to CHAMP model is +/- 1.071m and with respect to GRACE is +/- 1.017. (Unit is m)

Point	Observed signal N	CHAMP computed	GRACE computed	Diff. from GRACE	Diff. from CHAMP
KRTH	15.93	15.887	15.862	0.04	0.07
SKYR	40.87	41.003	41.187	-0.13*	-0.26
KARP	15.68	15.693	15.485	-0.01	-0.19
STRT	44.67	44.972	44.942	-0.30*	-0.27
IERA	19.50	19.497	19.463	0.00	-0.04
27 fault (earthquake)	39.30	39.671	39.653	-0.37*	-0.35
38	37.92	37.935	37.918	-0.02	0.00
47 earthquake focus	38.10	38.273	38.004	-0.17*	0.10
25 mine	38.33	38.437	38.412	-0.11*	-0.08
29 fault	38.94	39.002	38.996	-0.06	-0.06

Table 2. Distances between tested points, geoidal slopes in CHAMP, GRACE and local observed signals (spherical distance computed with Haversine formula: http://www.moveble-type.co.uk/scripts - the symbols in parentheses show the raw orientation of the distances)

Distance between points (Km)	Points	CHAMP geoidal diff. δN	GRACE geoidal diff. δN	***LOCAL geoidal diff. δN***
469.2	STRT-KRTH (N-S)	29.08	29.08	28.74
387.9	KRTH-KARP (S-SE)	0.194	0.377	0.250
760.0	STRT-KARP (N-SE)	29.27	29.45	28.99
256.3	KRTH–SKYR (S-NE)	25.12	25.26	24.94
161.4	47-25 (N-S)	-0.164	-0.408	-0.230
91.18	25-29 (S-NE)	0.565	0.584	0.610
78.61	47-29 (W-E)	-0.729	-0.992	-0.840

Table 3. Distance between tested points, difference of slope δN between observed signals and calculated with CHAMP and GRACE

Signal differences between pairs of points	Distance (Km)	Diff. from CHAMP (m)	Diff. from GRACE (m)	Observed Differences (m)
STRT – KARP	760.0	-0.28	-0.46	28.99
STRT – KRTH	469.2	-0.34	-0.34	28.74
KRTH – KARP	387.9	0.06	-0.13	0.25
KRTH – SKYR	256.3	-0.18	-0.32	24.94
47–29 fault-fault	78.6	0.11	-0.15	-0.84
29-25 fault-mine	91.2	0.05	0. 03	0.61
47-25 fault-mine	161.4	0.07	-0.18	-0.23

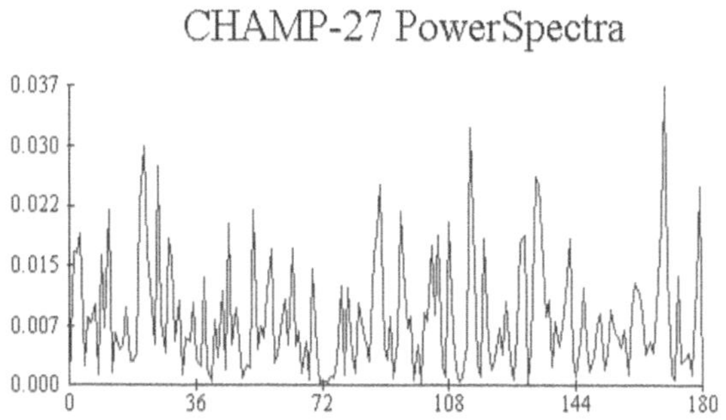

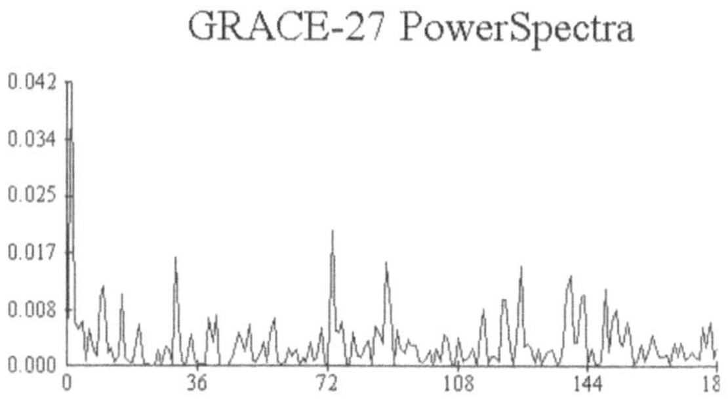

Fig. 5 FFT power spectra of frequency dependent de-trended geoid-signal after the CHAMP and GRACE coefficients ("source" geological fault – Earthquake 1999, depth 9-14 km)

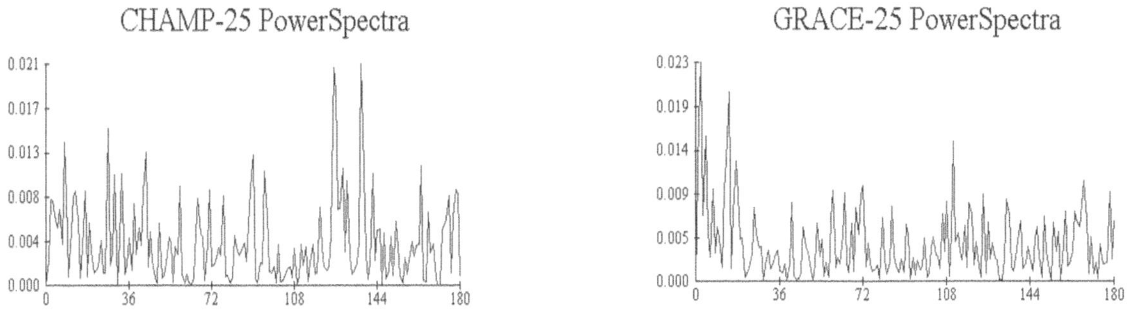

Fig. 6 FFT power spectra of frequency dependent de-trended geoid-signal after the CHAMP and GRACE coefficients ("source" mine-not in use)

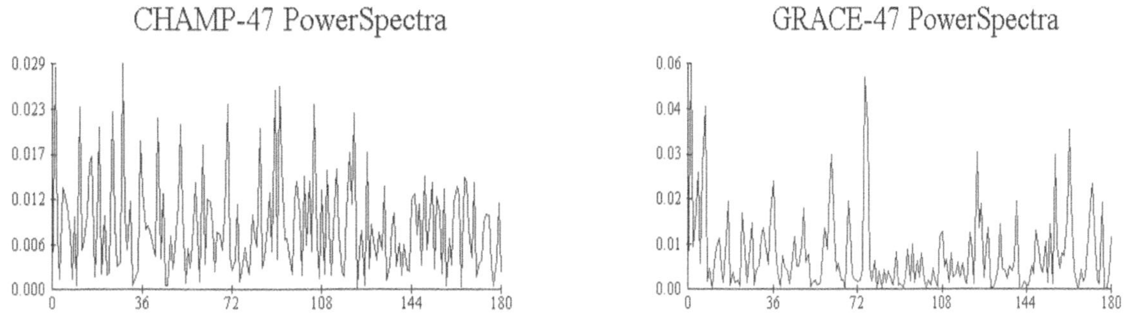

Fig. 7 FFT power spectra of frequency dependent de-trended geoid-signal after the CHAMP and GRACE coefficients ("source" – Atalanti tectonic fault)

4 Conclusions

It should be considered that the conclusions about how good is the adaptation of CHAMP and GRACE models in this area in respect to earlier ones, are not directly related to the spectral tests. The spectral tests shown in this investigation represent a very first research attempt to evaluate the information content of point undulations when these are de-trended by a well adapted model. In the future this spectral approach should be investigated in detail in theory and with several numerical experiments and spectral methods.

It must be emphasised that the small sample of test signals does not allow us to draw statistically significant conclusions. Nevertheless, it ones give a first insight into how well the CHAMP and GRACE models can adapt in the particular region: For example,

1) comparing these indicative evaluations with the previous investigations using earlier models, a first impression is that CHAMP and GRACE do reveal adequately long-wavelength variations coming from deep Newtonian sources that could not be revealed in earlier (Doufexopoulou 1984, Doufexopoulou & Papafitsorou 1986) or in the extended investigations with OSU91A and EGM96 (Pagounis 2000, Doufexopoulou & Pagounis 1998). The particularity of the region where the test points come from, exists in a large dispersion of frequency effects upon the gravity field. These start from low expansion degrees as revealed in (Doufexopoulou & Paradissis 1988). A possible reason can be the dominating long-wave-length effects from main tectonic plates with different age and origin (Makris, 1977; Parson&Daly, 1983; Doufexopoulou, 1985a) and the effect of a subduction zone within the area (Mcadoo, 1981). Indeed the results confirm that the first 100 degrees of CHAMP and GRACE models reveal most of the main tectonic formations that occur in the region, that has been studied extensively in respect to the geo-background by the local gravity field (e. g Doufexopoulou, 1985a; Doufexopoulou & Paradissis, 1988). The signal

residuals δN seems to contain local effects from near-to-surface sources.

2) The previous indications are supported by the observation that, of the 10 tested signals, the 5 showing the larger residuals in both models occur at points in the vicinity of local faults at depths near the Earth's surface. The location of these geologically determined faults has been revealed independently of geophysical inversion through error highs in local geoid interpolation (Doufexopoulou, 2002). The values corresponding to these points are indicated by an asterisk (*) in Table 1. This indication is also supported by the fact that points 47 and 25 are located above a local fault (essentially not known before 1999) and in the vicinity of an old mine. The differences between observed and computed signals at the 5 points indicate that the Newtonian sources which contribute to these values seem to originate from near-to-surface depths of "sources" at a rate of few km. Indeed, Point 47, which displays the signal with the largest difference, is located almost above the epicenter of the Athens earthquake 1999, the focal depth of which was estimated at a depth of about 9-14 km (http://www.earth.ox.ac.uk/~time/athens/fieldobs_orig.html). On the basis of the rms of the 10 signal residuals, GRACE is slightly better adapted than CHAMP.

Needless to say, these indications need to be confirmed by use of much bigger sample and geophysical and geological information need to be associated with each other so that a relation between spectra and estimated depths of "sources" can be established. However, if – as is expected – the present conclusions are confirmed, then the investigation of a regional/ local mass-density configuration from observed geoid geo-signals will be possible as an independent source of information, compared to other more rigorous and cost-effective methods.

Observing the FFT spectra of points 47, 27, & 25 (Fig. 4) a first conclusion that may be drawn is that the amplitude level is higher in CHAMP than in the GRACE de-trending and that the GRACE-based spectra are not noisy. This observation confirms again that GRACE is better adapted in the region than CHAMP. The spectra contain information on resolved local frequencies which need to be analysed in detail. The 3 points are within a small area (the distance between points 47 and 25 is 161 km). The spectra at each particular point are quite different in respect of both models. The spectrum of Point 47 (a tectonic fault) has more energy at lower frequencies comparing to the spectrum of Point 27 which is at a local geological fault.

References

Abers, G.A.., and R. McCaffrey (1994). Active arc-continent collision: Earthquakes, gravity anomalies and fault kinematics in the Huon-Finisterre collision zone, Papua New Guinea, *Tectonics*, 13, pp. 227-245.

Blais, J.A.R. Spectrum estimation using maximum entropy and multiresolution considerations, J. geod, 70 (1996) 6, Berlin, 349-356.

Bowin, C.O., (1983). Depth of the principal mass anomalies contributing to the Earth's geoidal undulations and gravity anomalies, *Marine Geodesy*, 7, pp. 61-100.

Reigber, Ch., Balmino, G., Schwintzer, P., Biancale, R., Bode, A., Lemoine, J.-M., Koenig, R., Loyer, S., Neumayer, H., Marty, J.-C., Barthelmes, F., Perosanz, F., Zhu, S. Y. (2002). A high quality global gravity field model from CHAMP GPS tracking data and Accelerometry (EIGEN-1S). *Geophysical Research Letters*, 29(14), 10.1029/2002 GL015064, 2002.

Doufexopoulou, M., (1985). A tectonic consideration of the geoid in the Greek area (in Greek with extended English summary). Techika Chronika. *Scientific Journal of Greek Association of Engineers*.

Doufexopoulou, M. (1985a). Study of the local gravity field in Greece for the geoid approximation, Ph. D dissertation, N.T.U.A, Athens, Greece (in Greek with English and French summary)

Doufexopoulou, M., Papafitsorou, A. (1986). Inversibility of geopotential models for local gravity field approximation. Proceedings of the IAG Conference "Figure of the Earth, Moon and other Planets", *Geodetic and Cartographic Institute VUGKT*, Prague.

Doufexopoulou, M., Paradissis, D., (1988). Doppler undulations and local gravity field, Proceedings of the IAG International Symposium " Instrumentation & Analysis for Integrated Geodesy" GGRI of Hung. Academy of Sciences (ed. J. Somognyi, G. Hein)

Doufexopoulou, M., Pagounis, V. (1998). Investigation of tectonic traces upon the regional geoid without the use of gravity. Proceedings of the second workshop on the geoid in Europe, *Reports of the Finnish Geodetic Institute*, 98:4, ISBN 951-711-222-X, Masala 1998.

Doufexopoulou, M. et al. (1999). A comparison of geophysical signatures between Greece and Italy as deduced by geodetic data, General Assembly of IUGG, Birmingham 1999. *International Association of Seismology and Physics of the Earth's Interior* (IASPEI). Poster ST14/W/57.

Doufexopoulou, M., Pagounis, V. (2001). A comparison between computed and observed undulation degree variances at the low and medium frequency spectrum. *International Geoid Service.* Bulletin N.13 –Special Issue / ISSN 1128 -3955 "Regional and local gravity field approximation", EGS Assembly Nice, March 2001

Doufexopoulou, M., (2002). Tracing signals from the Earth's crust by using only geodetic observations without inversion (in Spanish). *Proceedings of the Hispanoprtuguese Assembly of Geodesy and Geophysics*, February 2002.

Featherstone, W.E., (1997). On the use of geoid in Geophysics. A case study over the North and Shelf of Australia, *Exploration Geophysics*, 28 (1):52-57

Forsberg, R., (1984). Study of terrain reductions, density anomalies and geophysical inversion methods in gravity-

field modelling. *Technical report* AD-A-150788/8/XAB, Ohio State Univ. Research Foundation, Columbus (USA).
Kahn, M.A., (1977). Depth sources of gravity anomalies. *Geophysical Journal of the Royal Astronomical Society* 48, 197-209.
Kaipio, J.P., V Kolehmainen, M Vauhkonen and E Somersalo (1999). Inverse problems with structural prior information *Inverse Problems* 15 pp. 713-729.
Kakleas, Ch., Prodromou, I., (2000). Processing and Analysis of height observables, Diploma Thesis. National Technical University, Athens (in Greek)
Koening, R., Grunwaldt, L., Reigber, C., Schmidt, R., Schwintzer, P. (2002). SLR and the CHAMP Gravity Field Mission, *13th International Workshop on laser ranging*: Proceedings from the science session, October 2002, Washington DC.
Koumbis, C., (2001). Evaluation of interpolation methods for the geoid undulation N. Initial tracing of environmental geo-information (in Greek). *Diploma thesis*, National Technical University of Athens, Greece.
Makris, J., (1977). The crust and upper mantle of Aegean region obtained from deep seismic soundings. Publ. Inst. Geoph. Polish Academy of Sciences, A-4 (115)
Mcadoo, D., (1981). Geoid anomalies over deep sea trenches and the vicinity of subduction zones, J. G. R. Vol.. 86., No. B7
Oezener, H., A. Garagon Dogru (2003). The results of deformation studies on the Iznik – Meceke segment of Nafz by geodetic techniques, *Geophysical Research Abstracts*, Vol. 5, European Geophysical Society 2003.
Pagounis, V., N. (2000). The contribution of evaluating geodetic height data to local geodynamics. Application in Greece. *PhD dissertation* (in Greek), N.T.U.A, Athens.
Parson, B., Daly, S., (1983). The relationship between surface topography, gravity anomalies and temperature structure of convection. J. G. R., Vol. 88., B2
Pavlides, G. Papadopoulos, Ganas, Th. (1999) ATHENS METROPOLITAN EARTHQUAKE *Field report*.
Wallace, L.M., C. W. Stevens, E. Silver, R. McCaffrey, W. Loratung, S. Hasiata, R. Curley, R. Rosa, J. Taugaloidi, and H. Davies (2004). GPS Constraints on Active Tectonics and Arc-Continent Collision in Papua New Guinea: evidence for edge-driven microplate rotations, *Journal of Geophysical Research 109*, B05404, doi:10.1029/2003JB002481, 2004
http://www.earth.ox.ac.uk/~time/athens/fieldobs_orig.html
http://www.gfz-potsdam.de/pb1/media/champ/eigen-cg01c/reigber_dlr_nachrichten052004.pdf.

Deformations Monitoring by Integrating Local and Global Reference Systems

J. Zurutuza
GEOLan Donosti SL, Portuetxe 53 b Of. 314, 20.018 San Sebastián (Gipuzkoa).
jz@geolandonosti.com

M. J. Sevilla
Instituto de Astronomía y Geodesia (CSIC-UCM),
Facultad de Ciencias Matemáticas, Universidad Complutense de Madrid, 28040 Madrid, Spain.
sevilla@mat.ucm.es

Abstract. Although the transformation between threedimensional reference systems has been widely studied in geodesy, in this paper, a different way of integrating, and not transforming, both systems is used. On one hand, a straightforward computational method without iterations nor "a priori" known values is shown. On the other hand, some applications have been made and presented. The method, recently recognised by Grafarend, is known as the Procrustes solution and it is based on the Singular Value Decomposition of a matrix from which we obtain the Euler Elements. Finally real network located in Gipuzkoa has been used to monitor a simulated deformation via the Procrustes solution, which is compared with the Helmet's datum transformation classical approach.

Keywords: Procrustes, SVD, datum transformation, Euler elements, Orthogonal transformation.

1. Introduction

Still today, one of the most challenging tasks in geodesy is the datum transformation. We can admit that the datum transformation problem has never been as important as nowadays that a new global reference system has been adopted: ETRS89. Thus, obtaining correct and accurate transformation models from one system to other (global to global, global to local or local to local) is a main task in modern geodesy to warrant that there is no loose of accuracy in the data due to the transformation. Although the datum transformation has been widely studied by many authors (Krakiwsky and Thompson, 1974), this paper deals with the method known as the Procrustes problem (Schönemann, 1966, Sansò, 1973, Crosilla, 1983, Awange, 2003, Awange and Grafarend, 2005). The basic aim of the Procrustes solution is to fit one given set of points into another known set of points and get a measure of the quality. In the next pages, the different solutions will be treated in detail and an example is provided were transformation models are compared, using real data.

2. Three Dimensional Positioning Systems

In order to understand the Procrustes solution, the local and the global three dimensional positioning systems must be defined. The main characteristics of such reference systems are pointed in the next lines.

2.1. Local Positioning System

The local positioning system (Figure 1) owes its name to the following characteristic: every measure is referred to a local origin which is, usually, the centre of the instrument. The local positioning system (f_1, f_2, f_3) can be described as follows:

- Origin: Centre of the instrument.
- f_3 is the direction of the plumb line.
- f_1 in the horizontal plane.
- f_2 form a right handed system with f_3 and f_1

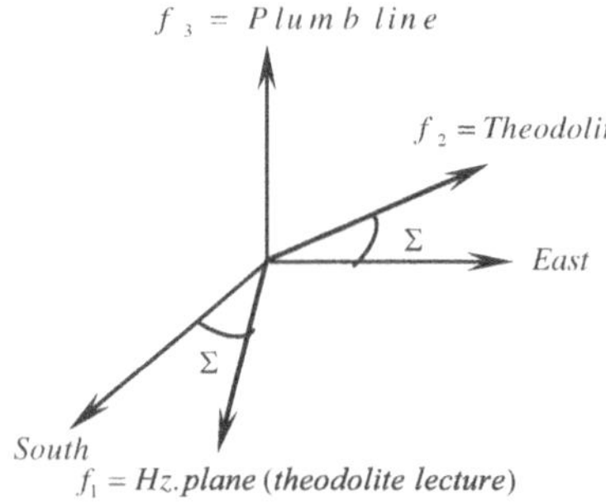

Fig. 1: Local positioning system.

2.2. Global Positioning System

The main characteristics of the global positioning system (F_1, F_2, F_3) are (Figure 2):

- Origin: Center of mass of the Earth.
- F_3 is the ITRS pole direction.
- F_1 is the ITRS Origin of longitudes.
- F_2 forms aright handed system with F_1 and F_3.

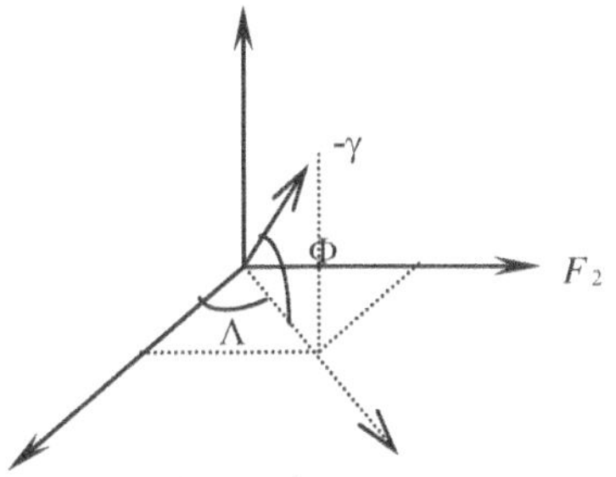

Fig. 2: Global positioning system.

3. Three Dimensional Datum Transformation Models

3.1. The Helmert Transformation

The classical Helmert datum transformation solution is given by a translation (vector $\mathbf{T}$), a scale factor k and a rotation (matrix $\mathbf{R}(\omega)$). The most important disadvantages of Helmert's solution are: they need "a priori" known values, they are iterative and need to be linearized. Commonly, the solution is presented by the equations

$$\begin{bmatrix} X \\ Y \\ Z \end{bmatrix}_{GPS} = \begin{bmatrix} T_U \\ T_V \\ T_W \end{bmatrix}_{Traslation} + (1+k) \begin{bmatrix} 1 & \omega_W & -\omega_V \\ -\omega_W & 1 & -\omega_U \\ \omega_V & \omega_U & 1 \end{bmatrix} \begin{bmatrix} U \\ V \\ W \end{bmatrix}_{Local} \quad (1)$$

The parameters of the transformations are obtained by the least squares mixed conditional model

$$\mathbf{Ax} + \mathbf{Bv} - \mathbf{t} = \mathbf{0}, \quad (2)$$

were $\mathbf{A}$ and $\mathbf{B}$ are the design matrices, $\mathbf{x}$ the parameter vector, $\mathbf{t}$ the vector of observations and $\mathbf{v}$ the residuals.

3.2. The Procrustes Transformations

The weighted extended orthogonal Procrustes (WEOP) transformation in given by the equation (Awange and Grafarend 2005)

$$\mathbf{x} = \mathbf{u}\mathbf{T}^{\mathrm{T}} + k\mathbf{X}\mathbf{R}^{\mathrm{T}} + \mathbf{E} \quad (3)$$

were $\mathbf{E}$ represents the residuals, $\mathbf{x}$, $\mathbf{X}$, the coordinates of several points in each system respectively, k the scale factor, $\mathbf{R}$ the rotation matrix and $\mathbf{u}$ is a vector of as many ones as known points in both systems. $\mathbf{T}$ is the translation vector.

To see how (3) is obtained we need to know how a transformation of a point is performed. To transform the coordinates of a point from a known reference system into another known reference system a translation vector, a rotation matrix and a scale factor shall be considered. Therefore, for one single point we have:

$$\begin{pmatrix} x \\ y \\ y \end{pmatrix} = \begin{pmatrix} T_1 \\ T_2 \\ T_3 \end{pmatrix} + k \begin{pmatrix} r_{11} & r_{12} & r_{13} \\ r_{21} & r_{22} & r_{23} \\ r_{31} & r_{32} & r_{33} \end{pmatrix} \begin{pmatrix} X \\ Y \\ Z \end{pmatrix} \quad (4)$$

In case of two points, (4) would be:

$$\begin{pmatrix} x_1 & x_2 \\ y_1 & y_2 \\ z_1 & z_2 \end{pmatrix} = \begin{pmatrix} T_1 & T_1 \\ T_2 & T_2 \\ T_3 & T_3 \end{pmatrix} + k \begin{pmatrix} r_{11} & r_{12} & r_{13} \\ r_{21} & r_{22} & r_{23} \\ r_{31} & r_{32} & r_{33} \end{pmatrix} \begin{pmatrix} X_1 & X_2 \\ Y_1 & Y_2 \\ Z_1 & Z_2 \end{pmatrix} \quad (5)$$

For n points, (4) would be written as:

$$\begin{pmatrix} x_1 & x_2 & \dots & x_n \\ y_1 & y_2 & \dots & y_n \\ z_1 & z_2 & \dots & z_n \end{pmatrix} = \begin{pmatrix} T_1 & T_1 & \dots & T_1 \\ T_2 & T_2 & \dots & T_2 \\ T_3 & T_3 & \dots & T_3 \end{pmatrix} + k \begin{pmatrix} r_{11} & r_{12} & r_{13} \\ r_{21} & r_{22} & r_{23} \\ r_{31} & r_{32} & r_{33} \end{pmatrix} \begin{pmatrix} X_1 & X_2 & \dots & X_n \\ Y_1 & Y_2 & \dots & Y_n \\ Z_1 & Z_2 & \dots & Z_n \end{pmatrix} \quad (6)$$

We consider the matrices of the coordinates:

$$\mathbf{x} = \begin{pmatrix} x_1 & y_1 & z_1 \\ x_2 & y_2 & z_2 \\ \dots & \dots & \dots \\ x_n & y_n & z_n \end{pmatrix} \text{ and } \mathbf{X} = \begin{pmatrix} X_1 & Y_1 & Z_1 \\ X_2 & Y_2 & Z_2 \\ \dots & \dots & \dots \\ X_n & Y_n & Z_n \end{pmatrix} \quad (7)$$

Taking into account the notation used in (3), and the Cartesian coordinates in (7), the transformation equation for n points can be rewritten as:

$$\mathbf{x}^{\mathrm{T}} = \mathbf{T}\mathbf{u}^{\mathrm{T}} + k\mathbf{R}\mathbf{X}^{\mathrm{T}} \quad (8)$$

By transposing (8) we get:

$$\mathbf{x} = \mathbf{u}\mathbf{T}^{\mathrm{T}} + k\mathbf{X}\mathbf{R}^{\mathrm{T}} \tag{9}$$

The data are nor error free and this fact is taken into account by adding the error matrix **E**. Thus, the Procrustes final equation is:

$$\mathbf{x} = \mathbf{u}\mathbf{T}^{\mathrm{T}} + k\mathbf{X}\mathbf{R}^{\mathrm{T}} + \mathbf{E} \tag{10}$$

The conditions imposed to determine the parameters of the transformation are:

- Norm of **E** must be a minimum.
- **R** is orthogonal

Since we need to solve a conditioned minimum, among all the possible solutions, the Lagrange method is considered. The norm used will be the Frobenius Norm:

Thus, the equation to be solved is (3) with the conditions:

$$\|\mathbf{E}\|_P = \min . \quad \mathbf{R}^{\mathrm{T}}\mathbf{R} = 1 \; and \quad \det(\mathbf{R}) = +1 \tag{11}$$

The Frobenius norm (seminorm) is defined by means of the trace of matrix as follows

$$\|\mathbf{E}\|_P^2 = \mathrm{tr}\left(\mathbf{E}^{\mathrm{T}}\mathbf{P}\mathbf{E}\right) \tag{12}$$

The weight matrix **P** of the observations is semidefinite positive because it might occur that not all the elements are known. In this cases, the solution is obtained by using null values for such element. **P** is computed by taking into account the dispersion matrixes of both sets of coordinates (see sec. 3.3).

To solve (3), subject to conditions (11) the first step is to compute the translation vector **T**. The norm to be minimized is

$$\begin{aligned}\mathrm{N}(k,\mathbf{T},\mathbf{R}) &= \mathrm{tr}\left(\mathbf{x}-\mathbf{u}\mathbf{T}^{t}-k\mathbf{X}\mathbf{R}^{\mathrm{T}}\right)^{\mathrm{T}}\mathbf{P}\left(\mathbf{x}-\mathbf{u}\mathbf{T}^{\mathrm{T}}-k\mathbf{X}\mathbf{R}^{\mathrm{T}}\right)=\\ &= \mathrm{tr}(\mathbf{x}^{\mathrm{T}}\mathbf{P}\mathbf{x}-\mathbf{x}^{\mathrm{T}}\mathbf{P}\mathbf{u}\mathbf{T}^{\mathrm{T}}-k\mathbf{x}^{\mathrm{T}}\mathbf{P}\mathbf{X}\mathbf{R}^{\mathrm{T}}-\\ &\quad -\mathbf{T}\mathbf{u}^{\mathrm{T}}\mathbf{P}\mathbf{x}+\mathbf{T}\mathbf{u}^{\mathrm{T}}\mathbf{P}\mathbf{u}\mathbf{T}^{\mathrm{T}}+k\mathbf{T}\mathbf{u}^{\mathrm{T}}\mathbf{P}\mathbf{X}\mathbf{R}^{\mathrm{T}}-\\ &\quad -k\mathbf{R}\mathbf{X}^{\mathrm{T}}\mathbf{P}\mathbf{x}+k\mathbf{R}\mathbf{X}^{\mathrm{T}}\mathbf{P}\mathbf{u}\mathbf{T}^{\mathrm{T}}+k^2\mathbf{R}\mathbf{X}^{\mathrm{T}}\mathbf{P}\mathbf{X}\mathbf{R}^{\mathrm{T}})\end{aligned} \tag{13}$$

For (13) to be a minimum, its partial derivative with respect to the unknowns must be null:

$$\begin{aligned}\frac{\partial \mathrm{N}}{\partial \mathbf{T}^{\mathrm{T}}} &= -\mathbf{x}^{\mathrm{T}}\mathbf{P}\mathbf{u}-\mathbf{x}^{\mathrm{T}}\mathbf{P}\mathbf{u}+2\mathbf{T}\mathbf{u}^{\mathrm{T}}\mathbf{P}\mathbf{u}+k\mathbf{R}\mathbf{X}^{\mathrm{T}}\mathbf{P}\mathbf{u}+k\mathbf{R}\mathbf{X}^{\mathrm{T}}\mathbf{P}\mathbf{u}=\\ &= -2\mathbf{x}^{\mathrm{T}}\mathbf{P}\mathbf{u}+2\mathbf{T}\mathbf{u}^{\mathrm{T}}\mathbf{P}\mathbf{u}+2k\mathbf{R}\mathbf{X}^{\mathrm{T}}\mathbf{P}\mathbf{u}=0\end{aligned} \tag{14}$$

Finally, the translation vector can be written as:

$$\mathbf{T} = \left(\mathbf{u}^{\mathrm{T}}\mathbf{P}\mathbf{u}\right)^{-1}\left(\mathbf{x}^{\mathrm{T}}-k\mathbf{R}\mathbf{X}^{\mathrm{T}}\right)\mathbf{P}\mathbf{u} \tag{15}$$

In the special case that **P**=**I**, the solution is :

$$\mathbf{T} = \frac{1}{n}\left(\mathbf{x}^{\mathrm{T}}-k\mathbf{R}\mathbf{X}^{\mathrm{T}}\right)\mathbf{u} = \frac{1}{n}\left(\mathbf{x}-k\mathbf{X}\mathbf{R}^{\mathrm{T}}\right)^{\mathrm{T}}\mathbf{u} \tag{16}$$

Once the translation vector has been computed, we do now proceed to obtain the scale factor.

Taking into account (15), the equation to be solved is now:

$$\begin{aligned}\mathrm{N}(k,\mathbf{R}) = \mathrm{tr}&\left[\mathbf{x}-\mathbf{u}\mathbf{u}^{\mathrm{T}}\mathbf{P}\left(\mathbf{x}^{\mathrm{T}}-k\mathbf{R}\mathbf{X}^{\mathrm{T}}\right)^{\mathrm{T}}\left(\mathbf{u}^{\mathrm{T}}\mathbf{P}\mathbf{u}\right)^{-1}-k\mathbf{X}\mathbf{R}^{\mathrm{T}}\right]^{\mathrm{T}}\mathbf{P}\\ &\left[\mathbf{x}-\mathbf{u}\mathbf{u}^{\mathrm{T}}\mathbf{P}\left(\mathbf{x}^{\mathrm{T}}-k\mathbf{R}\mathbf{X}^{\mathrm{T}}\right)^{\mathrm{T}}\left(\mathbf{u}^{\mathrm{T}}\mathbf{P}\mathbf{u}\right)^{-1}-k\mathbf{X}\mathbf{R}^{\mathrm{T}}\right]\end{aligned} \tag{17}$$

(17) is centred because the translation parameters have been included in the formula.

$$\begin{aligned}\mathrm{N}(k,\mathbf{R}) &= \mathrm{tr}\left[\left(\mathbf{x}-k\mathbf{X}\mathbf{R}^{\mathrm{T}}\right)-\left(\mathbf{u}^{\mathrm{T}}\mathbf{P}\mathbf{u}\right)^{-1}\mathbf{u}\mathbf{u}^{\mathrm{T}}\mathbf{P}\left(\mathbf{x}-k\mathbf{X}\mathbf{R}^{\mathrm{T}}\right)\right]^{\mathrm{T}}\mathbf{P}\\ &\quad\left[\left(\mathbf{x}-k\mathbf{X}\mathbf{R}^{\mathrm{T}}\right)-\left(\mathbf{u}^{\mathrm{T}}\mathbf{P}\mathbf{u}\right)^{-1}\mathbf{u}\mathbf{u}^{\mathrm{T}}\mathbf{P}\left(\mathbf{x}-k\mathbf{X}\mathbf{R}^{\mathrm{T}}\right)\right]\\ &= \mathrm{tr}\left\{\left[\mathbf{I}-\left(\mathbf{u}^{\mathrm{T}}\mathbf{P}\mathbf{u}\right)^{-1}\mathbf{u}\mathbf{u}^{\mathrm{T}}\mathbf{P}\right]\left(\mathbf{x}-k\mathbf{X}\mathbf{R}^{\mathrm{T}}\right)\right\}^{\mathrm{T}}\mathbf{P}\cdot\\ &\quad\cdot\left\{\left[\mathbf{I}-\left(\mathbf{u}^{\mathrm{T}}\mathbf{P}\mathbf{u}\right)^{-1}\mathbf{u}\mathbf{u}^{\mathrm{T}}\mathbf{P}\right]\left(\mathbf{x}-k\mathbf{X}\mathbf{R}^{\mathrm{T}}\right)\right\}\end{aligned} \tag{18}$$

A new matrix shall be defined:

$$\mathbf{C} = \mathbf{I}-\left(\mathbf{u}^{\mathrm{T}}\mathbf{P}\mathbf{u}\right)^{-1}\mathbf{u}\mathbf{u}^{\mathrm{T}}\mathbf{P} \tag{19}$$

With (19), (18) can now be written as:

$$\begin{aligned}\mathrm{N}(k,\mathbf{R}) &= \mathrm{tr}\left\{\left[\mathbf{C}\left(\mathbf{x}-k\mathbf{X}\mathbf{R}^{\mathrm{T}}\right)\right]^{\mathrm{T}}\mathbf{P}\left[\mathbf{C}\left(\mathbf{x}-k\mathbf{X}\mathbf{R}^{\mathrm{T}}\right)\right]\right\}\\ &= \mathrm{tr}\left\{\left(\mathbf{x}^{\mathrm{T}}-k\mathbf{R}\mathbf{X}^{\mathrm{T}}\right)\mathbf{C}^{\mathrm{T}}\mathbf{P}\mathbf{C}\left(\mathbf{x}-k\mathbf{X}\mathbf{R}^{\mathrm{T}}\right)\right\}\end{aligned} \tag{20}$$

or

$$\mathrm{N}(k,\mathbf{R}) = \mathrm{tr}\begin{pmatrix}\mathbf{x}^{\mathrm{T}}\mathbf{C}^{\mathrm{T}}\mathbf{P}\mathbf{C}\mathbf{x}-k\mathbf{x}^{\mathrm{T}}\mathbf{C}^{\mathrm{T}}\mathbf{P}\mathbf{C}\mathbf{X}\mathbf{R}^{\mathrm{T}}-\\ -k\mathbf{R}\mathbf{X}^{\mathrm{T}}\mathbf{C}^{\mathrm{T}}\mathbf{P}\mathbf{C}\mathbf{x}+k^2\mathbf{R}\mathbf{X}^{\mathrm{T}}\mathbf{C}^{\mathrm{T}}\mathbf{P}\mathbf{C}\mathbf{X}\mathbf{R}^{\mathrm{T}}\end{pmatrix} \tag{21}$$

Eq. (21) must be a minimum, so its derivative with respect the unknown must be null:

$$\begin{aligned}\frac{\partial \mathrm{N}(k,\mathbf{R})}{\partial k} &= -2\mathrm{tr}\left(\mathbf{x}^{\mathrm{T}}\mathbf{C}^{\mathrm{T}}\mathbf{PCXR}^{\mathrm{T}}\right)+ \\ &+2k\mathrm{tr}\left(\mathbf{RX}^{\mathrm{T}}\mathbf{C}^{\mathrm{T}}\mathbf{PCXR}^{\mathrm{T}}\right)=0\end{aligned} \tag{22}$$

Finally, we get the scale factor:

$$k=\frac{\mathrm{tr}\left(\mathbf{x}^{\mathrm{T}}\mathbf{C}^{\mathrm{T}}\mathbf{PCXR}^{\mathrm{T}}\right)}{\mathrm{tr}\left(\mathbf{X}^{\mathrm{T}}\mathbf{C}^{\mathrm{T}}\mathbf{PCX}\right)} \tag{23}$$

In the particular case that the weight matrix is diagonal unitary, the scale factor obtained is:

$$\begin{aligned}\mathbf{C}&=\mathbf{I}-\frac{1}{n}\mathbf{uu}^{\mathrm{T}} \\ k&=\frac{\mathrm{tr}\left(\mathbf{x}^{\mathrm{T}}\mathbf{CXR}^{\mathrm{T}}\right)}{\mathrm{tr}\left(\mathbf{X}^{\mathrm{T}}\mathbf{CX}\right)}\end{aligned} \tag{24}$$

Once the translation vector and the scale factor have been computes, the rotation matrix will be computed. To compute the rotations, (17) and (23) are taken into account. Thus, the new equation to be minimized is:

$$\begin{aligned}\mathrm{N}(\mathbf{R}) &= \mathrm{tr}\left(\left(\mathbf{x}^{\mathrm{T}}-\frac{\mathrm{tr}\left(\mathbf{x}^{\mathrm{T}}\mathbf{C}^{\mathrm{T}}\mathbf{PCXR}^{\mathrm{T}}\right)}{\mathrm{tr}\left(\mathbf{X}^{\mathrm{T}}\mathbf{C}^{\mathrm{T}}\mathbf{PCX}\right)}\mathbf{RX}^{\mathrm{T}}\right)\mathbf{C}^{\mathrm{T}}\mathbf{PC}\left(\mathbf{x}-\frac{\mathrm{tr}\left(\mathbf{x}^{\mathrm{T}}\mathbf{C}^{\mathrm{T}}\mathbf{PCXR}^{\mathrm{T}}\right)}{\mathrm{tr}\left(\mathbf{X}^{\mathrm{T}}\mathbf{C}^{\mathrm{T}}\mathbf{PCX}\right)}\mathbf{XR}^{\mathrm{T}}\right)\right) \\ &= \mathrm{tr}\left(\mathbf{x}^{\mathrm{T}}\mathbf{C}^{\mathrm{T}}\mathbf{PCx}\right)-\frac{\mathrm{tr}\left(\mathbf{x}^{\mathrm{T}}\mathbf{C}^{\mathrm{T}}\mathbf{PCXR}^{\mathrm{T}}\right)}{\mathrm{tr}\left(\mathbf{X}^{\mathrm{T}}\mathbf{C}^{\mathrm{T}}\mathbf{PCX}\right)}\mathrm{tr}\left(\mathbf{x}^{\mathrm{T}}\mathbf{C}^{\mathrm{T}}\mathbf{PCXR}^{\mathrm{T}}\right)- \\ &\quad -\frac{\mathrm{tr}\left(\mathbf{x}^{\mathrm{T}}\mathbf{C}^{\mathrm{T}}\mathbf{PCXR}^{\mathrm{T}}\right)}{\mathrm{tr}\left(\mathbf{X}^{\mathrm{T}}\mathbf{C}^{\mathrm{T}}\mathbf{PCX}\right)}\mathrm{tr}\left(\mathbf{RX}^{\mathrm{T}}\mathbf{C}^{\mathrm{T}}\mathbf{PCx}\right)+ \\ &\quad +\frac{\mathrm{tr}\left(\mathbf{x}^{\mathrm{T}}\mathbf{C}^{\mathrm{T}}\mathbf{PCXR}^{\mathrm{T}}\right)}{\mathrm{tr}\left(\mathbf{X}^{\mathrm{T}}\mathbf{C}^{\mathrm{T}}\mathbf{PCX}\right)}\frac{\mathrm{tr}\left(\mathbf{x}^{\mathrm{T}}\mathbf{C}^{\mathrm{T}}\mathbf{PCXR}^{\mathrm{T}}\right)}{\mathrm{tr}\left(\mathbf{X}^{\mathrm{T}}\mathbf{C}^{\mathrm{T}}\mathbf{PCX}\right)}\mathrm{tr}\left(\mathbf{RX}^{\mathrm{T}}\mathbf{C}^{\mathrm{T}}\mathbf{PCXR}^{\mathrm{T}}\right) \\ &= \mathrm{tr}\left(\mathbf{x}^{\mathrm{T}}\mathbf{C}^{\mathrm{T}}\mathbf{PCx}\right)-2\frac{\mathrm{tr}\left(\mathbf{x}^{\mathrm{T}}\mathbf{C}^{\mathrm{T}}\mathbf{PCXR}^{\mathrm{T}}\right)^2}{\mathrm{tr}\left(\mathbf{X}^{\mathrm{T}}\mathbf{C}^{\mathrm{T}}\mathbf{PCX}\right)}+\frac{\mathrm{tr}\left(\mathbf{x}^{\mathrm{T}}\mathbf{C}^{\mathrm{T}}\mathbf{PCXR}^{\mathrm{T}}\right)^2}{\mathrm{tr}\left(\mathbf{X}^{\mathrm{T}}\mathbf{C}^{\mathrm{T}}\mathbf{PCX}\right)} \\ &= \mathrm{tr}\left(\mathbf{x}^{\mathrm{T}}\mathbf{C}^{\mathrm{T}}\mathbf{PCx}\right)-\frac{\mathrm{tr}\left(\mathbf{x}^{\mathrm{T}}\mathbf{C}^{\mathrm{T}}\mathbf{PCXR}^{\mathrm{T}}\right)^2}{\mathrm{tr}\left(\mathbf{X}^{\mathrm{T}}\mathbf{C}^{\mathrm{T}}\mathbf{PCX}\right)}\end{aligned} \tag{25}$$

The condition for (25) to be a minimum is:

$$\mathrm{tr}\left(\mathbf{x}^{\mathrm{T}}\mathbf{C}^{\mathrm{T}}\mathbf{PCXR}^{\mathrm{T}}\right)=\max \tag{26}$$

Let $\mathbf{D}=\mathbf{x}^{\mathrm{T}}\mathbf{C}^{\mathrm{T}}\mathbf{PCX}$ and let $\mathbf{U\Sigma V}^{\mathrm{T}}$ be its Singular Value Decomposition. Then,

$$\begin{aligned}\mathbf{UU}^{\mathrm{T}}&=\mathbf{U}^{\mathrm{T}}\mathbf{U}=\mathbf{I} \\ \mathbf{VV}^{\mathrm{T}}&=\mathbf{V}^{\mathrm{T}}\mathbf{V}=\mathbf{I} \\ \mathbf{\Sigma}&=\mathrm{diagonal}\left(\sigma_1,\sigma_2,\ldots,\sigma_n\right)\end{aligned} \tag{27}$$

We can now write:

$$\mathbf{x}^{\mathrm{T}}\mathbf{C}^{\mathrm{T}}\mathbf{PCXR}^{\mathrm{T}}=\mathbf{DR}^{\mathrm{T}}=\mathbf{U\Sigma V}^{\mathrm{T}}\mathbf{R}^{\mathrm{T}} \tag{28}$$

Now:

$$\begin{aligned}\mathrm{tr}\left(\mathbf{x}^{\mathrm{T}}\mathbf{C}^{\mathrm{T}}\mathbf{PCXR}^{\mathrm{T}}\right)&=\mathrm{tr}\left(\mathbf{DR}^{\mathrm{T}}\right) \\ \mathrm{tr}\left(\mathbf{U\Sigma V}^{\mathrm{T}}\mathbf{R}^{\mathrm{T}}\right)&=\mathrm{tr}\left(\mathbf{\Sigma V}^{\mathrm{T}}\mathbf{R}^{\mathrm{T}}\mathbf{U}\right)\end{aligned} \tag{29}$$

let $\mathbf{K}=\mathbf{V}^{\mathrm{T}}\mathbf{R}^{\mathrm{T}}\mathbf{U}$, which is an orthogonal matrix with diagonal entries such that $\left|k_{ii}\right|\le 1$, then

$$\mathrm{tr}\left(\mathbf{\Sigma V}^{\mathrm{T}}\mathbf{R}^{\mathrm{T}}\mathbf{U}\right)=\mathrm{tr}\left(\mathbf{\Sigma K}\right)=\sum_{i=1}^{n}\sigma_i k_{ii}\le\sum_{i=1}^{n}\sigma_i \tag{30}$$

The maximum value of (30) is given by:

$$\begin{aligned}\max\left(\mathrm{tr}\mathbf{x}^{\mathrm{T}}\mathbf{C}^{\mathrm{T}}\mathbf{PCXR}^{\mathrm{T}}\right)&=\max\left(\mathrm{tr}\mathbf{\Sigma K}\right) \\ &=\sum_{i=1}^{n}\sigma_i=\mathrm{tr}\mathbf{\Sigma}\end{aligned} \tag{31}$$

But the maximum is:

$$\mathrm{tr}\left(\mathbf{\Sigma V}^{\mathrm{T}}\mathbf{R}^{\mathrm{T}}\mathbf{U}\right)=\mathrm{tr}\mathbf{\Sigma}, \tag{32}$$

so that

$$\mathbf{V}^{\mathrm{T}}\mathbf{R}^{\mathrm{T}}\mathbf{U}=\mathbf{I} \tag{33}$$

Multiplying by $\mathbf{V}$ and $\mathbf{U}^{\mathbf{T}}$, we get:

$$\begin{aligned}\mathbf{VV}^{\mathrm{T}}\mathbf{R}^{\mathrm{T}}\mathbf{U}&=\mathbf{V} \\ \mathbf{R}^{\mathrm{T}}\mathbf{U}&=\mathbf{V} \\ \mathbf{R}^{\mathrm{T}}\mathbf{UU}^{\mathrm{T}}&=\mathbf{VU}^{\mathrm{T}}\end{aligned} \tag{34}$$

From (34) we can write:

$$\mathbf{R}=\mathbf{UV}^{\mathrm{T}} \tag{35}$$

For the special case that **P=I**, we get:

$$\begin{aligned}\mathbf{D}&=\mathbf{x}^{\mathrm{T}}\mathbf{CX}=\mathbf{U\Sigma V}^{\mathrm{T}} \\ \mathbf{R}&=\mathbf{UV}^{\mathrm{T}}\end{aligned} \tag{36}$$

We can now compute the residuals by taking into account all the parameters computed:

$$\mathbf{E}=\left(\mathbf{I}-\left(\mathbf{u}^{\mathrm{T}}\mathbf{Pu}\right)^{-1}\mathbf{uu}^{\mathrm{T}}\mathbf{P}\right)\left(\mathbf{x}-\frac{\mathrm{tr}\left(\mathbf{x}^{\mathrm{T}}\mathbf{C}^{\mathrm{T}}\mathbf{PCXVU}^{\mathrm{T}}\right)}{\mathrm{tr}\left(\mathbf{X}^{\mathrm{T}}\mathbf{C}^{\mathrm{T}}\mathbf{PCX}\right)}\mathbf{XVU}^{\mathrm{T}}\right) \tag{37}$$

The Frobenius norm, which can be considered as a measure of the quality of the transformation equals:

$$\|\mathbf{E}\|_{\mathrm{P}}^{2}=\mathrm{tr}\left[\left(\mathbf{I}-\left(\mathbf{u}^{\mathrm{T}}\mathbf{P}\mathbf{u}\right)^{-1}\mathbf{u}\mathbf{u}^{\mathrm{T}}\mathbf{P}\right)\left(\mathbf{x}-\frac{\mathrm{tr}\left(\mathbf{x}^{\mathrm{T}}\mathbf{C}^{\mathrm{T}}\mathbf{P}\mathbf{C}\mathbf{X}\mathbf{V}\mathbf{U}^{\mathrm{T}}\right)}{\mathrm{tr}\left(\mathbf{X}^{\mathrm{T}}\mathbf{C}^{\mathrm{T}}\mathbf{P}\mathbf{C}\mathbf{X}\right)}\mathbf{X}\mathbf{V}\mathbf{U}^{\mathrm{T}}\right)\right]^{\mathrm{T}}\mathbf{P}$$
$$\left[\left(\mathbf{I}-\left(\mathbf{u}^{\mathrm{T}}\mathbf{P}\mathbf{u}\right)^{-1}\mathbf{u}\mathbf{u}^{\mathrm{T}}\mathbf{P}\right)\left(\mathbf{x}-\frac{\mathrm{tr}\left(\mathbf{x}^{\mathrm{T}}\mathbf{C}^{\mathrm{T}}\mathbf{P}\mathbf{C}\mathbf{X}\mathbf{V}\mathbf{U}^{\mathrm{T}}\right)}{\mathrm{tr}\left(\mathbf{X}^{\mathrm{T}}\mathbf{C}^{\mathrm{T}}\mathbf{P}\mathbf{C}\mathbf{X}\right)}\mathbf{X}\mathbf{V}\mathbf{U}^{\mathrm{T}}\right)\right] \tag{38}$$

The values for the residuals and the Frobenius Norm, in the special case **P=I** are respectively:

$$\mathbf{E}=\left(\mathbf{I}-\frac{1}{n}\mathbf{u}\mathbf{u}^{\mathrm{T}}\right)\left(\mathbf{x}-\frac{\mathrm{tr}\left(\mathbf{x}^{\mathrm{T}}\mathbf{C}\mathbf{X}\mathbf{V}\mathbf{U}^{\mathrm{T}}\right)}{\mathrm{tr}\left(\mathbf{X}^{\mathrm{T}}\mathbf{C}\mathbf{X}\right)}\mathbf{X}\mathbf{V}\mathbf{U}^{\mathrm{T}}\right) \tag{39}$$

$$\|\mathbf{E}\|_{\mathrm{P}}^{2}=\mathrm{tr}\left(\mathbf{x}-\frac{\mathrm{tr}\left(\mathbf{x}^{\mathrm{T}}\mathbf{C}\mathbf{X}\mathbf{V}\mathbf{U}^{\mathrm{T}}\right)}{\mathrm{tr}\left(\mathbf{X}^{\mathrm{T}}\mathbf{C}\mathbf{X}\right)}\mathbf{X}\mathbf{V}\mathbf{U}^{\mathrm{T}}\right)^{\mathrm{T}}\left(\mathbf{I}-\frac{1}{n}\mathbf{u}\mathbf{u}^{\mathrm{T}}\right)\left(\mathbf{x}-\frac{\mathrm{tr}\left(\mathbf{x}^{\mathrm{T}}\mathbf{C}\mathbf{X}\mathbf{V}\mathbf{U}^{\mathrm{T}}\right)}{\mathrm{tr}\left(\mathbf{X}^{\mathrm{T}}\mathbf{C}\mathbf{X}\right)}\mathbf{X}\mathbf{V}\mathbf{U}^{\mathrm{T}}\right) \tag{40}$$

3.3. Computation of the Weight Matrix

The Weighted Procrustes solution takes into account the dispersions of both coordinate sets. Nevertheless, the weight matrix *P*, has not been defined yet. This section deals with the computation of the weight matrix.

The weight matrix must take into account the dispersion matrices of both sets of coordinates and, of course, the local dispersion matrix when it is transformed to the global frame. It has been proved (Awange and Grafarend., 2005) that the weight matrix equals to:

$$\Sigma_{vec\mathbf{E}^{\mathrm{T}}}=\Sigma_{vec\mathbf{x}^{\mathrm{T}}}+\left(\mathbf{I}_{n}\otimes k\mathbf{R}\right)\Sigma_{vec\mathbf{X}^{\mathrm{T}}}\left(\mathbf{I}_{n}\otimes k\mathbf{R}\right)^{\mathrm{T}}-$$
$$-2\Sigma_{vec\mathbf{x}^{\mathrm{T}},(\mathbf{I}_{n}\otimes k\mathbf{R})vec\mathbf{X}^{\mathrm{T}}} \tag{41}$$

In (41), Σ makes reference to the variance covariance matrix, *vec* is the vec operator and $\otimes$ is the Kronecker product.

3.4. Partial Procrustes Analysis

The partial Procrustes Analysis deals with the computation only of the rotation elements. No translation vector nor scale factor are computed (Awange and Grafarend, 2005). Thus, the solution is given by (35) and (36).

Once the rotations matrix is computed, the next step is to transform the matrix elements into the orientation elements: astronomic latitude (Φ) astronomic longitude (Λ) and the orientation unknown (Σ), as has been stated in section 3, Euler angles are obtained by the parameterization of the rotation matrix as follows:

$$\mathbf{R}(\Lambda,\Phi,\Sigma)=\mathbf{R}(\Sigma)\cdot\mathbf{R}\left(\frac{\pi}{2}-\Phi\right)\cdot\mathbf{R}(\Lambda) \tag{42}$$

Taking into account the general rotations matrix for each angle, substituting the orientation angles and multiplying in the order (42), we get:

$$\mathbf{R}=\mathbf{R}(\Sigma)\mathbf{R}\left(\frac{\pi}{2}-\Phi\right)\mathbf{R}(\Lambda)$$
$$=\begin{pmatrix}\cos(\Sigma)\sin(\Phi)\cos(\Lambda)-\sin(\Sigma)\sin(\Lambda) & \cos(\Sigma)\sin(\Phi)\sin(\Lambda)+\sin(\Sigma)\cos(\Lambda) & -\cos(\Sigma)\cos(\Phi)\\ -\sin(\Sigma)\sin(\Phi)\cos(\Lambda)-\cos(\Sigma)\sin(\Lambda) & -\sin(\Sigma)\sin(\Phi)\sin(\Lambda)+\cos(\Sigma)\cos(\Lambda) & \sin(\Sigma)\cos(\Phi)\\ \cos(\Phi)\cos(\Lambda) & \cos(\Phi)\sin(\Lambda) & \sin(\Phi)\end{pmatrix} \tag{43}$$

4. Case Study

The data sets (courtesy of the Gipuzkoako Foru Aldundia: b5m.gipuzkoa.net) correspond to the GPS network of Gipuzkoa, which is formed by a total of 21 stations which coordinates are known in the ITRF00 frame and, thus, in the ETRS89 Reference System by applying the Altamimi transformation parameters (Altamimi et al. 2002).

To verify the capability of the WEOP to detect deformations different movements have been simulated in 6 stations. The displacements vary from 1 to 6 cm in the X and in the Z components. The movements simulated have been: ITURREGI (1 cm), MENDIZORROTZ (2 cm), ONDDY (3 cm), BIANDITZ (4 cm), ZUBELTZU (5 cm) and SAN TELMO (6 cm). The analysis of the residuals should tell us if the WEOP is or not a valid model to detect deformations. Real data has been used.

The residuals will be compared with those obtained from the classical Helmert datum transformation.

Both systems are global. The network (Figure 3) was stabilised in 2002 and it covers completely the Territory of Gipuzkoa, which is located in the Basque Country (Northern Spain) and its surface is about 2000 km^2. The network consists in 21 station which belong to the "Active and Passive GPS Network of Gipuzkoa" and it was presented in the

“4ª Asamblea Hispanolusa de Geodesia y Geofísica”, February 2004, Figueira da Foz (Portugal) (Zurutuza et al., 2004). Each station has been occupied at least twice in separate sessions of not less than 8 hours of time span. A total of 7 dual frequency GPS receptors were used in each session. The 15 sessions were computed taking into account only the non trivial vectors. After the adjustment of the whole campaign, the global accuracy was estimated to be better than 2 cm with respect the active station. The coordinates of the Active Station were also computed and the accuracy estimated was better than 1,5 cm in the ITRF00 frame. Once the Reference Station was computed, the rest of the stations were adjusted constraining strongly the reference one.

The main processing parameters were:

a. Elevation Mask: 10º.
b. Precise IGS ephemeris.
c. NO ocean loading model..
d. Tropospheric model: Niell model (Niell, 1996)
e. Ambiguities: QIF strategy (Hugentobler et al., 2001).
f. Processing frequency: L_3.
g. Software package: Bernese V 4.2.

The first part of the problem shows the differences between the classical and the WEOP transformations, while the second part deals with the deformation detection.

In table 1 the Cartesian coordinates of both sets are presented. Table 2 shows the computed coordinates after the transformation

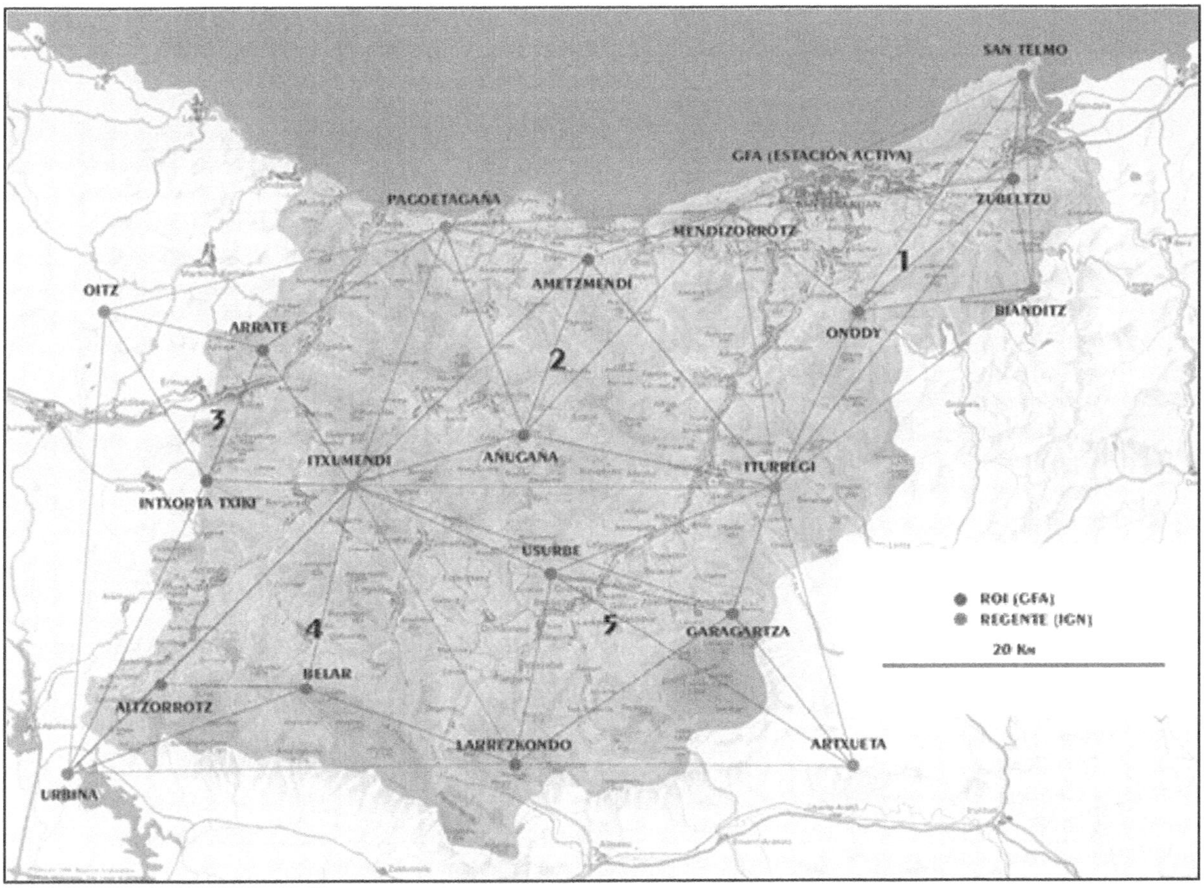

Fig 3: Test network: Active and Passive GPS network of Gipuzkoa.

	ITRF00 Coordinates (m)			ETRS89 Coordinates (m)		
S tation.	X	Y	Z	X	Y	Z
A ITZORROTZ	4667813.0044	-207091.2377	4328176.0110	4667813.1903	-207091.4595	4328175.8077
A METZMENDI	4648581.5707	-176745.6283	4349215.1022	4648581.7589	-176745.8491	4349214.8997
A NUGANA	4656972.4871	-181645.8861	4340292.5540	4656972.6747	-181646.1073	4340292.3512
A RRATE	4652438.4412	-199058.1997	4344726.7660	4652438.6280	-199058.4207	4344726.5633
A RTXUETA	4673825.5758	-160365.1493	4324588.3343	4673825.7640	-160365.3714	4324588.1311
BELAR	4668121.4190	-197465.2942	4328524.5982	4668121.6054	-197465.5160	4328524.3950
BIANDITZ	4651583.8997	-145430.0779	4348133.1014	4651584.0495	-145430.2989	4348132.8590
G ARAGARTZA	4664840.1331	-169153.4995	4332432.9138	4664840.3211	-169153.7211	4332432.7108
G FA	4644833.0652	-160817.6784	4353575.0764	4644833.2544	-160817.8991	4353574.8741
INTXORTA TXIKI	4658356.6606	-202888.2927	4338445.1462	4658356.8470	-202888.5140	4338444.9433
IT URREGI	4660182.6604	-164249.4372	4338251.9550	4660182.8389	-164249.6586	4338251.7422
ITXUMENDI	4658830.9201	-193130.1434	4338396.6208	4658831.1070	-193130.3647	4338396.4179
LARREZKONDO	4672777.6644	-182116.7758	4323931.8770	4672777.8515	-182116.9978	4323931.6737
MENDIZORROTZ	4646636.3737	-167697.9967	4351962.2560	4646636.5425	-167698.2175	4351962.0336
O ITZ	4650697.7378	-210481.1412	4346696.1158	4650697.9241	-210481.3621	4346695.9131
ONDDY	4652015.6655	-158468.8661	4346786.2290	4652015.8246	-158469.0871	4346785.9965
PAGOETAGANA	4646548.0436	-185892.9832	4351053.8710	4646548.2314	-185893.2039	4351053.6686
S AN TELMO	4640529.6294	-145676.2974	4358781.6667	4640529.7596	-145676.5179	4358781.4047
U RBINA	4671792.8232	-213717.2423	4323528.7964	4671793.0086	-213717.4642	4323528.5930
USURBE	4664072.7258	-179486.9185	4333355.0737	4664072.9133	-179487.1401	4333354.8707
ZU BELTZU	4645449.6443	-146437.3727	4353693.2581	4645449.7842	-146437.5934	4353693.0059

Table 1: Test network: Active and Passive GPS network of Gipuzkoa.

		WEOP detected error (m)			Helmert detected error (m)		
Station	Error (cm):	X	Y	Z	X	Y	Z
AITZORROTZ		-0.0108	0.0003	-0.0104	-0.0107	0.0003	-0.0105
AMETZMENDI		0.0153	-0.0001	0.0156	0.0153	-0.0001	0.0156
ANUGANA		0.0082	0.0001	0.0084	0.0082	0.0001	0.0084
ARRATE		0.0019	-0.0002	0.0022	0.0018	-0.0002	0.0022
ARTXUETA		0.0107	0.0002	0.0097	0.0107	0.0001	0.0097
BELAR		-0.0058	0.0002	-0.0058	-0.0057	0.0002	-0.0058
BIANDITZ	4	-0.0097	-0.0003	-0.0103	-0.0097	-0.0003	-0.0103
GARAGARTZA		0.0104	0.0004	0.0103	0.0104	0.0005	0.0103
GFA		0.0254	0.0000	0.0257	0.0254	0.0000	0.0257
INTXORTA TXIKI		-0.0033	-0.0001	-0.0031	-0.0033	-0.0001	-0.0031
ITURREGI	1	0.0058	0.0000	0.0054	0.0058	0.0000	0.0054
ITXUMENDI		0.0016	-0.0001	0.0016	0.0015	-0.0001	0.0016
LARREZKONDO		-0.0005	0.0005	-0.0007	-0.0005	0.0005	-0.0007
MENDIZORROTZ	2	0.0012	-0.0002	0.0013	0.0012	-0.0003	0.0013
OITZ		-0.0028	-0.0006	-0.0026	-0.0028	-0.0006	-0.0026
ONDDY	3	-0.0069	-0.0002	-0.0071	-0.0069	-0.0001	-0.0071
PAGOETAGANA		0.0116	-0.0002	0.0121	0.0117	-0.0002	0.0121
SAN TELMO	6	-0.0243	-0.0001	-0.0243	-0.0243	-0.0002	-0.0242
URBINA		-0.0165	0.0005	-0.0161	-0.0164	0.0005	-0.0161
USURBE		0.0057	0.0002	0.0055	0.0057	0.0001	0.0055
ZUBELTZU	5	-0.0173	-0.0001	-0.0174	-0.0173	-0.0001	-0.0174

Table 2: Residuals: simulated values – displaced values.

As can be seen in tables 1 and 2, the results are the same to the mm level. This is very important because it means that the WEOP solution is as good as the classical datum transformation. This paper deals with the datum transformation problem though the global to global solution has been tested it is also al solution for the local to global or local to local datum transformation.

4.1. Results

A very important part of the results has already be shown in table 2. As well as the residuals, it is also interesting to check the parameters obtained:

- **Helmert:**

Variance: 0.000101895

Tx	-2.2890 m	0.8479
Ty	4.5539 m	0.7480
Tz	2.4625 m	0.8717
ωx	-5.08316E-07	1.01058E-07
ωy	-5.7154E-07	1.7058E-07
ωz	5.50716E-07	1.04616E-07
k.	2.23824E-08	8.57186E-08

- **WEOP:**

Matrix Norm: 0.075588

Rotation matrix:

$$\begin{pmatrix} 0.99999999999969 & -0.00000055053971 & -0.00000057193875 \\ 0.00000055054000 & 0.99999999999972 & 0.00000050783095 \\ 0.00000057193847 & -0.00000050783126 & 0.99999999999971 \end{pmatrix}$$

k: 1.00000002191551

Tx: -2.2886 m

Ty: 4.55087 m

Tz: 2.46627 m

5. Conclusions

A different method, using real data, for the classical datum transformation has been used (Grafarend and Awange, 2003).

The main advantages are:

a. non iterative method, no “a priori” results must be known and it is a straight forward computational method.
b. Method capable of integrating the local and the global reference systems by computing the vertical deflections.
c. No need of linearization and easy to compute.
d. Promising results obtainable in the ITRF determination and densification.

From the authors point of view, the method worth deeper investigation, as well as the inclusion of the stochastic model, which has been already proposed (Awange and Grafarend, 2005)

Helmert vs Procurstes: difficulties comparing rotation results due to the small values obtained in global-to-global transformation.

Acknowledgements

The authors wish to acknowledge to the people of the Provincial Council of Gipuzkoa the data provided as well as the facilities given to get such data, specially to J. M. Aramburu, M. Elorza, and F. Gainzarain.

References

Altamimi, Z., P. Sillard and C. Boucher, 2002: ITRF2000: A new release of the International Terrestrial Reference Frame for Earth science applications. J. Geophys. Res. 107, No B10, 2214, 10.1029/2001JB000056.

Awange, J. L., 2003: Partial Procrustes solution of the three dimensional orientation problem from LPS/GPS observations. Geodesy – the Challenge of the 3rd Millenium. Springer, Heidelberg.

Awange, J. L. and E. W Grafarend, 2005. Solving Algebraic Computational Problems in Geodesy and Geoinformatics. Springer.

Crosilla, F (1983): Procrustes transformation as a tool for the construction os a criterion matrix for control networks. Manuscripta Geodaetica 8, pp. 343-370.

Grafarend, E. W. and J. L. Awange (2003): Nonlinear analysis of the three-dimensional datum transformation [conformal group $\mathbb{C}_7(3)$]. Journal of Geodesy, 77, pp. 66-76

Hugentobler, U., S. Schaer & P. Fridez (Edts.), 2001: Bernese GPS Software Version 4.2. Astronomical Institute. University of Berne.

Krakiwsky, E. J and Thompson, D. B., 1974: Mathematical models of combination of terrestrial and Satellite Networks. The Canadian Surveyor 28 (1974) 606-615.

Niell, A.E., 1996: Global mapping functions for the atmosphere delay at radio wavelengths. Journal of Geophysical Research, Vol. 101, No. B2, pp 3227-3246.

Sanso, F. (1973): An exact solution of the rototranslation problem. Photogrammetria n 29, pp. 203-216, Elsevier.

Schönemann, P. H., 1996: Generalized solution of the Orthogonal Procrustes Problem, Psychometrika 31 No.1 pp. 1-10.

Zurutuza, J., García, L. and Sevilla, M. J., 2004. Red de estaciones GPS y estación permanente en Gipuzkoa. 4ª Asamblea hispanolusa de Geodesia y Geofisica, Figueira da Foz, Portugal.

Uncertainty Modeling in Local Geodetic Networks

M. Chueca, S. Baselga, J.L. Berné, I.C. Maestro
Department of Cartographic Engineering, Geodesy and Photogrammetry,
Polytechnic University of Valencia, Camino de Vera, s/n. 46022 Valencia, Spain.

Abstract. A new theory and methodology that allows improving the knowledge of geodetic networks uncertainty has been developed.

This new approach provides the uncertainty figures not only in the observed stations and their correlatives but also at any point of the network space. To this end, the variances and covariances, known in several points, will be taken as a reference and, from that information and after the appropriate transformation, the corresponding value of any point in the network will be estimated.

The methodology is completed with the uncertainty modeling by means of a contour map of the expected error in the whole network. Finally, the theory is applied to a real network.

Keywords. Geodetic networks, confidence region, error estimation, deformation monitoring.

1 Introduction

In science and engineering, the solutions to most of the problems must include the degree of uncertainty associated with the solution. In this sense, one of the main contributions of the Gauss–Markov model, widely used in the resolution of geodetic networks, is that the solution that it provides is based on a reliable estimation of observational errors.

It thus becomes evident the importance of any effort devoted to know uncertainty in geodetic networks. Therefore, in deformation monitoring it is essential to accurately estimate error since the differences between coordinates in two different epochs could be caused just by the uncertainty in the solution, not by a deformation. It can be stated that uncertainty assessment is the most important aspect to consider in deformation analysis (Caspary et al., 1990, Hekimoglu et al., 2002).

The most widely used indicator for uncertainty estimation is the error ellipse that, as it is well-known, is defined by the covariance matrix of the adjustment coordinates.

In this sense, only the data stored in the diagonal, subdiagonal and supradiagonal of the covariance matrix is used for the estimation of the uncertainty in a network. The other data are ignored even though they possess similar statistical reliability. This approach could be understood as if there were infinite differentiated networks, and even solutions, for the same network, with different observations that share the same error values, which is absurd. It is then necessary to revise the current situation of the theory to define a robust simple algorithm by which only one single set of error figures will correspond to a given network. This paper intends to contribute to this effort by providing a complementary theoretical approach that allows to overcome this contradiction and to offer a clarifying interpretation.

A first step forward in this direction was the introduction of the so-called relative error ellipses (Richardus, 1984). Later, three different approaches (Chueca et al., 2001) were presented for the interpretation of error affecting a network: departing from the error hyperellipsoid as the most general error estimate (that, however lacks visual interpretation) through the *section* of this error hyperellipsoid by its coordinate planes one obtained vertex ellipses and correlatives ellipses; through the *projection* on its coordinate planes sets of vertex and correlatives rosettes were obtained; and through the *projection of its main semi-axes* on the coordinate planes sets of points and correlatives rosettes can be obtained.

Whereas the interpretation of the correlative ellipses (Chueca et al., 2001) increased the number of error figures by a power of 2, this paper provides an increase in the number of known uncertainty regions with no more limits than those established by the designer.

2 Objectives

In this paper we propose a new approach that allows knowing the uncertainty not only at the observed points but also at any point located in the neighboring space that includes the geodetic network under study. For this purpose, a transformation will be developed to define the change in the variances and covariances of the parameters involved, i.e., a transformation that defines how error is transferred along the network space. To do this, the known values of the variances and covariances in two points of the network will be taken as a reference. The error value in any point will be accurately estimated from these points and after the corresponding transformation, by means of the expression of both ellipse semi-axes.

The derived knowledge provides better understanding of the best location for monitoring points in deformation analysis and the network densification problem (the also called Third-Order Design problem, TOD).

In the following section we present and demonstrate the theory used for error estimation as well as the methodology for error modeling by means of an isoline map of the expected error.

3 Theoretical Description

As it is well known, the error figure of a geodetic network can be interpreted as an hypercuadratic with significance level α (henceforth HEα). This figure is defined *a priori* from the weighted design matrix of the usual least-squares adjustment ($S = A^TPA$) and *a posteriori* from the covariance matrix, C. For that reason, it is clear that if we intend to know the error anywhere along the network we have to start from this hypersurface. Hence, let the *a priori* hyperquadratic HEα, with variance factor $s^2=1$, be the error figure of a geodetic network

$$\mathrm{HE}\alpha \equiv x^T Sx - f_\alpha^2 = 0 \qquad (1)$$

referred to the Euclidean rectangular n-vector (x_1, x_2,..., x_i, x_j,..., x_k, x_l,..., x_{n-1}, x_n) in space $\Re^n$, with f_α defining its scale factor.

Let us suppose that (x_i, x_j) and (x_k, x_l) are the variables corresponding to the respective corrections of the coordinates of two network points (A and B in Figure 1, hence $x_i = x_A$, $x_j = y_A$, $x_k = x_B$, $x_l = x_B$) referring to the rectangular Cartesian axes OXY.

We have to deduce the expression from the error figure of a random point, M, whose geometrical location could be any inscribed angle, θ, over AB.

If the coordinates of point M are accepted a priori as exactly known (this can be done since a first-order error in these coordinates would only affect the final error ellipse in second-order terms), then it will be possible to deduce the slopes δ and γ of the lines that join each reference point (A and B) to point M. In other words, M is the intersection of axes x_A' and y_B' (see Figure 1), resulting from rotating x_A an angle δ, and y_B an angle γ, both assumed exactly known. As a result, both rotations around A and B solve the question. We obtain the rotated coordinates for both the coordinate systems defined by A and B as

$$\begin{aligned} x_A' &= x_A \cos\delta + y_A \sin\delta \\ y_A' &= -x_A \sin\delta + y_A \cos\delta \\ x_B' &= x_B \cos\gamma + y_B \sin\gamma \\ y_B' &= -x_B \sin\gamma + y_B \cos\gamma \end{aligned} \qquad (2)$$

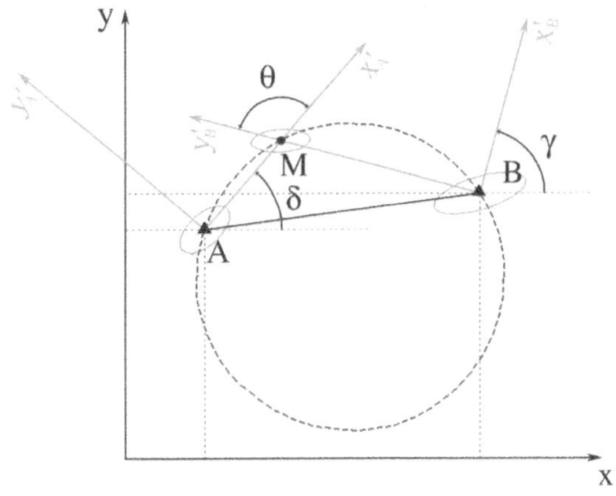

Fig. 1. Calculation of the error figure of a point M located over any inscribed angle where $\theta = \frac{1}{2}\pi + \gamma - \delta$.

In this way, according to the theory presented by Chueca et al. (2001), the point M becomes the node of the grid generated with the new axes x_A' and y_B'. As a consequence, the error figure in M is the correlative ellipse of the intersection of axes x'_A and y'_B, that can be obtained with the transformed HE$\theta\alpha$ (i.e. referred to the new axes) by cutting through the hyperplane

$$x'_m = 0 \quad \forall\, m \neq i, l \qquad (3)$$

that, ignoring null elements, yields

$$(x_A\cos\delta \;\; x_A\sin\delta \;\; -y_B\sin\gamma \;\; y_B\cos\gamma)\cdot C_M \cdot \begin{pmatrix} x_A\cos\delta \\ x_A\sin\delta \\ -y_B\sin\gamma \\ y_B\cos\gamma \end{pmatrix} = f_\alpha^2 \quad (4)$$

from which the uncertainty ellipse in M will be obtained referred to the oblique coordinate axes $Mx_M'y_M'$ in angle $\theta = \frac{1}{2}\pi + \gamma - \delta$, which solves the problem.

In the particular case $\theta = \frac{1}{2}\pi$, both reference points will lie at the ends of the diameter of the circumference that defines the inscribed angle, and the ellipse will only refer to some orthogonal axes, the rotations δ and γ being identical.

Returning to the general case, it is necessary to define rigorously and completely the error figure. Taking aside for the moment expression (4), the equation of the uncertainty figure at any point M with its center at the origin of the coordinates, with axes forming an angle θ, and significance level α, here called $HE\theta\alpha$, can be written as

$$HE\theta\alpha \equiv x^T Ax = (x \;\; y \;\; 1)\cdot \begin{pmatrix} a_{11} & a_{12} & 0 \\ a_{12} & a_{22} & 0 \\ 0 & 0 & a_{33} \end{pmatrix} \cdot \begin{pmatrix} x \\ y \\ 1 \end{pmatrix} = \quad (5)$$
$$= a_{11}x^2 + 2a_{12}xy + a_{22}y^2 + a_{33} = 0$$

The same ellipse now referred to orthogonal axes (named $HE\frac{1}{2}\pi\alpha$), that is, expressed in its canonical form, is

$$HE\tfrac{1}{2}\pi\alpha \equiv x^T A'x = (x \;\; y \;\; 1)\cdot \begin{pmatrix} a_{11}' & 0 & 0 \\ 0 & a_{22}' & 0 \\ 0 & 0 & a_{33}' \end{pmatrix} \cdot \begin{pmatrix} x \\ y \\ 1 \end{pmatrix} = \quad (6)$$
$$= a_{11}'x^2 + a_{22}'y^2 + a_{33}' = 0$$

By considering the metric conic invariants of (5) and (6):

$$a_{11}' + a_{22}' = \frac{I}{\sin^2\theta} = \frac{a_{11} + a_{22}}{\sin^2\theta} \quad (7)$$

$$a_{11}' \cdot a_{22}' = \frac{A_{33}}{\sin^2\theta} = \frac{\begin{vmatrix} a_{11} & a_{12} \\ a_{12} & a_{22} \end{vmatrix}}{\sin^2\theta} \quad (8)$$

$$a_{11}' \cdot a_{22}' \cdot a_{33}' = \frac{A}{\sin^2\theta} = \frac{\begin{vmatrix} a_{11} & a_{12} & 0 \\ a_{12} & a_{22} & 0 \\ 0 & 0 & a_{33} \end{vmatrix}}{\sin^2\theta} \quad (9)$$

it follows that the coefficients a_{11}' and a_{22}' are the roots of the quadratic equation

$$\rho^2 - \frac{I}{\sin^2\theta}\rho + \frac{A_{33}}{\sin^2\theta} = 0 \quad (10)$$

where A_{33} is the adjunct of element a_{33}, that can also be expressed as

$$\rho^2\sin^2\theta - (a_{11} + a_{22})\rho + (a_{11}a_{22} - a_{12}^2) = 0 \quad (11)$$

Its roots (ρ_1 and ρ_2) are

$$\rho = \frac{(a_{11} + a_{22}) \pm \sqrt{(a_{11} + a_{22})^2 - 4\sin^2\theta(a_{11}a_{22} - a_{12}^2)}}{2\sin^2\theta} \quad (12)$$

fulfilling that $\rho_1 = b^2$ and $\rho_2 = a^2$, been a and b the semi-axes of the ellipse.

In this way, the equation of the ellipse is:

$$\frac{1}{\rho_1}x^2 + \frac{1}{\rho_2}y^2 + \frac{A}{A_{33}} = 0 \quad (13)$$

where, by simplifying and taking into account the aforementioned invariants, the semi-axes of the ellipse in M (the aim of this work) can be deduced:

$$a^2 = -\frac{\rho_1}{a_{33}} \quad (14)$$

$$b^2 = -\frac{\rho_2}{a_{33}} \quad (15)$$

Since $a_{33} = -1$ for the ellipse is origin centered it yields

$$a^2 = \frac{(a_{11} + a_{22}) + \sqrt{(a_{11} + a_{22})^2 - 4(a_{11}a_{22} - a_{12}^2)\sin^2\theta}}{2\sin^2\theta} \quad (16)$$

$$b^2 = \frac{(a_{11} + a_{22}) - \sqrt{(a_{11} + a_{22})^2 - 4(a_{11}a_{22} - a_{12}^2)\sin^2\theta}}{2\sin^2\theta} \quad (17)$$

where if the ellipse equation is written as (4) the coefficients are

$$a_{11} = \cos^2\delta\, \sigma^2_{x_A} + \sin^2\delta\, \sigma^2_{y_A} + 2\cos\delta \sin\delta\, \sigma_{x_A y_A} \quad (18)$$

$$a_{22} = \cos^2\gamma\, \sigma^2_{y_B} + \sin^2\gamma\, \sigma^2_{x_B} - 2\cos\gamma \sin\gamma\, \sigma_{x_B y_B} \quad (19)$$

$$a_{12} = \cos\delta \cos\gamma\, \sigma_{x_A y_B} - \sin\delta \sin\gamma\, \sigma_{y_A x_B} + \cos\gamma \cos\delta\, \sigma_{y_A y_B} - \cos\delta \sin\gamma\, \sigma_{x_A x_B} \quad (20)$$

expressions that can be immediately checked in the classical canonical case with $a_{12} = 0$ and $\theta = \pi/2$. Furthermore it could be useful to transform the oblique case into the rectangular one, however this transformation will not affect the above semiaxes formulae for being invariant.

To finish, it is interesting to point out that this theory can be extended to the three-dimensional case obtaining thus the error ellipsoid of point M, located in the right inscribed angle over AB and on the plane of height z_M, by means of two plane rotations around the height axis.

4 Uncertainty Modeling

This section focuses on the second goal of present work: the uncertainty modeling in the network space. For this, the first aspect to solve is to find a suitable descriptor, i.e. a single parameter able to efficiently represent geographical error distribution, usually characterized by the three parameters that define the ellipse: orientation and both semi-axes.

After the analysis of these three variables and considering as a priority the dimensioning of the ellipse over its orientation, we concluded that the parameter that best defined by excess the uncertainty in the coordinates of a network point was the quadratic diagonal composition of the semi-axes of its error ellipse. This simple parameter that geometrically corresponds to the semi-diagonal of the rectangle confined to the ellipse (see Figure 2) is the parameter denoted with letter ι:

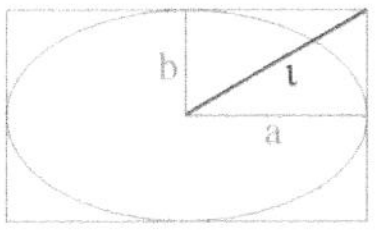

$$\iota = \sqrt{a^2 + b^2}$$

Fig. 2. Uncertainty parameter, ι, chosen for uncertainty modeling.

After the definition of the uncertainty parameter, it will suffice to calculate it on all the nodes of a quadrangular grid defined by the user. The values obtained will allow to define an uncertainty distribution model along the network that ensures the continuity and homogeneity of the medium.

The model obtained will represent the influence on accuracy derived on one hand from the own accuracy of the network points, and on the other hand from its configuration (its geometrical robustness). Therefore, the model should indicate greater accuracy if the point analyzed is closer to a more precise point or if it is in a position such that the triangle that it forms with the points of the intersection considered acquires greater geometrical robustness, i.e. an angle close to ½π.

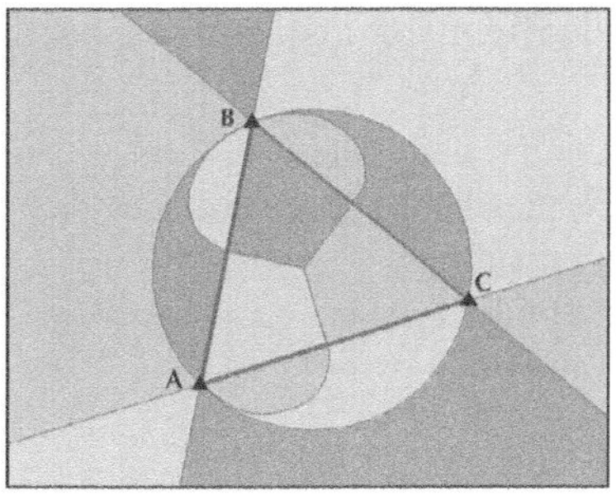

Fig. 3. Partition of the network space depending on the selected reference points A, B or C.

That is, the algorithm takes those points defining the most consistent triangle (with an angle closer to ½π) as criterion for the selection of the reference points (called A and B in the previous section). The partition of the network space in different regions (one for each of the possible pair of points, that is, the binomial coefficient, $_nC_2$) can be appreciated in Figure 3.

The first remarkable fact is that the proposed estimation of the error is based only in an angular solution (and only from two points) ignoring the possible, and often more realistic, use of trilateration. This will be discussed in the section 6.

Regardless any consideration about the datum definition (the method described here is independent of the consideration of any constraint in the network), the case of fixed networks require a particular treatment because the error is estimated from the covariance matrix which does not include the so-called "fixed" points (whose error is supposed to be null or known). Therefore, to avoid ignoring possible bisections from fixed network points when estimating the accuracy, the following consideration was adopted. It consists of partially "freeing" the fixed network points which are supposed to have an "optimum" error, i.e., an error region of small quantity (e.g., σ_x^2, σ_y^2 = 10^{-9} equivalent to a typical deviation smaller than one millimeter) and of homogeneous distribution (circular, σ_{xy} = 0). Following the nomenclature proposed by Schaffrin (1989), this type of points could be called "fiducial". In other words, it consists of using the well-known Baarda's absolute criterion minor matrix on these points (Baarda, 1973). In case the precision of fixed points could be reliably estimated one would have filled in the corresponding covariance matrix elements.

5 Application: Results and Discussion

This section will describe an example for illustrating the theory presented and demonstrating its main applications.

The example consists of a network with five points, from which two are taken as fixed points (A and D), and solved as trilateration but, of course, also applicable to the use of angular observations (no matter whether the origin of the reference ellipses are distances or angular observations). The lines indicated in Figure 4 have been observed, using a 1 cm precision EDM.

After the resolution of the network, by using the conventional adjustment by least squares, the adjusted coordinates of the points with their error figures were obtained (see Table 1).

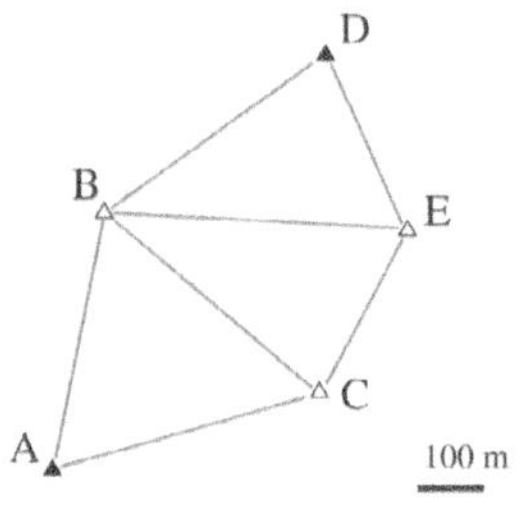

Fig. 4. Network used to demonstrate the application of the methodology.

For the application of the methodology described in this paper, the following procedure was followed. First, error was estimated using the theory described here, for all the nodes of a grid with a step of 50m. To do this, a computer program was developed on Mathematica® from Wolfram Research which generates a matrix with node coordinates and its estimated error (x, y, ι).

The second step consists of generating an error model, in the form of a map with lines of equal uncertainty, from the data obtained in the previous step and by the use of interpolation techniques. If the defined grid is dense enough it is possible to obtain directly a raster model of uncertainty without the need of any interpolation (remind the expressions 16-20 allow to estimate the error everywhere), otherwise it can be used an interpolation method to obtain (faster) the isolines of the model. In the second case, among the existing interpolation methods (minimum curvature, IDW, kriging, etc.), kriging was selected as it offers the additional advantage of considering data self-correlation (semivariogram). This feature improves the results since it facilitates the interpretation of the different uncertainty values that result from the different possible bisections for the estimated node (bear in mind that the algorithm chooses the one with best geometrical configuration) and to overcome the discontinuities derived from this reality. In addition, the kriging method provides the best unbiased linear estimator (BLUE).

The resulting models are shown in Figure 5. Observe that the smallest error is located along the ½π inscribed angle at the highest accuracy points, i.e. the fixed points A and D (Figure 5, left), as it should be.

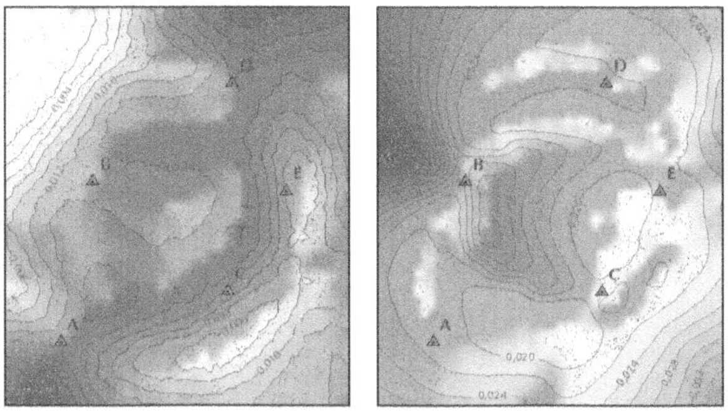

Fig. 5. Uncertainty models. Right: without considering the possible intersections from A and D, uncertainty ranges between 16 and 54 mm. Left: considering the possible intersections, uncertainty ranges between 0 and 28 mm.

A good technique to verify these results is to analyze the discrepancies existing between both models (Fixed – Free, see Figure 6). In this sense, if the theory developed by the authors is correct, this third model should:

1. Show values equal to or lower than zero since by considering the points as fixed error will always be reduced.
2. Show almost null values in the zones far from the fixed points (A and D), since the consideration of the fixed points will have little effect in these zones.
3. Show very low values in the region located between the three free points (B, C and E), since the improvement of an additional fixed point will be lower in that zone due to the fact that the configuration is good (i.e. precision) even without considering this point.

This behavior can be observed in the Figure 6.

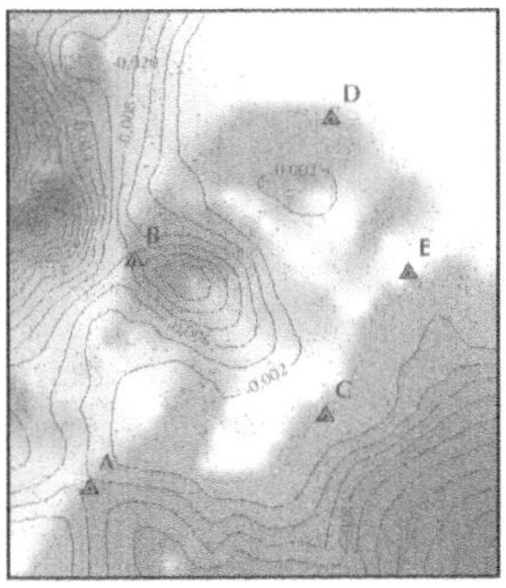

Fig. 6. Uncertainty discrepancy model depending on whether the bisections from the fixed points are considered or not. Uncertainty ranges from 0 to 28 mm.

On the other hand, observe the greater discrepancy in the areas in which the geometrical configuration of the bisection is significantly improved by considering the fixed points (A and D). In the case under study, this is even more evident in the area between point B and the left limit of the model, in which the configuration of one bisection is very poor unless the fixed points are considered. Finally, similarly, though at a lesser extent, happens on the opposite side of point B, where the inclusion of both fixed points (mainly of point A) involves a remarkable improvement of the bisection (as without this point, it would not reach 50^g).

To conclude, reasonable error estimates were obtained near the reference points. In general, error estimation using this methodology greatly approximates the value derived from the adjustment (the discrepancy is only of a few millimeters–see Table 1). Nevertheless what cannot be expected is that error estimation at any point be identical to that obtained in the adjustment, since then we would be obviating that in the first case, some of the information used (the *a posteriori* covariance matrix) differs from that used in the adjustment (the accuracy given by the manufacturer and used for weighting). This confirms the robustness of the theory presented in this paper.

Table 1. Comparison between uncertainty (in millimeters) derived from the conventional adjustment and from this new methodology.

Point	a	b	Adjust. error	Estimation discrepancy			
				Fixed network	Free network	Fixed network	Free network
B	14	3	17	24	13	7	4
C	17	12	20	17	15	3	5
E	15	12	19	17	2	2	17

6 Conclusions

In the last decades great efforts have been devoted to the optimization of geodetic networks. Different optimization problems have been formulated, and different approaches and methodologies have been proposed for its resolution, some of these showing to be more efficient than others. As Richardus (1984) stated, generally, "the rather sweeping rules old time surveyors and geodetic engineers used to apply (...) are still valid at present. Until new completely satisfactory criteria have been found, the existing methods will still be subject of discussion".

In this sense, the methodology proposed in this paper intends to be a step forward in the establishment of "satisfactory" criteria in terms of robustness, accuracy and reliability. Thus, this contribution may be understood as a transition from the classic considerations (the rules mentioned by Richardus) to a sharper and rigorous definition of precision in the network space.

This work allows to describe and, what is more important, *to quantify* in a rigorous way how the main variables acting on the precision of a point behave, that is how the precision of the supporting points and their configuration are affecting the network. Hence, those areas with greater "sensitivity", in which small changes in the position of one point involve great alterations in error estimation can be identified. In such zones the possible actions devoted to improve accuracy are limited (unless this action performs a change in the network configuration itself).

The procedure presented here can be very useful for network design, particularly when high accuracy is required. On the other hand, this system can also be useful in deformation control networks, as it permits to decide more accurately the optimal location of the reference points, which results in a considerable improvement of sensitiveness of the network. Finally another indirect advantage of this method is that the time interval between observation campaigns could be reduced.

However, a drawback of this technique for error estimation is that this approach is based only on the information of two network points (the points from which the bisection is supposed to start). This involves a multiplicity (in fact, real) in the uncertainty value; that is, for each point there will be as many error figures as pairs of points can be taken as reference. However, this drawback does not necessarily greatly affect our results if the reference points are adequately selected, as can be expected (that is, the regions shown in Figure 3 are taken into account). On the other hand, the interpolation method used (kriging) takes into account the information corresponding to the neighborhood of the point considered (by means of the semivariogram); thus, to some extent, it does consider this multiplicity. Nevertheless, efforts are currently being made towards a modified version of the methodology which uses a weighted estimation that takes into account all possible bisections and their geometrical configuration. It is evident that there is still much work to do in the field of error analysis for high precision geodetic networks. Advances may come from the application of a Bayesian approach to the problem or from the use of multivariate analysis. Fortunately, there exist different research possibilities but the correct diagnosis of uncertainty will still be one of the main issues in geodesy for the next years.

Acknowledgements

The authors would like to acknowledge A. Dermanis for his useful suggestions and comments.

References

Baarda, W. (1973). S-Transformations and criterion matrices, *Netherlands Geodetic Commission*, 5 (1). Delft.

Caspary, W.F., W. Haen and H. Borutta (1990). Deformation analysis by statistical methods. *Technometrics*, 32 (1).

Chueca, M., J.L. Berné, A.B. Anquela and S. Baselga (2001) Avances en la interpretación de resultados en redes locales. Recintos de error. Ed. Universidad Politécnica de Valencia, Valencia.

Hekimoglu, S., H. Demirel and C. Aydin (2002). Reliability of the conventional deformation analysis methods for vertical networks. *Proc. of XXII FIG Internacional Congress*, Washington, D.C. (USA).

Richardus, P. (1984). Project surveying. General adjustment and optimisation techniques with applications to engineering surveying. Ed. AA. Balkema, Rotterdam.

Schaffrin, B. (1989). Fiducial versus fixed points in the GPS network approach. *Proc. of the 5th International Geodetic Symposium on Satellite Positioning*, Vol. 1, pp. 500–511.

Session B

Special Measurement Tools, Data Integration and Specific Models

Chairman: E. Brückl

PS InSAR Integrated with Geotechnical GIS: Some Examples from Southern Lombardia

C. Meisina, F. Zucca
Department of Earth Sciences, University of Pavia, Via Ferrata n°1, 27100 Pavia, Italy.
cmeisina@manhattan.unipv.it

D. Fossati, M. Ceriani
Regione Lombardia - Struttura Rischi Idrogeologici

J. Allievi
Tele-Rilevamento Europa - T.R.E. S.r.l.

Abstract. The Oltrepo Pavese, which extends for almost 1100 km^2 in Lombardia Region (Northern Italy), has a complex geological-structural setting resulting from overthrusting of different tectonic units made up mainly of clays. All these characteristics make the Oltrepo Pavese particularly vulnerable to hydrogeological risk: shallow and deep landslides in the hill, swelling/shrinkage of the clayey soils and subsidence in the plain. In order to understand more about the hydrogeological hazard (related to landslides and other phenomena) and related risk in the Oltrepo Pavese area the Regione Lombardia decided to test the use of the Permanent Scatterers Technique.

The Permanent Scatterers (PS) Technique is an advanced technique for the processing of SAR data developed by Politecnico of Milano. The PS Technique overcomes the main limits of conventional approaches to surface deformation detection due to temporal and geometric decorrelation.

The PS data sets and their temporal series were included on a geological and geotechnical GIS supported data-base, where they were integrated with geological-structural, hydrogeological and geomorphologic data. Spatial and temporal clusters were highlighted and/or discovered. They are generally related to ground deformation due to well known phenomena (swelling/shrinkage, landslides, overpumping) or to unknown phenomena (uplift).

Keywords. InSAR, GIS, landslides, subsidence.

1 Introduction

Differential Synthetic Aperture Radar Interferometry (DInSAR) allows the measure of very small movements of the ground over time and large area coverage (100x100km for ERS data). However, DInSAR suffers for temporal and geometric decorrelation. The Permanent Scatterers technique developed by Politecnico of Milano (Ferretti et al. 2001) overcomes the main limitations of conventional DInSAR approaches. It identifies, quantifies and removed atmospheric distortions, leaving displacement as the only contribution to signal phase shift. DInSAR and particularly PS technique are powerful techniques for geological risk assessment and monitoring (e.g. landslides, subsidence, earthquakes) (Ferretti at al. 2000; Crosetto et al. 2002; Berardino et al. 2003; Colesanti at al. 2003; Canuti et al. 2004). Such an approach provides fast and updatable data acquisition over large areas, which can integrate conventional methods (e.g. field surveys, aerial photointerpretation). Of particular interest is the possibility to combine deformation measurements with geological data in a Geographical Information Systems.

The paper presents an application of the Permanent Scatterers technique for detecting and monitoring ground displacement related to landslides and other phenomena in the Oltrepo Pavese, which could be considered representative, in terms of geological hazard, of the Italian Apennines.

2 Study Area

2.1. Geological and geomorphological settings

The Oltrepo Pavese, which is situated in Northern Italy (Southern Lombardia), has an extension of about 1100 km^2.

Its Southern part corresponds to the northwestern sector of the Apennines. The area is at heights between 200 m and 1725 m a.s.l. and it is characterized by a complex geological and structural setting.

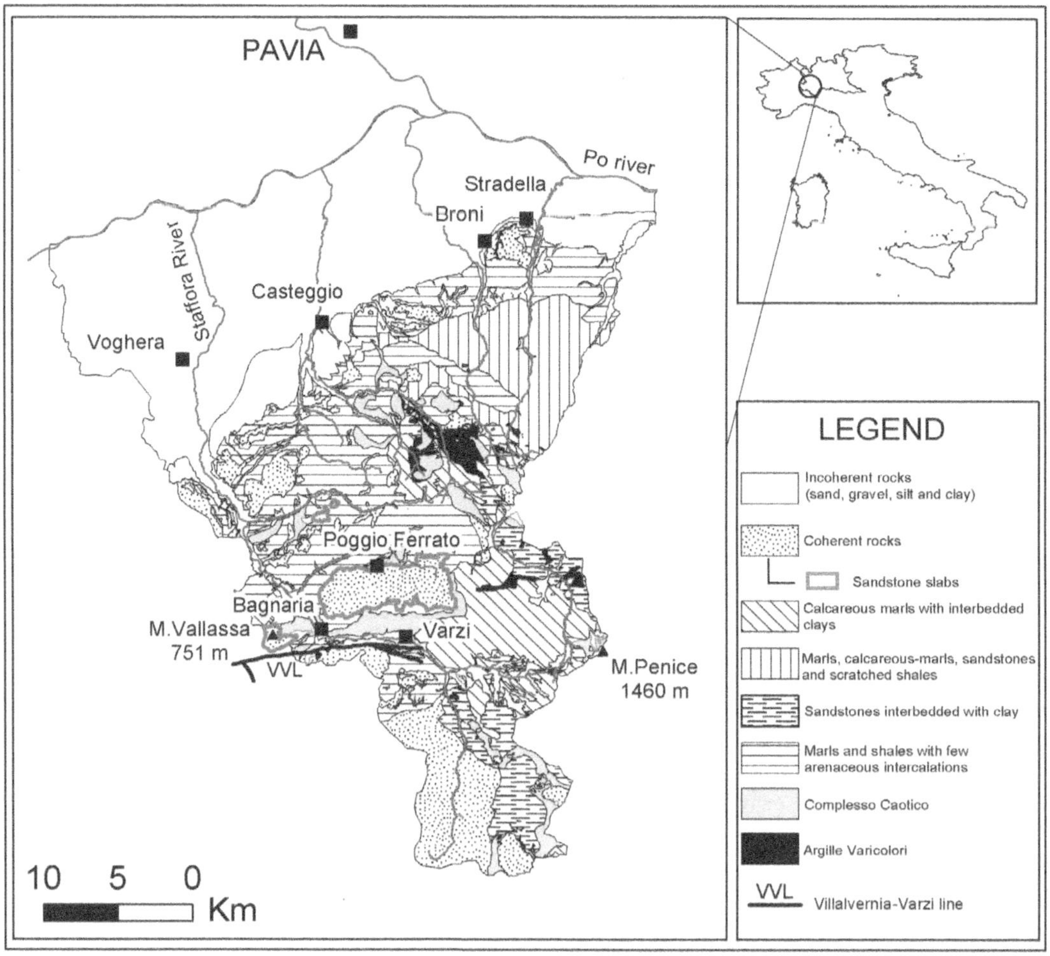

Fig. 1 Lithological map of the study area.

The geology is dominated by sedimentary formations, with a dominant clay component (Beatrizzotti et al. 1969; Braga et al. 1985) (Fig. 1). Clay shales, referred to as Argille Varicolori (varicoloured clays), or Complesso Caotico (chaotic complex), outcrop throughout the Oltrepo Pavese area, while calcareous flysch, made up of alternating marl, calcareous marl, and scratched shale predominated in the eastern part. Sandstones slabs lying on a deformable clayey substratum, are present as isolated area of relief in the central part. Silty and/or clayey deposits formed by weathering and down slope transportation cover the argillaceous bedrock units.

The northern part of Oltrepo Pavese corresponds to the Po River Plain. Morphologically, fluvial terraces and piedmont alluvial fans characterise this part. The recent river deposits consist of mainly coarse grained units. All along the piedmont margin, which separates the first foothills of the Apennines from the alluvial deposits of the Po plain, three orders of fluvial terraces deposits are present. The upper part of the older alluvial deposits is strongly weathered and it is composed of a large amount of clay.

The climate of Oltrepo Pavese has an average annual rainfall of around 700 mm in the plain and 998 mm in the hills. There are two distinct rainy seasons with the maxima in May and in October-November (Rossetti and Ottone 1979). The region has experienced, in the last two decades, many drought periods. The last major droughts were March 1989 - August 1993 (the most severe drought of the period) and May 1998 – September 2000.

2.2. Geological hazards

The main geological hazards are represented by landslides and subsidence due to shrinkage of clayey soils or ground water pumping.

The Oltrepo Pavese is characterized for the presence of a high number of mass movements, which cover up to 40% of the territory in the north-eastern sector. Up today 3707 landslide were mapped, the 89 % of which are active or dormant. A lot of them have been classified at high risk from the institutional authorities. The most frequent typologies are roto-translational slides associated with complex earth slides/earth flows. Most of the active landslides are relative small shallow landslides affecting the surficial cover, whereas dormant and inactive landslides are mainly slides or complex failures.

Swelling/shrinking soils occur extensively throughout the Oltrepo Pavese. The material source of expansive/shrinking soils is from the weathering of the sedimentary rocks in the hills and the alluvial deposits in the Apennine fringe and Po River Plain (Meisina, 2003).

A large number of residential buildings (more than 1000) have experienced damages. The majority of problems are related to single storey family residences. The buildings are founded on conventional concrete shallow strip footings, which generally extend to depths of between 1 m and 2 m below ground level. For the 46 % the causes the damages are due to landslide, for the 20 % the cause is the volume change of clay soils. The economic losses due to volume changes of clay soils have been estimated at around 20% of the building cost.

In this context the Lombardia government decided to test how PS technique may help to detect and to monitor geological hazard for risk assessment.

3 Methods

The project focuses on the combination of the PS analysis, able to provide displacement measurements on sparse points with geological and geotechnical data by the way of GIS.

3.1. Permanents Scatterers Technique

The Permanent Scatterers (PS) Technique is an advanced algorithm for the processing of data acquired by SAR sensors developed by Politecnico of Milano (Ferretti et al. 2001). It is an operational tool for ground deformation mapping at millimetric level on a high spatial density grid of phase stable radar targets (the so-called Permanent Scatterers, PS), acting as a "natural" geodetic network. In the Oltrepo Pavese the PS mainly correspond to man-made structures such as buildings. The PS approach allows to overcome the two most significant drawbacks of conventional Differential SAR Interferometry (DInSAR), namely decorrelation noise and atmospheric artifacts. The PS technique was already used to detect and monitor different geological phenomena such as: subsidence, landslides, seismic faults etc. and even to verify individual building stability (Ferretti et al. 2000; Colesanti et al. 2003; Farina et al. 2005).

Seventy-six ESA ERS1 and ERS2 images gathered along descending orbits and 26 images gathered along ascending orbits in the time span May 1992- November 2000/December 2001 have been exploited in Oltrepo Pavese. The analysis resulted in the identification of around 95500 descending PS and 3800 ascending PS. The minor number of ascending PS is due to the limited number of the images acquired along ascending orbit.

The most part of PS is in the plain area of Oltrepo Pavese, where density of man-made structures is the highest. No PS were detected in the mountain part of the area (the southern part) due to the limited number of urban areas and to the vegetation cover.

The LOS (Line of sight direction) displacement rates vary from +5 to -5 mm/yr.

Two type of PS analyses were performed: Standard Permanent Scatterers Analysis and Advanced Analysis. The Standard Permanent Scatterers Analysis (hereinafter SPSA) is suitable for mapping the territory at regional scale, in order to identify unstable areas, which deserve further detailed studies. PS are detected and their average velocity is then estimated by an automatic procedure, allowing to process large amounts of data relative to large areas in a limited period of time. Linear motion model is searched and information about linear velocity is extracted.

The Advanced Analysis (hereinafter APSA) is suitable for those small areas where a full exploitation of the information content of the satellite data is required. It is a very sophisticated and time-consuming analysis, which requires skilled technical staff. With the APSA the nonlinear movements can be estimated.

3.2 Geological and geotechnical databases

On the whole territory a standard PS analysis was performed and only the descending PS dataset was taken into account. The Permanent Scatterers interferometric analysis was integrated in a GIS with the geology (lithology and geostructures), the landslide and damaged building databases.

The landslide database consists in a landslide inventory produced from 1994 aerial photographs analysis supported by field checks. Mass movements were classified in relationship with their state of activity (active, dormant, inactive) and typology.

The damaged buildings database contains information concerning the type of soil foundation, the type of damage and the remedial works.

The SPSA analysis aimed to identify the general clusters of the ground movement in the area (uplift or subsidence), to verify or update the state of activity of the landslides, to update the landslide inventory with the identification of new unstable area and to verify the relationship between the PS and the damaged buildings.

Four significant test sites were selected and the APSA analysis was applied to. A geological model of each test area was obtained from the integration of the geological-geomorphological character of the site with the geotechnical data (penetrometer tests, boreholes and laboratory tests). The rainfall deficit, obtained by the arithmetic difference between the monthly water balance (difference between precipitation and potential evapotranspiration) and the average monthly water balance calculated over a 40-year-period, was studied. The ground water table fluctuations and the ground water pumping were also analysed and compared with the time series of the PS displacement occurring along the sensor-target line of sight (LOS) direction.

4 Results

4.1 General analysis

In the plain two areas of ground subsidence were clearly identified in the towns of Voghera and Broni (510 and 986 PS/km^2 respectively). In the first case the ground movement of the eastern part of Voghera was until now unknown. The subsidence of 10-30 mm from 1992 to 2000 (1-3 mm/yr) could be related to the presence in the subsoil of clay deposits thicker than 6-7 m. In the town of Broni clay shrinkage phenomena were already known and they are the cause of a lot of damage to buildings. The SPSA analysis detects a ground subsidence in the north-eastern part of about 10-22 mm from 1992 to 2000 (1-5 mm/yr).

Two areas of uplift were identified in the towns of Casteggio and Varzi (397 and 136 PS/km^2 respectively). In the Casteggio area the heave (1.3 mm/yr) could be related with the uplift of the miocenic substratum observed by geophysical prospecting (Pieri and Groppi, 1981). The SAR data are also in agreement with topographical measures of IGM (Italian Militar Geographic Institute) that registered a vertical movement in the same area (Arca and Beretta, 1985), even if with lower velocity (0.5 mm/yr for the time span 1897-1957).

In the Varzi zone the uplift (3 mm/yr) was also identified with the study of the alluvial sediment thickness along the Staffora riverbed. Between Bagnaria and Varzi the Staffora river is parallel to Villavernia-Varzi Line (VVL), defined as a dextral transpressive system, that constitutes one of the most important structural elements in the North Western Apennines (Cerrina Feroni et al. 2002) (Fig.1). The VVL divides the Staffora River, into two blocks: the northern block (downstream Bagnaria) with the minimum alluvial sediment thickness and the southern block (upstream Bagnaria) with the higher values of thickness. The different thickness of the alluvial sediments could be related to an uplifting of the northern block followed by the erosion of the alluvial sediments (rejuvenation process) (Piccio and Meisina, 2003). This is also in agreement with the soil vertical movement map of Arca and Beretta (1985).

The compliance analysis of the PS data sets and landslides in terms of distribution and state of activity allows to verify that the 7% of the descending PS are located on landslides (Fig.2). The great majority of moving PS fall within landslides classified as inactive from the geomorphological point of view. The PS analysis allows to modify the state of activity of 16 landslides (from inactive to active). A limited number of unstable areas not yet mapped have been detected through the use of the PS.

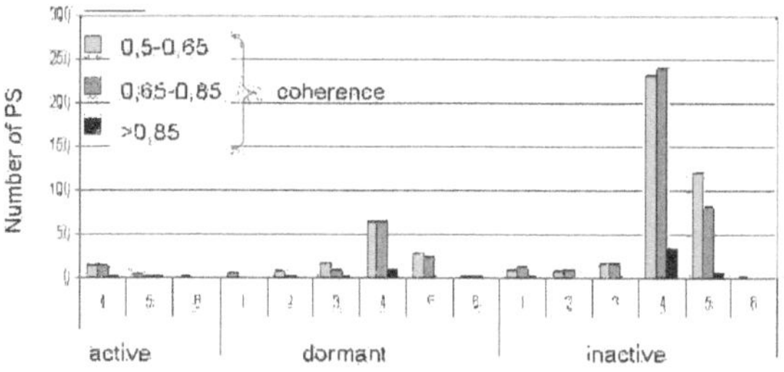

Fig.2 PS distribution vs landslide activity. Velocity classes 1: < -5 mm/yr, 2: -3/-5 mm/yr, 3: -1/-3 mm/yr, 4: -1/+1 mm/yr; 5: +1/+3 mm/yr; 6: +3/+5 mm/yr; 7: > +5 mm/yr.

Strong subsidence phenomena (-11 mm/yr) were detected on isolated area of relief, consisting of sandstones slabs lying on a deformable clayey substratum. This phenomenon develops on the borders where the slabs are highly fractured, with more sets of subvertical joints which separate sandstones blocks.

The PS distribution was also compared with the damaged building inventory. The 33 % of the damaged buildings are Permanent Scatterers. The PS velocity (<-1 mm/yr or > +1 mm/yr) allowed the confirmation of damage for the 20% of these buildings in the plain and the 40% in the hill. The damaged buildings with "stable" PS (velocity between +1 and –1 mm/yr) correspond to repaired houses before or within the monitored period. New unstable buildings not yet mapped were also detected on the basis of PS velocity (<-1 mm/yr or > +1 mm/yr).

4.2 Detailed analysis

Two examples of application of the APSA are presented in this section of the paper; they concern the Poggio Ferrato landslide and the town of Broni.

The landslide of Poggio Ferrato is located on the external northern border of a sandstone slab lying on a clay substratum (Fig.1 and 3). The landslide, is classified as an active complex failure: the movement started as a roto-translational slide with a NE-SW direction and became an earth flow with a NNW-SSE direction (Fig.3A). It has an elongated morphology with lengths of a thousand meters. In the last 50 years the crown has retrogressed by 250 m. In the same period the flow advanced about 855 m. A serious increase in the movements was recorded in December 1996, after a long period of heavy rainfall. The channel keeps on moving with displacement rates of 10 m/day and the flow advances about 160 m which caused cracks to open in nine buildings in the northern part of the village of Poggio Ferrato.

Site investigation consisted of 16 boreholes from 9 m to 40 m, trench pits and geophysical surveys. A geological model of the site was derived. The boreholes have found different thicknesses of the Sandstones (from 7 to 30 m) around Poggio Ferrato village (Fig.4). This has been explained by the existence of an important structural discontinuity, dipping SW and subvertical, which indicates differential settlement and sinking of sandstone blocks on the border of the slab (Fig. 3B and 4). Most of the damaged buildings are located near the outcrop of this discontinuity. The structural and geomorphological surveys and comparison with results obtained by some authors in similar geological contexts (Conti and Tosatti 1994) indicate the presence of deep-seated gravitational deformations, which could be classified as lateral spreads and block type slope movement (Braga et al. 2003). The stability conditions of the slope are governed by the groundwater circulating in the sandstone, sustained by the impermeable clayey and marly substratum. The waters emerging at the ground surface and flowing down the hill cause the softening of the clayey soils and contribute to the progress of slope movements.

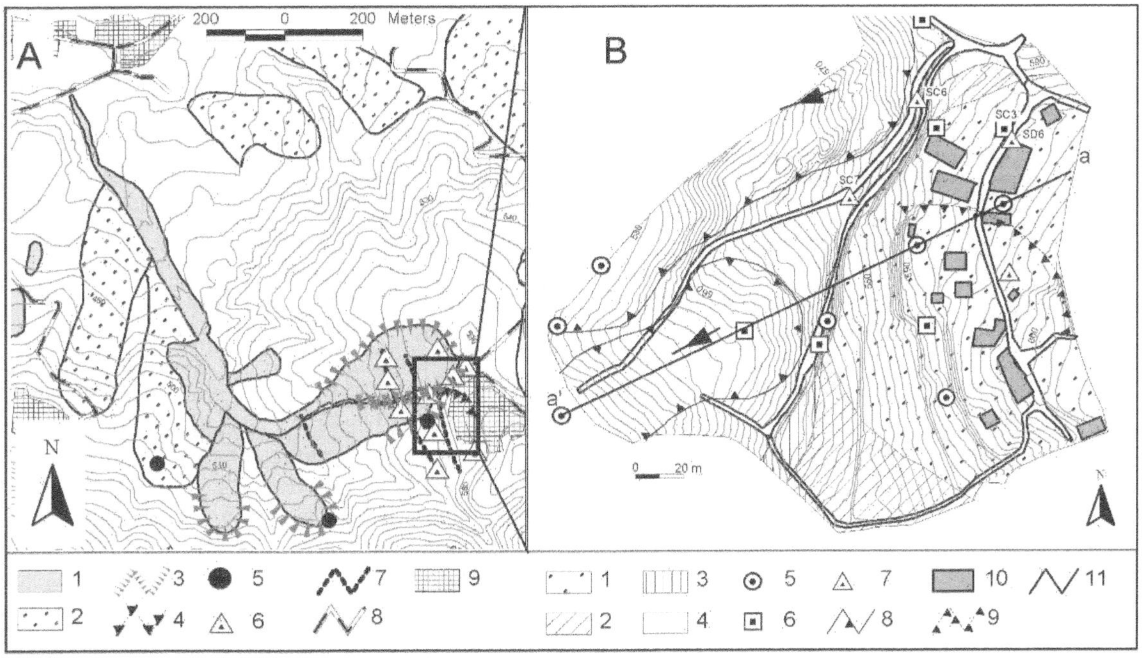

Fig.3 A: The active Poggio Ferrato landslide. 1: active landslide; 2: dormant landslide; 3: landslide scarp; 4: fracture; 5: spring; 6: trench pit; 7: seismic refraction profile; 8: road; 9: urban area. B. Detail of the old villane of Poggio Ferrato. 1: M. Vallassa Sandstone; 2: Monte Lumello Marls; 3: M. Piano Marls; 4: Chaotic Complex; 5: borehole; 6: borehole + open pipe piezometer; 7: borehole + Casagrande piezometer; 8: landslide scarp; 9: fracture; 10: building; 11: line of the geological sections.

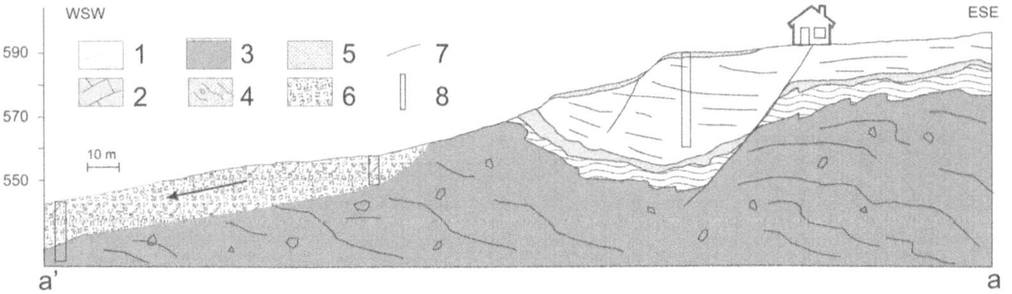

Fig.4 Longitudinal section of the depletion zone of the Poggio Ferrato landslide. 1: M. Vallassa Sandstone; 2: Monte Lumello Marls; 3: M. Piano Marls; 4: Chaotic Complex; 5: weathered sandstone; 6: landslide; 7: fracture; 8: borehole.

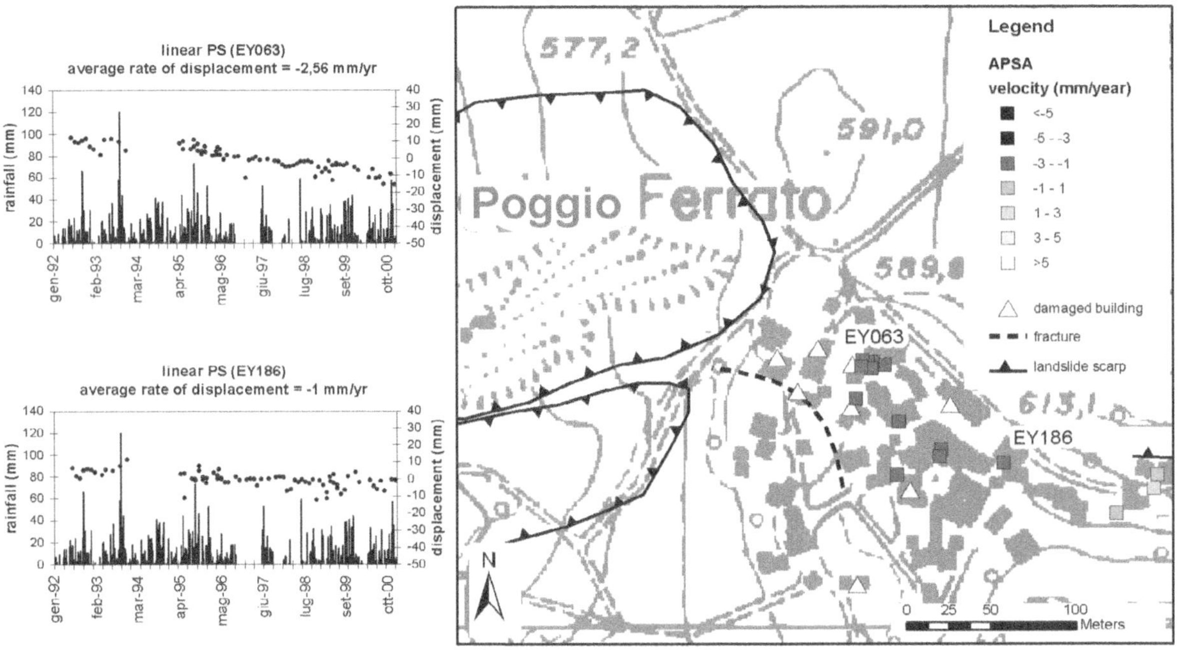

Fig.5 Aerial photo of the Poggio Ferrato landslide showing the distribution of the Permanent Scatterers.

The results of the advanced PS analysis are shown in Figure 5. A limited number of linear PS were obtained. The PS analysis shows a significant settlement of the sandstone slab in correspondence of the old village (25-35 mm in the period 1992-2000). The PS velocity ranges between –1 (southern part of the village) and –3 mm/year (north-eastern part). The SAR data support the geological hypothesis of a differential settlement and sinking of sandstone blocks on the border of the slab near Poggio Ferrato. The lack of PS in the north-western part of the village, that is the nearest to the landslide crown and where are concentrated the larger number of damaged buildings, could be related to the magnitude of the ground displacements, probably greater than 2.8 cm, which is the maximum displacement between two consecutive acquisitions. The historical series shows that the settlement is continuous in the period, there are not abrupt changes in correspondence of the reactivation of the landslide after December 1996.

The town of Broni, located in the Apennine fringe, is build upon alluvial fan deposits, which are very heterogeneous. A geological model of the area was obtained from the integration of the geological-geomorphological characteristics of the site with the geotechnical data (penetrometer tests, boreholes and laboratory tests). Silty clay and clayey silty deposits of high to very high swelling/shrinking potential constitute the first layer. They are soft soils with a tip resistence, measured with cone penetrometer test, of 1-2 MPa and their thickness varies from 4 m in the western part to 20 m in the northeastern. In the western sector thin lens of silt and sand are interbedded. The underlying layers consist, from top to bottom, of sand and gravel, silty clay, gravel and Tertiary marls.

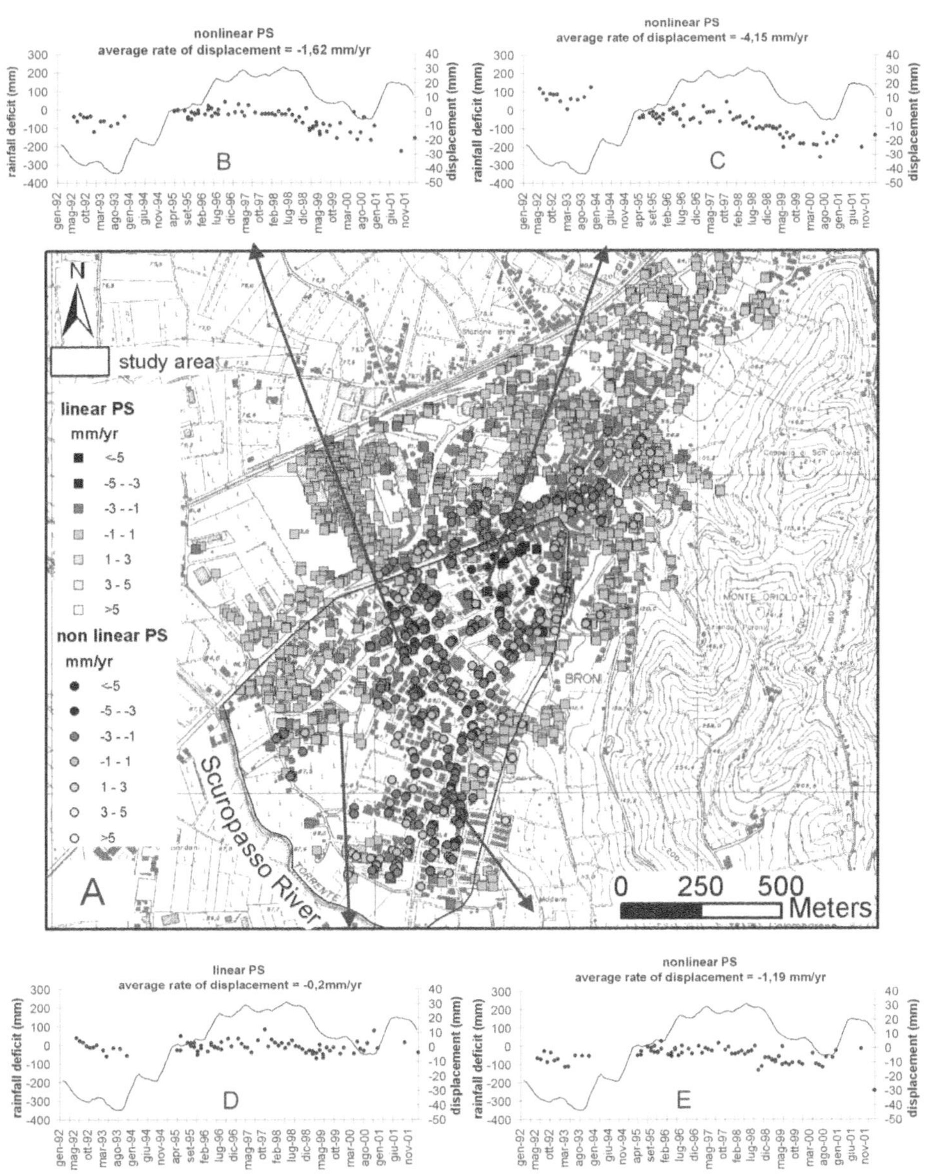

Fig.6 A: Linear and non linear advanced Permanent Scatterers. B, C, D, E: Time series relative to different Permanent Scatterers.

The alluvial deposits consist of two aquifers interbedded with discontinuous silty clay aquitard: the groundwater is pumped from the deepest, which is a semi-confined aquifer with a depth from 10 to 25 m. The water table fluctuation in drought period reaches 2.5-3 m. The depth of the active zone in the period 1992-2000 was 2.8 m. The moisture regime of soil could also be influenced by a seasonal perched water table.

Several buildings (single storey family residences), founded on conventional concrete shallow strip footings, were damaged in the last years; the cracks appeared at the end of 1980s (drought period) with a worsening in 1998-2000. Cracks are progressive and generally close up during the wet season in relationship with rainfall and open up again during the dry seasons.

The ASPA identifies vertical ground displacement in the northern and in the eastern part of Broni near the Apennine fringe, where nonlinear PS are present (Fig.6). This is in agreement with the soil stratification and clay thickness. The time series of nonlinear PS show abrupt variations after May 1998: we observe a rapid increase of the settlement that reaches 14-22 mm at the end of the drought period. The comparison between the vertical displacement of the Permanent Scatterers and the rainfall deficit curves shows a relationship between the curves. The phenomenon could be related to the shrinkage of the soil in the drought period. The lowering of the water table associated with pumping well could contribute to the ground settlement.

5 Conclusions

InSAR for the wide area coverage (100 x 100 km), the high spatial resolution (20 x 20 m) and the availability of a long historical SAR dataset (more than 14 years) may used as powerful instruments for monitoring and detection of surficial deformations. Its intrinsic limits due to temporal and geometric decorrelation may be overcame by the use of some techniques of persistent scatterers analysis as the Permanent Scatterers of Ferretti et al (2000-2001).

In this study the PS analysis at a large scale (SPSA) allows to verify the existence of known unstable areas (buildings, landslides and shrinkage phenomena) and to identify unknown phenomena (subsidence on the border of the sandstone slabs, new unstable areas, uplift zones, new damaged buildings). Discrepancies between the PS distribution/velocity and the landslides movements were also identified and allowed the modification of the state of activity of some landslides.

The PS analysis at detailed scale (APSA) allows to study the temporal evolution of the phenomena also of non-linear type.

The PS also allowed the production of deformation rate maps (e.g shrinkage in drought period) for the derivation of qualitative hazard zonation very useful for land planning.

References

Arca, S., Beretta, G.P. (1985). Prima sintesi geodetico-geologica sui movimenti verticali del suolo nell'Italia settentrionale (1897-1957). *Boll. Geodesia e Scienze Affini*, 2, pp.125-15

Beatrizzotti, G., Bellinzona, G., Beltrami, G., Boni, A., Braga, G., Marchetti, G., Mosna, S., (1969). Geological map of Italy Fg.71 – Voghera (2nd Ed) - Servizio Geologico d'Italia, Roma

Berardino, P., Costantini, M., Franceschetti, G., Iodice, A., Pietranera, L., Rizzo, V. (2003). Use of differential SAR interferometry in monitoring and modelling large slope instability at Maratea (Basilicata, Italy). *Eng.Geol.*, 68, pp. 31-51.

Braga, G., Meisina, C., Piccio, S. (2003). Study of a complex landslide movement for risk assessment: the example of the Poggio Ferrato landslide (Northern Italy). *International Conference Fast slope movements prediction and prevention for risk mitigation*, 1, pp.37-45, Picarelli Editor, Naples.

Braga, G., Braschi, G., Calculli, S., Caucia, F., Cerro, A., Colleselli, F., Grisolia, M., Piccio, A., Rossetti, R., Setti, M., Spalato, A., Soggetti, F., Veniale, F. (1985). I fenomeni franosi nell'Oltrepo Pavese: tipologia e cause. Geologia Applicata e Idrogeologia, 20, pp. 621-666.

Canuti, P., Casagli, N., Ermini, L., Fanti, R., Farina, P. (2004). Landslide activity as a geoindicator in Italy: significance and new perspectives from remote sensing. Environmental Geology, 45, pp. 907-919.

Cerrina Feroni, A, Otaria, G, Martinelli P, Martelli, L, 2002. Structural geological map of the Emilia Romagna Apennines. Firenze. SELCA.

Colesanti, C., Ferretti, A., Prati, C. & Rocca, F. (2003). Monitoring Landslides and Tectonic Motion with the Permanent Scatterers Technique. *Eng. Geol.*, 68, pp. 3-14.

Conti, S., Tosatti, G. (1994). Caratteristiche geologico-strutturali della Pietra di Bismantova e fenomeni franosi connessi (Appennino Reggiano). Quaderni di Geologia Applicata, 1, pp.31-49.

Crosetto, M., Crippa, B., Barzaghi, R. (2002). Quantitative subsidence monitoring using SAR interferometry. *Proceedings of IGARSS 2002, 20-24 June 2002, Toronto (Canada).*

Ferretti, A., Prati, C., Rocca, F. (2000). Nonlinear Subsidence Rate Estimation Using Permanent Scatterers in Differential SAR Interferometry. *IEEE Trans. Geosci. Remote Sens.*, 38, pp. 2202-2212.

Ferretti, A, Prati, C., Rocca, F. (2001). Permanent Scatterers in SAR Interferometry. *IEEE Trans. on geosci. Remote Sens.*, 39, pp. 8-20.

Ferretti, A., Prati, F., Rocca; N., Casagli, N., Farina, P. Young, B. (2005). Permanent Scatterers Technology: a powerful state of the art tool for historic and future monitoring of landslides and other terrain instability phenomena. *International Conference on Landslide Risk Management, Vancouver,* 31 May - 3 June 2005; pp. 1-9.

Meisina, C. (2003). Light buildings on swelling/shrinking soils: case histories from Oltrepo Pavese (north-western Italy). *Int. Conf. on Problematic Soils*, Nottingham, July 28-30, 2, pp. 365-374.

Meisina, C., Piccio, A. (2003). River dynamics and slope processes along a sector of the Villalvernia-Varzi line (Northern Italy). *Quaternary International*, 101-102, pp. 179-190.

Pieri, M., Groppi, G. (1981). Subsurface geological structure of the Po plain, Italy- Consiglio nazionale delle ricerche – Progetto finalizzato geodinamica – Sottoprogetto "Modello strutturale" - Pubblicazione n°414 – Agip.

Rossetti, R., and Ottone, C. (1979). Esame preliminare delle condizioni pluviometriche dell'Oltrepo Pavese e dei valori critici delle precipitazioni in relazione ai fenomeni di dissesto franoso. *Geologia Applicata e Idrogeologia*, 14, pp. 83-99

Tidal Errors and Deformations in Regional GPS Networks

L. Biagi
Politecnico di Milano – DIIAR c/o Polo Regionale di Como, via Valleggio, 11 – 22100 Como, Italy

G. Pietrantonio, F. Riguzzi
Istituto Nazionale di Geofisica e Vulcanologia, via di Vigna Murata, 605 - 00143 Roma, Italy

Abstract. At present, the tidal correction model described in the IERS 2003 Conventions should be applied in the processing of permanent networks: in brief, the model involves the computation of a first order tidal correction plus some corrective terms, as functions of space and time. In the BERNESE software, one of the main scientific softwares for GPS network processing, the IERS tidal model is implemented; in July 2004, a bug in the tidal correction routine was identified by one of the paper authors and corrected by the BERNESE staff: the error concerned a second order term, related to the semidiurnal tidal signal.

An experimental analysis on the error effects in the results stemming from processing of regional networks has been carried out. A network of 8 European permanent GPS stations has been selected: the network choice ensures heterogeneity in the length and the orientation of the baselines. One year of data has been considered and two elaborations of the data have been performed using the BERNESE 4.2 SW, by applying the same processing strategy and, respectively, the wrong (old) and the right (corrected) tidal correction routine. The differences in the network results and the deformations caused by the error are discussed.

Keywords. GPS permanent networks, BERNESE SW, Deformation control, Tidal modeling error

1 Introduction

It is well known that every celestial body exerts a gravitational attraction on the points of the Earth; a gravitational potential is usually related to the attraction. Attraction and potential are functions of the distance between the body and the point: so, they vary both inside the Earth and on its surface.

The tidal attraction of a celestial body B in a point P of the Earth is defined as the difference of the B gravitational attraction between the point itself and the Earth mass center (EMC).

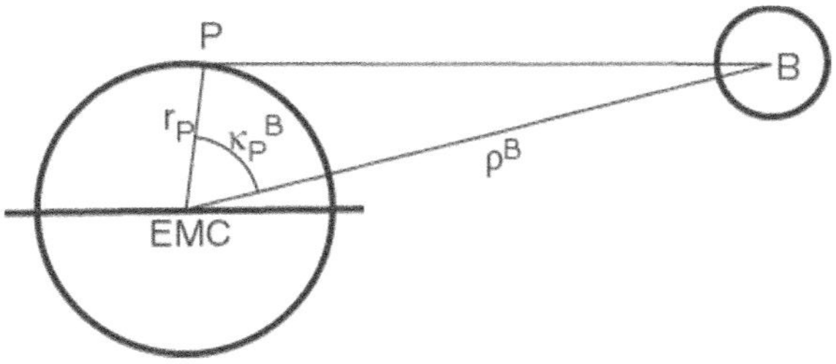

Fig. 1 The geometry of tidal potential

According to (Vanicek, 1980, Fig. 1), the related potential is:

$$W_E^B(P) = \frac{GM^B}{\rho^B}\sum_{n=2}^{\infty}\left(\frac{r_P}{\rho^B}\right)^n P_n\left(\cos\kappa_P^B\right)$$

where: G is the universal gravitational constant, M^B is the body mass, ρ^B is the distance between the body and the EMC, r_p is the distance between the point P and the EMC, κ_P^B is the geocentric angle between B and P . It is assumed that, for each geodetic and geodynamic application, only Moon and Sun tidal attractions are significant.

In the time domain the tidal attraction shows typical bands centered on infinite, monthly, diurnal and semidiurnal periods.

Tidal attractions cause periodic deformations of the Earth's shape; obviously, in the analysis of time series of geodetic observations (VLBI, SLR, GPS, ...) finalized to the estimation of positions and long term displacements on the Earth's surface, the tidal deformation on each single epoch position estimate needs to be removed: in other terms, an accurate model of tidal deformations has to be available to process raw geodetic observations. The response of the Earth's shape to the tidal attraction depends on Love and Shida numbers. They express the actual displacement in a point as the ratio between the response of an ideally rigid and an ideally fluid

Earth. The Love number is related to the radial (vertical) displacement in the point, while The Shida number is related to the tangential (horizontal) displacements. Due to the Earth's anelasticity, both Love and Shida numbers are complex functions both of the point position and of the tidal signal frequency.

In the last decades a lot of effort has been invested to properly model the tidal deformation: among the others, we mention the works of (Wahr, 1982, Mathews et al., 1995).

At present, the suggested formula for the tidal effects modeling on the Earth's surface is described in IERS conventions 2003 (McCarthy and Petit, 2004). Without entering into the details, the numerical tidal modeling is performed in four steps:

0. zero order estimate, taking into account the latitude dependences of Love and Shida numbers;
1. corrections for out of phase dependences of Love and Shida numbers on the diurnal and semidiurnal frequencies of the tidal signal;
2. final corrections for the Love number dependences on the diurnal and long period frequencies;
3. computation of the permanent tide (not applied for conventionally tide free coordinates) and the polar motion tidal effects.

Several numerical routines have been implemented according to the IERS conventions: among them, the FORTRAN routine of Veronique Dehant is freely available on the website ftp://omaftp.oma.be/dist/astro/dehant/IERS; in the next months, a C translation of the Dehant routine will be freely available on the website http://geomatica.como.polimi.it.

The moon-sun induced Earth's tide can reach values up to some decimeters; its time series show the same frequency bands of the tidal attraction. As an example, in Fig. 2 the tidal effects for the Matera IGS GPS permanent station are shown; in Table 1 the relevant statistics are summarized.

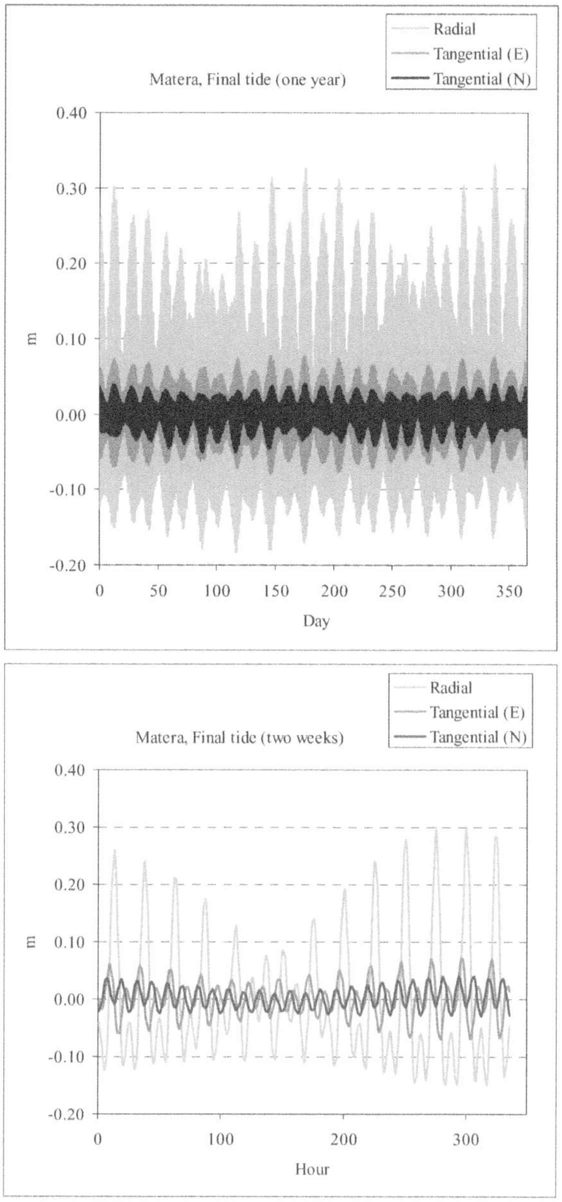

Fig. 2 An example of tidal effect.
Matera IGS station. Top: one year (from July, 1st, 2003 to June, 30th, 2004) of hourly tidal corrections. Bottom: a detail on the first two weeks.

2 The Tidal Modeling in GPS Data Adjustment

In the following we will shortly summarize the tidal effect, and the way to model it, on GPS observations, either code or phase. To do that, we'll consider only the distance term in the GPS observation equation, disregarding the atmospheric delays, clock errors, internal delays and integer ambiguities: for a complete description of them see for example (Teunissen, Kleusberg, 1998). Moreover, in the following we do not consider (assuming as already corrected for) the offsets of the satellite/receiver antenna phase centers with respect to the satellite mass center/ground station positions.

The GPS "distance" observation from a receiver R to a satellite S at a given epoch t $\rho_R^S(t)$ can be written as

$$\rho_R^S(t) = \left\| \mathbf{x}^S(t-\tau_R^S) - \mathbf{x}_R(t) \right\|$$

where $\mathbf{x}^S(t-\tau_R^S)$ is the satellite mass center position at the emission epoch $t-\tau_R^S$, $\mathbf{x}_R(t)$ is the position of the station at the reception epoch t. Taking into account the tide, we can write:

$$\mathbf{x}_R(t) = \mathbf{x}_{0,R}(t) + \boldsymbol{\delta}\mathbf{x}_{T\,R}(\mathbf{x}_{0,R}, t)$$

where $\mathbf{x}_{0,R}(t)$ is the "tide free" position of the station, $\boldsymbol{\delta}\mathbf{x}_{T\,R}(\mathbf{x}_{0,R}, t)$ is the tidal effect (displacement) in the station at the reception epoch.

The distance observation equation can be linearized with respect to the tidal effect:

$$\rho_R^S(t) \cong \left\| \mathbf{x}^S(t-\tau_R^S) - \mathbf{x}_{0,R}(t) \right\| + \mathbf{e}_R^S(t) \cdot \boldsymbol{\delta}\mathbf{x}_{T\,R}(\mathbf{x}_{0,R}, t)$$

where $\mathbf{e}_R^S(t)$ is the unitary vector directed from the satellite to the station.

Given approximate values for the station position, $\tilde{\mathbf{x}}_{0,R}$, given a tidal (e.g. the IERS2003) model, we can compute the tidal correction to the observation:

$$\delta\rho_{T\,R}^{\;S}(t) = \mathbf{e}_R^S(t) \cdot \boldsymbol{\delta}\mathbf{x}_{T\,R}(\mathbf{x}_{0,R}, t)$$

Next, we can compute a tide free distance observation by subtracting the tidal correction from the original observation:

$$\rho_{0,R}^S(t) \cong \rho_R^S(t) - \delta\rho_{T\,R}^{\;S}(t) = \left\| \mathbf{x}^S(t-\tau_R^S) - \mathbf{x}_{0,R}(t) \right\|$$

Finally, a set of tide free observations can be adjusted to estimate a tide free position of the station.

In GPS network adjustment, some softwares construct differences between the raw observations from two receivers to one or two satellites in order to estimate the network coordinates; particularly, the single differences between two receivers are given by the

$$\rho_{R1,R2}^S(t) = \rho_{R1}^S(t) - \rho_{R2}^S(t)$$

Double differences from two receivers to two satellites are given by the

$$\rho_{R1,R2}^{S1,S2}(t) = \rho_{R1,R2}^{S1}(t) - \rho_{R1,R2}^{S2}(t)$$

The corresponding differenced tidal corrections (dtc) are given, respectively, by the:

$$\delta\rho_{T\,R1,R2}^{\;S}(t) = \delta\rho_{T\,R1}^{\;S}(t) - \delta\rho_{T\,R2}^{\;S}(t)$$

and by the

$$\delta\rho_{T\,R1,R2}^{\;S1,S2}(t) = \delta\rho_{T\,R1,R2}^{\;S1}(t) - \delta\rho_{T\,R1,R2}^{\;S2}(t)$$

To compute the tide free single/double differences, two ways are possible:

1) to compute the tide free raw observations before differencing them;
2) to difference the raw observations and then to compute and subtract the dtc.

Table 1. Statistics of tidal corrections.
Matera IGS PS; one year (from July, 1st, 2003 to June, 30th, 2004) of hourly tidal corrections: 8760 data points. E: mean value; σ: RMS of data; M: maximum; m: minimum. S0: step 0 estimates; S1: step 1 corrections; S2: step 2 corrections; PM: polar motion effect; PT: permanent tide effect; F: final estimate.

	Radial (mm)				Tangential, East comp. (mm)				Tangential, North comp. (mm)			
	E	σ	m	M	E	σ	m	M	E	σ	m	M
S0	-15	108	-187	331	0	29	-78	77	-23	17	-77	16
S1	<1	<1	-1	1	<1	<1	-1	1	<1	<1	1	1
S2	<1	9	-15	16	<1	<1	-1	1	<1	<1	0	1
PM	<1	4	-7	6	<1	<1	-1	1	0	0	0	0
PT	16	0	16	16	0	0	0	0	25	0	25	25
F	0	102	-183	332	0	29	-79	78	0	17	-51	41

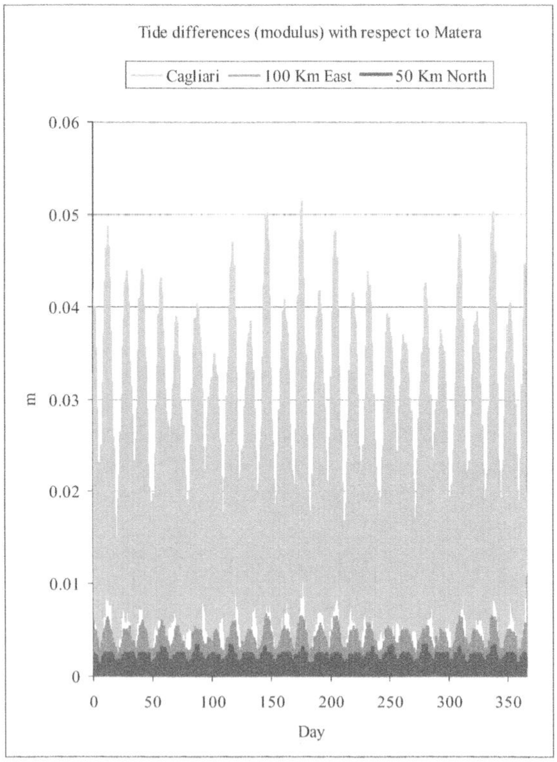

Fig. 3 An example of difference between tidal corrections for two stations.
One year (from July, 1st ,2003 to June, 30th, 2004) of hourly tidal corrections. The plotted time series contain the modulus of the vector of the tidal correction differences. Reference station: Matera IGS PS. Second station: Baseline 1: 50 km toward North, max difference 3 mm; B2: 100 km toward East, max difference 7 mm; B3 Cagliari IGS PS (680 km), max difference 5 cm.

Both the approaches provide the same final results: the choice is purely a numerical matter. A last remark is worth: the dtc depend both on the observation geometry (absolute and relative positions of the stations and the satellites) and on the difference between the tidal displacements of the stations. In very local networks (baseline lengths smaller than 20 Km), tidal differences are always negligible at the mm level; on baselines between 20 and 100 Km, particularly if oriented in East-West direction, the differences can reach the cm; on regional networks, they are never negligible. An example of tidal correction differences between two stations is given in Fig. 3.

3 The Tidal Modeling Bug in BERNESE GPS SW

In the BERNESE SW (Beutler et al., 2000, Hugentobler et al., 2005) the tidal modeling is performed, according the IERS Conventions, by the subroutines tide96 for the 4.2 release and tide2000 for the 5.0 release. A bug in the tidal modeling was discovered by one of the authors (Biagi, personal e-mail to Urs Hugentobler, 2004), and corrected by the AIUB staff (BSWMail 190, 2004). The bug concerned the step 2 computation; particularly, the correction to the step 1 estimate of the radial component of the tidal effect was affected.

In the following, the errors introduced in the time series by the tidal modeling error are investigated. In this section, at first the tidal modeling errors on a single station (absolute errors) are presented; in the following, the differences between the tidal modeling errors for two stations (relative errors) are discussed. In the next section, errors in position estimates caused by the tidal modeling errors in the GPS data adjustment for a network are shown.

As an example, the Matera and the Cagliari IGS permanent stations have been chosen: their distance is about 680 Km (Fig. 6, Table 4).

In order to investigate long term periodicity of the errors, two years have been analyzed, from January, 1st, 2003 to December, 31st, 2004. Adopting a computation interval of one hour, both the right and the wrong tidal corrections have been computed: absolute tidal errors for both stations have been derived by subtracting the right from the wrong tidal correction. Absolute errors are shown in Fig. 4 for Matera; the relevant statistics are summarized in Table 2 for both stations. The absolute tidal errors show two clear periods: the first diurnal (Fig. 4, bottom), the second annual (Fig. 4, top), both with an amplitude of about 25-30 mm. The errors have zero mean, with maxima and minima of about 30 mm.

Table 2. BERNESE absolute tidal errors on radial component.
Two years (from January, 1st, 2003 to December, 31st, 2004) of differences between wrong and right tidal corrections: 17520 epochs. E: mean error; σ: RMS of the errors; m: minimum; M: maximum.

	E (mm)	σ (mm)	m (mm)	M (mm)
Matera	0	14	-30	30
Cagliari	0	13	-29	30

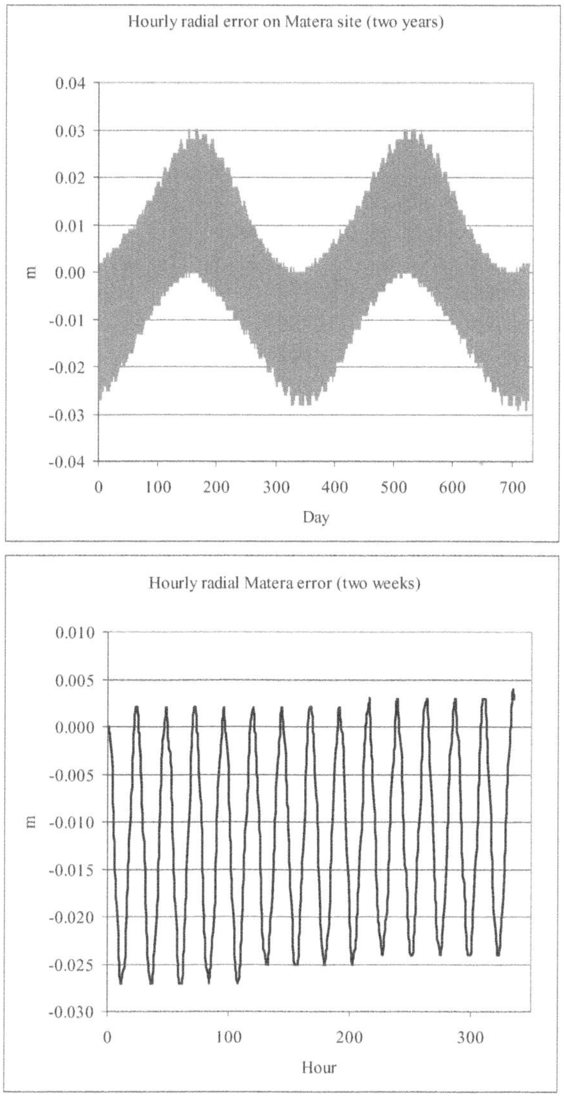

Fig. 4 (up). BERNESE absolute tidal errors on radial component. Matera IGS station. Top: two years (from January, 1st ,2003 to December, 31st, 2004) of differences between right and wrong tidal corrections. Bottom: detail of the first two weeks. Not the same scale for the graphs.

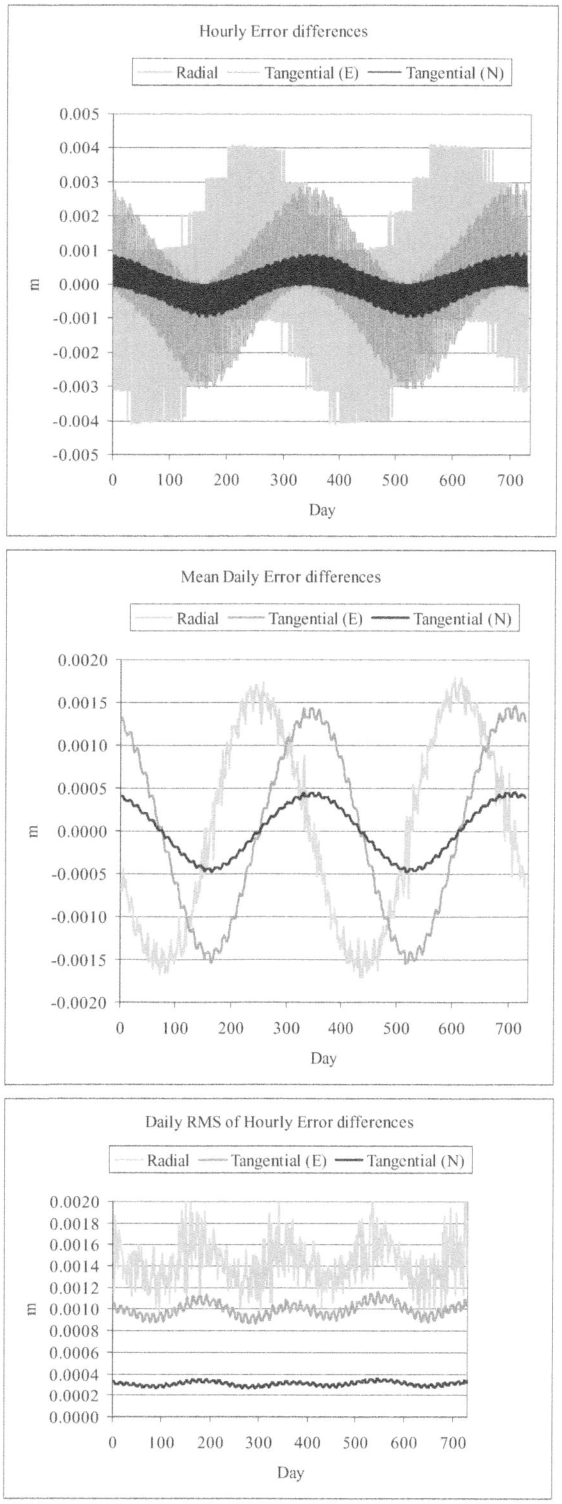

Fig. 5 (on the right). Relative tidal errors between Matera and Cagliari IGS stations (distance 680 km).
The three components of the relative tidal error vectors have been converted from cartesian ITRF2000 components to local (with respect to Cagliari position) North, East, Up. Two years (from January, 1st, 2003 to December, 31st, 2004) of data. Top: hourly tidal error differences. Center: mean daily error differences. Bottom: RMS of the 24 hourly data for each day in the two years. Not the same scale for the three graphs.

Table 3. Relative tidal errors between Matera and Cagliari IGS stations (distance 680 km). Two years (from January, 1st, 2003 to December, 31st, 2004) of data. The three components of the relative tidal error vectors have been converted from cartesian ITRF2000 components to local (with respect to Cagliari position) North, East, Up. H: hourly differences. D: mean daily differences. E: mean error; σ: RMS of the errors; m: minimum; M: maximum.

	E (mm)	σ (mm)	m (mm)	M (mm)
N, H	0	0.4	-0.9	0.9
E, H	0	1.4	-3.1	3.0
U, H	0	1.8	-4.1	4.0
N, D	0	0.3	-0.5	0.5
E, D	0	1.0	-1.4	1.5
U, D	0	1.1	-1.8	1.7

In the GPS data adjustment, BERNESE SW constructs and uses double differences: so, only tidal error differences influence the final results. Moreover, typically the GPS data of a permanent network are processed on a daily basis, in order to estimate daily coordinates. So, mean daily error differences seem to be particularly interesting. Fig. 5 and Table 3 summarize the hourly and mean daily error differences between Matera and Cagliari. The error vectors have been converted from cartesian ITRF2000 components to local (with respect to Cagliari position) North, East and Up directions. The relative errors are significant at the mm level; their daily means are not on average null, and show a clear annual period: this may involve significant effects also on GPS daily solutions. Moreover, the greater errors are in the local Up and East directions: taking into account the orientation of the baseline between Matera and Cagliari (Fig. 6), in GPS data processing this should imply errors mainly in the estimates of heights and baseline lengths.

4 Data Set and Processing

A permanent GPS network (Fig. 6) of 8 stations has been selected ensuring heterogeneity in the length and orientation of the baselines: Matera (MATE), Cagliari (CAGL), Villafranca (VILL), Wettzell (WTZR), Roma (INGR and M0SE), Roseto degli Abruzzi (RSTO) and Perugia (UNPG).

According to the periodicity in the tidal correction errors previously evidenced, one year of data has been considered, from July, 1st, 2003 to June, 30th, 2004. We processed the GPS observations twice by the BPE automatic procedure of the BERNESE SW, 4.2 (Beutler et al., 2000). In both the network adjustments, the EUREF (EPN, 2002) guidelines for regional network adjustment were applied. The datum was fixed by using IGS final ephemerides, IERS C04 EOP, and by constraining the Matera coordinates to their ITRF2000 estimates, daily propagated. The main characteristics of the implicitly defined baselines between Matera and the others stations are summarized in Table 4.

In the first processing, the wrong (old) tidal model was applied to correct GPS observations; in the second processing the right (corrected) tidal correction was applied. The differences (tidal error effects) between the results are discussed.

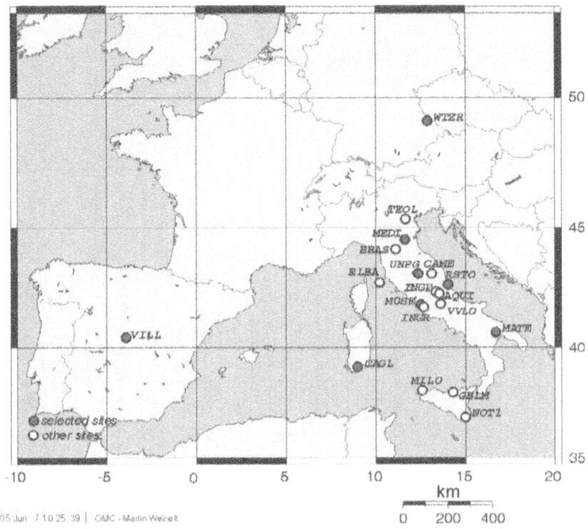

Fig. 6 The test network (7 baselines considered, from 320 (MATE-RSTO) to 1740 Km (MATE-VILL) long.

Table 4. Main characteristics of the network baselines.
L: baseline length; A: baseline azimuth, clockwise from North direction; Dh: height difference between the stations.

Matera to	L (km)	A (°)	Dh (m)
RSTO	317	306	433
MOSE	379	286	415
UNPG	453	300	185
MEDI	597	307	486
CAGL	682	259	297
WTZR	990	336	124
VILL	1741	270	112

At first, we focus our analysis on the tidal error effects on the coordinate estimates. The differences between the estimated coordinates show a clear annual signal, as anticipated from the analysis of the mean daily errors. The magnitudes of the errors linearly depends on the baseline length, and are of the order of some parts per billion. Typically, the horizontal errors are oriented along the baseline direction: this fact should involve a scale effect on the baseline length estimate, that will be analyzed at the end of the section; for the majority of the stations, horizontal and vertical errors are comparable: an exception is provided by the Wettzell station, for which the vertical errors are significantly smaller. In Fig. 7 and Table 5 the results for some stations, representative of the different classes of distances from Matera, are shown.

It has to be mentioned that the tidal errors are negligible with respect to the dispersion of the daily results: an example (Medicina estimates) is given by Fig. 8; moreover, the mean value of the errors over one year of results doesn't significantly differ from zero. Nevertheless, due to the periodicity of the tidal errors, their effect isn't expected to vanish in coordinates and velocities estimated from time series shorter than one year.

As regards the baseline lengths (Fig. 8, bottom; Fig. 9; Table 6), the same considerations can be repeated: the differences are not significant with respect to the dispersion of the daily results; the yearly estimated trends of the baseline lengths in the two cases (right and wrong tidal corrections) do not change significantly (Fisher test on differences between estimated initial length and length time derivative, with significance α both =5% and =1%); however, tidal error effects are highly correlated in time and show a clear annual signal, with a magnitude of 5×10^{-9} of the baseline length.

To conclude, it's worth noting that a time shift of about 2.5 months exists between the error signals in the height and in the baseline length estimates.

5 Conclusions

A bug has been discovered in the BERNESE routine for tidal modeling and correction. In the daily adjustment of a regional GPS network, the bug introduced errors in height and scale estimates that were not significant with respect to the dispersions of the results; moreover, on an annual basis, they had zero mean.

However, the errors show a high correlation in time; particularly, a clear annual signal is present, both in the height estimates and in the scale of GPS networks, that reaches 5×10^{-9}.

Acknowledgements

We wish to thank the Italian Space Agency and the "GPS networks and Geodesy" group of INGV for providing GPS observations. We received useful suggestions by M. Crespi. L. Biagi's work has been supported in the framework of the "Satellite Positioning Services for the e-government" Italian PRIN (MIUR founded) 2004 project.

Table 5. Tidal error effects on BERNESE daily estimates. One year (from July, 1st, 2003 to June, 30th, 2004) of data. Reference station: Matera IGS station. V-: Villafranca IGS station; W-: Wettzel IGS station; M-: Medicina IGS station. R- Università di Roma La Sapienza station.
Dh, DN, DE: vertical, local North and East components of the errors. E: mean error; σ: RMS of the errors; m: minimum; M: maximum.

	E (mm)	σ (mm)	m (mm)	M (mm)
V-Dh	0.4	3.8	-9.6	8.9
V-DN	0.1	1.0	-3.9	3.7
V-DE	0.4	4.7	-12.2	8.7
W-Dh	0.0	1.0	-2.7	3.3
W-DN	0.0	2.0	-3.9	3.7
W-DE	0.0	1.1	-2.5	3.1
M-Dh	0.1	0.9	-2.3	1.8
M-DN	0.0	0.9	-1.8	1.8
M-DE	0.1	0.9	-1.6	1.5
R-Dh	0.1	0.9	-2.2	5.0
R-DN	-0.1	0.3	-1.2	0.5
R-DE	0.0	0.7	-1.3	1.7

Table 6. Tidal error effects on BERNESE daily baseline length estimates. One year (from July, 1st, 2003 to June, 30th, 2004) of data. Reference station: Matera IGS station. E: mean error; σ: RMS of the errors; m: minimum; M: maximum.

	E (mm)	σ (mm)	m (mm)	M (mm)
RSTO	-0.1	0.7	-1.8	2.4
MOSE	-0.1	0.8	-2.0	2.3
UNPG	0.1	1.0	-2.2	2.8
MEDI	0.0	1.3	-2.6	2.6
CAGL	0.6	1.2	-2.3	3.5
WTZR	-0.1	2.2	-5.5	4.5
VILL	-0.4	3.8	-7.0	10.5

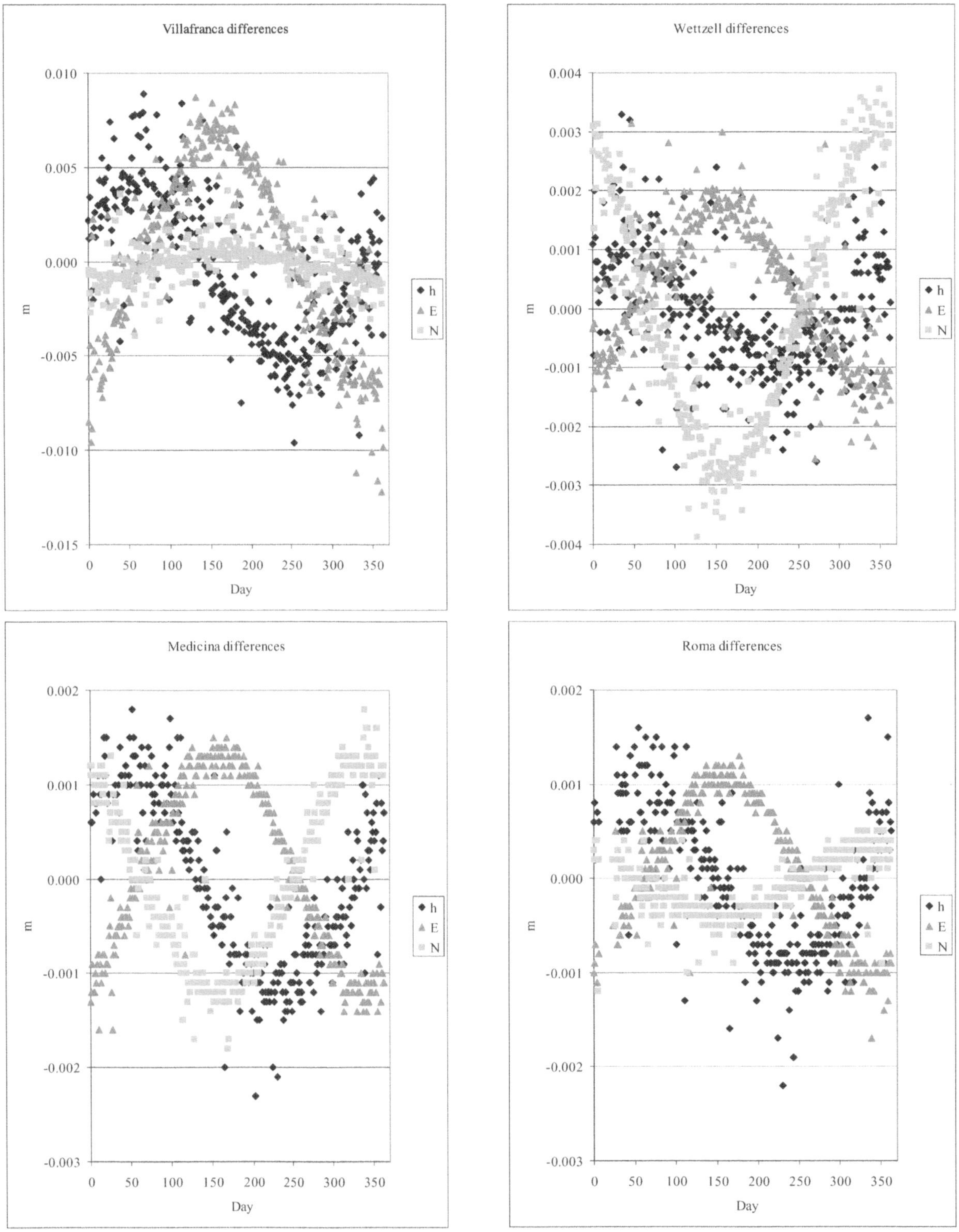

Fig. 7 Tidal error effects on BERNESE daily estimates. One year (from July, 1st ,2003 to June, 30th, 2004) of data. Reference station: Matera IGS station. Top: Villafranca IGS station; Wettzel IGS station; bottom: Medicina IGS station; Università di Roma La Sapienza station. Not the same scale for the graphs.

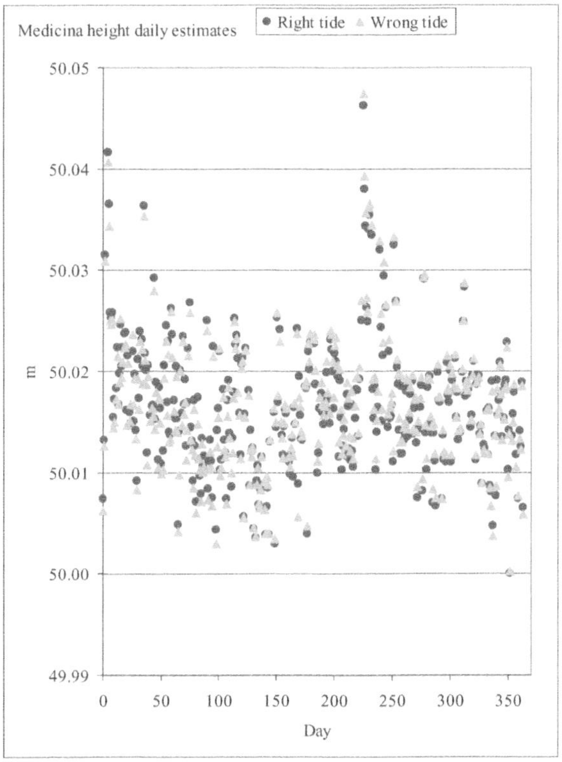

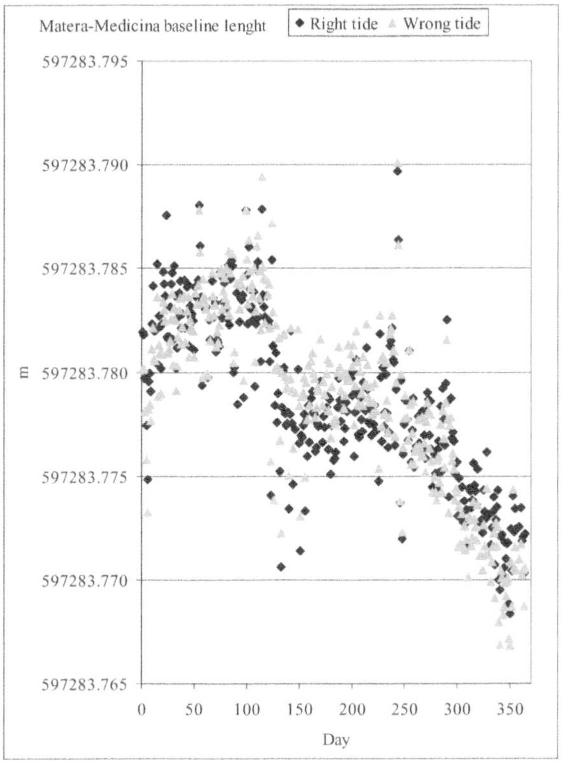

Fig. 8. Medicina IGS station daily estimates. One year (from July, 1st, 2003 to June, 30th, 2004) of data. Reference station: Matera IGS station. Top: height estimates (mean height estimate: 50.015 m; RMS: 7 mm; min. value: 50 m; max. value: 50.046 m). Bottom: baseline length estimates.

Trend: $l(t) = l_0 + v \cdot t,\ v = -1.1cm/y,\ \sigma_v = 0.6mm/y$

Not the same scale for the graphs.

References

Beutler, G., Brockmann, E., Dach, R., Fridez, P., Gurtner, W., Hugentobler, U., Johnson, J., Mervant, L., Rothacher, M., Schaer, S., Springer, T., Weber, R. (2000). *Bernese Software 4.2.* Astronomical Institute, University of Berne.

EPN (2002). *Guidelines for EPN analysis centres.* Document and related mails on-line: http://www.epncb.oma.be/

Hugentobler, U., Dach, R., Fridez, P. (2005). *Bernese GPS Software 5.0, Draft.* Astronomical Institute, University of Berne.

Hugentobler, U., BSWMail 190 (2004), on-line: http://www.aiub.unibe.ch/download/bswmail/

Mathews, P. M., Buffett, B. A., Shapiro, I. I. (1995). *Love numbers for a rotating spheroidal Earth: new definitions and numerical values.* Geophys. Res. Lett., 22, pp 579-582.

McCarthy, D. D., Petit, G. eds. (2004). *IERS Conventions (2003).* IERS Technical Note No. 32, Verlag des Bundesamts für Kartographie und Geodäsie, Frankfurt am Main.

Teunissen, P. J. G., Kleusberg A., eds (1998). *GPS for Geodesy, 2nd edition.* Springer Verlag Berlin.

Vanicek, P. (1980). *Tidal Corrections to Geodetic Quantities.* NOAA Technical Report, NOS 83 NGS 14, US Department of Commerce, Rockville, Maryland.

Wahr, J. M. (1982). *Computing tides, mutations and tidally induced variations in the Earth rotation rate for a rotating, elliptical Earth* in Geodesy and Global Geodynamics, Moritz H., Sünkel H. eds, Mitteilungen der geodätischen Institute der Technischen Universität, Folge 41, Graz.

Astronomical Institute, University of Berne AIUB FTP site: ftp.unibe.ch

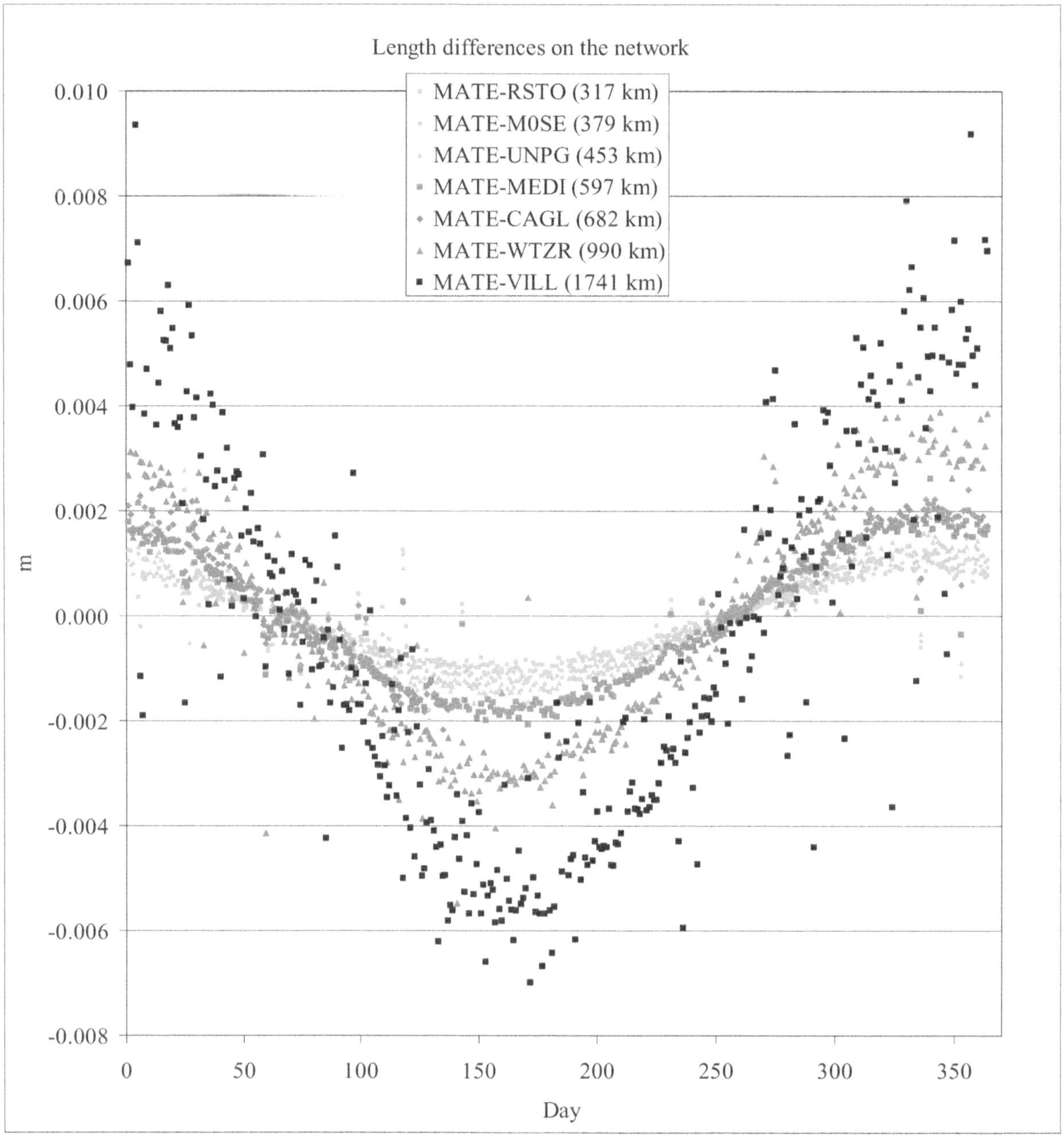

Fig. 9. Baseline lengths differences. One year (from July, 1st, 2003 to June, 30th, 2004) of data: a clear scale effect is evidenced.

The Treatment of Time Continuous GPS Observations for the Determination of Regional Deformation Parameters

Ludovico Biagi[(1)] and Athanasios Dermanis[(2)]

[(1)] DIIAR, c/o Polo regionale di Como, Politecnico di Milano,
[(2)] Department of Geodesy and Surveying, The Aristotle University of Thessaloniki

Abstract

The calculation of invariant deformation parameters, entering in the constitutional equations of crustal dynamics, requires information on the geometry of the crust in the study region, which is continuous in both the time and spatial domain. The use of continuously observing GPS stations provides geodetic data which are practically time-continuous and must be only spatially interpolated. A strategy is developed for the treatment of dense series of horizontal coordinates from a regional GPS network, which are typically exhibiting a time-linear behavior. The role of the choice of reference system is examined for the removal of trend before the spatial interpolation as well as the determination of the motion of the region as whole with respect to the ITRF or of the relative motion of tectonically homogeneous sub regions. Rigorous formulas are presented for various horizontal deformation parameters and their intrinsic time derivatives, without the usual infinitesimal approximations. Finally the problem of quality assessment for the derived parameters is investigated completely ignoring the questionable formal statistical characteristic of the original geodetic data. A realistic numerical example demonstrates the suggested techniques, involving spatial interpolation by the classical finite-element method. A software package in standard C language has been developed in order to implement the proposed algorithms.

1 Introduction

Although crustal deformation is naturally a three-dimensional (3D) phenomenon, there is a long tradition of a two-dimensional (2D) treatment of the relevant data, primarily by the finite element method using horizontal triangles within which deformation is assumed to be homogeneous. This restriction is partly due to the difficulty of obtaining high quality leveling data, or nowadays in the lower quality of heights from GPS observations. The main reason though is that our discrete positional information is fundamentally 2D as restricted to the surface of the earth. Interpolation can provide continuous information in the surface or horizontal sense, but extension to the 3D unknown deformation requires an extrapolation in the surface-normal or height sense, which is not justified for the available data.

The derivation of horizontal deformation, ignoring height information is somewhat unnatural since it is not concerned with the deformation of the material points constituting the crust, but rather with the deformation of their projections on a horizontal reference surface. The results are more realistic when the surface is essentially flat and height variations do not alter height differences in the area, i.e. the whole area is uplifted or down-lifted as a whole. Only in this case horizontal deformation is the horizontal trace of the actual 3D deformation.

A realistic treatment must study the deformation of the 2D natural earth surface as embedded in 3D space and provide the "trace" of the unknown 3D deformation in the tangent plane for each surface point. This generalized treatment will be presented elsewhere. For the time being we mention an obvious extension of the finite element method in the "right direction". In Dermanis (1994) we have presented a version of the finite element method where, instead of coordinates the horizontal distances of the triangle sides at epochs are used. If the horizontal distances are replaced by spatial distances taking also height information into account, more realistic deformation parameters can be computed, which are referring to the tangential rather than the horizontal trace of the actual 3D deformation.

In any case, here we will follow the traditional restriction of the horizontal treatment, since our aim is to update some existing computational techniques (Dermanis & Livieratos, 1983, Dermanis & Rossikopoulos,1988, Rossikopoulos, 2001), in a direction which takes into account the nature of presently available data from permanent GPS stations monitoring a deforming area. The daily network solutions, apart from some missing or obvious outlier days, manifest a linear evolution in time, even for long periods up to 2 years. This is not surprising, since the motions of "floating" tectonic plates are rotations around a fixed axis. Furthermore they are so slow that the displacements, circular in principle, appear to be linear in the "infinitesimal" time interval of few years, especially for points far away from the plate-rotation axis, which may well be located outside the plate itself. The use of a "constant velocity" model for the station coordinates as functions of time provides a surplus of data that allows us to address the data quality assessment problem, completely ignoring the unrealistic (specially when derived from GPS observations) formal covariance matrices accompanying the daily network solutions.

2 Horizontal Deformation Parameters

Deformation is primarily described by the deformation function $\mathbf{f}$, which relates the coordinates $\mathbf{x} = \mathbf{f}(\mathbf{x}_0)$ of any material point at some particular epoch t to the coordinates $\mathbf{x}_0$ of the same material point at some reference epoch t_0. In the differential equations of motion, which connect the deformation of a material body under external forces with its response characteristics (constitutional equations), the deformation function enters through its local linear approximation, the deformation gradient $\mathbf{F} = \frac{\partial \mathbf{f}}{\partial \mathbf{x}_0} = \frac{\partial \mathbf{x}}{\partial \mathbf{x}_0}$. The linear mapping $\mathbf{F}$ describes how each curve changes its spatial direction and rate of length variation, by mapping a tangent vector to a curve at epoch t_0 to the tangent vector to its new position at epoch t. The deformation gradient depends on the chosen reference systems. Under system changes described by $\tilde{\mathbf{x}} = \mathbf{S}\mathbf{x}$, $\tilde{\mathbf{x}}_0 = \mathbf{S}_0\mathbf{x}_0$, the deformation gradient is represented by the new matrix $\tilde{\mathbf{F}} = \mathbf{S}\mathbf{F}\mathbf{S}_0^T$. In order to get an insight into the nature of the deformation gradient independently of the coordinate systems involved, we will resort to the singular value decomposition (SVD)

$$\mathbf{F} = \mathbf{Q}^T \mathbf{\Lambda} \mathbf{P} \tag{1}$$

where $\mathbf{P}$, $\mathbf{Q}$ are orthogonal matrices and $\mathbf{\Lambda}$ a diagonal matrix having diagonal elements the singular values $\lambda_i = \Lambda_{ii}$ of $\mathbf{F}$. These are related to the diagonalization of two symmetric matrices, the right Cauchy strain matrix

$$\mathbf{C} = \mathbf{F}^T \mathbf{F} = \mathbf{P}^T \mathbf{\Lambda}^2 \mathbf{P} \tag{2}$$

and the left Cauchy strain matrix

$$\mathbf{B} = \mathbf{F}\mathbf{F}^T = \mathbf{Q}^T \mathbf{\Lambda}^2 \mathbf{Q}. \tag{3}$$

Thus the columns of $\mathbf{P}^T$ are the eigenvectors of $\mathbf{C}$, the columns of $\mathbf{Q}^T$ are the eigenvectors of $\mathbf{B}$ and the diagonal elements $\lambda_i^2 = (\mathbf{\Lambda}^2)_{ii} = \Lambda_{ii}^2$ are the common eigenvalues of $\mathbf{C}$ and $\mathbf{B}$, arranged in order of magnitude from larger to smaller. Related to the SVD is the polar decomposition

$$\mathbf{F} = \mathbf{R}\mathbf{U} = \mathbf{V}\mathbf{R}, \tag{4}$$

which involves the right stretch matrix

$$\mathbf{U} = \mathbf{P}^T \mathbf{\Lambda} \mathbf{P} = \mathbf{C}^{1/2} \tag{5}$$

the left stretch matrix

$$\mathbf{V} = \mathbf{Q}^T \mathbf{\Lambda} \mathbf{Q} = \mathbf{B}^{1/2} \tag{6}$$

and the rotation matrix

$$\mathbf{R} = \mathbf{Q}^T \mathbf{P}. \tag{7}$$

Most common in engineering applications is the use of the strain matrix

$$\mathbf{E} = \tfrac{1}{2}(\mathbf{C} - \mathbf{I}) = \mathbf{P}^T [\tfrac{1}{2}(\mathbf{\Lambda}^2 - \mathbf{I})] \mathbf{P} \tag{8}$$

with eigenvalues $\frac{1}{2}(\lambda_i^2 - 1)$, or its infinitesimal approximation

$$\mathbf{E}_{\text{inf}} = \tfrac{1}{2}(\mathbf{J} + \mathbf{J}^T) = \mathbf{E} - \tfrac{1}{2}\mathbf{J}\mathbf{J}^T \tag{9}$$

expressed in terms of the displacement gradient

$$\mathbf{J} = \frac{\partial \mathbf{u}}{\partial \mathbf{x}_0} = \frac{\partial(\mathbf{x} - \mathbf{x}_0)}{\partial \mathbf{x}_0} = \frac{\partial \mathbf{x}}{\partial \mathbf{x}_0} - \frac{\partial \mathbf{x}_0}{\partial \mathbf{x}_0} = \mathbf{F} - \mathbf{I}, \tag{10}$$

in equation (9) the quadratic terms $\frac{1}{2}\mathbf{J}\mathbf{J}^T$ are neglected.

It can be easily shown that under a change of reference frames the above matrices are replaced by $\tilde{\mathbf{F}} = \mathbf{S}\mathbf{F}\mathbf{S}_0^T$, $\tilde{\mathbf{C}} = \mathbf{S}_0\mathbf{C}\mathbf{S}_0^T$, $\tilde{\mathbf{B}} = \mathbf{S}\mathbf{B}\mathbf{S}^T$, $\tilde{\mathbf{R}} = \mathbf{S}\mathbf{R}\mathbf{S}_0^T$, $\tilde{\mathbf{Q}} = \mathbf{Q}\mathbf{S}^T$, $\tilde{\mathbf{P}} = \mathbf{P}\mathbf{S}_0^T$ and $\tilde{\mathbf{\Lambda}} = \mathbf{\Lambda}$. The only invariants are the singular values $\tilde{\lambda}_i = \lambda_i$. The matrix $\mathbf{P}$ transforms the direction of the axes of the coordinate system at epoch t_0 to the directions of the eigenvectors of $\mathbf{C}$, which are physical invariants called *principal directions*. Points at these directions are displaced along the same directions so that their distances from the reference point change by factors λ_i. After this deformation the area around the reference point is rotated as described by the matrix $\mathbf{R}$.

In the horizontal treatment the SVD becomes

$$\mathbf{F} = \mathbf{R}(-\theta_Q)\mathbf{\Lambda}\mathbf{R}(\theta_P) \tag{11}$$

explicitly

$$\begin{bmatrix} F_{11} & F_{12} \\ F_{21} & F_{22} \end{bmatrix} = \tag{12}$$

$$= \begin{bmatrix} \cos\theta_Q & -\sin\theta_Q \\ \sin\theta_Q & \cos\theta_Q \end{bmatrix} \begin{bmatrix} \lambda_1 & 0 \\ 0 & \lambda_2 \end{bmatrix} \begin{bmatrix} \cos\theta_P & \sin\theta_P \\ -\sin\theta_P & \cos\theta_P \end{bmatrix}$$

The singular values and the angle θ_P, computed by the well known solution of the 2D diagonalization problem $\mathbf{C} = \mathbf{R}(-\theta_P)\mathbf{\Lambda}^2\mathbf{R}(\theta_P)$ are

$$\lambda_1^2 = \frac{C_{11} + C_{22}}{2} + \sqrt{\left(\frac{C_{11} - C_{22}}{2}\right)^2 + C_{12}^2} \tag{13}$$

$$\lambda_2^2 = \frac{C_{11} + C_{22}}{2} - \sqrt{\left(\frac{C_{11} - C_{22}}{2}\right)^2 + C_{12}^2} \tag{14}$$

$$\tan 2\theta_P = \frac{C_{12}}{\frac{1}{2}(C_{11} - C_{22})}, \tag{15}$$

where $C_{11} = F_{11}^2 + F_{21}^2$, $C_{22} = F_{12}^2 + F_{22}^2$, $C_{12} = F_{11}F_{12} + F_{21}F_{22}$. If the strain matrix is used then

$$\lambda_1^2 = 1 + (E_{11} + E_{22}) + \sqrt{(E_{11} - E_{22})^2 + 4E_{12}^2} \tag{16}$$

$$\lambda_2^2 = 1 + (E_{11} + E_{22}) - \sqrt{(E_{11} - E_{22})^2 + 4E_{12}^2} \tag{17}$$

$$\tan 2\theta_P = \frac{2E_{12}}{E_{11} - E_{22}}. \tag{18}$$

Instead of the singular values which are scale factors along the principal direction, more appealing to the engineering visualization of deformation seem to be the dilatation Δ (percentage of area change) and shear γ, which relates to the distortion of a square into an oblique parallelogram with bases along the same straight lines, skewed by an angle having γ as its tangent.

A shear along the x-axis is represented by the shear matrix

$$\mathbf{\Gamma} = \begin{bmatrix} 1 & \gamma \\ 0 & 1 \end{bmatrix} \tag{19}$$

and a shear in the direction with angle ϕ by the matrix

$$\mathbf{\Gamma}_\phi = \mathbf{R}(-\phi)\mathbf{\Gamma}\mathbf{R}(\phi). \tag{20}$$

The 4-parameters deformation gradient $\mathbf{F}$ can be represented by a shear γ in direction ϕ, a scale factor s and an additional rotation by an angle ψ

$$\mathbf{F} = s\mathbf{R}(\psi)\mathbf{\Gamma}_\phi = s\mathbf{R}(\psi)\mathbf{R}(-\phi)\mathbf{\Gamma}\mathbf{R}(\phi). \tag{21}$$

Since $|\mathbf{\Gamma}| = 1$ and all orthogonal rotation matrices have determinant 1 too, the scale factor is directly determined from the determinant

$$s = \sqrt{|\mathbf{F}|} = \sqrt{F_{11}F_{22} - F_{12}F_{21}}, \tag{22}$$

and is related to the singular values by $s^2 = |\mathbf{F}| = |\mathbf{P}^T||\mathbf{\Lambda}||\mathbf{Q}| = |\mathbf{\Lambda}| = \lambda_1\lambda_2$ and

$$s = \sqrt{\lambda_1\lambda_2}. \tag{23}$$

The shear can then be determined from

$$tr\mathbf{C} = tr(\mathbf{F}^T\mathbf{F}) = s^2 tr\{\mathbf{R}(-\phi)\mathbf{\Gamma}^T\mathbf{\Gamma}\mathbf{R}(\phi)\} =$$

$$= s^2 \mathrm{tr}\{\mathbf{\Gamma}^T\mathbf{\Gamma}\mathbf{R}(\phi)\mathbf{R}(-\phi)\} = s^2 \mathrm{tr}\{\mathbf{\Gamma}^T\mathbf{\Gamma}\} =$$

$$= s^2 \mathrm{tr}\begin{bmatrix} 1 & \gamma \\ \gamma & 1+\gamma^2 \end{bmatrix} = s^2(2+\gamma^2) = \mathrm{tr}(\mathbf{Q}^T\mathbf{\Lambda}^2\mathbf{Q}) =$$

$$= \mathrm{tr}(\mathbf{\Lambda}^2\mathbf{Q}\mathbf{Q}^T) = \mathrm{tr}(\mathbf{\Lambda}^2) = \lambda_1^2 + \lambda_2^2 \tag{24}$$

and

$$\gamma = \sqrt{\frac{C_{11} + C_{22}}{s^2} - 2} = \sqrt{\frac{\lambda_1^2 + \lambda_2^2}{\lambda_1\lambda_2} - 2} = \frac{\lambda_1 - \lambda_2}{\sqrt{\lambda_1\lambda_2}} \tag{25}$$

To determine the direction of shear ϕ we use the SVD of $\mathbf{\Gamma}$:

$$\mathbf{\Gamma} = \mathbf{R}_\Gamma'^T\mathbf{M}\mathbf{R}_\Gamma = \mathbf{R}(-\theta_\Gamma')\begin{bmatrix} \mu_1 & 0 \\ 0 & \mu_2 \end{bmatrix}\mathbf{R}(\theta_\Gamma) \tag{26}$$

The diagonalization problem

$$\boldsymbol{\Gamma}^T\boldsymbol{\Gamma} = \mathbf{R}_\Gamma{}^T\mathbf{M}^2\mathbf{R}_\Gamma \tag{27}$$

gives

$$\tan 2\theta_\Gamma = \frac{2}{-\gamma}. \tag{28}$$

Comparison with the SVD of $\mathbf{F}$

$$\mathbf{F} = s\mathbf{R}(\psi - \phi - \theta'_\Gamma)\mathbf{M}\mathbf{R}(\theta_\Gamma + \phi) =$$

$$= \mathbf{R}(-\theta_Q)\boldsymbol{\Lambda}\mathbf{R}(\theta_P) \tag{29}$$

establishes the relation

$$\phi = \theta_P - \theta_\Gamma. \tag{30}$$

3 Propagation of Variances and Covariances

Starting from the complete covariance matrix of the elements of the deformation gradient $\mathbf{F}$ it is possible to compute covariance matrices for other deformation parameters in steps, as follows

$$\mathbf{F} \to \mathbf{C} \to \{\lambda_1, \lambda_2, \theta_P\} \to \{\gamma, \Delta, \phi\}$$

utilizing at each step the following relevant partial derivatives:

$$\frac{\partial \begin{bmatrix} C_{11} \\ C_{12} \\ C_{22} \end{bmatrix}}{\partial \begin{bmatrix} F_{11} \\ F_{12} \\ F_{21} \\ F_{22} \end{bmatrix}} = \begin{bmatrix} 2F_{11} & 0 & 2F_{21} & 0 \\ F_{12} & F_{11} & F_{22} & F_{21} \\ 0 & 2F_{12} & 0 & 2F_{22} \end{bmatrix} \tag{31}$$

$$\frac{\partial \begin{bmatrix} \lambda_1 \\ \lambda_2 \\ \theta_P \end{bmatrix}}{\partial \begin{bmatrix} C_{11} \\ C_{12} \\ C_{22} \end{bmatrix}} = \begin{bmatrix} \frac{1}{4\lambda_1}\left(1+\frac{K}{B}\right) & \frac{C_{12}}{2\lambda_1 B} & \frac{1}{4\lambda_1}\left(1-\frac{K}{B}\right) \\ \frac{1}{4\lambda_2}\left(1-\frac{K}{B}\right) & -\frac{C_{12}}{2\lambda_2 B} & \frac{1}{4\lambda_2}\left(1+\frac{K}{B}\right) \\ -\frac{C_{12}}{4B^2} & \frac{K}{2B^2} & \frac{C_{12}}{4B^2} \end{bmatrix} \tag{32}$$

where $K = (C_{11} - C_{22})/2$ and $B = \sqrt{K^2 + C_{12}^2}$,

$$\frac{\partial \begin{bmatrix} \gamma \\ \Delta \\ \phi \end{bmatrix}}{\partial \begin{bmatrix} \lambda_1 \\ \lambda_2 \\ \theta_P \end{bmatrix}} = \begin{bmatrix} \frac{\lambda_1+\lambda_2}{2\lambda_1\sqrt{\lambda_1\lambda_2}} & -\frac{\lambda_1+\lambda_2}{2\lambda_2\sqrt{\lambda_1\lambda_2}} & 0 \\ \lambda_2 & \lambda_1 & 0 \\ -\frac{\gamma\lambda_2}{4B} & \frac{\gamma\lambda_1}{4B} & 1 \end{bmatrix} \tag{33}$$

4 Interpolation of Displacements Using the Finite Element Method

The simplest way to interpolate displacements determined at the vertices of a control network, is by assuming that it is homogenous within properly formulated network triangles, serving as finite elements. The interpolation model has the form

$$\begin{bmatrix} U \\ V \end{bmatrix} = \begin{bmatrix} J_{11} & J_{12} \\ J_{21} & J_{22} \end{bmatrix} \begin{bmatrix} X \\ Y \end{bmatrix} + \begin{bmatrix} g_x \\ g_y \end{bmatrix}. \tag{34}$$

The elements J_{ik} of the displacement gradient matrix, which are constant within each triangle, are recovered from the known displacement values (U_i, V_i) at the triangle vertices having coordinates (X_i, Y_i), $i = A, B, C$, using the relations

$$J_{11} = \frac{(Y_B - Y_C)U_A + (Y_C - Y_A)U_B - (Y_B - Y_A)U_C}{D}$$

$$J_{12} = \frac{(X_C - X_B)U_A - (X_C - X_A)U_B + (X_B - X_A)U_C}{D}$$

$$J_{21} = \frac{(Y_B - Y_C)V_A + (Y_C - Y_A)V_B - (Y_B - Y_A)V_C}{D}$$

$$J_{22} = \frac{(X_C - X_B)V_A - (X_C - X_A)V_B + (X_B - X_A)V_C}{D} \tag{35}$$

where

$$D = (X_B - X_A)(Y_C - Y_A) - (X_C - X_A)(Y_B - Y_A). \tag{36}$$

From the displacement gradient $\mathbf{J}$, the deformation gradient $\mathbf{F} = \mathbf{I} + \mathbf{J}$ can be computed, from which all relevant deformation parameters can be derived. The complete covariance matrix of the network velocities and coordinates can be used in order to compute the complete covariance matrix of the elements of $\mathbf{F}$, by covariance propagation for all triangles, and proceed on for the covariances of other deformation parameters. To perform the covariance propagation from the coordinates and displacements to the elements F_{ij}, we need the relevant partial derivatives. In fact we need only those with respect to the displacements, since the ones referring to coordinates are too small to be of any practical importance. The required derivatives for this first step of covariance propagation ($U_i, V_i \rightarrow \mathbf{F}$) are

$$\frac{\partial[F_{11}\,F_{12}\,F_{21}\,F_{22}]^T}{\partial[U_A\,V_A\,U_B\,V_B\,U_C\,V_C]^T} =$$

$$= \begin{bmatrix} \frac{Y_B-Y_C}{D} & 0 & \frac{Y_C-Y_A}{D} & 0 & \frac{Y_A-Y_B}{D} & 0 \\ \frac{X_C-X_B}{D} & 0 & \frac{X_A-X_C}{D} & 0 & \frac{X_B-X_A}{D} & 0 \\ 0 & \frac{Y_B-Y_C}{D} & 0 & \frac{Y_C-Y_A}{D} & 0 & \frac{Y_A-Y_B}{D} \\ 0 & \frac{X_C-X_B}{D} & 0 & \frac{X_A-X_C}{D} & 0 & \frac{X_B-X_A}{D} \end{bmatrix} \quad (37)$$

5 The Case of Linearly Moving Permanent GPS Stations

In a permanent GPS network, daily or weekly solutions provide coordinates which exhibit a linear trend with superimposed noise. Each separate single-epoch solution is accompanied by a complete covariance matrix for all coordinates of all network points, which presumably reflect the attained accuracy, while coordinate estimates for different epochs are formally uncorrelated since they are computed from different observations which are presumably uncorrelated. On the other hand such estimate statistics are commonly recognized to be overoptimistic and therefore unrealistic, since unknown biases affecting the observations are not taken into account. For this reason we propose to discard the available covariance matrices and to perform an independent variance-covariance component estimation utilizing the residuals resulting after fitting a linear model to each point coordinate.

Some simplifying assumptions are necessary to arrive to a computationally reasonable and efficient algorithm, which of course may be subject to criticism. In any case we claim that at least they offer an attractive alternative to the use of the formal unrealistic variances and covariance coming from the network per epoch adjustments. Explicitly we assume that network performance does not vary essentially with time while different epoch estimates remain uncorrelated. The first assumption is supported from the analysis of coordinate series from daily solutions. For the second one we note that a small correlation is present between consecutive days. A refined procedure seeking also an estimate of this correlation will be the object of future work.

We outline the suggested procedure taking into account only horizontal coordinates X, Y, the extension to three coordinates being straightforward. The linear trend model is

$$X_{ki} \equiv X(P_k, t_i) = X_{0k} + U_k(t_i - t_0) + v_{X_k t_i}$$

$$Y_{ki} \equiv Y(P_k, t_i) = Y_{0k} + V_k(t_i - t_0) + v_{Y_k t_i} \quad (38)$$

for network points P_k, $i = 1, 2, \ldots, M$, at epochs t_i, $i = 1, 2, \ldots, N$, where X_{0k}, Y_{0k} are the initial coordinates of point P_k at the chosen reference epoch t_0 and U_k, V_k the corresponding constant velocity components. We explicitly assume that

$$E\{v_{Xki}\} = 0\,,\quad E\{v_{Yki}\} = 0\,,$$

$$E\{v_{X_k t_i} v_{X_m t_j}\} = \delta_{ij}\sigma_{X_k X_m}\,,$$

$$E\{v_{Y_k t_i} v_{Y_m t_j}\} = \delta_{ij}\sigma_{Y_k Y_m}\,,$$

$$E\{v_{X_k t_i} v_{Y_m t_j}\} = \delta_{ij}\sigma_{X_k Y_m}\,. \quad (39)$$

In the first iteration we use a common unknown variance $\sigma_{X_k X_m} = \sigma_{Y_k Y_m} = \sigma_{X_k Y_m} = \delta_{km}\sigma^2$ and zero correlation between different coordinates, which allows separate linear regressions yielding residuals

$$\hat{v}_{X_k t_i} = X(P_k, t_i) - \hat{X}_{0k} - t_i\hat{U}_k$$

$$\hat{v}_{Y_k t_i} = Y(P_k, t_i) - \hat{Y}_{0k} - t_i\hat{V}_k\,. \quad (40)$$

For each point P_k the coordinate and velocity estimates are derived from the following algorithm

$$\bar{t} = \frac{1}{N}\sum_i t_i\,, \qquad m_t^2 = \frac{1}{N}\sum_i (t_i - \bar{t})^2$$

$$\bar{X}_k = \frac{1}{N}\sum_i X_{ki}\,, \quad m_{tX,k} = \frac{1}{N}\sum_i (t_i - \bar{t})(X_{ki} - \bar{X}_k)\,,$$

$$\bar{Y}_k = \frac{1}{N}\sum_i Y_{ki}\,, \quad m_{tY,k} = \frac{1}{N}\sum_i (t_i - \bar{t})(Y_{ki} - \bar{Y}_k)\,,$$

$$\hat{U}_k = \frac{m_{tX,k}}{m_t^2}, \qquad \hat{V}_k = \frac{m_{tY,k}}{m_t^2},$$

$$\hat{X}_{0k} = \bar{X}_k - \hat{U}_k \bar{t}, \quad \hat{Y}_{0k} = \bar{Y}_k - \hat{V}_k \bar{t}. \tag{41}$$

The residuals are utilized for the derivation of unbiased estimates

$$\hat{\sigma}_{X_k X_m} = \frac{1}{N-2} \sum_i \hat{v}_{X_k t_i} \hat{v}_{X_m t_i},$$

$$\hat{\sigma}_{Y_k Y_m} = \frac{1}{N-2} \sum_i \hat{v}_{Y_k t_i} \hat{v}_{Y_m t_i},$$

$$\hat{\sigma}_{X_k Y_m} = \frac{1}{N-2} \sum_i \hat{v}_{X_k t_i} \hat{v}_{Y_m t_i}, \tag{42}$$

which constitute an estimate for the common covariance matrix for each epoch to be used in the second iteration:

$$\hat{\mathbf{\Sigma}} = \begin{bmatrix} \hat{\sigma}^2_{X_1} & \cdots & \hat{\sigma}_{X_1 X_M} & \hat{\sigma}_{X_1 Y_1} & \cdots & \hat{\sigma}_{X_1 Y_M} \\ \vdots & \ddots & \vdots & \vdots & \ddots & \vdots \\ \hat{\sigma}_{X_M X_1} & \cdots & \hat{\sigma}^2_{X_M} & \hat{\sigma}_{X_M Y_1} & \cdots & \hat{\sigma}_{X_M Y_M} \\ \hat{\sigma}_{Y_1 X_1} & \cdots & \hat{\sigma}_{Y_1 X_M} & \hat{\sigma}^2_{Y_1} & \cdots & \hat{\sigma}_{Y_1 Y_M} \\ \vdots & \ddots & \vdots & \vdots & \ddots & \vdots \\ \hat{\sigma}_{Y_M X_1} & \cdots & \hat{\sigma}_{Y_M X_M} & \hat{\sigma}_{Y_1 Y_M} & \cdots & \hat{\sigma}^2_{Y_M} \end{bmatrix} \tag{43}$$

In the second iteration the linear regressions are performed simultaneously ordering the data as follows

$$\mathbf{b} = [\mathbf{b}_{X_1}^T \cdots \mathbf{b}_{X_M}^T \mathbf{b}_{Y_1}^T \cdots \mathbf{b}_{Y_M}^T]^T \tag{44}$$

with

$$\mathbf{b}_{X_k} = \begin{bmatrix} X(P_k, t_1) \\ \vdots \\ X(P_k, t_N) \end{bmatrix}, \quad \mathbf{b}_{Y_k} = \begin{bmatrix} Y(P_k, t_1) \\ \vdots \\ Y(P_k, t_N) \end{bmatrix}. \tag{45}$$

In the collective linear model $\mathbf{b} = \mathbf{Ax} + \mathbf{v}$, the residuals have covariance matrix of the form $\mathbf{C} = \mathbf{\Sigma} \otimes \mathbf{I}_N$. We utilize the estimate $\hat{\mathbf{\Sigma}}$ and set

$$\mathbf{C} = E\{\mathbf{v}\mathbf{v}^T\} = \kappa^2 (\hat{\mathbf{\Sigma}} \otimes \mathbf{I}_N) \tag{46}$$

including an accommodating variance factor κ^2 and the adjustment is performed using a weight matrix

$$\mathbf{P} = (\hat{\mathbf{\Sigma}} \otimes \mathbf{I}_N)^{-1} = \hat{\mathbf{\Sigma}}^{-1} \otimes \mathbf{I}_N = \mathbf{W} \otimes \mathbf{I}_N \tag{47}$$

where

$$\mathbf{W} = \begin{bmatrix} \mathbf{W}_X & \mathbf{W}_{XY} \\ \mathbf{W}_{XY}^T & \mathbf{W}_Y \end{bmatrix} = \hat{\mathbf{\Sigma}}^{-1}. \tag{48}$$

It can be shown that the solution breaks down into individual linear regressions leading to the same estimates as in the first step. The only difference is in the covariance propagation step which is performed by first computing the estimate of the variance factor

$$\begin{aligned} \hat{\mathbf{v}}^T \mathbf{P} \hat{\mathbf{v}} = & \sum_{k,m,i} (\mathbf{W}_X)_{km} \hat{v}_{X_k t_i} \hat{v}_{X_m t_i} + \\ & + \sum_{k,m,i} (\mathbf{W}_{XY})_{km} \hat{v}_{X_k t_i} \hat{v}_{Y_m t_i} + \\ & + \sum_{k,m,i} (\mathbf{W}_{XY}^T)_{km} \hat{v}_{Y_m t_i} \hat{v}_{X_k t_i} + \\ & + \sum_{k,m,i} (\mathbf{W}_Y)_{km} \hat{v}_{Y_k t_i} \hat{v}_{Y_m t_i}, \end{aligned} \tag{49}$$

$$\hat{\kappa}^2 = \frac{\hat{\mathbf{v}}^T \mathbf{P} \hat{\mathbf{v}}}{2(N-2)M} \tag{50}$$

and then the variances and covariances from

$$\begin{aligned} \hat{\sigma}_{\hat{X}_{0k} \hat{X}_{0m}} &= \hat{\kappa}^2 \, \hat{\sigma}_{X_k X_m} \frac{m_t^2 + \bar{t}^2}{N m_t^2}, \\ \hat{\sigma}_{\hat{X}_{0k} \hat{Y}_{0m}} &= \hat{\kappa}^2 \, \hat{\sigma}_{X_k Y_m} \frac{m_t^2 + \bar{t}^2}{N m_t^2}, \\ \hat{\sigma}_{\hat{Y}_{0k} \hat{Y}_{0m}} &= \hat{\kappa}^2 \, \hat{\sigma}_{Y_k Y_m} \frac{m_t^2 + \bar{t}^2}{N m_t^2}, \\ \hat{\sigma}_{\hat{U}_k \hat{U}_m} &= \hat{\kappa}^2 \, \hat{\sigma}_{X_k X_m} \frac{1}{N m_t^2}, \\ \hat{\sigma}_{\hat{V}_k \hat{V}_m} &= \hat{\kappa}^2 \, \hat{\sigma}_{Y_k Y_m} \frac{1}{N m_t^2}, \end{aligned} \tag{51}$$

$$\hat{\sigma}_{\hat{U}_k\hat{V}_m} = \hat{\kappa}^2\,\hat{\sigma}_{X_kY_m}\frac{1}{Nm_t^2}\,, \tag{52}$$

$$\hat{\sigma}_{\hat{X}_{0k}\hat{U}_m} = -\hat{\kappa}^2\,\hat{\sigma}_{X_kX_m}\frac{\bar{t}}{Nm_t^2}\,,$$

$$\hat{\sigma}_{\hat{X}_{0k}\hat{V}_m} = -\hat{\kappa}^2\,\hat{\sigma}_{X_kY_m}\frac{\bar{t}}{Nm_t^2}\,,$$

$$\hat{\sigma}_{\hat{Y}_{0k}\hat{U}_m} = -\hat{\kappa}^2\,\hat{\sigma}_{Y_kX_m}\frac{\bar{t}}{Nm_t^2}\,,$$

$$\hat{\sigma}_{\hat{Y}_{0k}\hat{V}_m} = -\hat{\kappa}^2\,\hat{\sigma}_{Y_kY_m}\frac{\bar{t}}{Nm_t^2}\,. \tag{53}$$

Note that the above formulas are greatly simplified when the reference epoch t_0 is chosen so that $\bar{t} = 0$. The obtained velocity component estimates $\hat{U}_k$, $\hat{V}_k$ at each network point P_k must be interpolated to obtain a continuous velocity field $\mathbf{v} = \mathbf{v}(\mathbf{x}_0)$, where $\mathbf{x}_0 = [X_0\,Y_0]^T$ and $\mathbf{v} = [U\,V]^T$ from which the velocity gradient

$$\mathbf{L}_0 = \frac{\partial \mathbf{v}}{\partial \mathbf{x}_0} \tag{54}$$

referring to the reference epoch t_0 can be calculated. The displacements $\mathbf{u}$ between epoch t_0 and any other epoch t are given by $\mathbf{u}(t_0,t) = (t - t_0)\mathbf{v}$ and the displacement gradient is simply $\mathbf{J}(t_0,t) = (t - t_0)\mathbf{L}_0$. Thus the deformation gradient becomes

$$\mathbf{F}(t_0,t) = \mathbf{I} + \mathbf{J}(t_0,t) = \mathbf{I} + (t - t_0)\mathbf{L}_0\,. \tag{55}$$

Once the deformation gradient is calculated the deformation analysis can be performed as previously described. When the finite element method is used, instead of the displacements $\mathbf{u}$ one may interpolate the velocities $\mathbf{v}$, in which case the resulting matrix is not $\mathbf{J}$ but $\mathbf{L}_0$. For this reason we have used the same symbols (U, V) for both displacement and velocity components, while the equations (35) for the computation of the elements of $\mathbf{J}$ can be used for the computation of the corresponding elements of $\mathbf{L}_0$.

It is also possible to perform the deformation analysis for any pair of epochs t_1, t_2 considering that

$$\mathbf{F}(t_1,t_2) = \frac{\partial \mathbf{x}_2}{\partial \mathbf{x}_1} = \frac{\partial \mathbf{x}_2}{\partial \mathbf{x}_0}\frac{\partial \mathbf{x}_0}{\partial \mathbf{x}_1} = \frac{\partial \mathbf{x}_2}{\partial \mathbf{x}_0}\left(\frac{\partial \mathbf{x}_1}{\partial \mathbf{x}_0}\right)^{-1} =$$

$$= \mathbf{F}(t_0,t_2)\mathbf{F}(t_0,t_1)^{-1} =$$

$$= [\mathbf{I} + (t_2 - t_0)\mathbf{L}_0][\mathbf{I} + (t_1 - t_0)\mathbf{L}_0]^{-1}\,. \tag{56}$$

Taking into account that for a "small" matrix $\mathbf{\Delta}$ $(\mathbf{I} + \mathbf{\Delta})^{-1} \cong \mathbf{I} - \mathbf{\Delta}$, we may use the approximation

$$\mathbf{F}(t_1,t_2) \approx \mathbf{I} + (t_2 - t_1)\mathbf{L}_0\,. \tag{57}$$

The velocity gradient at any epoch t is given by

$$\mathbf{L} = \mathbf{L}(t) = \frac{\partial \mathbf{v}}{\partial \mathbf{x}} = \frac{\partial \mathbf{v}}{\partial \mathbf{x}_0}\frac{\partial \mathbf{x}_0}{\partial \mathbf{x}} = \mathbf{L}_0\mathbf{F}(t_0,t)^{-1} =$$

$$= \mathbf{L}_0[\mathbf{I} + (t - t_0)\mathbf{L}_0]^{-1} \approx \mathbf{L}_0 \tag{58}$$

and it can be used to derive the corresponding symmetric "stretch" or "rate-of-strain" matrix

$$\mathbf{D} = \frac{1}{2}(\mathbf{L} + \mathbf{L}^T) \tag{59}$$

and the antisymmetric "spin" matrix

$$\mathbf{W} = \tfrac{1}{2}(\mathbf{L} - \mathbf{L}^T) = \begin{bmatrix} 0 & w \\ -w & 0 \end{bmatrix}, \quad w = \frac{L_{12} - L_{21}}{2}\,. \tag{60}$$

6 A Numerical Example

For a numerical illustration of the proposed methodology, we consider a network of permanent GPS stations (fig. 1) covering a region separated in two parts by a fault. Coordinate time series have been simulated for a time span of two years, applying linear velocities of the order of some cm/year to the station coordinates. Moreover, two rigid transformations, linear with respect to time, with the same order of magnitude of the velocities, have been added to the data: a first one on the right (R) network alone, to simulate a tectonic effect of that region with respect to the other; the second one on both networks, to simulate a reference frame effect. At the end, a realistic noise (of the order of some mm) has been added to the coordinates.

We shall derive the motion (displacements in X, Y and rotation) for the whole region, and each one of the separate parts (subregions) with respect to the global reference system (WGS84 or ITRF) used in GPS data analysis, as well as the relative motion of one subregion with respect to the other. Within each subregion a deformation analysis is performed using the finite element interpolation method, which leads to estimates of deformation parameters and their standard deviations and correlation. These statistics are based on the previously described variance-covariance analysis and are independent from the unrealistic formal statistics derived within the adjustment of the GPS observations.

The motion of a network covering a deforming region is derived by defining its corresponding "Tisserand" reference system, which best represents the network as a whole, by preserving the "center of mass" of the network points (considered as material points with unit mass) and by the vanishing of their relative angular momentum (Dermanis & Kotsakis, 2005). In the planar case the original coordinates X, Y are transformed into new coordinates $\tilde{X}$, $\tilde{Y}$ by a (counter-clockwise) rotation θ and displacements d_X, d_Y (all functions of time) by requiring that

$$\sum_k \tilde{X}_k = \sum_k \tilde{Y}_k = 0 \tag{61}$$

$$\tilde{h} = \sum_k (\tilde{X}_k \tilde{V}_k - \tilde{Y}_k \tilde{U}_k) = 0 \tag{62}$$

for the conservation of the center of mass and the vanishing of the relative angular momentum $\tilde{h}$, respectively. For linearly varying coordinates the solution is given by

$$\theta(t) = \theta_0 + \frac{h_0}{2\sqrt{S_v^2 S_0^2 - S_{0v}^2}} \arctan \frac{S_v^2 t + S_{0v}}{\sqrt{S_v^2 S_0^2 - S_{0v}^2}} \tag{63}$$

$$\begin{bmatrix} d_X(t) \\ d_Y(t) \end{bmatrix} = -\begin{bmatrix} \cos\theta(t) & \sin\theta(t) \\ -\sin\theta(t) & \cos\theta(t) \end{bmatrix} \begin{bmatrix} m_{0X} + t\, m_U \\ m_{0Y} + t\, m_V \end{bmatrix} \tag{64}$$

where

$$m_{0X} = \frac{1}{N}\sum_k X_{0k}, \qquad m_{0Y} = \frac{1}{N}\sum_k Y_{0k},$$
$$m_U = \frac{1}{N}\sum_k U_k, \qquad m_V = \frac{1}{N}\sum_k V_k; \tag{65}$$

$$\Delta X_{0k} = X_{0k} - m_{0X}, \qquad \Delta Y_{0k} = Y_{0k} - m_{0Y},$$
$$\Delta U_k = U_k - m_U, \qquad \Delta V_k = V_k - m_V, \tag{66}$$

$$S_0^2 = \frac{1}{N}\sum_k (\Delta X_{0k}^2 + \Delta Y_{0k}^2), \tag{67}$$

$$S_{0v} = \frac{1}{N}\sum_k (\Delta X_{0k}\Delta U_k + \Delta Y_{0k}\Delta V_k), \tag{68}$$

$$S_v^2 = \frac{1}{N}\sum_k (\Delta U_k^2 + \Delta V_k^2) \tag{69}$$

$$h_0 = \frac{1}{N}\sum_k (\Delta X_k \Delta V_k - \Delta Y_k \Delta U_k). \tag{70}$$

Using the computed parameters $\theta(t)$, $d_X(t)$, $d_Y(t)$, the station coordinates within each subregion can be transformed from the original global frame to the corresponding Tisserand frame, so that any common displacement and rotational trend is removed before performing the deformation analysis.

A noticeable property of such a reference system transformation (where a generally different time dependent transformation is performed on the coordinates of each epoch), is that the property of linear coordinate variation is not preserved from a strictly theoretical point of view. In fact this linear model of reference epoch coordinates and constant velocities, so widely used (e.g. in the formulation of the ITRF) has no solid theoretical foundation. Despite common belief, there exists no such thing as an intrinsic (coordinate-free) linear in time deformation of the network! From a practical point of view though we may limit ourselves to small displacements and rotations, in which case we can use instead of (63) the approximation $\theta(t) \approx \theta' t$ (with $\theta_0 = 0$), where

$$\theta' \equiv \frac{d\theta}{dt} \approx \frac{h_0}{S_0^2}. \tag{71}$$

with the same degree of approximation the displacements are $d_X(t) = t\, d'_X$, $d_Y(t) = t\, d'_Y$, with constant displacement velocities

$$d'_X \equiv \frac{d(d_X)}{dt} \approx -m_U - \theta' m_{0Y}$$

$$d'_Y \equiv \frac{d(d_Y)}{dt} \approx -m_V + \theta' m_{0X} \tag{72}$$

The values of the time derivatives θ', d'_X, d'_Y characterize the motion of the subregion as a whole, as represented by its own Tisserand frame. The computed values for the two regions in the left (L) of the fault and in the right (R) of the fault as well as for the whole network (L+R) are presented in the following table

	d'_X (m/y)	d'_Y (m/y)	θ' ("/y)
$L+R$	0.0369	0.0067	–0.64
L	0.0349	0.0102	–0.74
R	0.0285	0.0353	–0.48
$R–L$	–0.0064	0.0251	0. 26

Table 1: Displacement components and angular velocity of network regions and subregions

Of particular importance are the first row (L+R) which represents the motion of the whole region with respect to the global network, as well as the last one (R–L), which represents the relative motion of the right subregion with respect to the left one. The network of the whole region moves in a north-east direction with a velocity of 3.75 cm/year (eastern component 3.7 cm/year - northern component 0.7 cm/year) and rotates around its own mass center clock-wise with angular velocity of 0.64"/year. Considering Tisserand system of the left region as constant, the network of the right region moves in a north-west direction with a velocity of 2.6 cm/year (western component 0.6 cm/year - northern component 2.5 cm/year) and rotates around its own mass center counter-clock-wise with angular velocity of 0.26"/year.

After the removal of the trend due to the definition of the reference frame the velocities of all stations, (each one referring to the Tisserand frame of its own subregion) are depicted in fig. 1.

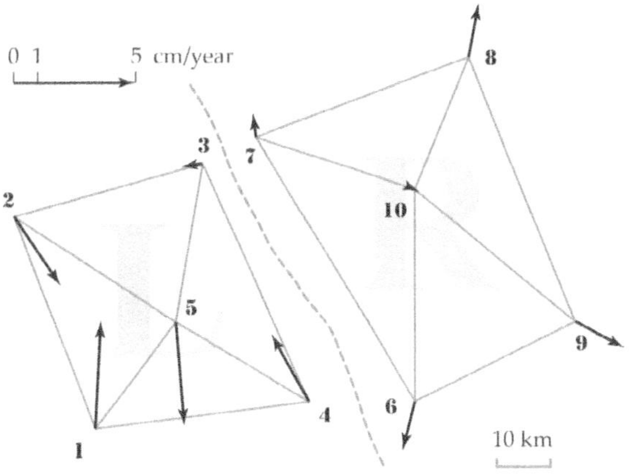

Fig. 1: The network of permanent GPS stations and the remaining velocities after the removal of trend due to the reference system definition

pt	X_0 (m)	Y_0 (m)	U (mm/y)	V (mm/y)
1	–11999.9176	–23000.0431	2.3	45.1
2	–27000.0610	16999.9033	19.2	–28.8
3	7999.9034	27000.0286	3.4	–43.2
4	28000.0646	–17999.8997	–15.7	28.3
5	3000.0107	–2999.9891	–9.2	–1.4
6	–1999.9211	–34000.0047	–5.2	–20.5
7	–32000.0370	15999.9259	–1.4	10.2
8	7999.9281	31000.0184	4.1	22.5
9	28000.0442	–18999.9351	20.5	–11.2
10	–2000.0141	5999.9955	2.0	–0.9

Table 2: Estimated reference-epoch coordinates and velocity components

point	σ_{X_0} (mm)	σ_{Y_0} (mm)	σ_U (mm/y)	σ_V (mm/y)
1	0.12	0.16	0.20	0.28
2	0.11	0.12	0.19	0.20
3	0.17	0.16	0.29	0.27
4	0.13	0.17	0.23	0.29
5	0.14	0.11	0.25	0.20
6	0.17	0.11	0.30	0.20
7	0.12	0.18	0.20	0.31
8	0.17	0.13	0.29	0.23
9	0.18	0.12	0.31	0.20
10	0.16	0.13	0.28	0.22

Table 3: Standard deviations of the estimated reference epoch coordinates and velocity components

The results of the finite element deformation analysis are listed in tables 4 and 5 and are depicted in figures 2 and 3. Figure 2 depicts the values of $1+k(\lambda_i-1)$, as the semi-axes of an ellipse compared to a unit circle. The semi-axes are oriented according to their computed directions (angle θ_P and θ_P+90°). The exaggeration factor is $k=0.25\times10^6$. Figure 3 depicts the direction of maximal shear by an oriented (angle ϕ) line segment having $\gamma\times10^{-6}$ as its length. Dilatation Δ is depicted by a circle with radius $R=\sqrt{1+k\Delta}$ compared to a unit circle, with an exaggeration factor of $k=0.25\times10^6$.

The standard deviations and correlations of the deformation parameters are given in tables 6, 7 and 8. In tables 7 and 8, the upper triangular part contains the correlations between λ_1, λ_2 and θ_P, while the lower triangular part contains the correlation between γ, Δ and ϕ.

triangle	$(\lambda_1-1)\times10^6$	$(\lambda_2-1)\times10^6$	θ_P (°)
1-5-2	-0.862	-2.031	8.313
2-5-3	-0.526	-1.465	-16.303
3-5-4	0.083	-1.536	-13.623
4-5-1	-0.395	-2.240	6.335
6-10-7	0.841	0.489	2.178
7-10-8	1.010	0.659	66.990
8-10-9	0.838	0.482	-72.339
9-10-6	0.813	0.444	-20.763

Table 4: Principal extensions and their direction

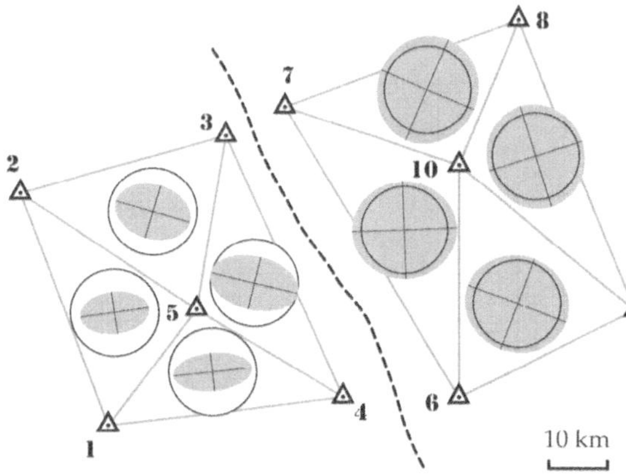

Fig. 2: Principal extensions and their directions.

triangle	$\gamma\times10^6$	$\Delta\times10^6$	ϕ (°)
1-5-2	1.169	-2.893	53.313
2-5-3	0.939	-1.991	61.303
3-5-4	1.619	-1.454	58.623
4-5-1	1.846	-2.635	51.335
6-10-7	0.351	1.330	47.178
7-10-8	0.351	1.669	111.990
8-10-9	0.355	1.320	117.339
9-10-6	0.370	1.257	65.763

Table 5: Dilatation, maximum shear and its direction

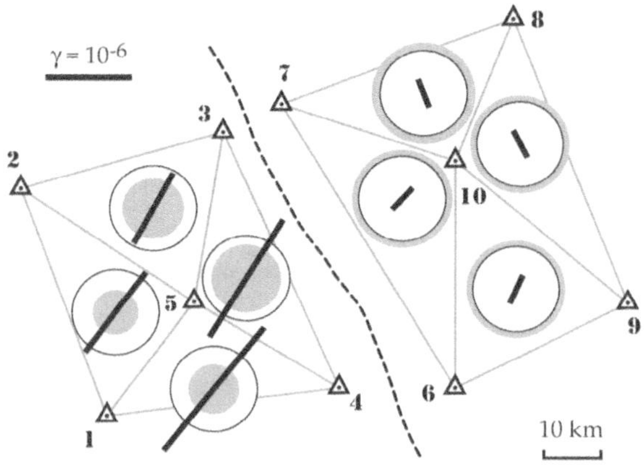

Fig. 3: Dilatation, maximum shear and its direction

triang.	σ_{λ_1} ($\times10^6$)	σ_{λ_2} ($\times10^6$)	σ_{θ_P} (deg)	σ_γ ($\times10^6$)	σ_Δ ($\times10^6$)	σ_ϕ (deg)
1-5-2	0.012	0.010	0.360	0.015	0.017	0.360
2-5-3	0.009	0.008	0.412	0.012	0.013	0.412
3-5-4	0.016	0.006	0.249	0.018	0.017	0.249
4-5-1	0.006	0.013	0.265	0.015	0.015	0.265
6-10-7	0.012	0.007	1.278	0.014	0.014	1.278
7-10-8	0.013	0.007	1.150	0.016	0.013	1.150
8-10-9	0.009	0.013	1.203	0.016	0.015	1.203
9-10-6	0.012	0.007	0.964	0.014	0.014	0.964

Table 6: Standard deviations of the deformation parameters

		1 - 5 - 2			2 - 5 - 3			3 - 5 - 4			4 - 5 - 1				
		λ_1	λ_2	θ_P	λ_1	λ_2	θ_P	λ_1	λ_2	θ_P	λ_1	λ_2	θ		
1	γ		0.14	0.59	0.36	0.22	-0.64	-0.72	0.07	0.03	-0.28	0.21	0.78	λ_1	1
5	Δ	0.16		0.63	-0.12	-0.15	-0.12	-0.13	-0.06	-0.23	0.25	0.85	0.45	λ_2	5
2	ϕ	0.03	0.80		0.08	-0.25	-0.31	-0.54	-0.06	-0.29	0.04	0.75	0.70	θ_P	2
2	γ	0.09	0.09	0.24		0.08	0.19	0.15	0.24	0.30	-0.02	0.04	0.02	λ_1	2
5	Δ	0.46	0.17	-0.10	0.08		0.28	-0.02	0.75	0.83	-0.14	-0.41	0.30	λ_2	5
3	ϕ	-0.44	-0.53	-0.31	-0.05	0.31		0.69	0.54	0.54	0.26	-0.18	-0.43	θ_P	3
3	γ	-0.49	-0.55	-0.47	0.24	-0.15	0.44		-0.04	0.29	0.68	-0.34	-0.54	λ_1	3
5	Δ	-0.43	-0.55	-0.53	-0.01	0.33	0.85	0.73		0.57	-0.21	-0.04	-0.00	λ_2	5
4	ϕ	0.18	-0.12	-0.29	-0.36	0.75	0.54	0.06	0.49		-0.03	-0.52	0.22	θ_P	4
4	γ	0.22	-0.63	-0.67	-0.26	0.18	0.28	0.56	0.54	0.46		0.01	-0.03	λ_1	4
5	Δ	-0.55	0.59	0.70	0.33	-0.26	-0.06	0.02	-0.07	-0.48	-0.67		0.26	λ_2	5
1	ϕ	0.32	0.82	0.70	-0.20	0.21	-0.43	-0.49	-0.51	0.22	-0.25	0.22		θ_P	1
		γ	Δ	ϕ	γ	Δ	ϕ	γ	Δ	ϕ	γ	Δ	ϕ		
		1 - 5 - 2			2 - 5 - 3			3 - 5 - 4			4 - 5 - 1				

Table 7: Correlations between deformation parameters for the left sub-region

		6 - 10 - 7			7 - 10 - 8			8 - 10 - 9			9 - 10 - 1				
		λ_1	λ_2	θ_P	λ_1	λ_2	θ_P	λ_1	λ_2	θ_P	λ_1	λ_2	θ_P		
6	γ		0.01	0.60	0.39	0.28	0.77	-0.14	-0.71	0.60	0.13	-0.43	0.53	λ_1	6
10	Δ	0.49		0.52	-0.52	0.02	0.19	-0.49	0.23	0.16	0.39	0.72	0.43	λ_2	10
7	ϕ	0.25	0.78		-0.35	0.39	0.45	-0.39	-0.22	0.44	0.35	0.13	0.44	θ_P	7
7	γ	0.39	-0.05	-0.46		-0.20	0.10	0.57	-0.61	0.14	-0.03	-0.73	0.24	λ_1	7
10	Δ	0.70	0.20	-0.13	0.55		-0.20	0.36	0.10	-0.28	-0.05	0.01	-0.02	λ_2	10
8	ϕ	0.57	0.76	0.45	0.17	-0.01		-0.68	-0.55	0.76	0.07	-0.13	0.44	θ_P	8
8	γ	0.65	0.18	-0.05	0.60	0.85	0.04		-0.05	-0.47	-0.12	-0.40	-0.15	λ_1	8
10	Δ	-0.53	-0.63	-0.42	-0.26	-0.00	-0.86	-0.31		-0.57	0.29	0.66	-0.66	λ_2	10
9	ϕ	0.44	0.60	0.44	0.24	-0.02	0.76	0.18	-0.76		-0.20	0.04	0.23	θ_P	9
9	γ	0.29	0.27	0.25	0.29	0.30	0.13	0.11	-0.00	-0.20		0.03	0.21	λ_1	9
10	Δ	-0.44	0.26	0.37	-0.30	-0.39	-0.01	-0.63	0.29	-0.15	0.51		-0.27	λ_2	10
1	ϕ	0.24	0.67	0.44	0.21	0.22	0.44	0.43	-0.63	0.23	0.32	0.05		θ_P	1
		γ	Δ	ϕ	γ	Δ	ϕ	γ	Δ	ϕ	γ	Δ	ϕ		
		6 - 10 - 7			7 - 10 - 8			8 - 10 - 9			9 - 10 - 1				

Table 8: Correlations between deformation parameters for the right sub-region

Conclusions

Despite some simplified assumptions on the variance-covariance structure of the data, a consistent and rigorous approach has been formalized to infer network deformations from continuous time series of coordinates. The proposed approach allows also the evaluation of both the accuracies and the correlations of the estimated parameters.

The numerical example shows that by the proposed empirical approach, a realistic estimate of the network covariance matrix is possible. The Tisserand approach, applied to the linear trends estimated from the original coordinate series, provides the estimate of the global motion of a region (reference frame effects) as well as the relative motion of two regions (relative tectonic effects).

The estimated standard deviations and correlations of the deformation parameters provide useful information in the deformation analysis; for example, in the case study, all the direction parameters of network L are characterized by smaller standard deviations than the corresponding of network R: this is due to the nature of the deformations in the R region, that are practically homogeneous.

With regard to the principal extension parameters, the smaller correlations within each triangle are between λ_1 and λ_2; with regard to dilatation, maximum shear and the related angle, the same holds for γ and Δ. Generally, adjacent triangles show the highest correlations.

All the algorithms have been implemented in a software package, written in standard C language. After some minor refinements it will become available to the scientific community.

Acknowledgements

L. Biagi's work has been supported by the "Satellite Positioning Services for the e-government" Italian PRIN 2004 project.

References

Dermanis, A. & E. Livieratos (1983): Applications of Deformation Analysis in Geodesy and Geodynamics. *Reviews of Geophysics and Space Physics*, vol. 51, no. 1, 41-50.

Dermanis, A. and D. Rossikopoulos (1988): Modeling Alternatives in Four-Dimensional Geodesy. Proceedings of the International Symposium "*Instrumentation, Theory and Analysis for Integrated Geodesy*", Sopron, Hungary, May 16-20, 1988, Vol. 2, 115-145.

Dermanis, A. (1994): A method for the determination of crustal deformation parameters and their accuracy from distances. *Journal of the Geodetic Society of Japan*, vol. 40, no. 1, 17-32.

Rossikopoulos, D. (2001): Modeling Alternatives in Deformation Measurements. In: A. Carosio & H. Kutterer (eds.) "*First International Symposium on Robust Statistics and Fuzzy Techniques in Geodesy and GIS*", ETH Zurich, Inst. of Geodesy and Photogrametry, No. 295.

Dermanis, A and C. Kotsakis (2005): Estimating crustal deformation parameters from geodetic data: Review of existing methodologies, open problems and new challenges. In these proceedings.

High Precision GPS Deformation Monitoring using Single Receiver Carrier Phase Data

Noor Raziq, Philip Collier
Department of Geomatics,
The University of Melbourne, VIC 3010, Australia

Abstract. The use of GPS in single receiver mode may offer significant potential for deformation monitoring. Single receiver positioning eliminates the need for base stations located on nearby stable points. Single receiver positioning offers logistical, technical and economic advantages and may be the only viable option in some monitoring applications. Not only may single receiver positioning be applicable for the detection of structural deformation, it may also provide a means of monitoring the long and short-term stability of Continuously Operating Reference Stations (CORS). There is a proliferation of CORS networks around the world, but users need to be assured that these stations are stable if they are to be used with confidence for high precision applications.

In conventional differential carrier phase based GPS positioning, advantage is taken of the fact that many of the error sources are spatially correlated. The formation of between-receiver and between satellite differences helps to eliminate or at least reduce the impact of these errors. However, many of these errors also exhibit a strong temporal correlation that can likewise be exploited to minimise their impact. This paper presents results of temporal correlation analysis conducted on various sources of errors that affect single receiver positioning after a discussion of different types of GPS errors. These results will form the basis of a new GPS processing strategy that uses the temporal correlation between GPS errors to achieve high accuracy GPS positioning using a single receiver.

Keywords. Precise Point Positioning (PPP), Deformation Monitoring,,

1 Introduction

The use of the Global Positioning System in single receiver mode for high precision applications (often referred to as Precise Point Positioning [PPP]) is a key area of interest for GPS researchers (Zumberge et al. 1997; Witchayangkoon and Segantine 1999; Kouba and Héroux 2001; Gao and Shen 2002; Beran et al. 2003). To achieve accuracy comparable to that achievable in a relative mode is an exciting prospect for some deformation monitoring applications where the base lines are long thus making positions determined using relative positioning less accurate and reliable. The success of GPS relative positioning lies in the fact that two nearby receivers are affected by similar observational errors. In this situation, most of these spatially correlated errors can either be eliminated or at least minimized.

In the case of single receiver positioning it is no longer possible to exploit the spatial correlation of GPS errors. In addition, a number of other error sources, normally ignored in relative positioning, must be considered in the single receiver case. Examples include satellite phase centre offset, phase windup error, relativistic effects, solid earth tides, ocean loading and polar tides.

One way of dealing with errors that impact on single receiver positioning is to utilise independent data from external sources to model and remove their influence. For example, the International GPS Service [IGS] (2004) provides accurate satellite ephemeris, clock and polar motion parameters. The influence of earth tides can be computed using models presented in IERS Conventions (McCarthy and Petit 2003) and ocean loading can be accounted for by accessing information from research groups such as Onsala Space Observatory (2004). Since most of these errors have a predominant sub-daily period, the use of external data for modelling purposes is most effective when observing sessions extend over at least one full day, thereby allowing daily averages to make up for any shortcomings in the model data. PPP computations using the Gypsy/Oasis II software or the Bernese software, with 24 hour data sets and external data to model the major errors, typically achieve position solutions accurate to a few centimetres (Kouba

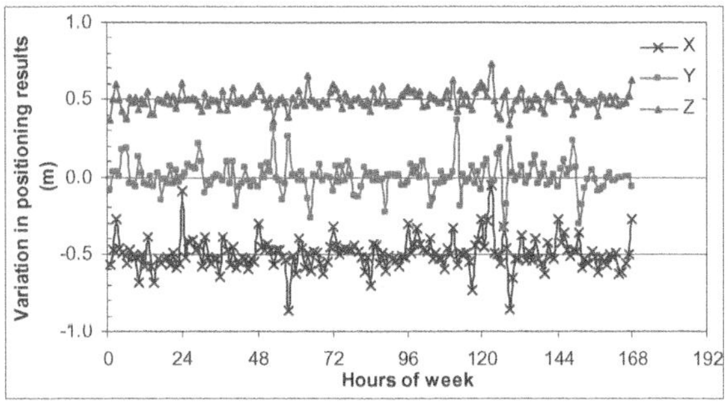

Figure 1. Variation in Bernese PPP hourly solutions for Melbourne, RMIT CORS station (For clarity results along X and Z axes are offset by +0.5 and -0.5 m respectively)

2003). However, when shorter duration datasets are used or epoch-by-epoch solutions are required, the use of external data for error modelling is much less successful. Position solution quality in such cases deteriorates to the decimetre level if not the metre level as shown in Figure 1.

An alternative to the use of external data to model errors in the GPS point positioning solution is to utilise the fact that many GPS errors are temporally correlated. In the same way as spatial error correlation increases as receiver separation decreases, temporal correlation between errors tends to increase as the time between observations decreases. Therefore, taking differences between successive observations between a single receiver and a satellite, particularly if the observations are separated by a short time interval, will eliminate or minimise the influence of many errors. Using this technique GPS errors will be minimized utilizing a limited amount of GPS data. High precision single receiver GPS positions will thus be available on a sub-daily basis compared to currently available daily results. These highly precise GPS positions will improve the understanding of geodetic deformation on a sub-daily basis.

The objective of this paper is to evaluate the effectiveness of temporal correlation in the minimisation of the influence of a number of GPS errors for the purposes of single receiver positioning. A brief introduction of GPS error sources will precede the discussion and analysis of temporal correlation.

2 GPS Errors

In relative GPS positioning, the main systematic errors are satellite clock and orbit errors, ionospheric and tropospheric refraction, receiver clock errors, phase centre variation and receiver multipath. In addition, errors that emerge as significant for single receiver positioning can be classified as satellite attitude effects, relativistic effects and site displacement effects. These errors will now be discussed in some detail.

2.1 Satellite Attitude Effects

In an Earth-Centred, Earth-Fixed frame of reference satellites move with respect to a receiver on surface of the Earth. This relative motion has two distinct affects on GPS positioning.

2.1.1 *Satellite Antenna Phase Centre Variation*

The phase centre of a GPS satellite antenna is offset relative to the satellite's centre of mass. All GPS measurements and the broadcast ephemeris refer to the antenna phase centre. The IGS precise ephemeris, on the other hand, refers to the centre of mass of a satellite. This offset can be incorporated into GPS processing algorithms. However, in reality, the orientation of the offset vector varies due to the motion of the solar panels as they orient themselves to face the Sun. Variation in the direction of the offset vector is a function of the position of the satellite and the Sun in an Earth-centred, Earth-fixed frame of reference and can be modelled mathematically.

In a satellite centred frame of reference, as shown in Figure 2, the offset between the satellite's centre of mass and the antenna phase centre can be assumed to be constant. The IGS adopts values of 0.279 m and 1.023 m along X and Z axes respectively for Block II and Block IIA satellites, and an offset of zero for Block IIR satellites. For compatibility reasons, when using IGS precise products, these offsets must be applied.

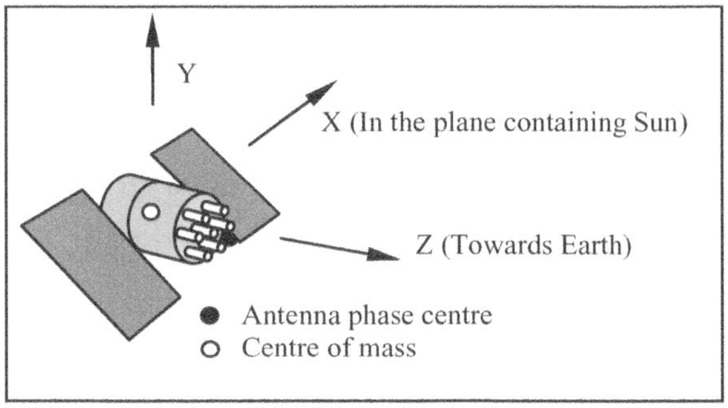

Figure 2: Antenna phase centre offset for GPS satellites

2.1.2 *Phase Windup Error*

GPS signals are right-handed circularly polarized and may be visualized as a rotating electric field. At a given instant, the measured carrier phase at the receiver is the geometrical difference between the electric field at the receiver antenna and a reference direction of the antenna. A change in the orientation of receiver antenna means the relocation of the reference direction and hence a change in the measured range. Similarly, a change in the orientation of the transmitter antenna, changes the direction of electric field, which subsequently affects the direction of the electric field at the receiving antenna. The result of a change in orientation of either antenna therefore affects the measured range. This effect is commonly known as *phase windup error*.

Phase wind up can cause a maximum error of one cycle in carrier phase observations. The details on modelling this error can be found in papers such as (Wu et al. 1993).

2.2 Relativistic Effects

Relativity affects satellite orbits, satellite signals, the satellite clock and the receiver clock (Hofmann-Wellenhof et al. 2001). The effect on satellite orbits takes the form of perturbations in the order of 3 x 10^{-10} m/sec^2 (Zhu and Groten 1998). This effect is less than one third of other perturbation effects such as the earth's tidal potential (1x 10^{-9} m/sec^2) and can be safely neglected.

The effects of relativity on GPS signals, satellite clock and receiver clock, however, can be considerable. These effects are further explained below.

2.2.1 *Relativistic Effects on GPS Signals*

The speed of a GPS signal deviates from its nominal value of c (the speed of light in vacuum), due to the effect of the Earth's gravitational field. A propagation correction is required to correct for this change, when calculating the distance between a GPS satellite and a receiver.

This range correction can affect relative positioning by 0.001ppm (Zhu and Groten 1998). The effect, however, is more pronounced in point positioning where the value of the range correction can be several centimetres.

2.2.2 *Relativistic Effects on Satellite and Receiver Clocks*

Relativity affects satellite clocks in two ways. The major effect is in the form of a constant clock offset. This offset is incorporated into satellite clocks by slightly tuning the frequency of onboard clocks at the time of production and is known as "the factory offset". The frequency of onboard satellite clocks is reduced by multiplying the nominal frequency by a factor of 446.47 x 10^{-12}.

However, GPS orbits are not perfectly circular as they are assumed to be for computing this offset. Due to orbit eccentricity, this frequency offset is not constant and can be considered to consist of a constant and a variable part. The variable part of this effect can be termed as a residual relativistic effect for convenience. The value and variation of the residual relativistic effect is a function of orbit eccentricity. Residual relativistic effects cancel out in relative positioning as part of the satellite clock error but must be considered for single receiver positioning.

GPS time is realized in an Earth-centred Earth-fixed frame of reference. Due to the rotation of the Earth, the receiver clock, at the surface of Earth, is in motion with respect to the Earth's centre of mass. The relativistic effect caused by this rotation of the receiver clock is known as *the Sagnac Effect*. The magnitude of the Sagnac Effect depends upon the velocity of the GPS receiver. The Sagnac Effect is important for both relative and single receiver positioning. The relevant clock correction is usually applied by the GPS receiver and can subsequently be ignored.

2.3 Site Displacement Effects

The surface of the Earth experiences different types of movement. These movements can be regular when different strata move along fault lines on a local scale or along tectonic boundaries on a continental scale. They can also be abrupt in the form of earthquakes and other major seismic events. Both forms of movement are irreversible, and hence can be called deformations. There are, however, other types of movement that have a periodic nature and are reversible. These movements include tidal movements due to the attraction of the Moon, Sun and other celestial bodies, movements due to the loading effect of the atmosphere and oceans and movements due to variation in the rotation of the Earth.

In differential carrier phase positioning, over short baselines, the effect of these periodic movements tends to cancel. For long baselines and high precision applications, however, these movements must be taken into account. For single receiver positioning, their effect can reach up to several decimetres and requires careful consideration.

3 Temporal Correlation of GPS Errors

All systematic errors in GPS positioning are correlated in time to some degree. This correlation, if properly used, can lead to the minimisation of the impact of these errors.

One way of analysing a time series for temporal correlation is to form an auto-correlation function. An auto-correlation function derived from a discrete time series quantifies the tendency for similarity between nearby data points. The stronger the tendency for nearby points to be similar in magnitude and sign, the stronger and more lasting is the correlation. If, on the other hand, nearby data points do not tend to be similar in magnitude and sign, the correlation is weak and the auto-correlation function tends quickly to a value of zero.

The shape of an auto-correlation function derived from a given time series gives useful information about temporal correlation within the series. A gradual drop of the auto-correlation curve indicates strong and sustained temporal correlation. A sudden drop of the curve, on the other hand, indicates minimal correlation in the time series.

Figure 3a shows a time series with evident correlation. The auto correlation function of this series is given in Figure 3b which shows a gradual drop to zero. Figure 4a, on the other hand, shows a time series with very low correlation. The auto-correlation function for this series is plotted in Figure 4b which drops to zero almost immediately, as expected.

The auto-correlation function $r_{xx}(k)$ of a time series X_t at a time lag k is computed using (2).

$$r_{xx}(k) = \frac{\sum_{t=k+1}^{T} \left\{ \left(X_t - \overline{X}\right)\left(X_{t-k} - \overline{X}\right) \right\}}{\sum_{t=1}^{T} \left(X_t - \overline{X}\right)^2} \quad \cdots\cdots\cdots(2)$$

where $\overline{X} = \frac{1}{T} \sum_{t=1}^{T} X_t$

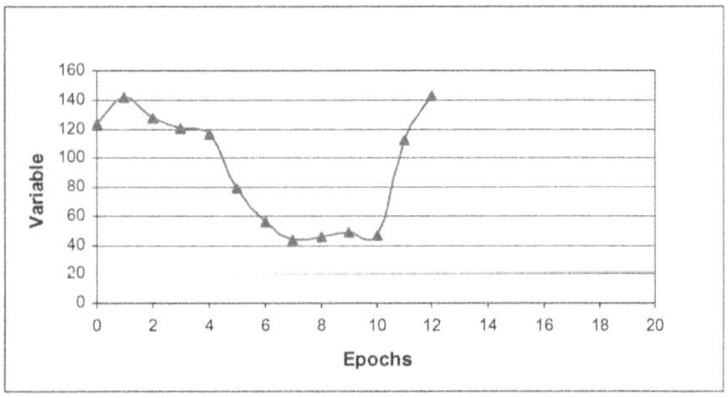

Figure 3a. A time series with certain degree of correlation

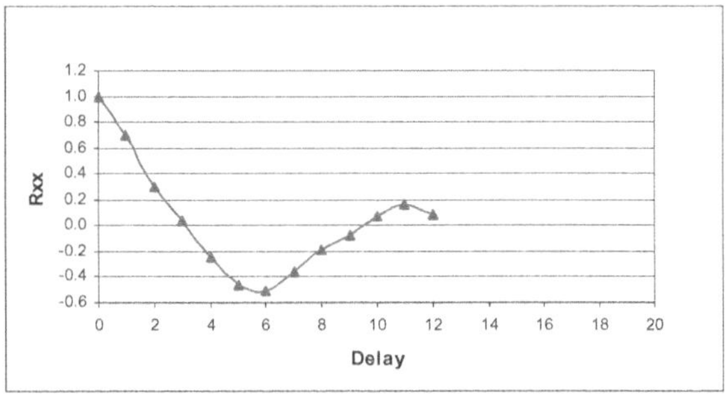

Figure 3b. Auto-correlation function of series in Figure 3a

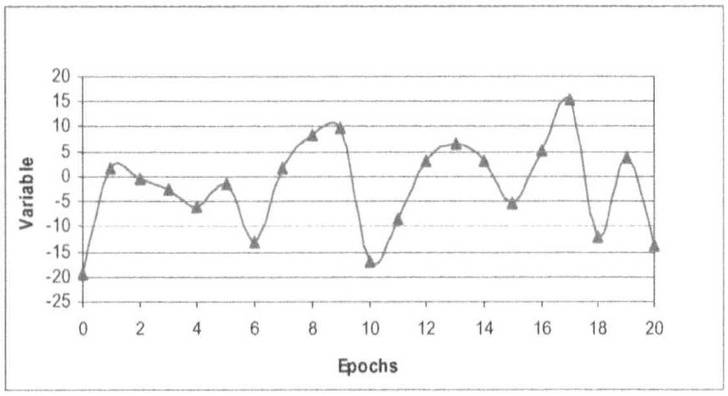

Figure 4a. A time series with no correlation

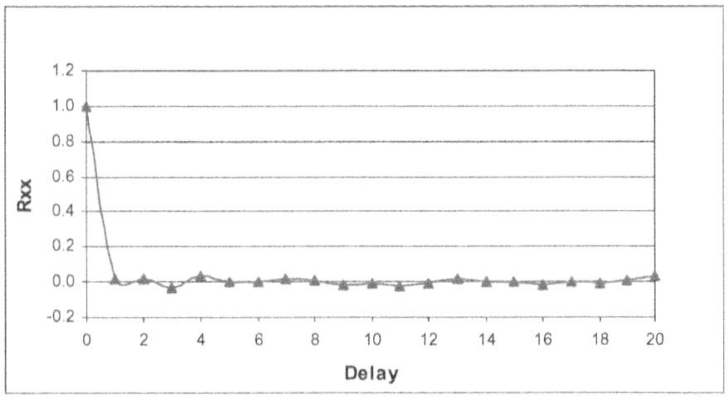

Figure 4b. Auto-correlation function of series in Figure 4a

For the purpose of studying temporal correlation of different GPS errors, it is convenient to classify these errors into three categories

1. Errors that tend to move the physical location of the point where GPS measurements are made e.g. receiver antenna phase centre offset and

variation, solid earth tides, polar tides, ocean loadings etc.

2. Errors affecting the position of satellites e.g. satellite orbit errors, satellite antenna offset variation.

3. Errors affecting the range between a GPS satellite and receiver e.g. satellite and receiver clock errors, ionospheric and tropospheric effects, residual relativistic effect, the Sagnac effect and the propagation range correction.

For errors in category 1, the possibility of modelling these errors normally exists. Their effects can, therefore, be largely accounted for in the computed positions and they will, be ignored in the following discussion.
Errors in categories 2 and 3 will be analysed for temporal correlation in the subsequent sections. Time series of these errors are plotted followed by the corresponding auto-correlation functions. In analysing these results, strong temporal correlation, in an auto-correlation function indicates that time differencing between adjacent observations will effectively minimise the impact of the error under consideration. On the other hand, low temporal correlation demonstrates that time differencing will be largely ineffective as a means of error minimisation.

3.1 Satellite Orbit Error

To study the temporal correlation of satellite orbit errors, ephemeris data from the International GPS service (2004) for different GPS satellites was used. The IGS final ephemeris is considered as truth and the difference between IGS ultra-rapid predicted orbits and IGS final orbits along WGS84 coordinate axes is plotted against time for one full day in Figure 5a. The epoch interval of GPS data used is 15 minutes. The corresponding auto-correlation function in Figure 5b shows strong temporal correlation, even though the epoch interval is fairly large. Shorter epoch intervals will most likely result in stronger temporal correlation.

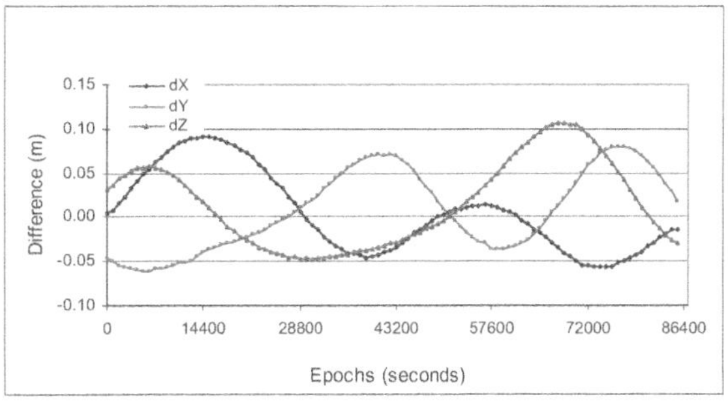

Figure 5a. A 24 hours time series of satellite orbit error

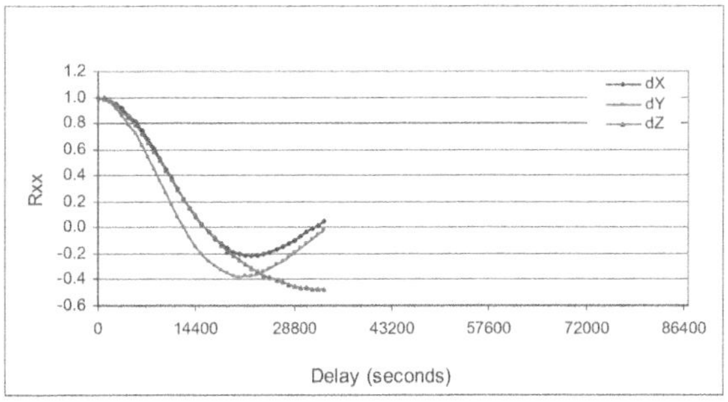

Figure 5b. Auto-correlation function of satellite orbit error

3.2 Satellite Antenna Phase Centre Variation

The IGS precise ephemeris refers to the centre of mass of a GPS satellite. All GPS measurements and broadcast orbits, on the other hand, refer to the phase centre of the satellite antenna. Some agencies, for example the National Geo-spatial Intelligence Agency [NGA] (2005), provide precise orbit results referred to both the satellite centre of mass and antenna phase centre.

GPS orbits referred to both the satellite centre of mass and the antenna phase centre can be obtained from the NGA web service. Differences between X,

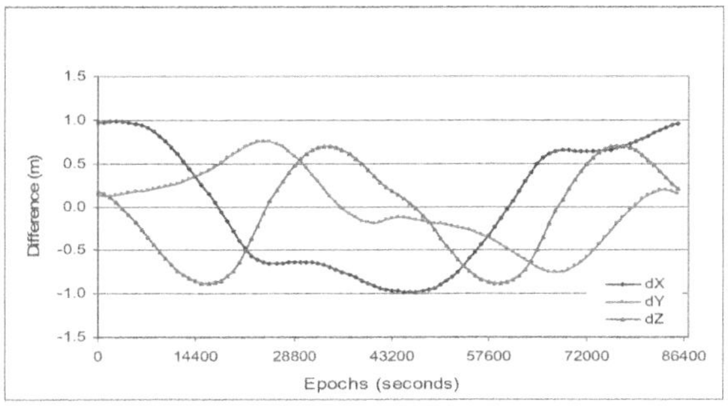

Figure 6a. Difference between orbits referred to the centre of mass and the antenna phase centre of a GPS satellite

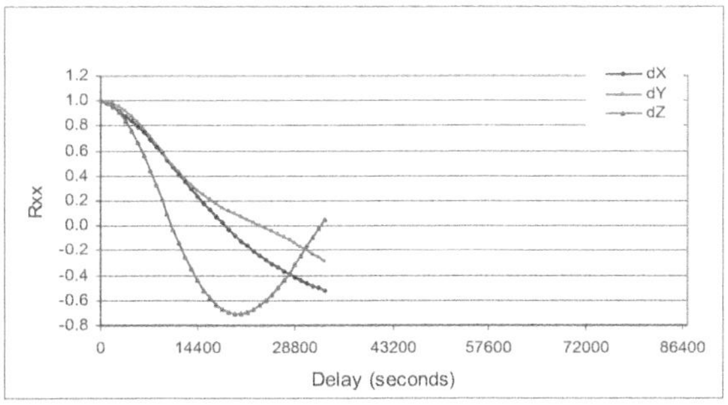

Figure 6b. Auto-correlation function of series in Figure 6a

Y and Z coordinates of GPS satellites referred to the centre of mass and the antenna phase centre for a 24 hour period are plotted in Figure 6a with an epoch interval of 15 minutes. Figure 6b shows the auto-correlation functions for these differences.

The correlation functions show a gradual drop indicating the presence of strong temporal correlation in the effect caused by satellite antenna phase centre variation.

3.3 Satellite Clock Error

GPS satellites carry very stable and precise atomic clocks. These clocks, however, can drift from GPS system time.

Satellite clock errors and their rates are part of the broadcast message as received by a GPS receiver. Research organisations such as IGS (International GPS Service 2004) also provide precisely computed satellite clock errors. Examples of satellite clock errors for two satellites are plotted in Figure 7a, using IGS data with an epoch interval of 15 minutes. The auto-correlation functions for these two series are given in Figure 7b.

The curves have a gradual slope which indicates strong temporal correlation in satellite clock error.

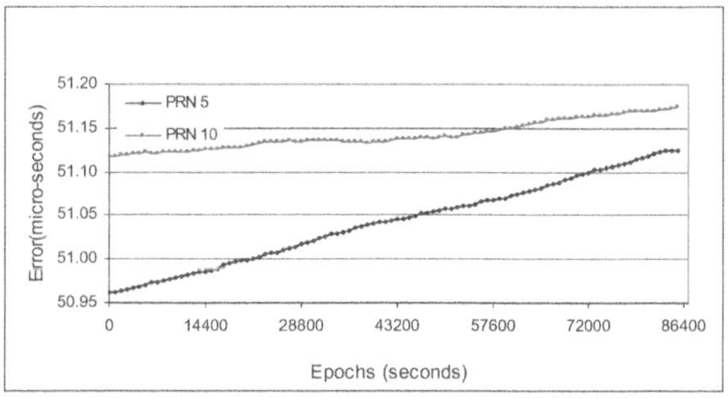

Figure 7a. Satellite clock error for two GPS satellites

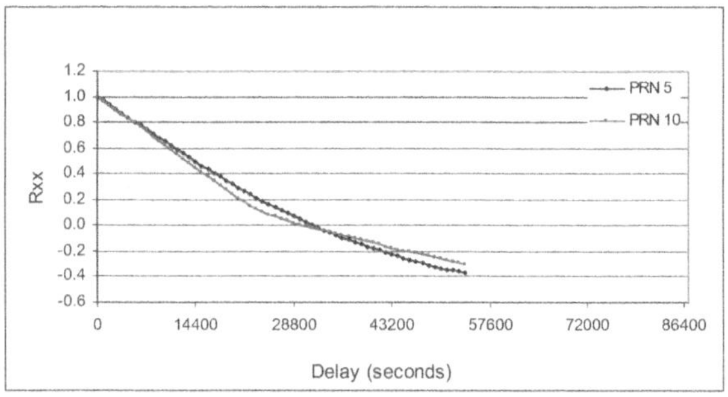

Figure 7b. Auto-correlation function of series in Figure 7a

3.4 Ionospheric Effect

The ionosphere is a dispersive medium for GPS signals and increases the phase velocity. The result is a less than actual range measured between a GPS satellite and a receiver.

This effect of the ionosphere is the result of free electrons produced due to ultra-violet rays from the Sun. A measure of free electron content can thus be used to predict the effect of the ionosphere on GPS signals. Different ionospheric models such as the Klobuchar model (Hofmann-Wellenhof et al. 2001) determine the ionospheric effect from total electron content.

Ionospheric modelling is the focus of research in different agencies. The Centre for Orbit Determination in Europe [CODE] (2004) is one such organisation providing Vertical Total Electron Content (VTEC). VTEC for a location near the city of Melbourne, Australia is plotted in Figure 8a for a period of six days, using the final ionosphere products from CODE. The sampling interval of the data is 2 hours. Figure 8b shows the auto-correlation functions for these time series. The plots indicate that a very strong temporal correlation exists for the effect of ionosphere.

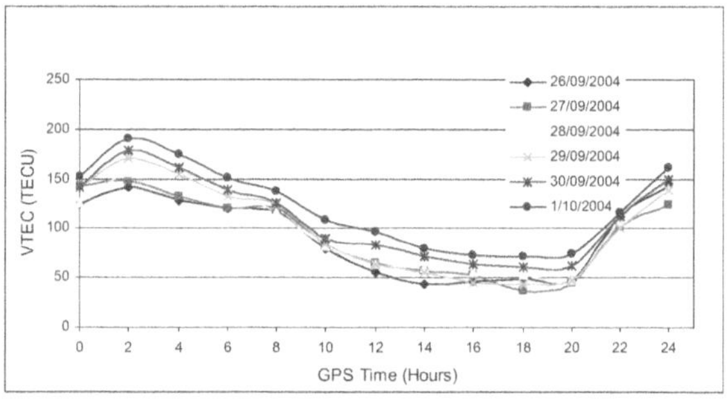

Figure 8a. 24 hour VTEC variation at Lat -37.5 Lon 145

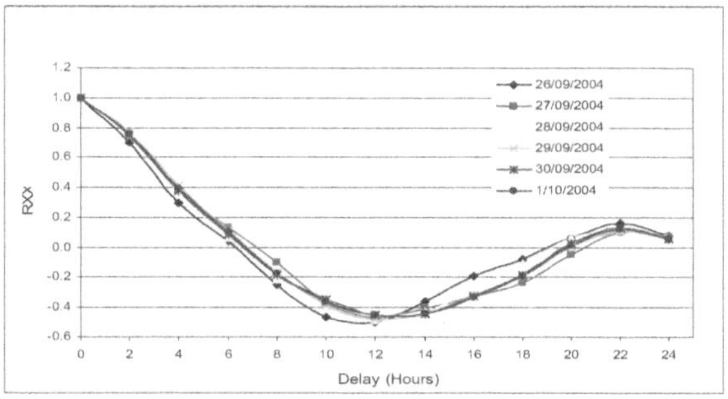

Figure 8b. Auto-correlation function of series in Figure 8a

3.5 Tropospheric Effect

The troposphere is a non-dispersive medium that delays GPS carrier phase signals. The range measured between GPS receiver and satellite is thus longer than the actual range.

The effect of the troposphere consists of a dry and a wet part and is a function of temperature, pressure and humidity of the atmosphere. Efforts have been made to model the effect of the troposphere and finding a suitable model is still the focus of much research.

The Saastamoinen model (Hofmann-Wellenhof et al. 2001) as given in (3) is used below for assessing the temporal correlation of the tropospheric error.

$$\Delta^{Trop} = \frac{0.002277}{\cos z}\left[p + \left(\frac{1255}{T} + 0.05\right)e - \tan^2 z\right] \ldots (3)$$

In Equation (3), Δ^{Trop} is the tropospheric delay in metres, z is the zenith angle of the GPS satellite, p and e are atmospheric pressure and partial water vapour pressure in millibars and T is temperature in Kelvin.

The Zenith Path Delay (ZPD), which is the effect of the troposphere when a GPS satellite is at the zenith, is calculated for Melbourne using the Saastamoinen model and hourly meteorological data from the Bureau of Meteorology, Australia.

ZPD for four separate days is plotted in Figure 9a with a sampling interval of 1 hour. The auto-correlation functions resulting from these time series are shown in Figure 9b and indicate that the temporal correlation is strong for the effect of the troposphere. It should be noted, though, that results are highly variable from day to day.

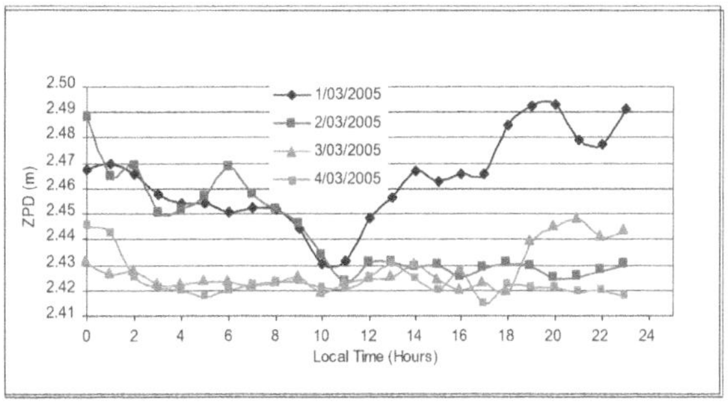

Figure 9a. 24 hour ZPD variation in Melbourne, Australia

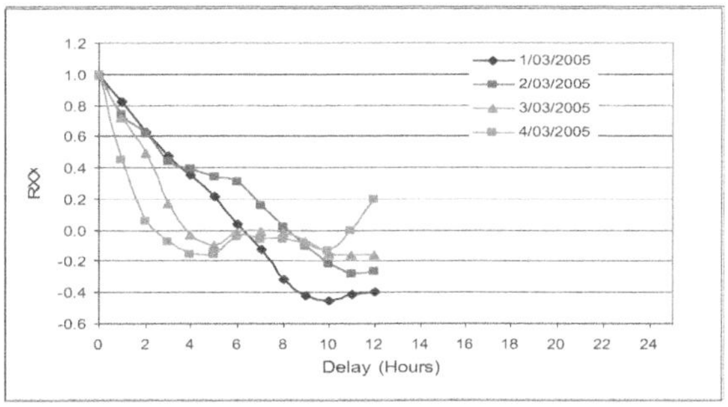

Figure 9b. Auto-correlation function of series in Figure 9a

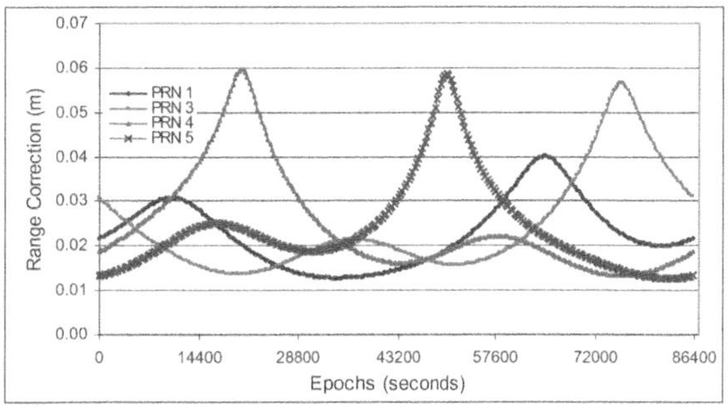

Figure 10a. Propagation correction due to relativity to signals from GPS satellites

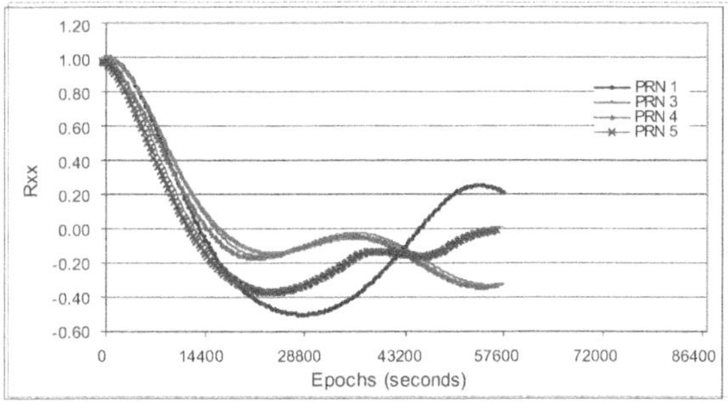

Figure 10b. Auto-correlation function of series in Figure 10a

3.6 Relativistic Effect on GPS Signals

The propagation correction due to relativistic effects δ^{rel} on the measured range between a GPS satellite and receiver is given by Equation (4) (Hofmann-Wellenhof et al. 2001).

$$\delta^{rel} = \frac{2\mu}{c^2} \ln \frac{\rho^j + \rho_i + \rho_i^j}{\rho^j + \rho_i - \rho_i^j} \ldots\ldots\ldots\ldots (4)$$

In Equation (4), μ is the Earth gravitational constant, ρ^j and ρ_i are the geocentric distances to the GPS satellite and receiver respectively, ρ_i^j is the distance between GPS satellite and receiver and c is the speed of light in a vacuum.

Figure 10a shows time series for the propagation correction calculated for four GPS satellites using (4). Figure 10b shows the auto-correlation functions of these series. A strong temporal correlation is evident from these functions.

3.7 Relativistic Effect on Satellite Clock

The residual effect of relativity due to GPS orbit eccentricity can be computed using equation (5) (Parkinson and Ashby 1996)

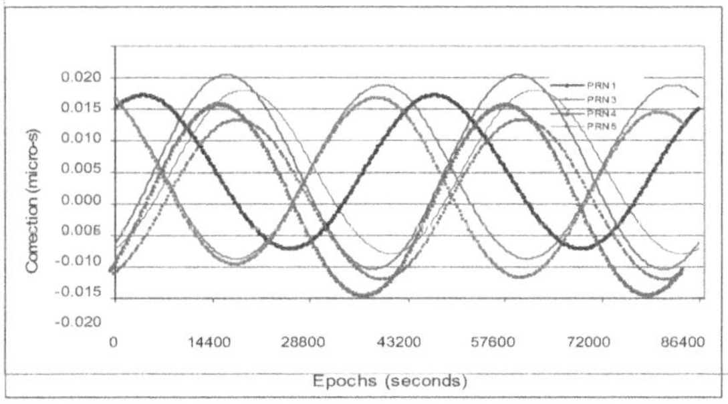

Figure 11a. The residual effect of relativity on satellite clock due to orbit eccentricity

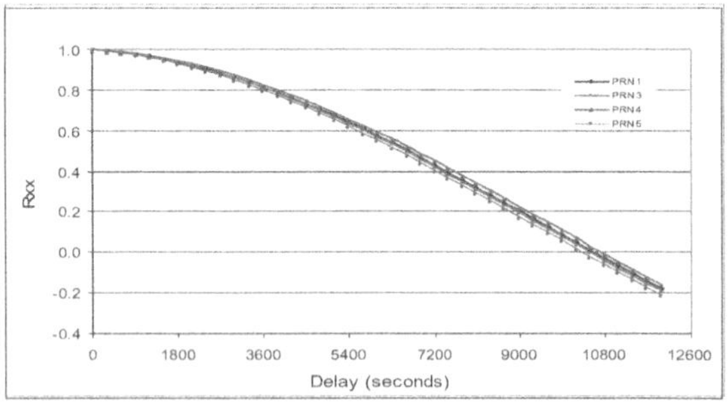

Figure 11b. Auto-correlation function of series in Figure 11a

$$\Delta t' = +\frac{2\vec{r}\cdot\vec{v}}{c^2} \quad (5)$$

In Equation (5), $\Delta t'$ is the relativity clock correction in seconds, and $\vec{r}$ and $\vec{v}$ are the position and velocity vectors of the GPS satellite relative to the Earth centre at the instant of transmission.

Figure 11a shows the variation of this effect over 24 hours for four GPS satellites. The auto-correlation function of these series is given in Figure 11b indicating a strong temporal correlation.

4 Conclusions and Future Work

The results in section 3 indicate that most systematic errors affecting GPS positioning are strongly correlated in time. For two sets of observations separated by a short duration, the effect of different sources of errors is nearly the same and hence can be eliminated using time differences between observations. Development of a GPS processing algorithm that will use temporal correlation for eliminating the effects of different error sources is the focus of future work. Using this algorithm, it would be possible to compute precise GPS positions irrespective of the base line length and duration of GPS data collection sessions. This would help to observe and interpret sub-daily geodetic deformation activities in an absolute frame of reference.

References

Beran, T., Kim, D., and Langley, R. B. "High-Precision Single-Frequency GPS Point Positioning." 16th International Technical Meeting of the Satellite Division of the U.S. Institute of Navigation, ION GPS / GNSS 2003, Portland, Oregon, USA.

Centre for Orbit Determination in Europe. (2004). "The AIUB IGS Page." [online], <http://www.aiub.unibe.ch/igs.html> (20 August 2004).

Gao, Y., and Shen, X. (2002). "A New Method Of Carrier Phase Based Precise Point Positioning." Navigation: Journal of the Institute of Navigation, 49(2).

Hofmann-Wellenhof, B., Lichtenegger, H., and Collins, J. (2001). Global Positioning System : Theory and Practice, Springer-Verlag, Wein.

McCarthy, D. D., and Petit, G. (2003). "IERS Technical Note #32 – IERS Conventions (2003)." U.S. Naval Observatory (USNO) Bureau International des Poids et Mesures (BIPM).

International GPS Service. (2004). "IGS Products." [online], <http://igscb.jpl.nasa.gov/components/prods.html>, (12 June 2004)

Kouba, J. (2003). "A Guide to using International GPS Service (IGS) Products." [online], <ftp://igscb.jpl.nasa.gov/igscb/resource/pubs/GuidetoUsingIGSProducts.pdf>, (15 August 2004).

Kouba, J., and Héroux, P. (2001). "Precise Point Positioning Using IGS Orbit and Clock Products." GPS Solutions, 5(2), 12-28.

National Geospatial-Intelligence Agency. (2005). "NGA GPS Satellite Antenna Phase Center Precise Ephemeris Page." [online], <http://earth-info.nga.mil/GandG/sathtml/APCexe.html> (12 January 2005)

Onsala Space Observatory. (2004). "Ocean Tide Loading Provider." [online], <http://www.oso.chalmers.se/~loading/> (15 August 2004).

Parkinson, B. W., and Ashby, N. (1996). "Introduction to Relativistic Effects on the Global Positioning System." The global positioning system : theory and applications: volume I, American Institute of Aeronautics and Astronautics, Washington, DC, 679-682.

Witchayangkoon, B., and Segantine, P. C. L. (1999). "Testing JPL's PPP Service." GPS Solutions, 3(1), 73-76.

Wu, J. T., Wu, S. C., Hajj, G. A., Bertiger, W. I., and Lichten, S. M. (1993). "Effects of antenna orientation on GPS carrier phase." Manuscripta Geodaetica, 18, 91-98.

Zhu, S. Y., and Groten, E. (1998). "Relativistic effects in GPS." Lecture Notes in Earth Sciences, S. Bhattacharji, G. M. Friedman, H. J. Neugebauer, and A. Seilacher, eds., Springer-Verlag Berlin, Heidelberg, New York, London, Paris, Tokyo, 41-46.

Zumberge, J. F., Heflin, M. B., Jefferson, D. C., Watkins, M. M., and Webb, F. H. (1997). "Precise point positioning for the efficient and robust analysis of GPS data from large networks." Journal of Geophysical Research, 102(B3), 5005-5018.

Automated Form Recognition of Laser Scanned Deformable Objects

Christian Hesse and Hansjörg Kutterer
Geodetic Institute,
University of Hannover, Nienburger Straße 1, 30167 Hannover, Germany

Abstract. The use of terrestrial laser scanners is a new and promising way for monitoring the deformations of artificial and natural objects. These new sensors offer distance measurements without artificial reflectors. The data acquisition rates range from 1000 Hz up to 600 kHz, the obtained accuracies are better than 1 cm. The result of such a scan is a highly dense point cloud which enables more precise object models.

Two main problems exist if dense point clouds from laser scanner measurements are used for deformation analysis: the huge data volume which has to be handled, and the lack of fully automated analysis methods. As in a typical deformation scenario the shape of the objects changes due to bending and flexing, a thorough but information preserving parameter reduction is needed by a set of a few characteristic parameters. This is often possible by describing the objects or parts of them using quadratic forms.

A statistically founded identification of the particular quadratic form is presented which solves this task automatically. Thus, a dedicated functional model for the adjustment of the point data can be set up. This procedure is illustrated using numerical examples which are based on synthetic point clouds as well as on real data which were observed with terrestrial laser scanners.

1 Terrestrial laser scanning

The use of laser scanning techniques for 3D point determination is well known from airborne surface acquisition where a oscillating laser beam is measuring the angle and the distance to the ground in a section orthogonal to the plane's flight direction. In engineering geodesy this type of sensors has been available since a few years. Up to now, terrestrial laser scanners have become important tools for solving tasks such as cultural heritage or as-built documentation or object monitoring.

Similar to airborne devices, terrestrial laser scanners use a laser beam to determine the slope distance s to an object's surface, either by time-of-flight or phase-shift method. Usually the laser beam is moved in a raster-shaped way over the object with two rotating or oscillating mirrors which are arranged perpendicularly to each other. The vertical angle φ and horizontal angle λ are also registered. Hence, terrestrial laser scanners are polar measuring devices. For a better handling, the polar coordinates are usually transformed into cartesian coordinates. The result of such a scan is a so called point cloud, the data volume of which can easily exceed 100 million points. Not only the geometric coordinates can be derived from measurements with terrestrial laser scanners but also the intensity i of the reflected laser signal what leads to the expression "4D laser scanning". As terrestrial laser scanners are able to acquire point clouds with a single point accuracy of better than 1 cm and a data acquisition rate of up to 500.000 points per second, they are getting more and more interesting for kinematic and quasi-kinematic monitoring tasks in engineering geodesy. The point spacing at the object can be up to a few millimetres, depending on the distance to the surface. This enables a representative continuous modelling of the object's surface.

The scanning speed varies depending on the range measurement method. In general, phase-shift scanners offer data rates of more than 100 kHz but have a limited measurement range of less than 70 m. Time-of-flight devices are able to realize maximum ranges of about 1 km with an acquisition rate between 1000 to 5000 points per second.

When scanning from multiple positions, homologous points are necessary for transforming scans into a common coordinate system. Such points can be realized either by flat targets with scanner-specific patterns or by spheres. Most scanner manufacturers prefer spheres because of their rotation-symmetric geometry.

Note that laser scanners produce a huge amount of data in a short time. High speed scanners deliver approximately 50 GByte of ASCII-encoded data per hour. The handling and especially the reduction of this volume without loosing relevant information is a challenging task. Here, the focus lies on parameter reduction by point cloud fitting using three-dimensional (3d) quadratic forms.

Hence, the paper is organized as follows. First, some basic ideas of point cloud fitting are introduced. Afterwards, the statistically founded, advanced procedure is explained which has been used for this study. The main part of the paper is concerned with the presentation and discussion of numerical results derived from synthetic and real world data.

2 Point cloud fitting

2.1 Conventional procedure

The need of reducing the complexity of point clouds by fitting simple mathematical forms is common to all applications of laser scanning. Therefore it is helpful that artificial objects of interest such as buildings are usually designed using computer software like CAD software. The shape of these design elements is clearly based on well-known mathematical models.

A typical fitting procedure starts with segmenting the cloud with each segment representing a simple geometric part of the object. The mathematical models for these parts are set up manually. Their parameters are estimated by means of a least-squares adjustment. However, it is usually ambiguous to define a proper mathematical form. Linear or quadratic forms can be used as well as higher degree forms. This study emphasizes the identification of quadratic forms in a highly automated way.

2.2 Quadratic forms

A 3d quadratic form or quadric, respectively, is given in matrix notation as

$$\boldsymbol{x}^T \boldsymbol{M} \boldsymbol{x} + \boldsymbol{m}^T \boldsymbol{x} + \alpha = 0 \qquad (1)$$

with the coordinate vector

$$\boldsymbol{x}^T = \begin{bmatrix} x_1 & x_2 & x_3 \end{bmatrix}, \qquad (2)$$

the symmetric coefficient matrix

$$\boldsymbol{M} = \begin{bmatrix} a_1 & a_4 & a_5 \\ a_4 & a_2 & a_6 \\ a_5 & a_6 & a_3 \end{bmatrix}, \qquad (3)$$

the coefficient vector

$$\boldsymbol{m} = \begin{bmatrix} a_7 \\ a_8 \\ a_9 \end{bmatrix} \qquad (4)$$

and the scalar

$$\alpha = a_{10}. \qquad (5)$$

Note that not all elements of $\boldsymbol{M}$ and $\boldsymbol{m}$ are zero.

A simple classification of the correct quadratic form can either be done by a principle axis transformation based on an eigenvalue decomposition (Anton 1994, Bronstein and Semendjajew 1991, Drixler 1993) or by interpreting certain second degree invariant parameters of a crossover substitution (Anton 1994, Kutterer and Schön 1999).

In total, there are 18 possible quadratic imaginary or real forms (Holz and Wille 2002). However, imaginary forms are not relevant for geodetic deformation analysis, since they do not occur at real objects. See Table 1 for the singular and regular quadratic forms of interest.

Table 1: Singular and regular quadratic forms

	Class	Form
Singular forms	1	Plane
	2	Two perpendicular planes
	3	Parabolic cylinder
	4	Elliptical cylinder
	5	Hyperbolic cylinder
	6	Cone
Regular forms	7	Hyperbolic paraboloid
	8	Elliptic paraboloid
	9	Sphere
	10	Rotation ellipsoid
	11	Ellipsoid
	12	Double walled hyperboloid
	13	Single walled hyperboloid

2.3 Advanced procedure

2.3.1 Introduction

The type of the respective quadratic form can be identified by analyzing the estimated form parameters. Afterwards, the likely model can be validated through statistical test decisions. These testing possibilities can be extended for use in deformation monitoring. The change of some form parameters can, e.g., be analyzed or compared with the shapes in CAD models or of different observed states in order to quantify some deviations.

The presented method of a statistically founded automatic shape recognition consists of four steps (see Fig. 1) which are explained in the following sub-sections.

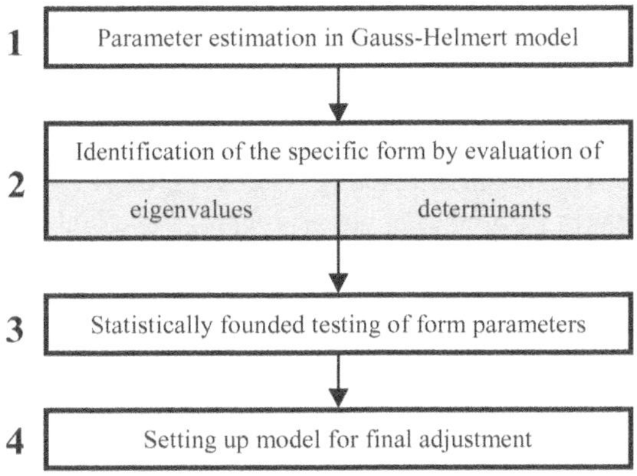

Fig. 1: Procedure for the identification of quadratic forms

The procedure starts with a parameter estimation in the Gauss-Helmert model. The identification of the specific quadratic form by evaluation of characteristic parameters in the second step can be realized in two ways, the eigenvalue and the determinants approach (Table 2). The eigenvalue approach seems to be straightforward. However, the respective test statistics (Step 3) are much simpler to derive for the determinant approach (Kutterer and Schön 1999) as the expressions are only of polynomial type. For this reason and due to the restrictions of space only the determinant approach is discussed in the following.

Table 2: Eigenvalue and determinant approach

	Eigenvalues	Determinants
Parameters	λ_1, λ_2, λ_3, m	Δ, δ, B, J
Derivatives	Rational and irrational	Only polynomial
Requirements	Principle axis transformation	None

2.3.2 Estimation of the form parameters

In order to gain information about the type of the quadratic form, it is necessary to estimate the vector of the ten form coefficients (parameter vector)

$$\boldsymbol{x}^T = \begin{bmatrix} a_1 & a_2 & a_3 & a_4 & a_5 & a_6 & a_7 & a_8 & a_9 & a_{10} \end{bmatrix} \quad (6)$$

using the least-squares method. With these parameters the Matrix $\boldsymbol{M}$, the vector $\boldsymbol{m}$ and the scalar α can be set up. Actually, Eq. (1) represents a non-linear functional Gauss-Helmert model according to

$$f\left(E(\boldsymbol{l}), \boldsymbol{x}\right) = \boldsymbol{0} \quad (7)$$

with the observation vector

$$\boldsymbol{l}^T = \left[\ldots \ \boldsymbol{l}_i^T \ \ldots\right]_{i=1,\ldots,n}, \quad \boldsymbol{l}_i^T = \left[x_{1,i} \ \ x_{2,i} \ \ x_{3,i}\right] \quad (8)$$

which is composed of the point coordinates and its expectation vector $E(\boldsymbol{l})$ (Niemeier 2001). A linearization with sufficiently close approximate values $\boldsymbol{x}_0$ yields

$$\boldsymbol{A}\boldsymbol{\Delta x} + \boldsymbol{B}\boldsymbol{v} + \boldsymbol{w} = \boldsymbol{0} \quad (9)$$

with

$$\boldsymbol{A} = \left.\frac{\partial \boldsymbol{f}}{\partial \boldsymbol{x}}\right|_{x_0, l}, \quad B = \left.\frac{\partial \boldsymbol{f}}{\partial \boldsymbol{l}}\right|_{x_0, l}$$

and $\boldsymbol{\Delta x} = \boldsymbol{x} - \boldsymbol{x}_0$ denoting the parameter updates and $\boldsymbol{w}$ the vector of discrepancies. As Eq. (1) is a homogeneous equation system, a sufficient additional constraint has to be introduced in order to prevent the trivial solution (all estimated elements equal to zero).

As the approximate values $\boldsymbol{x}_0$ are typically not known a priori in 3d point cloud fitting (or need time consuming procedures for determination), the estimation in the Gauss-Helmert model needs several iteration steps for convergence – if it actually converges.

This problem can be overcome by applying a constraint to only a part of the unknown form parameters. Hence, they can be separated into the ones that are affected by the constraint ($\boldsymbol{x}_1$) and the ones that are not ($\boldsymbol{x}_2$). The extended functional Gauss-Helmert model than reads as

$$\begin{aligned} \boldsymbol{A}_1\boldsymbol{\Delta x}_1 + \boldsymbol{A}_2\boldsymbol{\Delta x}_2 + \boldsymbol{B}\boldsymbol{v} + \boldsymbol{w} &= \boldsymbol{0} \\ \boldsymbol{R}\boldsymbol{\Delta x}_1 + \boldsymbol{w}_R &= \boldsymbol{0} \end{aligned} \quad (10)$$

with the matrix of the constraints $\boldsymbol{R}$, the corresponding discrepancy vector $\boldsymbol{w_R}$. The stochastic model is defined by

$$\boldsymbol{\Sigma}_{ll} = \sigma_0^2 \boldsymbol{Q}_{ll} = \sigma_0^2 \boldsymbol{P}^{-1} \tag{11}$$

with $\boldsymbol{\Sigma}_{ll}$ the variance-covariance matrix (vcm) of the observation, $\boldsymbol{Q}_{ll}$ their cofactor matrix and $\boldsymbol{P}$ their weight matrix. According to Drixler (1993) Eqs. (10) and (11) can be transformed into a special eigenvalue problem via an equivalent Gauss-Markov model. This yields

$$\begin{aligned} \left(\boldsymbol{A}_1^T \boldsymbol{H} \boldsymbol{A}_1 - \lambda \boldsymbol{I}\right)\hat{\boldsymbol{x}}_1 &= \boldsymbol{0} \\ \hat{\boldsymbol{x}}_1^T \hat{\boldsymbol{x}}_1 &= 1 \end{aligned} \tag{12}$$

with the identity matrix $\boldsymbol{I}$, the matrix

$$\boldsymbol{H} = \boldsymbol{P}_{ww} - \boldsymbol{P}_{ww}\boldsymbol{A}_2\left(\boldsymbol{A}_2^T \boldsymbol{P}_{ww} \boldsymbol{A}_2\right)^{-1} \boldsymbol{A}_2^T \boldsymbol{P}_{ww} \tag{13}$$

and the matrix

$$\boldsymbol{P}_{ww} = \left(\boldsymbol{B}\boldsymbol{Q}_{ll}\boldsymbol{B}^T\right)^{-1}. \tag{14}$$

The estimated parameters $\hat{\boldsymbol{x}}_1$ are obtained as the eigenvector which corresponds to the minimum eigenvalue. Here, no approximate values $\boldsymbol{x_0}$ are needed. However, the weight matrix $\boldsymbol{P}_{ww}$ is not known in the beginning. Fortunately, initialization as $\boldsymbol{P}_{ww} = \boldsymbol{I}$ and iterative computation performs quite well and converges typically after a very few iteration steps.

The parameters $\hat{\boldsymbol{x}}_2$ are consequently obtained as

$$\hat{\boldsymbol{x}}_2 = -\left(\boldsymbol{A}_2^T \boldsymbol{P}_{ww} \boldsymbol{A}_2\right)^{-1} \boldsymbol{A}_2^T \boldsymbol{P}_{ww} \boldsymbol{A}_1 \hat{\boldsymbol{x}}_1 . \tag{15}$$

Finally, the cofactor matrix $\boldsymbol{Q}_{\hat{x}\hat{x}}$ and the vcm $\boldsymbol{\Sigma}_{\hat{x}\hat{x}}$ of the estimated parameters are obtained using the law of variance propagation (Drixler 1993).

2.3.3 Determinant approach

In the determinant approach the so called extended form matrix

$$\boldsymbol{M}^* = \begin{bmatrix} \boldsymbol{M} & \boldsymbol{m} \\ \boldsymbol{m}^T & \alpha \end{bmatrix} \tag{16}$$

has to be set up which consists of the form matrix $\boldsymbol{M}$, the form vector m and the scalar α. It can be shown that there are four motion invariant parameters δ, Δ, B and J through which the particular quadratic form (Table 1) can be identified.

The first parameter

$$\delta = \left|\boldsymbol{M}^*\right| \tag{17}$$

with $|.|$ denoting the determinant distinguishes between regular forms ($\delta \neq 0$) and irregular forms ($\delta = 0$).

The second parameter

$$\Delta = \left|\boldsymbol{M}\right| \tag{18}$$

divides the forms into centre point forms ($\Delta \neq 0$) and non centre point forms ($\Delta = 0$). For the first two parameters the decision is just binary depending on the question if their values are zero or not.

The geometric interpretation of the third and fourth parameter is not as simple, since both have complementary meanings in several cases and can reach positive, negative or zero values. Their formal representations are

$$B = \left|M_{11}^*\right| + \left|M_{22}^*\right| + \left|M_{33}^*\right| \tag{19}$$

and

$$J = \left|M_{11}\right| + \left|M_{22}\right| + \left|M_{33}\right| \tag{20}$$

with $\boldsymbol{M}_{ii}$ the ith main minor of $\boldsymbol{M}$. For a complete classification scheme see Fig. 7. Since all classification parameters can be ascribed to the determinants of (16) or one of its main minors, this procedure of classifying quadratic forms is called the determinant approach. For further reference see Knopp (1959) or Strubecker (1966).

For the determination of the correct quadratic form it is essential to assess the significance of the respective classification steps according to Fig. 1. The combined vector of the four invariant parameters of the determinant approach reads as

$$\boldsymbol{\Omega} = \left[\delta,\ \Delta,\ B,\ J\right] \tag{21}$$

The corresponding (joint) vcm can be obtained be means of the law of variance propagation

$$\Sigma_{\Omega\Omega} = \boldsymbol{F}_\Omega \Sigma_{\hat{x}\hat{x}} \boldsymbol{F}_\Omega^T \tag{22}$$

with

$$\boldsymbol{F}_\Omega = \left[\frac{\partial \delta}{\partial a_i}, \frac{\partial \Delta}{\partial a_i}, \frac{\partial B}{\partial a_i}, \frac{\partial J}{\partial a_i}\right]^T_{i=1,\dots,10} . \tag{23}$$

If just the significance of each single parameter is of interest, the parameters can be tested component by component using a univariate test statistics. Using theoretical variances, the test statistics follow the normal distribution as it is indicated in the following. The tests are carried out in a two-sided way.

$$T_1 = \frac{\delta}{\sigma_\delta} \sim N(0,1) \Big| \quad H_0 : E(\delta) = 0 \qquad (24)$$

$$T_2 = \frac{\Delta}{\sigma_\Delta} \sim N(0,1) \Big| \quad H_0 : E(\Delta) = 0 \qquad (25)$$

$$T_3 = \frac{B}{\sigma_B} \sim N(0,1) \Big| \quad H_0 : E(B) = 0 \qquad (26)$$

$$T_4 = \frac{J}{\sigma_J} \sim N(0,1) \Big| \quad H_0 : E(J) = 0 \qquad (27)$$

In order to decide between the forms sphere, rotation ellipsoid and ellipsoid, who only differ from each other by the relations of their parameters, it is necessary to introduce some additional test values

$$T_5 = \begin{pmatrix} \kappa_1 \\ \kappa_2 \end{pmatrix}^T \Sigma_{\kappa\kappa}^{-1} \begin{pmatrix} \kappa_1 \\ \kappa_2 \end{pmatrix} \sim \chi_2^2 \Big| \quad H_0 : \begin{pmatrix} \kappa_1 \\ \kappa_2 \end{pmatrix} = \mathbf{0} \qquad (28)$$

with the values κ_i derived from the eigenvalues $\lambda_i, i \in \{1,2,3\}$ as

$$\begin{aligned} \kappa_1 &= \lambda_1 - \lambda_2 \\ \kappa_2 &= \lambda_1 - \lambda_3 \end{aligned} \qquad (29)$$

$$T_6 = \frac{\kappa_i}{\sigma_{\kappa_i}} \sim N(0,1) \Big| \quad H_0 : E(\kappa_i) = 0, \quad i = 1,2$$

(30)

For an exact derivation of the formulas refer to Drixler (1993) and Kutterer and Schön (1999). In order to have a comparable test decision it is necessary to harmonize the error probabilities with respect to the degrees of freedom. As the main result of this procedure a full classification of the 3d quadratic forms is possible. Hence, they can be automatically subdivided in the classes listed in (Table 1). For the complete test tree see Fig. 7 at the end of the document.

3 Performance test of the algorithm

3.1 Possible results of classification

The quadratic forms with relevance for engineering geodesy are listed in Table 1. These types are possible results of the classification algorithm, presented in section 2.3.

3.2 Test on artificial forms

In order to test the performance of the classification algorithm, several artificial forms were created. Afterwards they were overlaid with an additional noise in order to simulate realistic conditions without systematic influences. An additional rotation and translation out of the zero state was applied. Since the quality of the form classification mainly depends on the noise level of the coordinates, four different levels of white noise were added to the unbiased points ($\sigma_{x1}=\sigma_{x2}=\sigma_{x3}=\sigma$). The complete list of parameters used for setting up the artificial test forms given in Table 3.

Table 3: Parameters for artificial test forms

	Parameter	Value
Ellipsoid	Number of points	121
	Dimension x_1; x_2; x_3	5.0; 5.0; 5.0 [m]
	Noise $\sigma_{x1}=\sigma_{x2}=\sigma_{x3}$	5 mm; 10 mm; 100 mm; 500 mm
	Translation x_1; x_2; x_3	10; 10; 10 [m]
	Rotation α; β; γ	35; 40; 65 [gon]
Single Walled Hyperboloid	Number of points	121
	Dimension x_1; x_2; x_3	5.0; 5.0; 5.0 [m]
	Noise $\sigma_{x1}=\sigma_{x2}=\sigma_{x3}$	0 mm; 5 mm; 10 mm; 20 mm
	Translation x_1; x_2; x_3	10; 10; 10 [m]
	Rotation α; β; γ	35; 40; 65 [gon]

The classification process was repeated 100 times for each form to take non random effects of the point noise into account, which can lead to a false classification. The first form chosen for the test was a triaxial ellipsoid of class 11 (Fig. 2a).

Fig. 2: Ellipsoid (a) and Hyperboloid (b)

All classes in the following figures and Table 4 are labeled to Table 1.

Without noise, the form was classified correctly as an ellipsoid in 100% of all cases. With an applied noise level of 5 mm and 10 mm the classification algorithm showed the rotation ellipsoid as the most probable estimation in about 10-15% of the cases. Finally with a noise of 100 mm the classification determined the double walled hyperboloid (class 12) and the rotation ellipsoid (class 10) as the most probable solutions. The explanation for this can be, that the rotation ellipsoid only differs from the triaxial ellipsoid by the relation of their coefficients. A second reason can be, that the test points are only computed for a small section of the whole form. This is the case when forms are scanned from one or two points of view. In this case with a comparably high noise the classification fails (Fig. 3, 0.1 m noise).

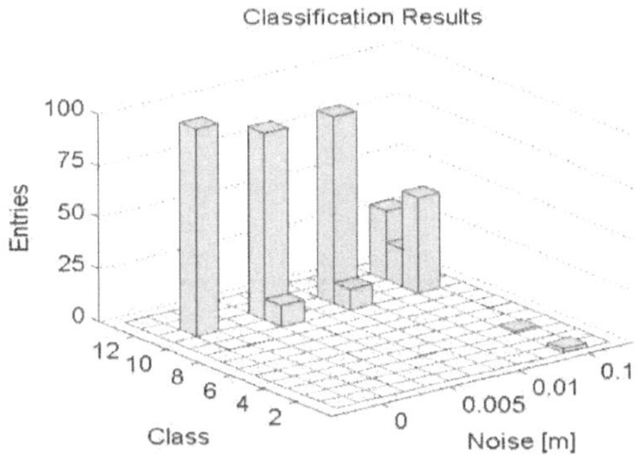

Fig. 3: Classification results of ellipsoid

The second quadratic form which was chosen for the test was a single walled hyperboloid of class 12 (Fig. 2b). The parameters of this form can be obtained from Table 3.

Compared to the classification of the ellipsoid the approach for the single walled hyperboloid seems more stable related to higher noise levels. The results show clearly, that with a noise of 10 mm and 100 mm the classification delivers correct results. Even with 200 mm noise it fails only in 3% of all cases. At a noise level of 500 mm, which is 10% of the overall object dimension, the algorithm determines a plane as the most probable estimation in 92% of the cases. Correct classification is yielded only in 4% of the cases (Fig. 4). This shows that the border between misclassification and correct classification is much sharper for single walled hyperboloids. For ellipsoids there is a transition from correct to false results with increasing noise.

The detailed results of 100 single classifications for the computed forms ellipsoid and hyperboloid with changing 3D point noise are shown in Table 4.

Table 4: Distribution of 100 classifications into classes, correct result marked bold

Class	Ellipsoid, noise [mm]				Hyperboloid, noise [mm]			
	0	5	10	100	10	100	200	500
1	-	-	-	3	-	-	-	92
2	-	-	-	-	-	-	-	-
3	-	-	-	-	-	-	-	-
4	-	-	-	1	-	-	-	-
5	-	-	-	-	-	-	2	3
6	-	-	-	-	-	-	-	-
7	-	-	-	-	-	-	1	3
8	-	-	-	-	-	-	-	-
9	-	-	-	-	-	-	-	-
10	-	10	15	46	-	-	-	-
11	**100**	**90**	**85**	**18**	-	-	-	-
12	-	-	-	32	**100**	**100**	**97**	**4**
13	-	-	-	-	-	-	-	-
∑	100	100	100	100	100	100	100	100

Legend for class numbers

1: Plane; 2: Two perpendicular planes; 3: Parabolic cylinder; 4: Elliptical cylinder; 5: Hyperbolic cylinder; 6: Cone; 7: Hyperbolic paraboloid; 8: Elliptic paraboloid; 9: Sphere; 10: Rotation ellipsoid; 11: Ellipsoid; 12: Double walled hyperboloid; 13: Single walled hyperboloid.

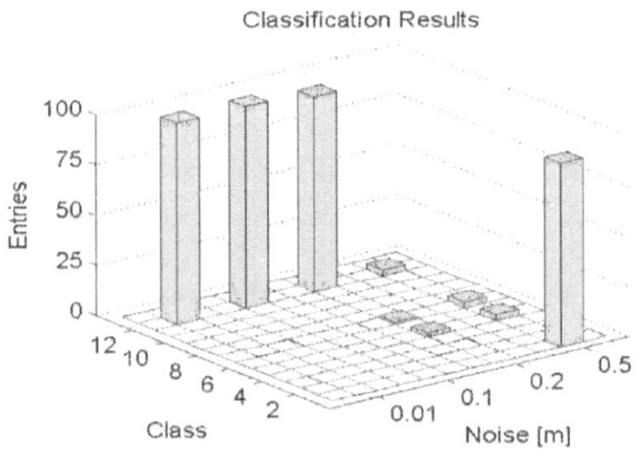

Fig. 4: Classification results single walled hyperboloid

3.3 Tests on empirical forms

To test the algorithm on real measurements two laser scans of real objects were classified. First a typical target sphere was considered for classification, which is used as a homologous point, because the correct detection of these points is evident for each laser scanning task. In this case, the topic of interest is, if the functional model of a sphere is correct, or if there are any systematic effects. This is important because changing angles of incidence from the centre to the border of the sphere cause an incorrect adjusted centre point and radius when using a simple least squares approach. The functional model can be verified by a classifycation of the underlying form.

Table 5: Parameters of empirical test objects

	Parameter	Value
Target Sphere	Default form	Sphere
	Number of points	904
	Default diameter	158.1 mm
	Diameter tolerance	± 0.1 mm
	Noise $\sigma_x=\sigma_y=\sigma_z$	approx 5 mm
Quarter Pipe	Default form	Cylinder
	Number of points	2695
	Length	11.5 m
	Radius	1.9 m
	Noise $s_x=s_y=s_z$	4 mm;

Since the classification algorithm is deterministic and the noise is given through the measurements, it is not necessary to iterate the process when using empirical measurements. The parameters of the scanned objects, their point cloud dimensions and the approximate noise values taken from the laser scanner software are shown in Table 5.

The target sphere was scanned from a single point of view with low noise settings in the scanning software. Fig. 5 shows the sphere in front and side view. The automatic form classification identified this object as an ellipsoid, which is not surprising.

Fig. 5: Target sphere front (a) and side view (b)

As mentioned before, systematic effects are arising at the side of curved objects due to harmful angles of incidence. This results in so called comet tails at the sides of a sphere, ellipsoid or similar forms. This fact makes the set up of an appropriate adjustment model difficult.

The second empirical example to be classified was a quarter pipe like form that is part of a lock (Fig. 6) located in Uelzen (Germany) where a deformation analysis project took place. This structure was computed as a section of a cylinder in CAD and constructed with solid concrete. The expected form is a elliptic cylinder, but in contrary to this assumption the algorithm classified the object as a cone. In order to verify the form, 4 radii were adjusted in 4 sections along the longitudinal axis of the object. From Table 6 can be seen, that except the first section the radii are decreasing from left to right, which is a strong indicator for a cone.

Table 6: Radii at 4 sections with standard deviation

Section	R1 [m]	R2 [m]	R3 [m]	R4 [m]
Radius	3.737	3.742	3.723	3.705
S_R	0.003	0.003	0.004	0.004

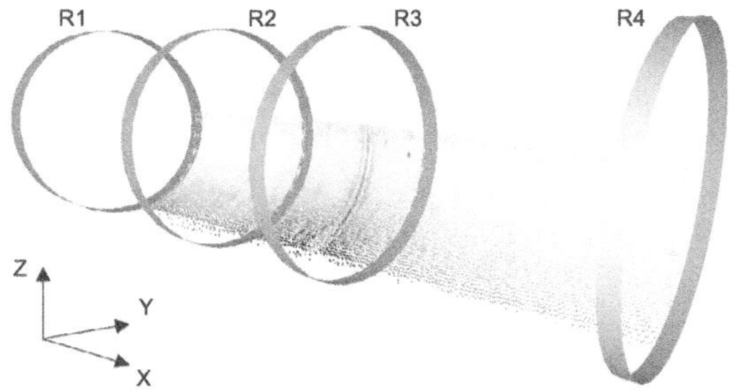

Fig. 6: Quarter pipe with adjusted radii

This example shows that the automated shape classification algorithm is able to deliver results which disprove false assumptions of a human user and thereby significantly improves the quality of parameter reduction.

4 Conclusions

Terrestrial laser scanning has a great potential for engineering geodesy. Especially the high speed data acquisition in addition to very dense point clouds offer a promising way for reliable monitoring of object deformations.

To handle the large amount of data and to automate the analysis of point clouds, a parameter reduction for the object model is necessary. This can be done by segmenting the point cloud into several parts and replace the resulting subclouds with three dimensional quadratic forms.

By the presented algorithm a statistically founded decision about the underlying quadratic form can be done. Thus the functional model for a final adjustment can be set up correctly avoiding over parametrization. Such a classification can be accomplished without preparatory work, since the introduced test values are motion invariant. Prior rotations or translations are not necessary. A fully automated form classification and adjustment is possible as presented.

The initial estimation of the ten form parameters without approximation values has been shown as well as the automated form classification process. Statistic measures for the test values were given as well. The algorithm was tested with two synthetic examples and two real scans. The computed forms were reliably classified under the condition that the measurement noise is less than some centimetres.

The major problems occurring due to the use of real laser scanner point clouds were exemplified by a target sphere. Such spheres are mandatory when scanning the same object from two or more points of view. Especially the changing angles of incidence from the centre to the border of the sphere cause systematic effects on the measured data. These effects prevent the detection of the correct form. The advantage of the automated form classification was proved at the Lock Uelzen as well. A large quarter pipe of solid concrete has decreasing radii along its longitudinal axis. An adjustment with a cylindric approach in the functional model would not consider this. The classification stated correctly a cone as the most probable estimation.

The next steps are to use the presented approach for deformation analysis based on form comparison methods. This may help to detect the change of an object shapes automatically with terrestrial laser scanners. For an improvement of the analysis process it is necessary to obtain knowledge on the error model of the whole sensor and knowledge on systematic influences arising from the object's surface, meteorological effects or through the measurement configuration itself.

References

Anton, H. (1994). Elementary Linear Algebra. Anton Textbooks. John Wiley & Sons, Inc., New York.

Bronstein, I. N., K. A. Semendjajew (1991). Handbook of Mathematics, 20th edition. Van Nostrand Reinhold, New York.

Drixler, E. (1993). Analyse der Lage und From von Objekten im Raum. DGK Reihe C, Heft Nr. 409, München.

Holz, H., D. Wille (2002). Repetitorium der linearen Algebra II. Binomi Verlag, Springe.

Knopp, K. (1959). H. v. Mangoldts Einführung in die höhere Mathematik – Band II: Differentialrechnung, unendliche Reihen, Elemente der Differentialgeometrie und der Funktionentheorie. S. Hirzel Verlag, Leipzig.

Kutterer, H., S. Schön (1999). Statistische Analyse quadratischer Formen – der Determinantenansatz. AVN, Heft 10/1999, Wichmann Verlag, Heidelberg.

Niemeier, W. (2001). Ausgleichungsrechnung. Walter de Gruyter Verlag, Berlin.

Strubecker, K. (1966). Einführung in die höhere Mathematik – Band IV: Grundzüge der linearen Algebra, Differential- und Integralrechnung der Funktionen von mehreren Veränderlichen. R. Oldenbourg, München.

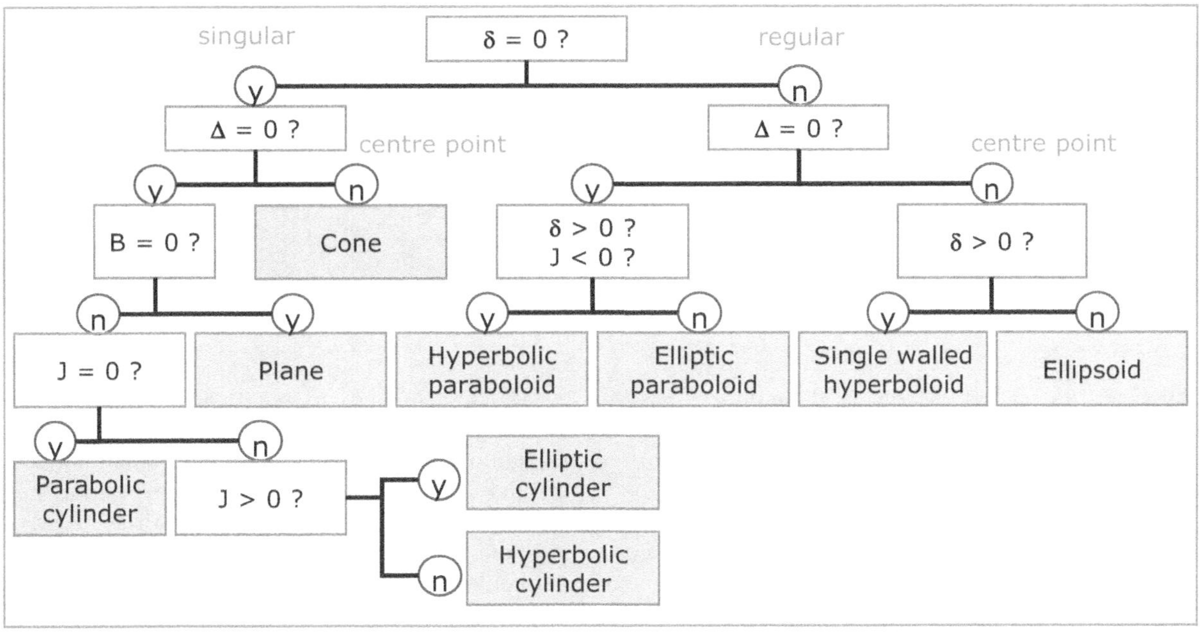

Fig. 7: Complete test tree for determinant approach

The Impact of Vertical Refraction at Local Three-Dimensional Engineering Control Networks

J. Casaca, M.J. Henriques, J. Coelho
Division of Applied Geodesy, National Laboratory for Civil Engineering,
Av. do Brasil 101, 1700-066 Lisbon, Portugal

Abstract. Local heterogeneous three-dimensional control networks, combining electronic tacheometry with geometric levelling, have been used in Portugal to monitoring displacements of large embankment dams. The most important limits to the accuracy of these networks are due to atmospheric phenomena, such as refraction and turbulence.

The refraction, which introduces curvature and torsion of the path of electromagnetic energy and affects its propagation velocity, and the turbulence, which causes vibration of the images of optical sightings and oscillation of EDM's signals, originate significant systematic errors on angles, distances and height differences. A careful choice of the observation periods – close to sunrise and to sunset, specially at hot sunny days – avoids the main effects of turbulence, but not of refraction.

In order to assess the magnitude and variability of the vertical refraction angle, an experiment, with regular geometric and trigonometric levelling measurements, was carried out, from 1996 to 2004, at Odeleite dam, in the South of Portugal. More than four hundred vertical refraction angles were measured and fitted to a normal distribution, which will be adopted as a prior distribution to future inferences (design and adjustment) on the vertical error distribution, at similar three-dimensional engineering control networks.

Keywords. Local three-dimensional networks, trigonometric levelling, vertical refraction angle, vertical uncertainty.

1 Introduction

The monitoring of displacements at large dams, for the analysis of the structural behaviour and to safety control, has been carried out since the beginning of the XXth century with geodetic surveying methods (precision triangulation and geometric levelling), together with other techniques, such as pendulums and extensometric bars.

At concrete and masonry dams, submilimetric accuracy requirements, as well as morphological constraints, impose the use of independent horizontal and vertical (geometric levelling) control networks. At embankment dams, less strict accuracy requirements allow the use of local three-dimensional control networks, combining horizontal angles, slope distances, height differences (measured by geometric or trigonometric levelling) and GPS derived coordinate differences.

According to the method of variation of coordinates, the vector of the displacements (ΔX) of the network's vertices between two measurement epochs, may be directly computed, by best linear unbiased estimation, from the vector (ΔY) of the variations of the observables between the two epochs:

$$\Delta X = (A^T W A)^{-1} A^T W \Delta Y \tag{1}$$

where the weight matrix (W) is the inverse of the variance-covariance matrix of the vector of the observables:

$$V(\Delta Y) = \begin{bmatrix} \Sigma_A & 0 & 0 & 0 \\ 0 & \Sigma_S & 0 & 0 \\ 0 & 0 & \Sigma_H & 0 \\ 0 & 0 & 0 & \Sigma_{GPS} \end{bmatrix} \tag{2}$$

where the diagonal blocks are the variance-covariance matrices of correlated observable groups (horizontal angles, distances, height differences and coordinate differences).

The prior knowledge of the variance-covariance matrix $V(\Delta Y)$, or of its diagonal blocks, is necessary for fundamental tasks: i) network design; ii) quality control, by means of statistical tests; iii) estimation of displacements.

The characterization of the distribution of the vertical refraction angle plays an important part on the estimation of the variance and covariance components of the altimetric diagonal blocks (Σ_H).

2 Trigonometric Levelling

At a short distance from a station, where the geopotential is close enough to the local sphere of radius R, the orthometric heights of the station point (H_S) and a target point (H_T) may be related by the formula for trigonometric levelling (Cooper, 1987):

$$H_T = H_S + S\cos Z + (a_I - a_R) + C \tag{3}$$

where S and Z are the slope distance and the zenith angle between the tacheometer and the reflector, respectively, a_I and a_R are the heights of the instrument and the reflector above the station and target points, respectively, C is a correction for both the curvature of the optical path and the curvature of the geopotential arc between the tacheometer and the reflector.

The combined correction for curvature and refraction (Cooper, 1987) is:

$$C = S^2 \sin^2 Z \frac{1-k}{2R} \tag{4}$$

where R is the mean radius of curvature of the reference ellipsoid at the station and k is the refraction coefficient, which is the ratio of R and the mean curvature radius of the optical path between the tacheometer and the reflector.

If the optical path is supposed to be on a straight line, the refraction coefficient (k) vanishes and the curvature correction becomes:

$$C_0 = \frac{S^2 \operatorname{Sin}^2 Z}{2R} \tag{5}$$

Since the observed zenith angle (Z_0) is affected by an observation error, which is a sum of several small random and systematic errors, where the vertical refraction angle (β) prevails, the formula for trigonometric levelling may be written as:

$$H_T = H_S + S\cos(Z_0 + \beta) + (a_I - a_R) + C_0 \tag{6}$$

and, assuming that $\cos\beta \approx 1$ and $\sin\beta \approx \beta$, the vertical refraction error may be expressed as:

$$\beta = \frac{S\cos Z_0 + C_0 + (H_S - H_T) + (a_I - a_R)}{S\sin Z_0} \tag{7}$$

The relation (7) allows the computation of the vertical refraction angle from field measurements: the slope distances (S) and the zenith angles (Z_0) are measurable with an electronic tacheometer and the height differences ($H_S - H_T$) are measurable with geometric levelling, which is assumed to be errorless within the scope of this work.

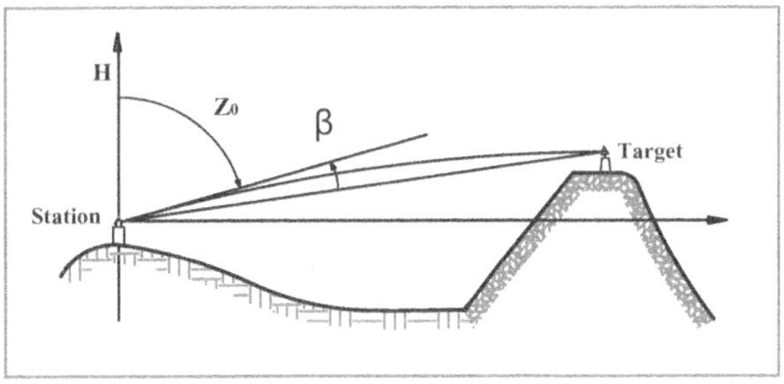

Fig. 1 Vertical refraction angle β

3 The Theoretical Model

The vertical refraction angle (β) of the optical path between the tacheometer and the reflector (Fig. 1), may be related (Casaca *et al*, 1994) to the atmospheric vertical gradient of the light's refractive index of the atmosphere (n) by:

$$\beta \approx -\sin Z \int_0^S \left(\frac{1}{n}\frac{\partial n}{\partial H}\right)\left(1 - \frac{s}{S}\right) ds \tag{8}$$

where the integration variable (s) is the optical path's length and S is the total path length.

The integral (8) may be simplified taking into account the relation between the refractive index and the atmosphere's temperature (T), pressure (P) and humidity. Taking the vertical gradient of the atmospheric pressure as –13.3222 Pa m^{-1} (–0.1 mm Hg m^{-1}) and ignoring the humidity, the vertical refraction angle (in mgon) may be expressed, for an horizontal sighting, as:

$$\beta \approx -0.06 \int_0^S \frac{\partial T}{\partial H}\left(1 - \frac{s}{S}\right) ds \tag{9}$$

A vertical temperature's distribution of the logarithmic type:

$$T = a\ln(\eta) + b \tag{10}$$

where a and b are empirical coefficients, that eventually vary along the optical path, and η is the height above the ground along the path, may be used to derive the temperature's vertical gradient:

$$\frac{\partial T}{\partial H} = \frac{a}{\eta} \tag{11}$$

Relations (8) and (9) show that the vertical refraction angle does not depend directly from the distance between the station and the target points, but it integrates atmospheric information, along the optical path, weighted by a triangular function that varies from 1 at the station to 0 at the target.

Fig. 2 Odeleite dam

4 The Odeleite Experiment

The river Odeleite is a tributary, on the right bank, of the Guadiana river, in the South of Portugal. The Odeleite dam (Fig. 2) is actually composed of a main dam and a secondary dam. Several object points, materialized at the crests of the main dam (12 points) and the secondary dam (11 points), were subject to geometric and trigonometric levelling between 1996 and 2004. The distances from the station points to the object (target) points ranged between 40m and 300m. The points on the main dam's crest were measured at 19 epochs and the points on the secondary dam's crest were measured at 18 epochs. A total of 426 vertical refraction angles were computed using relation (7).

Table 1 presents the sample means and standard deviations (SD) estimated for the main dam (228 angles), for the secondary dam (198 angles) and for all the 426 angles.

Table 1. Sample Means and Standard Deviations of Vertical Refraction Angles

	Mean	SD
Main dam	– 0.56 mgon	+ 0.52 mgon
Secondary dam	– 0.56 mgon	+ 0.44 mgon
Total	– 0.56 mgon	+ 0.48 mgon

The sample of 426 vertical refraction angles was classified (Fig. 3) and subject to a chi-square test of fit to a normal distribution with parameters (mean value and standard deviation) identical to the sample parameters. The null hypothesis (identical distribution) was accepted for a 0.95 significance level.

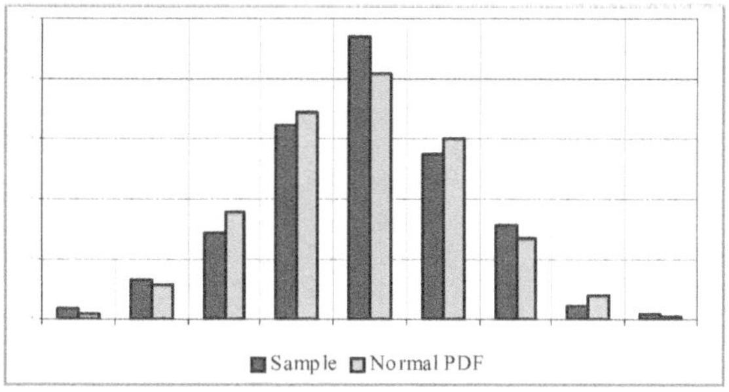

Fig. 3 Classified sample of vertical refraction angles

5 Altimetric Uncertainty

At Odeleite dam three-dimensional network, the distribution of the vertical refraction angle (β), is well modelled by a normal distribution with parameters mean value and standard deviation:

$$\mu(\beta) = -0.6\,\text{mgon}, \quad \sigma(\beta) = 0.5\,\text{mgon} \qquad (12)$$

where the mean vertical refraction angle is a consequence of the mean atmospheric temperature and pressure vertical gradients along the network's optical paths and the standard deviation represents the dispersion due to the instrumental uncertainty and the variation of atmospheric conditions between different observation epochs and between different optical paths at the same observation epoch.

From a statistical point of view, the –0.6 mgon bias, although affecting heights, does not affect vertical displacements. Consequently, the distribution of the errors of the vertical displacements between two observation epochs, due to instrumental uncertainty and vertical refraction, may be modelled by a central normal distribution with standard deviation:

$$\sigma(\Delta H) = \sqrt{2}\, S \sin Z_0\, \sigma(\beta) \qquad (13)$$

Vertical standard deviations computed with relation (13) may be used to derive the vertical diagonal block (Σ_H) of the variance-covariance matrix V(ΔY) of the observables (2).

Considering, as an example, three short distances and horizontal sightings, relation (13) leads to the *prior* vertical displacement standard deviations presented in Table 2, which illustrates the linear growth of the vertical displacement standard deviation (σ(ΔH)) with the distance (S) between the station and the target points.

Table 2. Standard Deviations of Vertical Displacements

S	50m	150m	300m
$\sigma(\Delta H)$	+ 0.6mm	+ 1.7mm	+ 3.3mm

6 Conclusion

The results of Odeleite experiment agree with the theoretical model and provide a prior distribution to the vertical uncertainty that is adequate to the design of local three-dimensional networks and to the quality control of the observations.

Bibliographic References

Casaca, J. *et* Henriques, M. J. (1994). Numerical Modelling of Atmospheric Refraction. In: Proc. of the International Symposium on Deformation, Istanbul (pp. 116-125).

Cooper, M. (1987). Control Surveys for Civil Engineers. Collins, London.

Monitoring and Analysing Structural Movements with Precise Inclination Sensors

B. Erol, S. Erol, R. N. Çelik
Department of Geodesy and Photogrammetry Engineering
Istanbul Technical University, Civil Engineering Faculty, Maslak, 34469, Istanbul, Turkey

Abstract. A multistorey high–rise building has been monitored using microradian precision inclination sensors and the structural movements have been analyzed as consequence of 40 days observation period. In the first step, the time series obtained from inclination sensors' data were processed with Least Squares Spectral Analysis technique. In documenting the periodicity of the data, the correlation functions were also used as a straightforward way of evaluating the time series. Afterwards the results from the analysis were interpreted and linked to the other observed data such as temperature changes, wind load effects, population in the building and instant earthquake information in the observation period. Further on the results were compared with processed data from consecutive GPS campaigns processes. The study has shown that precise inclination sensors are efficient tools for continuously monitoring and investigating structural movements of large engineering structures.

Keywords. Structural Deformations, Inclination Sensors, Least Squares Spectral Analysis

1 Introduction

Continuously monitoring the deformations of engineering structures has a particular importance especially in active tectonic regions. The improved precision of measurement systems as a consequence of developing technology provides better and more reliable results in analysis of deformations in structures and monuments. Global Positioning System (GPS) as a high precision point positioning technology has a wide spread use in deformation measurements apart from conventional geodetic surveying techniques today. However, despite many strengths and advantages of GPS, there are still some weaknesses, which affect its performance in high precision works as in deformation measurements, in practice. In order to overcome the weaknesses of the GPS system, combining and/or supporting this technique with other auxiliary systems is suggested (Erol et al. (2004)).

Also determining the most efficient solution to approximate functions, which define the deformations in mathematical point of view, is a complex problem to be solved by geodesists. Especially, when dealing with continuously monitoring processes, spectral analysis approaches provide an efficient tool to decompose this physical phenomenon, which is a dynamic system indeed.

The present study aims continuously monitoring and analysing structural deformations of a multistorey building close to North Anatolian Fault Zone (NAFZ) using microradian precision digital inclination sensors and comparing the obtained results with GPS solutions. In this study, Least Squares Spectral Analysis technique has been applied to decompose and understand the structural deformations regarding inclination sensors' data in addition to correlation functions.

2 Time Series Analysis

Deformation monitoring of engineering structures is based on measurements that are done using special and highly precise techniques and equipments. Appropriate analysis of these measurements in time series form with mathematical and statistical methods helps to clarify the amount and periodicity of the deformations in the structure. During the analysis of these time series, also the correlations between the effects of deformations and physical forces on the structure are also interpreted.

Due to the nature of measurements, the observed time series are assumed to be composed of two main constituents as *signal* and *noise*. The noise, which is the disturbing part on observations, can be random and/or systematic. In the concept of noise, the uncorrelated random noise with constant spectral density is assumed as ideal. This kind is also called as *white noise*. However, in practice, the observables mostly include non-white random noise, which is a band–limited random function of time. Systematic noise also may have systematic and non-systematic parts and can be modeled with certain mathematical forms as explained in section 2.2 (see Pagiatakis (1999) for more information).

Non–systematic noise can include datum shift, trend etc. and renders the series non–stationary. While analyzing the time series, generally the possible trend, which may fit linear, quadratic, exponential and so on expressions, is identified and removed in the first step in order to avoid non–systematic noise (Pytharoli et al. (2004), Pagiatakis (1999), Hekimoğlu (1981)). Usually the trial–and–error technique is used in the determination of best fitting model in order to test the models statistically (Fotopoulos (2003), Hardy (1971) can be referred for various applications of determination of a best–fit–model to a data set). De-trended data consist of superimposed signals (see Figure 1).

Depending on the structure of the analyzed data, in some cases a particular signal of a time series can be dominant and be modeled using uncomplicated techniques such as parametric model, different order polynomials fits and so on. However, most of the time, measurements have irregular distributions with unequal intervals and in such cases spectral analysis techniques are used to analyze the data.

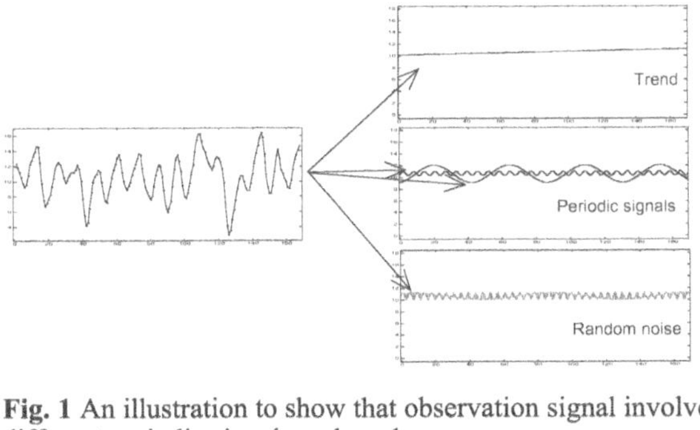

Fig. 1 An illustration to show that observation signal involves different periodic signals and random errors

2.1 Autocorrelation and cross correlation functions

Spectral analysis technique is applied to de-trended data. After documenting the periodicity of the data, periods and amplitudes of each periodic signal in the series are estimated. Autocorrelation functions are helpful while determining the periodicities of the series (Box et al. (1994)). This function helps to determine the change of linear correlation coefficient ($-1 \leq r \leq 1$) of the time series $F(\tau)$ and $F(\tau+k)$, in which the difference is k time interval (lag) ($k=0,1,2\ldots K$). An estimate of the kth lag autocorrelation (ρ_k) is r_k given in equation (1). The value of r_k is also called as sample autocorrelation function (Box et al. (1994)).

$$r_k = \frac{c_k}{c_0} \tag{1}$$

$$c_k = \frac{1}{N}\sum_{t=1}^{N-k}(z_t - \bar{z})(z_{t+k} - \bar{z}), \quad k = 0,1,2,\ldots, K \tag{2}$$

$$c_0 = \frac{1}{N}\sum_{t=1}^{N}(z_t - \bar{z})^2 \tag{3}$$

where c_k is the estimate of the autocovariance coefficient (γ_k) at lag k; N is the number of observations in the time series; z_t is the observation at time t; $\bar{z}$ is sample mean of the time series and k is time lag. Figure 2 shows an illustration for autocorrelation analysis.

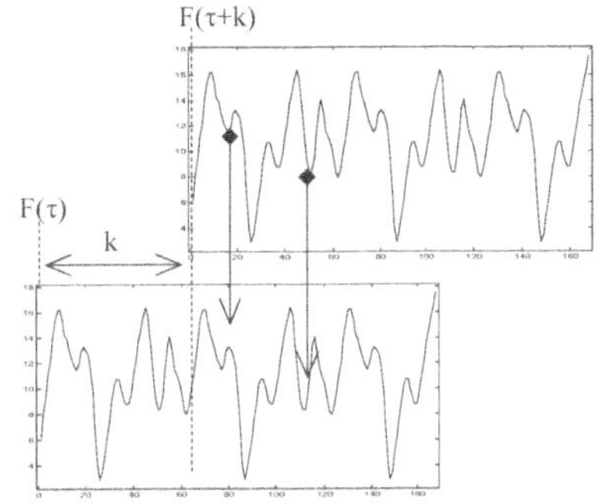

Fig. 2 Autocorrelation of $F(\tau)$ and $F(\tau+k)$ signal parts

The graph of autocorrelation coefficient–ρ_k versus lag k (correlogram) is also called the autocorrelation function of the process. The sinusoidal oscillations in autocorrelation function reveal a periodicity in time series (Figure 3a). On the contrary, a linear and non–oscillating trend in the function may indicate that there is not any periodic signal in the series (Figure 3b) (Hekimoğlu (1981), Box et al. (1994), Pytharouli et al. (2004)).

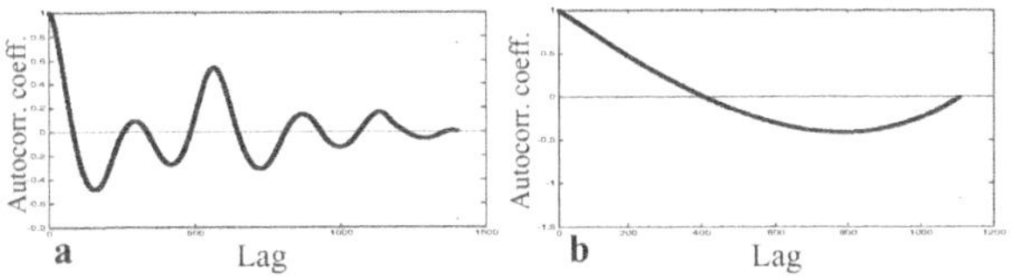

Fig. 3 Autocorrelation function samples; observation **a** includes, **b** does not include, periodic component.

The autocorrelation function is used to identify the correlations in a single time series. In a similar way, cross correlation function is also used for estimating the correlations but this time between two different time series, $F(\tau)$ and $G(\tau)$ (Hekimoğlu (1981)), Box et al. (1994)). An estimate $r_{xy}(k)$ of cross correlation coefficient ($\rho_{xy}(k)$) at k lag is as given in equation (4);

$$r_{xy}(k) = \frac{c_{xy}(k)}{s_x s_y} \quad , \quad k = 0, \pm 1, \pm 2, \ldots \tag{4}$$

$$c_{xy}(k) = \begin{cases} \frac{1}{n}\sum_{t=1}^{n-k}\left(x_t - \bar{x}\right)\left(y_{t+k} - \bar{y}\right) & k = 0,1,2,\ldots \\ \frac{1}{n}\sum_{t=1}^{n+k}\left(y_t - \bar{y}\right)\left(x_{t-k} - \bar{x}\right) & k = 0,-1,-2,\ldots \end{cases} \tag{5}$$

where, n is the number of pairs of the values to be analyzed such as (x_1,y_1), (x_2,y_2), … (x_n,y_n); $\bar{x}$, $\bar{y}$ are the sample means of x and y series respectively; $c_{xx}(k)$ is the estimate of cross covariance coefficient ($\gamma_{xx}(k)$) at k lag; s_x and s_y are estimates for σ_x and σ_y respectively.

In Figure 4, the graph of cross correlation coefficients versus lags is shown. In the example $G(\tau)$ is correlated with $F(\tau+k)$. The extremum point of the graphic shows the best match location of two time series with k lag.

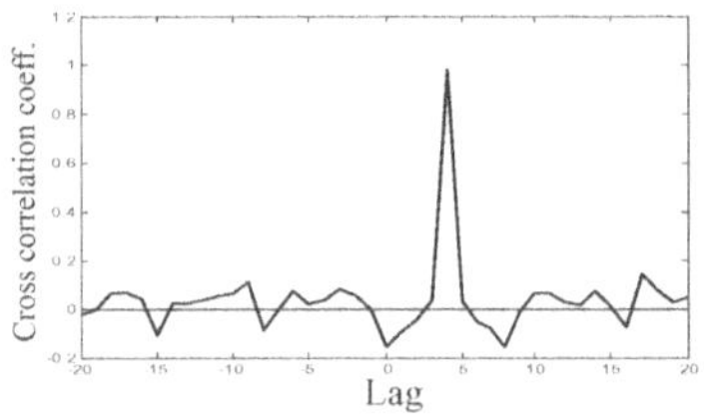

Fig. 4 Sample cross correlation function: the extremum point expresses the highest correlation between two series.

2.2 Least Squares Spectral Analysis (LSSA)

In least squares spectral analysis technique, the observed time series f is considered as a function of time t_i, $i=1,2,\ldots n$. Here, the time series may or may not have equally spaced values. The main objective of LSSA is to determine and clarify the periodic signals in f, especially when f includes both random and systematic noise.

The time series f can be modeled with function g as in equation (6).

$$g = Ax \tag{6}$$

where, $A=[A_s][A_n]$ is the coefficient matrix and called as Vandermonde matrix; $x^T=[x_s|x_n]^T$ is the vector of unknown parameters. The s and n indices refer to the *signal* and *noise* respectively. Matrix A involves the functional form of both signal and systematic noise component. Also it should be considered that the difference between the signal and noise cannot be distinguished easily, therefore the portioning of A and x is determined arbitrarily (Pagiatakis (1999)). The elements of matrix A are expressed as in equation (7).

$$A = \left[\cos\omega_i t, \sin\omega_i t\right] \qquad i = 1,2,\ldots,k \tag{7}$$

The parameters are determined according to least–squares method and the technique seeks to minimize the difference between f and g functions. The standard least–squares representation is as given in equation (8) (see Vaniček and Krakiwsky (1986)).

$$\hat{r} = f - \hat{g} = f - A\left(A^T A\right)^{-1} A^T f \tag{8}$$

where, $\hat{g}$ is the orthogonal projection of f and constitutes a column vector of matrix A (for more information refer Wells et al. (1985) and Pagiatakis (1999)). In other words, f is decomposed into two components as a signal $\hat{g}$ and noise $\hat{r}$ ($\hat{r} \perp \hat{g}$).

In order to find out something similar to spectral value, $\hat{g}$ is compared with the original series. This can be accomplished by orthogonal projection onto $\hat{g}$ and comparing the norm of this projection to the norm of f. This is related with orthogonality of vector spaces in Hilbert space. Hence, a measure is obtained (*as a percentage*) that how much of $\hat{g}$ is contained in f. This measure is expressed as in equation (9).

$$s = \frac{\left\langle f, \hat{g} \right\rangle |f|}{|f|} = \frac{\left\langle f, \hat{g} \right\rangle}{|f|^2} = \frac{f^T \hat{g}}{f^T f} \tag{9}$$

where, the sign of $\langle \; \rangle$ means inner product.

This spectral value for each of a set of spectral frequencies w_i (i=1,2,3…) is expressed as in equation (10).

$$s(w_i) = \frac{f^T \hat{g}(w_i)}{f^T f} \quad , \quad i = 1,2,\ldots,k \tag{10}$$

where, $w=2\pi f$ *radial frequency* and its value is subjective, however upper and lower limits are considered while determining its value. While the lowest is the inverse of the number of observations, the highest one is the fundamental frequency. This fundamental frequency is expressed with equation (11).

$$f_N = \frac{1}{2\Delta t} \tag{11}$$

where, Δt is the interval between two observations.

Equation (12) is used to derive power spectra of the series.

$$PSD_{LS} = \frac{Q_n}{f}\left[\frac{s}{1-s}\right] \tag{12}$$

where, PSD_{LS} is the power spectrum density of a series from the least squares spectrum s, Q_n is the quadratic norm of noise component, f is frequency (see Pagiatakis (2000)). The significant peaks are identified with comparing the power spectrum density to a critical value in order to an assumed confidence level α (generally 99%) in the series. The aspects of this significance test are given by the theorems 3 and 4 of Pagiatakis 1999 (also see Pagiatakis 2000 for practice and discussion). According to them, c_α critical value is expressed as in equation (13).

$$c_\alpha = \left(1 - \alpha^{2/(n-u)}\right) * 100 \tag{13}$$

where, n is the number of observations, u is the number of unknowns. The higher spectral values with respect to the critical value are assumed significant in the order of α confidence level.

The basics and comprehensive applications of LSSA can be found in Omerbashich (2003), Omerbashich and Vaniček (2000), Pagiatakis (1999, 2000), Vaniček (1969, 1971), Wells et al. (1985), Wert (2004).

Also for the theoretical aspects and applications of well established statistical tests, Fotopoulos (2003), Koch (1999), Vaniček and Krakiwsky (1986) can be seen.

3 Case Study: Monitoring and Analysing of the Building Movements

The structural and time dependent deformations of high buildings are monitored and investigated using different surveying techniques. These deformations occur under the environmental forces, which can be basically grouped as (according to Öztan (1985)):

- sliding and/or vertical deformations of the structure with the effect of its own mass
- loading of population living in the structure
- effects of natural phenomena such as sun rays, strong winds, tide effects for the structures in the coastline, earthquakes and so on

The geodetic measurements with the aim of deformation monitoring are done with certain periodicity and serve to determine the uniform and continuous deformations of the structure very well. However a sudden change, which occurs in the structure between the two periods of measurement. This change might not be understood for a long time to come while employing the conventional geodetic metrology techniques alone and this can be dangerous. Therefore, to avoid such a risk the geodetic deformation measurement techniques should be supported by a precise continuous monitoring technique especially, for the structures in critical regions such as land slide areas, active fault regions etc.

In this study, precise digital inclination sensors were used in addition to GPS technique while monitoring and determining a multistorey high rise building (17 floor), which is located very close to an active fault zone and was affected by East Marmara Earthquake in 1999 in Turkey.

3.1 Monitoring System and Description of Data

The study constitutes two measurement techniques namely GPS technique and inclination sensors. Three GPS campaigns were done with six month intervals. The control network for GPS monitoring, which is established during the first campaign, is seen in Figure 5. The stations of the control network were located on stable sites around the building and also four deformation points were located at the top of the piers of the building (see Figure 5). Dual frequency geodetic GPS receivers have been used during the campaigns and the observation times at each station varies between 40–60 minutes with 5 second recording interval. The deformations were investigated with millimeter accuracy (relative) point positions from three GPS campaigns.

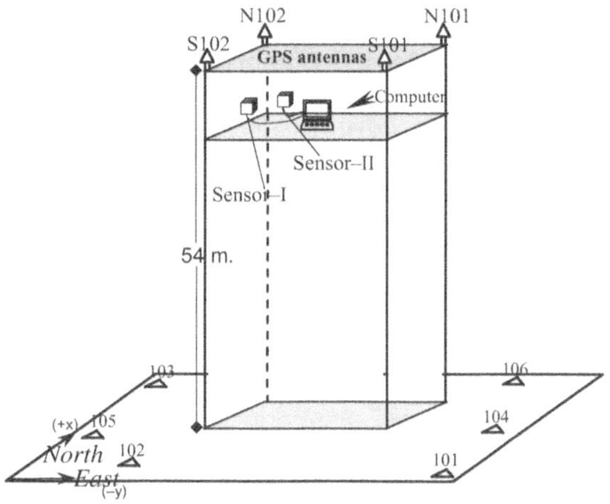

Fig. 5 The high–rise building, locations of inclination sensors at its top floor and the deformation network points

Other than GPS technique, the structural movements of the building have been monitored continuously between second and third GPS campaigns using microradian precision Leica NIVEL20 double axis inclination sensors. The aim of monitoring the movements with precise inclination sensors in this project has three parts, which are a) analysing the long term elastic deformation behaviors of the building, b) recording probable static deformations under effects of natural forces such as an earthquake, strong winds and so on, and c) testing the usability and performance of these high precision measurement equipments to support and control the GPS technique in deformation measurements.

The inclination sensors were set on steel console platforms, specifically designed for NIVEL20, on two columns of the top floor of the building. The two perpendicular sides of the building were chosen intentionally to locate the sensors' platforms (sensor I to the West–side and sensor II to the North–side) in order to be able to observe the deflections through both directions. The inclination data in milliradian were transmitted to a terminal computer (see Figure 5) and recorded to a data–base in every 5 minutes during the 40–day campaign. However, even though that the sampling interval was constant, the extreme changes in inclinations out of the interval were also recorded by sensors. In Figure 6a–d, the inclination graphics are seen. Also the temperature of the environment in Celsius have been measured and recorded simultaneously during the observations (see Figure 6e for temperature graph).

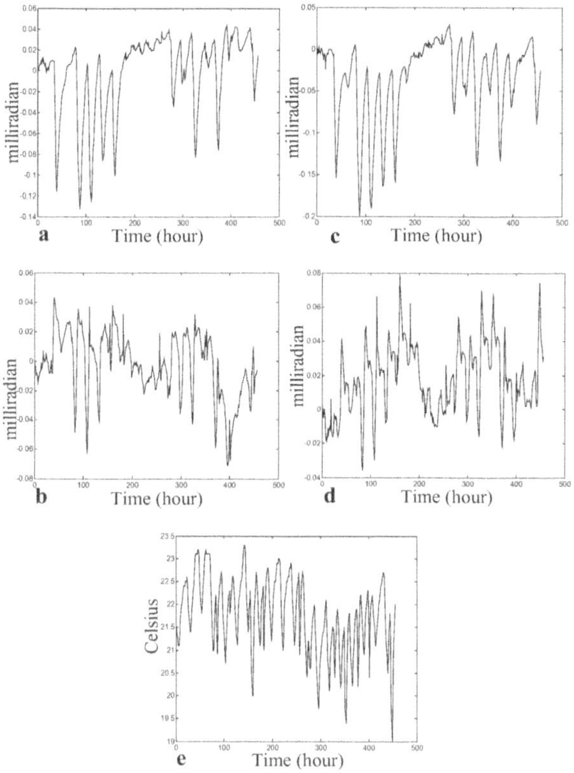

Fig. 6 the graph show inclinations versus time: sensor I (*north–south* direction / x axis) **a**; sensor I (*east–west* direction / y axis) **b**; sensor II (*north–south* direction / x axis) **c**; sensor II (*east–west* direction / y axis) **d** and temperature values in Celsius during observations in **e**.

3.2 Accuracies of Data Sets and Error Sources

As is well known, monitoring the deformations of engineering structure requires high precision measurement techniques. Under consideration of this fact, special precautions are applied to minimize the errors in observables. In case of using GPS technique, the error sources, which have an important role on deformation analysis, are given below (see Fotopoulos (2003), Featherstone (1998), Erol et al. (2004));

- Vertical Dilution of Precision (VDOP) (geometry of satellite system)
- satellite ephemeris and GPS baseline length (geometry of satellite system)
- ionosphere and trophosphere (the environmental effects)
- multhipath (the environmental effects)
- antenna orientation and phase center (instrumental and user)
- measuring antenna height (user)
- centering of the antenna (user)

Mentioned error sources are significant especially in height component of GPS derived position. Nevertheless, special precautions are applied against to each error source. Even after applying the precautions there remain errors, which distort high precision heights. Therefore, in addition to these precautions, GPS technique is recommended to support with an auxiliary measurement technique in order to improve GPS derived height component (see Erol et al. (2004a, 2004b)).

Most of these error sources were considered in this study. Some special precautions were applied during GPS campaigns such as using forced centering equipments at each station of deformation network to avoid centering errors of antennas.

The inclination sensors are calibrated under laboratory conditions during production in order to avoid instrumental error sources (Anonym (1998)). In addition to this, there are special precautions to be considered while setting the monitoring system. The stability of the sensors has enormous importance during observations. Therefore special carrier console platforms, which are made of special alloy and do not change their shape due to atmospheric changes, were used for mounting the sensors. Special covers protected them from dust, water and other environmental effects during the observations.

The relative accuracy of GPS derived positions of deformation points was better than ±3 millimeter in this study. The static GPS measurement method has been applied in each campaign and the probable deformations have been investigated in order to differences between coordinate sets.

The measuring accuracy of the double axis inclination sensors is in microradian. The axes directions of each inclination sensor were set parallel (the y axis to east–west direction and x axis to north–south direction) at the beginning of observation period to simplify comparing and interpreting the sensors' data with each other.

3.3 Autocorrelation and Cross correlation Analysis of Inclination Data

Autocorrelation functions of inclinations recorded by sensor–I along east–west and north–south directions (see Figure 7 a–b) and by sensor–II along east–west and north–south directions (see Figure 7c–d) reveals the presence of periodicity.

When the correlograms in Figure 7 are considered, it is seen that the autocorrelation coefficient has peaks in every 24 hour – time lag. Therefore the highest period can be assumed as 24 hour for each of the series (see Figure 7).

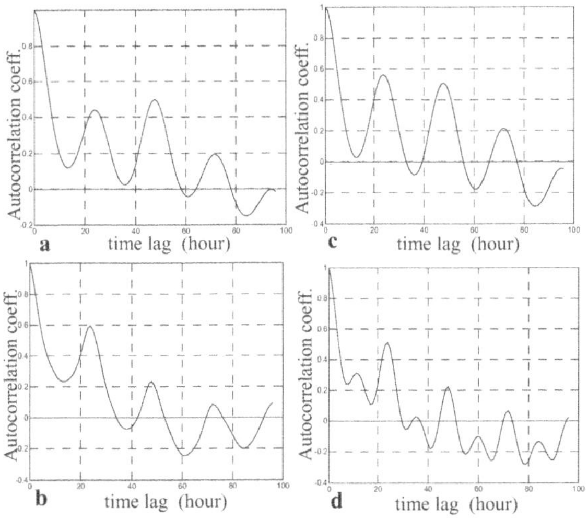

Fig. 7 the autocorrelation functions of inclination observations of *sensor* I – *north–south* direction (x axis) **a** and *east–west* direction (y axis) **b** , and of *sensor* II – *north–south* direction (x axis) **c** and *east–west* (y axis) direction **d**.

The cross correlation functions between the parallel axis inclinations of both sensors are shown in Figure 8. From the correlogram in Figure 8a, the high correlation between inclination signals on x–axes of sensor–I and sensor–II can be seen. However, the correlogram in Figure 8b, which reflect the correlations between inclination signals on y–axes of the sensors, does not show high correlation values due to the changes between time–lags. According to the tests applied, the reason for this is that the instrumental errors affected the y–axis values of sensor–II.

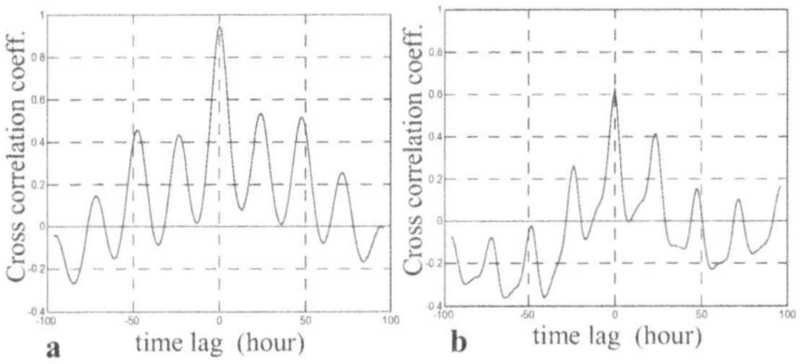

Fig. 8 the cross correlation functions between time series (inclination observations) of sensor–I and sensor–II: the data on *north–south* (x axis) directions of both sensors **a**; the data on *east–west* (y axis) directions of the sensors **b**.

The correlation function clarified the periodicity of the movements of the building. However, a more rigorous investigation of the movements was performed with LSSA technique in the following.

3.4 Analysis of Structural Deformations with Inclination Data Using LSSA

The least squares spectral analysis of sensor–I and sensor–II data has provided identical results as expected. Therefore, only the analysis results of sensor–I data have been shown here.

Analysing the inclination data by LSSA method was done using Least Squares Spectral Analysis program Version 5.02 (Anonym (2002)).

At the first stage of data evaluation using the LSSA technique, the whole observation period shown in Figure 6a and 6c, has been taken into account. Figure 9 shows the results of this evaluation. In Figure 9a (x–axis series) and 9b (y–axis series), the percentage variances (%var) and in Figure 9c and 9d, the respective power densities in mrad^2/frequency are shown. The dashed–line expresses the critical Least Squares Spectrum value at 99% confidence level, while detecting the significant peaks. In evaluating of the spectrum, the significant peak has been identified with 24–hour periods.

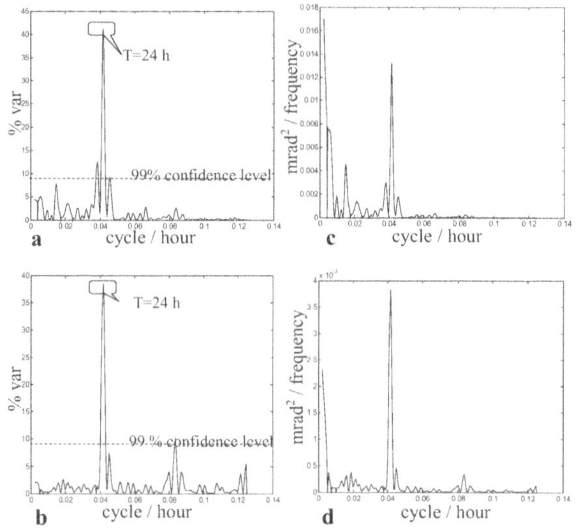

Fig. 9 the Least Squares Spectrum and power densities of the whole observation period: **a** and **b** graphics are least squares spectrum of inclination observations in percentage variance (%var) at *north–south* direction (x–axis) and *east–west* direction (y–axis) respectively. The dashed–line shows the critical %var value at 99% confidence level for detecting significant peaks. The respective power density graphics in mrad^2/frequency are given in **c** and **d**.

For more rigorous investigation of the data spectrum, it has been divided into weekly periods and LSSA approach has been applied to each of them in the second stage of analysis. The found results at previous stage have been supported by the results of weekly evaluation of the data. The significant peaks with a 24–hour period were recognized in the results. The least squares spectrum and power densities of the analysis can be seen in Figure 10.

In order to recognize higher frequency periodicities in the spectrum, the 24–hour period has been forced. In this third stage of data evaluation with LSSA, another significant period of 12–hour along north–south direction (x–axis) has been recognized. Also two significant periods of 12–hour and 8–hour have been identified along east–west direction (y–axis). The least squares spectrum values (Figure 11a and 11b) and respective power density graphs (Figure 11c and 11d) show these significant higher frequency signals clearly.

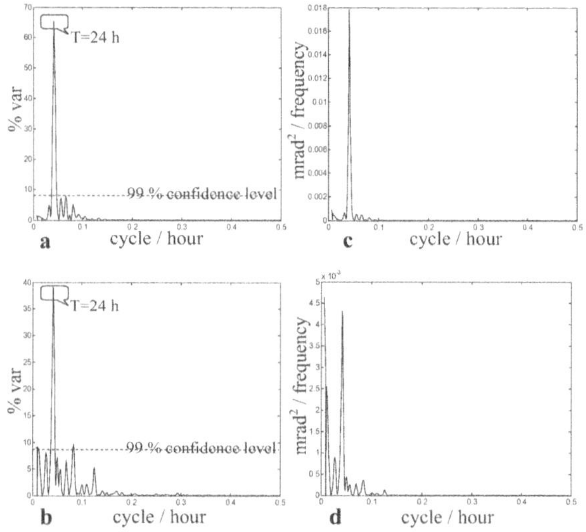

Fig. 10 the Least Squares Spectrum and power densities of the weekly observation periods: **a** and **b** graphics are least squares spectrum (%var) of inclination observations at *north–south* direction (x axis) and *east–west* direction (y axis) respectively. The respective power density graphics (mrad^2/frequency) are given in **c** and **d**.

From the results of evaluation of the data using LSSA technique, the inclination movements in the sense of vertical deflections of the building seemed to have two repeating components along north–south direction with 12–hour and 24–hour periods and three repeating components along east–west direction with 8–hour, 12–hour and 24–hour periods. The so–called periodicities appeared in the results were interpreted in the following.

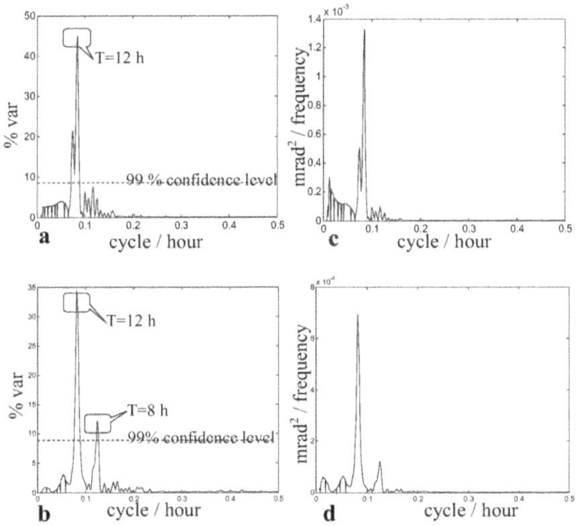

Fig. 11 the Least Squares Spectrum and power densities of the weekly observation periods after the signal with highest period is forced: **a** and **b** graphics are least squares spectrum of inclination observations in %var at *north–south* direction (x axis) and *east–west* direction (y axis) respectively. The respective power density graphics in mrad²/frequency are given in **c** and **d**.

3.5 Combined Interpretation of the Results of Inclination Sensors and of GPS Data

According to studies that used GPS campaigns for building deformation surveys it seemed possible to mention about the presence of movements (Çelik (2003)). However, it was not easy to judge these movements as the static deformations of the building and to differentiate them from the natural elastic movements. Therefore, it was found useful to combine the results of GPS data analysis with inclination sensor observations.

Figure 12 shows the magnitudes of the movements from the coordinate differences of deformation points between GPS campaigns. The root mean square error (rms) of the coordinates has been determined to be approximately ±3 millimeter. Three times of rms value has been accepted as significance level with 99% confidence (Figure 12).

The monitoring of the building with inclination sensors was performed between the second and the third GPS campaigns. According to least squares spectral analysis of inclination data, the elastic movements of the building were investigated and interpreted with the help of the consultations with structure engineering and material specialists. Also the information on environmental factors such as temperature and wind velocity and earthquake information are also considered during the interpretations. The results of LSSA of the data are summarized in Table 1a and 1b.

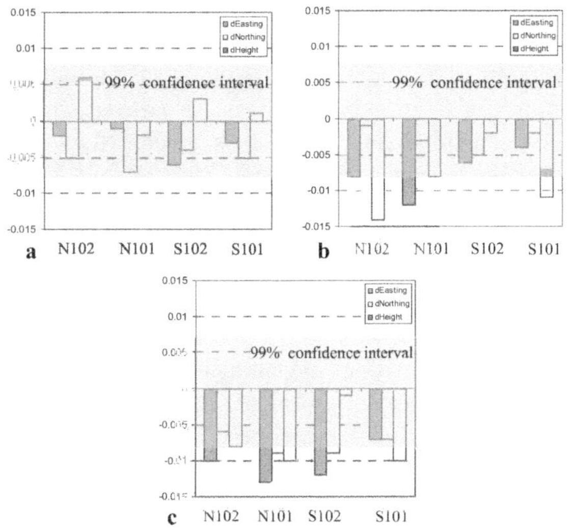

Fig. 12 The coordinate differences at deformation points from GPS campaigns: GPS^1–GPS^2 in **a**, GPS^2–GPS^3 in **b** and GPS^1–GPS^3 in **c** (the superscripts show the campaign number). Units are in meter.

Table 1a. north–south direction (x–axis) inclinations

Period (hour)	Frequency (cyc / hour)	Amplitude (mrad)	Description (efect)
8	–	–	–
12	0.083208	0.012	–
24	0.041816	0.024	warming–cooling

Table 1b. east–west direction (y–axis) inclinations

Period (hour)	Frequency (cyc / hour)	Amplitude (mrad)	Description
8	0.125751	0.007	–
12	0.082188	0.012	–
24	0.041649	0.013	warming–cooling

According to results in Table 1a and 1b, the elastic movements with 24–hour period along the x and y axes were interpreted as the natural behavior of a reinforced concrete high rise building under the warming–cooling phenomena during a whole day and repeat itself with a constant magnitude (see also Özöner (Erol) (2000)). However, the higher harmonics with 12–hour periods along the x and y axes and 8–hour period along the y axis were interpreted as the consequence of 24–hour periodicity in the result of non–linear response of the building. It was decided that any of the environmental effects under consideration may not have 8–hour fundamental period.

When the elastic movement with 24–hour period is considered, 0.024 mrad amplitude corresponds to 1.3 mm horizontal displacement on the x axis direction and 0.013 mrad amplitude corresponds to 0.7 mm horizontal displacement on the y axis direction on the top of the building (1mrad=0.001 mm/m and the building is 54 meter high). According to these results, it is evident that GPS technique stays insufficient to resolve the deformations observed by the inclination sensors.

With this respect, the benefit of combining GPS technique and inclination sensors in investigation of deformations of engineering structures is twofold. While the GPS technique is very useful to determine the static deformations in centimeter level, it can not resolve the elastic movements in sub-centimeter. Therefore, the input from inclination sensors data analysis makes contribution for understanding the type (statistic, elastic or semi–elastic) and character (continuous or instantaneous) of the deformations in the result of the analysis of GPS campaigns. On the other hand, the precision of inclination sensors is useful to clarify the amount and periodicity of the elastic deformations even at sub–millimeter level. However, analysis results of the inclination sensors data should be interpreted along with the output of GPS analysis in order to be able to reach more realistic results about the total deformations of the structure.

4 Conclusion

In this investigation, the movements of a high rise building with 17 floors has been observed using digital double axis inclination sensors in microradian precision and analysed using correlation functions and LSSA technique. The results have been compared and combined with analysis results of GPS data of three measurement campaigns.

From the least squares spectral analysis of the movements with inclination data, the hidden periodic components were figured out with 24–hour and 12–hour periods along *north–south* direction and an additional 8–hour period along *east–west* direction. The 24–hour periodic movements with 0.024 mrad and 0.013 mrad amplitudes on x and y axes directions respectively were interpreted to be natural as depending on warming–cooling effect of the material of the structure during the whole day. Since the warming and cooling phenomena of building is not purely sinusoidal, the higher harmonics with 12–hour and 8–hour periods were explained as the result of non–linear response of the building and occurred by mathematically splitting of the unique fundamental signal with 24–hour periodicity in the LSSA results. This interpretation is also supported by the analysis results using correlation functions. In the auto–correlation functions of inclinations on each axis, the 24–hour periodicity of the inclining movements were identified too. Therefore, correlation functions provided a straightforward control for LSSA results.

The irregularity of the sampling intervals in the inclination data set was the reason for using LSSA technique in the study. However, this technique was found successful for evaluation of inclination data in this study particularly, as a tool to find out the hidden periodicities in a time series having equal or non–equal intervals as its difference from other spectral analysis techniques.

The deformations of the building were analyzed from coordinate differences of deformation points between GPS campaigns as well. When the accuracy of coordinates derived from observations was considered and the magnitudes of the movements from coordinate differences between the campaigns were compared with horizontal displacements from inclination sensors, the insufficiency of GPS technique to recognize the movements in sub–millimeter level was evident. However, the benefit of combining these two techniques was twofold. While deciding the deformations in the result of GPS analysis whether static or elastic, inclination sensor analysis results, which clarify the amount and the periodicity of deformations, made contribution. On the other hand, while interpreting and understanding the analysis results of inclination data, output from GPS technique provided a general look in investigation of the entire (static and elastic) deformations of the building.

In the result of the combined analysis of the inclination data and GPS technique, the GPS derived coordinate differences were perceived as the result of elastic movements of the structure instead of constant static deformations contrary to expectations.

Acknowledgement. We would like to acknowledge Prof. Dr. Spiros Pagiatakis for giving us his valuable time during our investigation. Profs. E. Brückl and M. Llubes are kindly acknowledged for their contributions as reviewers in the improvement of the article. NIVEL20 sensors have been provided by Leica Sistem Bilgisayar Ltd. Company to this study.

References

Anonym. (1998). NIVEL20 Instruction Manuel. *Handbook 308 592 EN-IX.98*, Leica Geosystems AG, Switzerland.

Anonym. (2002) LSSA: Least Squares Spectral Analysis program. Version 5.02. *Copyright©2002*, University of New Brunswick.

Box, G. E. P., G. M. Jenkins, G. C. Reinsel (1994). Time Series Analysis- Forecasting and Control. Prentice-Hall. New Jersey. ISBN 0-13-060774-6. pp. 21-46.

Canıtez, N., C. Gürbüz., (1987). Preliminary Processes in Spectrum Computations. Spectral Analysis and Geophysical Applications. Canıtez, N., U. Yaramancı and H. Özdemir (eds). The Chamber of Geophysical Engineers of Turkey, pp.146-169. (in Turkish)

Çelik, R. N., (2001). Real Time Large Structure Monitoring Using NIVEL20. In: *Proc. of the NATO Advanced Research Workshop on Strong Motion Instrumentation for Civil Engineering Structures*, Istanbul, Turkey, Kluwer Academic Publishers, ISBN 0792369165, Boston , Science Series.

Çelik, R. N., (2003). Technical Report on Investigation of Time Dependent Structural Deformations of Halkalı Condominium-A503 Block after 1999 Marmara Earthquake. Istanbul Technical University. (in Turkish)

Çelik, R. N., B. Erol, C. Güney and S. Erol (2003). Monitoring and Investigating the Movements of a High Building Using Measurements and Precise Inclination Sensors. In: *Proc. of International Symposium of Modern Technologies, Education and Professional Practice in Globalizing World*, Sofia, Bulgaria, November 6-8, pp. 79-88.

Erol, S., B. Erol and T. Ayan (2004a). General Overview of the Deformation Monitoring Techniques and A Case Study: Analysing Deformations Using GPS/Levelling. *In: Proc. ISPRS 2004 XXI General Assembly*, Istanbul, Turkey, 12-23 July.

Erol, S., B. Erol, M., Tait and T. Ayan (2004b). 1D and 3D Analysis of Deformations in Engineering Structures Using GPS and Terrestrial Data. *In: Proc. FIG INGEO 2004*, Bratislava, Slovakia, 11-13 Nov.

Featherstone, W. E., M. C. Denith and J. F. Kirby (1998). Strategies for the accurate determination of orthometric heights from GPS, *Survey Review*, 34–267, pp. 278-296.

Fotopoulos, G., (2003). An Analysis of Optimal Combination of Geoid, Orthometric and Ellipsoidal Height Data. Ph.D. Thesis, *UCGE Report 20185*, Dept. of Geomatics Engineering, The University of Calgary.

Hardy, R. L., (1971). Multiquadric Equations of Topography and Other Irregular Surfaces. Journal of Geophysical Research, 76-8, pp. 1905-1915.

Hekimoğlu, Ş. (1981). Random Processes. *Research Reports Series. 1981/1*, Faculty of Geosciences, Karadeniz Technical University. (in Turkish)

Koch, K. R., (1999). Parameter Estimation and Hypothesis Testing in Linear Models. Springer Verlag, ISBN 3-540-65257-4, pp-271-307.

Omerbashich, M., (2003). Earth–Model Discrimination Method. *Ph.D. Thesis*, Graduate Academic Unit of Geodesy and Geomatics Engineering, The University of New Brunswick, pp. 31-37.

Omerbashich, M. and P. Vaniček, (2000). Least Squares Spectral Analysis of Gravity Data from the Canadian Superconducting Gravimeter: an ongoing Project Report. *Presentation at GAM Conference*, GEOIDE Network, internet site: http://gge.unb.ca/Personnel/Vanicek/

Özöner (Erol), B., (2000). Monitoring Deformations of Large Buildings by Precision Inclination Sensors. MSc. Thesis, Istanbul Technical University, Institute of Science and Technology. (in Turkish)

Öztan, O., (1985). Methods for Continuous Monitoring of Structural Deformations. *Map Journal*, 95, pp. 71-86. (in Turkish)

Pagiatakis, S. D., (1999). Stochastic Significance of Peaks in the Least-Squares Spectrum. *Journal of Geodesy*, 73, pp. 67-78.

Pagiatakis, S. D., (2000). Application of the Least Squares Spectral Analysis to Superconducting Gravimeter Data Treatment and Analysis. European Center of Geodynamic and Sysmology, 17, 103–113 .

Pytharouli, S., V. Kontogianni, P. Psimoulis and S. Stiros, (2004). Spectral Analysis Techniques in Deformation Analysis Studies. In: *Proc. of the INGEO 2004 and FIG Regional Central and Eastern European Conference on Engineering Surveying*, Bratislava, Slovakia, November 11-13.

Vaniček, P., E. J. Krakiwsky, (1986). Geodesy: the concepts. *Elsevier Science Pub. Co.*, Amsterdam.

Vaniček, P., (1969). Approximate Spectral Analysis by Least Squares Fit. *Astrophysics and Space Science 4*, 387-391, D. Reidel Publishing Company, Dordretcht Holland, Kluwer Academic Publishers.

Vaniček, P., (1971). Further Development and Properties of the Spectral Analysis by Least-Squares. *Astrophysics and Space Science 12*, 10-33, D. Reidel Publishing Company, Dordretcht Holland, Kluwer Academic Publishers.

Wells, D. E., P. Vaniček and S. D. Pagiatakis (1985). Least Squares Spectral Analysis Revisited. *Technical Report 84*, Department of Surveying Engineering, University of New Brunswick.

Wert, T. D., (2004). Tidal Height Retrieval Using Globally Corrected GPS in the Amundsen Gulf Region of the Canadian Arctic. MSc Thesis, Graduate Academic Unit of Geodesy and Geomatics Engineering, University of New Brunswick, pp. 101-109.

Session C

Geodynamical Applications, Crustal Deformation

Chairmen: H. G. Henneberg and A. Dermanis

Tilting and Horizontal Movement at and across the Northern Border of the Adria Plate

Chiara Pinato Gabrieli, Carla Braitenberg*, Ildiko' Nagy
Dipartimento di Scienze della Terra, Università di Trieste, Via Weiss 1, 34100

David Zuliani
Centro Ricerche Sismologiche, Istituto Nazionale di Oceanografia e Geofisica Sperimentale, Via Treviso 55, I-33100 Udine

Abstract. We report on the analysis of the seven permanent GPS stations of the recently installed Friuli Regional Deformation Network (FReDNet) of NE-Italy for the time period that goes from the moment of installation (between June 2002 and July 2003, according to the station) to 31 December 2004. We first make a comparison of two available solutions, calculated by the Agenzia Spaziale Italiana (ASI, Centro di Geodesia Spaziale "G. Colombo") and by the Centro Ricerche Sismologiche (CRS, Istituto Nazionale di Oceanografia e Geofisica Sperimentale). The rms (root mean square) difference between the two solutions results to be between 4 and 8 mm, depending on the component. The greatest difference is found for the vertical component, the best agreement is found for the NS component. We then proceed with the analysis of the time series, using the solutions of CRS, as this agency has longer time series available. Our attention goes to three classes of signals, which is the high frequency component (several days to weeks), the seasonal and annual signal, and the interpolated linear trend, representative of the average velocity of the monuments. The mutual correlation coefficients of the stations for the same component are between 0.3 and 0.7, with the highest values found for the vertical component, which demonstrates that there is a common signal, which is in phase. The correlation coefficient does not show a distance relation of the stations, which indicates that the origin of the common signal could be due to pressure loading or an incomplete atmospheric correction. In order to determine reliable velocities, we determine the sinusoidal annual and semi-annual oscillations, which are present in all stations. The amplitudes of the yearly oscillation vary between 0.3 mm and 1.4 mm for the horizontal components, and are definitely larger for the vertical component (between 2 and 5 mm). Those of the semi-annual period range between 13% and 95% of the annual amplitude. At last we determine the relative velocities of all stations with respect to the Trieste station, which can be considered tectonically stable, as it is located on the aseismic Karst. We find that the pedemountain stations have a relative southward motion of about 2-3 mm/year and an uplift rate of 1-2 mm/year. The stations on the Friuli plane have nearly no horizontal movement relative to Trieste, and show a relative subsidence. The observations are in agreement with the expectations from neotectonics, which give subsidence of the Friuli plane and uplift of the pedemountain area. They also agree with north-westward tilting observed with the long-base tiltmeters of the Grotta Gigante station (Trieste Karst). The southward movement of the pedemountain stations fits the picture of the relative movement of the Eurasian and Adria plates, the Trieste and Friuli-plane stations belonging to the Adria plate, the remainder stations belonging to the Eurasian plate.

Keywords. Horizontal and vertical crustal movement, tiltmeters, GPS, Friuli (NE-Italy), Adria plate

1 Introduction

The Friuli area belongs to the eastern portion of the subalpine mountain range located to the south of the Peri-Adriatic lineament. Starting from late Oligocene to present a series of north-dipping shallow angle faults have been activated, that extend from the Garda lake (west) to Slovenia (east) (e.g. Doglioni, 1992). In the western sector of the Friuli area the quaternary front of the subalpine range corresponds to the Aviano and Sequals overthrusts; in the oriental sector, the front of the

* corresponding author: berg@units.it

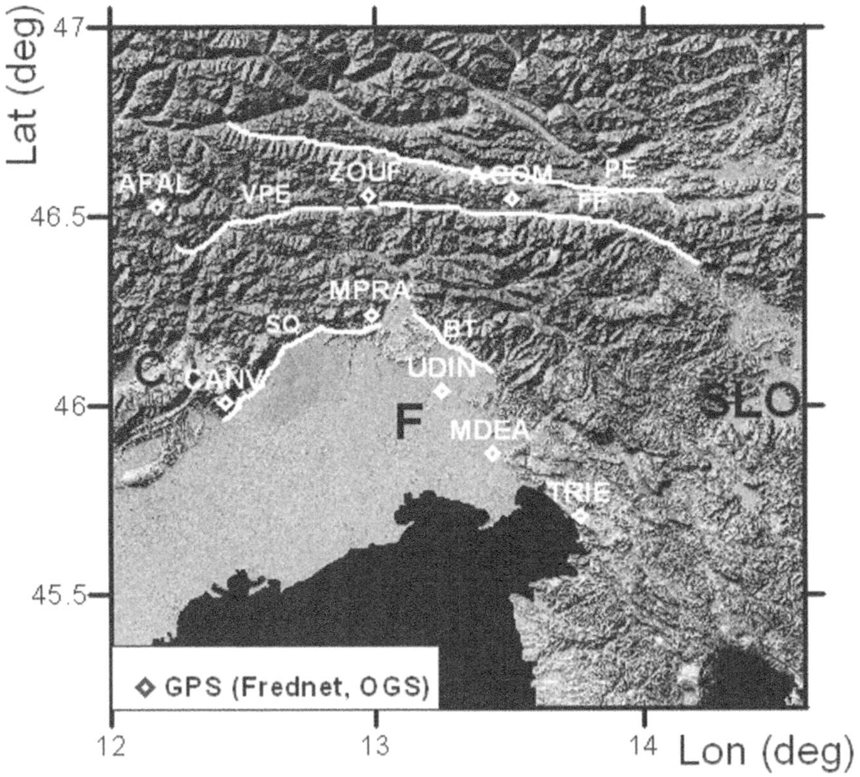

Fig. 1 Map of NE-Italy and surrounding areas with GPS stations. SLO: Slovenia, F: Friuli, C: Cansiglio. Fault lines: SQ: Sequals, BT: Buia-Tricesimo, VPE: Val Pesarina, FE: Fella Sava, PE: Periadriatic.

mountain range corresponds to a series of overthrusts that extend into the Friuli plane, and belong to overthrusts of the previous Dinaric front (Zanferrari et al., 2000). The external margins of the southern Alps and of the Dinarides show signs of recent deformation, that is maximum in the area of the Friuli pedemountain arch. The external margins represent the present tectonic evolution of the area, characterized by overthrusts with a southward movement towards the Adria plate. Presently, the pre-alpine Friuli sector is uplifting, while the frontal plane is subsiding (Zanferrari et al., 1982). The Friuli area is located in a ribbon of high seismicity (Fig. 3) that affects the pedemountain range extending from the Garda lake (to the west) to Slovenia (to the east) and bends south-eastwards along the Dinarides front. The alignment of the seismic ribbon follows the orientation of the major tectonic lineaments, which are seen in Fig.1. The seismicity along the E-W oriented faults has preponderantly thrust mechanisms on subhorizontal faults dipping northwards, whereas the events towards the east on the NW-SE oriented faults have dextral mechanisms. The Friuli and Cansiglio have been, since historical times, the areas of highest seismic potential. In the 20th century the greatest destructive earthquakes were the Friuli event (May 6, 1976) and the Cansiglio event (October 18, 1936) of eastern Veneto. Slejko et al. (1987) quantified the regional seismicity in terms of number of events with magnitude greater than 3.3 in 1000 km^2 for the years 1850-1984 and found values of 1 and greater for Friuli and Cansiglio, a value of 0.5 for the Garda lake and Ljubljana (Slovenia) and lower values for the remaining portions. The hypocenters affect the upper crust and are principally located between 5 and 15 km depth, with a peak distribution at 9 km (Braitenberg, 2000). The crustal deformation in Friuli has been monitored by underground tilt and extensometric instrumentation since 1977 (Zadro and Braitenberg, 1999, Braitenberg, 1999). The long data series have allowed to make a good statistic on the observed co-seismic signals and to study the ambient influence due to temperature, atmospheric pressure and hydrology on the crustal deformation measurements. A long-base tiltmeter station has been active since 1960 on the Trieste Karst to the south of the main seismic area, giving

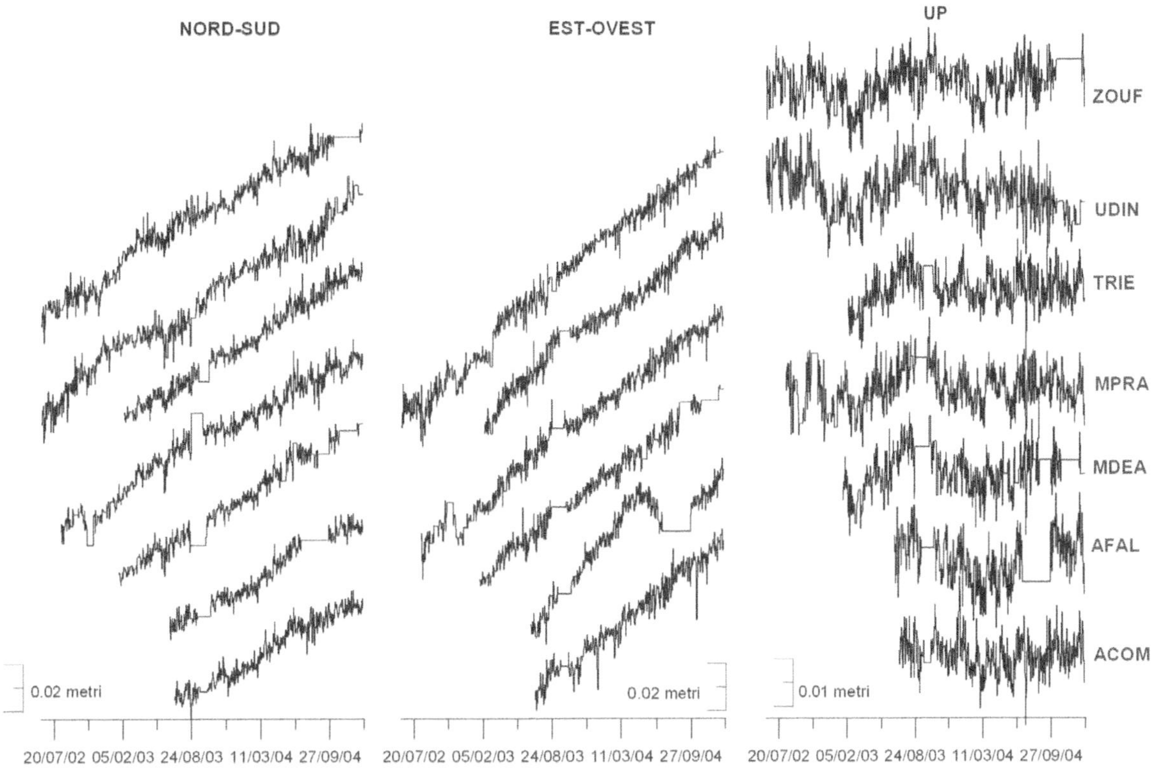

Fig. 2 The daily GPS solutions (CRS) for the seven FReDNet stations (ITRF00 reference frame). Interruptions are replaced by constant values.

reliable observations of a NW-ward tilting of the Karst block over the recent 40 years (Braitenberg et al., 2005). A tilt should imply a differential vertical movement of the stations aligned along the tilting direction. In the case that the observed tilt is a regional signal and the resolution sufficient, a GPS station to the NW of the Trieste Karst should be subsiding with respect to a station located in the Trieste Karst. Recently (starting with June 2002) a network of continuous GPS stations was installed with the aim of monitoring the surface crustal movement and obtain the geodetic strain rate (Battaglia et al., 2003). All the GPS locations, except the UDIN station, have been carefully chosen in order to be linked to solid rock. The UDIN station was set up in the immediate vicinity of the CRS main building and is mounted on an industrial warehouse, therefore it is less stable than the other stations. The locations of the stations are shown in Fig. 1. The three stations TRIE, MDEA, UDIN, are to the south of the seismic ribbon, and are located on the Adria plate, with the TRIE station being on the Karst, the two stations MDEA and UDIN on the Friuli plane. The station MPRA is just to the north of the plane and at the southern margin of the seismic ribbon, stations ZOUF and ACOM are just to the north of the central Friuli seismic area, bordering the Fella-Sava and Val Pesarina lineament. AFAL is farther to the west, dislocated from the main Friuli seismicity, about 50km north of the destructive Cansiglio earthquake of 1936.

In the present study we intend to determine the relative velocities of all stations with respect to the Trieste station, which can be considered as tectonically stable as there is no seismicity (e.g. Slejko et al., 1987). The pedemountain stations are located to the north of the transition from the Adria plate to the European plate margins. The determination of the relative velocities is preceded by a study which aims at determining the nature of the high frequency (days to weeks) signal and to clean the data from the annual and semi-annual signal (Pinato Gabrieli, 2005). This analysis was made on the time series from the time of installation of the stations to 31 December 2004. The stations were installed between June 2002 and July 2003, as reported in Table 1.

Table 1. Time intervals covered by the GPS observations of the FReDNet stations for the solutions of the CRS agency.

Station	Month of Installation
Acom (M. Acomizza)	July 2003
Afal (Alpe Faloria)	June 2003
Mdea (Medea)	January 2003
Mpra (M. Prat)	August 2002
Trie (Trieste)	February 2003
Udin (Udine)	June 2002
Zouf (Zouf Plan)	June 2002

2 Data Presentation

The GPS solutions have been calculated with daily sampling rate adopting the ITRF00 reference frame. The CRS solutions are computed using GAMIT/GLOBK (M.I.T and U.C.S.D, 2002) software and the time series cover the time intervals given in Table 1.

On routinely basis, the FReDNet GPS stations are also calculated with daily sampling by ASI, which uses the Microcosm vs. 2001 (GEODYN II, NASA, Eddy et al., 1990) software. The available solutions start at a later date (February 2004) than the ones calculated by CRS. A comparison of the two data sets over the common time interval has shown that the rms-difference between the two data series is 3-4 mm for the NS, 4-5 mm for the EW and 7-8 mm for the UP component. The rms difference is slightly higher than that expected given the formal error given by the two agencies (CRS: average 1.5 mm for the horizontal, 4.5 mm for the Up components; ASI: 1mm for the NS and UP, and 0.7 mm for the EW components). The difference in the solutions is probably due to the different software used and to the specific choice of the reference stations used in the calculations. Due to the short common time interval, presently the origin for the encountered differences is unknown and requires further investigation. In the remainder of the study we have used the CRS solutions, as the available data interval is the longest. The observed data from all stations are presented in Fig.2. The data series are characterized by a near to linear movement of the monuments, a near to yearly variation and a high-frequency variation. The high frequency content is a signal which is common to all stations, as can be verified from the mutual correlation coefficient between the stations. We calculate the correlation coefficients on the detrended time series, and obtain the values reported in Table 2.

Table 2. Mutual correlation coefficients of the FReDNet stations. The relatively high values of the coefficients demonstrate that there is a common signal to the stations.

	CORRELATION COEFFICIENTS		
	NORTH-SOUTH	EAST-WEST	UP
MDEA-MPRA	0.47	0.56	0.65
MDEA-TRIE	0.52	0.62	0.63
ACOM-MDEA	0.48	0.42	0.43
ACOM-MPRA	0.38	0.53	0.49
TRIE-MPRA	0.31	0.65	0.67
MPRA-ZOUF	0.55	0.61	0.54
MDEA-ZOUF	0.36	0.58	0.59
ACOM-ZOUF	0.58	0.46	0.57
ZOUF-TRIE	0.43	0.55	0.60
ACOM-TRIE	0.59	0.46	0.55
UDIN-ACOM	0.17	0.22	0.13
UDIN-MDEA	0.34	0.51	0.37
UDIN-MPRA	0.06	0.49	0.48
UDIN-TRIE	0.44	0.48	0.47
UDIN-ZOUF	0.14	0.42	0.36
ACOM-AFAL	0.63	0.01	0.52
MDEA-AFAL	0.48	-0.12	0.47
MPRA-AFAL	0.26	-0.03	0.47
TRIE-AFAL	0.50	-0.21	0.48
ZOUF-AFAL	0.41	-0.20	0.59
UDIN-AFAL	-0.05	0.27	0.02

Generally the highest values are found for the Up component (0.5-0.7), followed by the E-W component (0.4-0.6), the NS component having lower values (0.3-0.6). Exceptions to these general findings are the correlations with the EW component of AFAL, which has very low coefficients with all stations. The latter component though also presents anomalously large deviation from the linear trend, which could be due to a problem on the stability of this station (antenna, monument or other), reflected also in the correlation coefficients. Also all correlations with station UDIN are generally lower, presumably due to additional noise on this station caused by its relatively less stable position. The generally high mutual

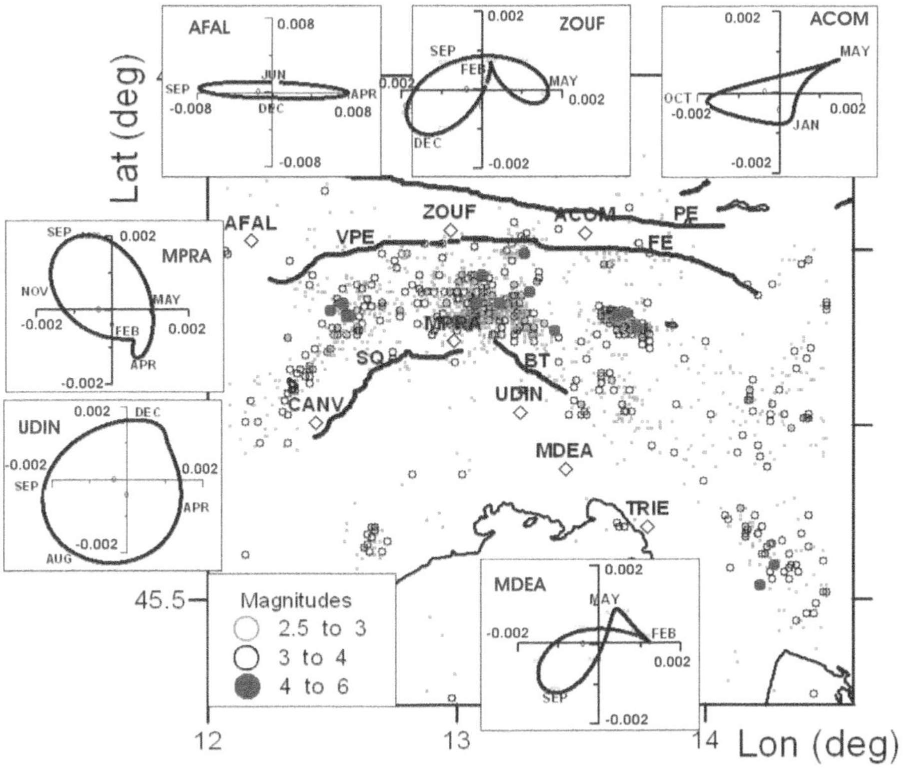

Fig. 3 Graphical representation of the seasonal movements of the monuments in the horizontal plane. The extremal values are labeled with the month of the year. Seismicity (CRS-OGS seismologic network) for the years 1977-2004

correlation coefficients demonstrate that the noise-like high frequency content has a large portion common to all stations. The fact that the coefficient is independent of the mutual distances of the stations, is an indication that the signal is not of tectonic origin but due to atmospheric effects. Infact, atmospheric disturbances and pressure variations are nearly homogeneous over a scale of 100 km (distance scale of the network); these could be the cause of the common signal. Atmospheric disturbances can either influence the range calculation or the movement of the monument (pressure loading, thermoelastic effects, soil condition influenced by humidity and hydrologic effects). By calculating the relative movements of all the stations with respect to one of the stations, this high frequency signal is reduced, and a more reliable analysis of the station velocities can be made. We choose the Trieste station as a reference station due to its position on the tectonically stable Karst block. Therefore the relative movements we obtain can be considered as being referred to the Adria plate.

3 Seasonal Variation and Relative Velocities

The presence of a seasonal or yearly signal is a known phenomenon, and has been observed in underground measurements (e.g. Zadro and Braitenberg, 1999; Braitenberg et al., 2001) as well as in satellite geodetic observations (e.g. Dong et al., 2002, Williams, 2003, for northern Italy see Caporali, 2003). The driving force of the signal is ascribed to the thermoelastic effects (Pawidodirdjo et al., 2004), hydrologic loading (water table or snow: Heki, 2003), atmospheric pressure loading, ocean loading, or a combination of the above effects (Zerbini et al., 2004). Especially in the case that the available time-series cover only a few years, the annual signal may distort the interpolated linear velocities of the strain, tilt, or crustal movement observations (Blewitt and Lavallèe, 2002) and must be corrected for. We model the signal with a combination of an annual and semi-annual sinusoidal variation which we determine by least square approximation together with the linear interpolation.

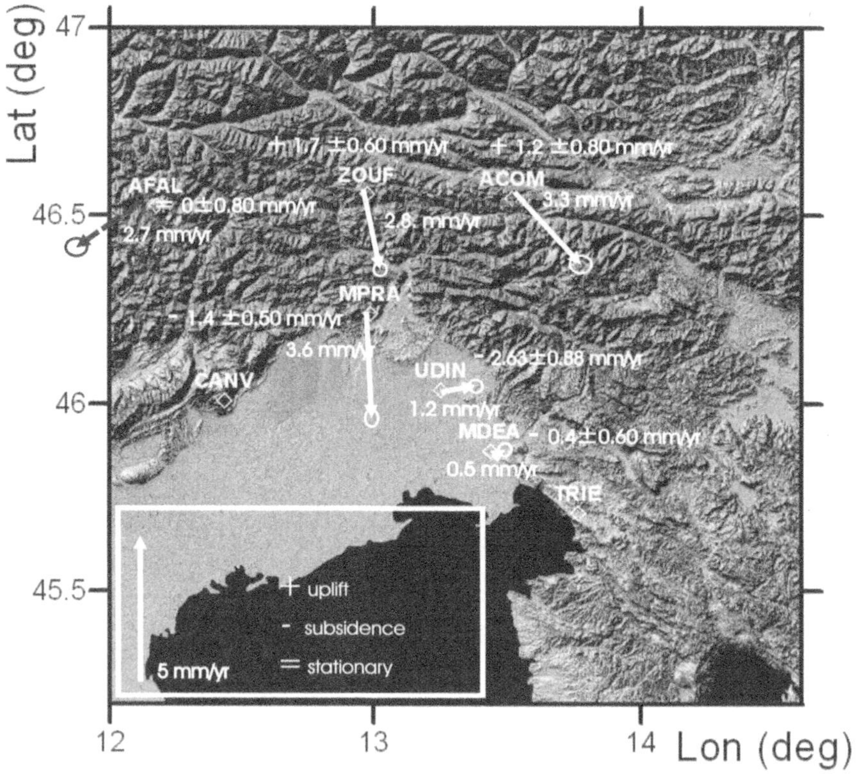

Fig. 4 – Horizontal and vertical velocities of the FReDNet stations relative to the Trieste GPS station. Error ellipses indicate 2 rms error.

We find that the seasonal variations are successfully extracted by this model, and determine amplitudes and phases of the annual and semi-annual variation. The data gaps (see Fig. 2) are less than 8 weeks long, except one (AFAL, max. 10 weeks) and therefore should have a negligible effect on the analysis (Blewitt & Lavallèè, 2002). The amplitude of the semi-annual variation is always smaller than the annual variation, but not neglectable. The amplitudes of the yearly oscillation vary between 0.3 mm and 1.4 mm for the horizontal components, and are definitely larger for the vertical component (between 2 and 5 mm). Those of the semi-annual period range between 13% and 95% of the annual amplitude. The amplitude of the annual period of temperature is about 10°, and the semi-annual period is not present in the temperature variation. The seasonal movement we find is graphically summarized in Fig.3, which shows the movement in the horizontal plane for every station. At the extremal positions the corresponding month in the year is added. The stations that show an elliptical movement are those for which the annual period is preponderant with respect to the semi-annual period. In Fig. 4 the final linearly interpolated velocities of the FReDNet stations relative to the Trieste station are shown, with the 2 rms error-ellipses. The vertical velocities are given as numbers next to the location of the station. We find a very interesting pattern of the velocities, which is expedient in the frame of plate-tectonics: the stations that are located on the Adria plate have very little horizontal movement (UDIN and MDEA), whereas the stations to the north of the Adria plate boundary (MPRA, ZOUF, ACOM, AFAL) have a definite relative southward movement in the order of a few mm/year. As mentioned before, the EW component of the AFAL station may be affected by an instrumental problem, and the observed west-ward movement cannot be considered to be reliable. Due to its position, also the station UDIN is not as reliable as the remainder stations. Although the vertical components are noisier than the horizontal components, the result we find is compatible with the general geologic picture: the stations in the mountain are uplifting (AFAL, ZOUF, ACOM), whereas the stations in the plane (UDIN, MDEA) are subsiding with respect to the Karst block. Station MPRA is very close to the transition from the Friuli plane to the prealpine overthrusts, which could be the reason to its anomalous movement, in that it is southward moving, but subsiding.

4 Discussion and Conclusion

Our study is concerned with a detailed analysis of the continuous GPS stations of the FReDNet network situated in the seismic Friuli area in NE-Italy. The GPS solutions are computed with daily sampling rate and the time series cover time intervals that range from 17 to 30 months. We have compared the solutions calculated by the two agencies CRS and ASI over the common time interval of 11 months and find an rms deviation of 3 to 8 mm, depending on the component. The greater value pertains to the Up component, and the NS component has the smallest rms difference between the two solutions for all stations. The rms difference is only slightly greater than the mean rms given by the agencies for the solutions. The time series contain a high frequency signal (period of several days) which we find by cross correlation analysis to be common to all stations, and which presumably is due to pressure loading or a residual in the range correction.

Another evident signal is the annual and semi-annual variation, which we have studied by least mean square approximation with two sinusoids of period of 365.25 days and 182.6 days, respectively. At this stage of the analysis we have determined the amplitude and phases of the oscillation with the aim of obtaining a reliable estimate of the linearly interpolated station velocities. Therefore the investigation on the cause of the oscillation remains to a follow-up study. The semi-annual oscillation cannot be due to a temperature effect, as the temperature variation does not contain semi-annual variations but is exclusively annual.

The third evident signal contained in the observations is the relative movement of the continuous stations with respect to the Trieste station. We find a clear difference in the horizontal velocities of the stations which are located on the Adria plate (TRIE, MDEA, UDIN) and the stations which are located on the European plate (MPRA, ZOUF, ACOM, AFAL). The difference in velocities is confirmed by the findings of D'Agostino et al., 2005. The analysed time period is relatively short for a reliable velocity detection. Nevertheless, the differences in the velocities of the pedemountain stations and the stations in the plane are so great that we are confident that the availability of longer time series will not alter the results. What will presumably change in the future will be the values of the relative velocities. The recent TRANSALP geophysical section which crosses the Adria-Eurasian plate boundary 100 km westwards of the Friuli area revealed the European plate dipping southwards beneath the Adria plate. The crustal structure in the collision zone could though not be resolved (Transalp Working Group, 2002; Kummerow et al., 2004). It remains controversial whether the Adria Moho south of the Alpine orogen remains flat (Kummerow et al., 2004) or is dipping northwards towards the orogen (Transalp Working Group, 2002). Teleseismic tomography (Lippitsch et al., 2003) indicates subduction of the European plate below the Adria plate west of the TRANSALP profile, and a change in the subduction pattern towards the east, with the Adria plate subducting below the European plate. The fault plane solutions of the seismicity are an indication of the northward movement of the Adria plate with respect to the European plate, with shallow angle N-dipping thrust faults in the central Friuli area, dextral strike slip events at the eastern border of the Adria plate and sinistral strike slip events at the western border (e.g. Caporali et al., 2005). The transition from the Adria plate to the European plate can be set at the northern margin of the Friuli plane, north of which the pedemountain overthrusts are found and the seismic belt begins. Our analysis finds that the head-on NS oriented convergence rate of the Eurasia and Adria plate is of about 2-3 mm/year, averaged over the interval of 2.5 years (2002 to end 2004). The two stations on the Friuli plane (UDIN, MDEA) and the Trieste station have nearly no relative horizontal movement. The stations on the plane are subsiding with respect to the Trieste station with a rate of 0.4 mm/yr or greater (UDIN being not totally reliable due to its location). The subsidence of the plane has further geomorphologic evidences on the basis of the following observation: a karstic cave at the depth of -180 m has been found 30 km northwest of Trieste by drilling a borehole (Albrecht and Mosetti, 1987). The cave must have been formed above sea level, as subsurface water erosion was responsible for its formation. Another observation was that of a sequence of submerged marine terraces in the gulf of Trieste. The uppermost terrace at the level of -20 m was attributed

(Antonioli et al., 2004) to the MIS 5.5 (marine isotope stage, about 125 ka). This marine terrace was correlated by Albrecht and Mosetti (1987) to a sediment horizon of the drill-log seated at the depth of 75 m. If the timing of those features is correct, this observation would imply a relative subsidence of 0.4 mm/year (55 m depth difference in 125 000 years), which is in the order of magnitude of our observations. A northward tilting, in accordance with the geological evidences, comes from the Trieste long-base tiltmeter station, which has recorded an average NW tilting of 41 nrad/year of the Karst block over a time window of the recent 40 years (1966-2004) (Braitenberg et al., 2005).

The good agreement between the GPS relative velocities and the movements expected from the geological consideration gives us confidence that our results are reliable and can be used for further studies concerned with the estimate of seismic risk based on the observed geodetic strain rate.

Acknowledgments

We thank Agenzia Spaziale Italiana (ASI, Centro di Geodesia Spaziale "G. Colombo" for the daily GPS solutions of the FReDNet stations. For discussions we thank Dr. Maurizio Battaglia, Dr. Lina Ferraro and Prof. Maria Zadro. This work was made with the contribution of the Istituto Nazionale di Geofisica e Vulcanologia and the Italian Department for Civil Defence.

References

Albrecht, P., and F. Mosetti, (1987). Karst evolution and sea level. *Mem. Soc. Geol. It.*, 40, 383-387.

Antonioli, F., G.B. Carulli, S.Furlani, R. Auriemma, and R. Marocco (2004). The enigma of submerged marine notches in northern Adriatic Sea. *Quaternaria*, VIII, 263-275.

Battaglia M., D. Zuliani, D. Pascutti, A. Michelini, I. Marson, M.H. Murray, and R. Burgmann (2003). Network Assesses Earthquake Potential in Italy's Southern Alps. *Eos*, vol. 84, No. 28, 262-264.

Blewitt G., and D. Lavallèe (2002). Effect of annual signals on geodetic velocity. *J. Geophys. Res.*, vol. 107, No. B7, 2145.

Braitenberg, C. (1999) The Friuli (NE Italy) tilt/strain gauges and short term observations. *Annali di Geofisica* ,42, 1-28.

Braitenberg, C. (2000) Non-random spectral components in the seismicity of NE Italy, *Earth Planet. Science Lett.*, 179/2, 379-390.

Braitenberg, C., I. Nagy, M. Negusini, C.,Romagnoli, M. Zadro, and S. Zerbini (2001) Geodetic measurements at the northern border of the Adria plate, Millennium Issue of the *Journal of Geodynamics*, Vol. 32, N. 1/2, 267-286.

Braitenberg C., G. Romeo, Q. Taccetti, and I. Nagy (2005) The very-broad-band long-base tiltmeters of Grotta Gigante (Trieste, Italy): secular term tilting and the great Sumatra-Andaman Islands earthquake of December 26, 2004, *J. of Geodynamics*, in press.

Caporali, A. (2003). Average strain rate in the Italian crust inferred from a permanent GPS network – I. Statistical analysis of the time-series of permanent GPS stations. *Geophys. J. Int.*, 155, 241-253.

Caporali A., C. Braitenberg, and M. Massironi (2005). Geodetic and Hydrological Aspects of the Merano Earthquake of July 17, 2001, *J. of Geodynamics*, 39, 317-336.

D'Agostino N., Cheloni D., Mantenuto S., Selvaggi G., Michelini A., Zuliani D. (2005). Strain accumulation in the southern Alps (NE Italy) and deformation at the northeastern boundary of Adria observed by CGPS measurements, *Geophysical Research Letters*, 32, L19306.

Department of Earth, Atmospheric and Planetary Sciences Massachussets Institute of Technology, Scripps Institution of Oceanography University of California at San Diego (2002). Documentation for the GAMIT GPS Analysis Software, Release 10.0, December 2000, web site: www-gpsg.mit.edu/~simon/gtgk.

Doglioni, C. (1992). Relationship between Mesozoic extentional tectonics, stratigraphy and Alpine inversion in the Southern Alps. *Eclogae geol. Helv.*, 85, 105-126.

Dong D., Fang Y., Bock M., Cheng M.K., Miyakazi S. (2002). Anatomy of apparent seasonal variations from GPS-derived site position time series, *Journal of Geophysical Research*, 107, No. B4.

Eddy W.F., McCarthy J.J., Pavlis D.E., Marshall J.A., Luthke S.B., Tsaouissi L.S., Leung G., Williams D. A. (1990). GEODYN II system operation manual, vol. 1-5, Contractor report, S.T System Corp., Lanham, Maryland, USA.

Heki, K. (2003). Snow load and seasonal variation of earthquake occurence in Japan. *Earth and Planetary Science Letters*, 207, 159-164.

Kummerow, J., R. Kind, O. Oncken, P. Giese, T. Ryberg, K. Wylegalla, F. Scherbaum, TRANSALP Working Group (2004). A natural and controlled source seismic profile through the Eastern Alps: TRANSALP, *Earth Planet. Sci. Lett.*, 225, 115-129.

Lippitsch, R., E. Kissling, and J. Ansorge (2003). Upper mantle structure beneath the Alpine orogen from high-resolution teleseismic tomography, *Journal of*

Geophysical Research, 108, B8, 2376, doi: 10.1029/2002JB002016

Pawidodirdjo L., Y. Ben-Zion, and Y. Bock (2004). Observation and Modelling of Thermoelastic Strain in SCIGN Daily Position Time Series. *Eos Trans. AGU*, 85(47), Fall Meet. Suppl., Abstract G53A-0109.

Pinato Gabrieli, C. (2005) Movimenti crostali in Friuli Venezia Giulia da osservazioni geodetiche spaziali e clino-estensimetriche. Laurea thesis from University of Trieste, Tutor: Carla Braitenberg.

Slejko D., G.B. Carulli, F.Carraro, D.Castaldini, A. Cavallin, C. Doglioni, V. Iliceto, R. Nicolich, A. Rebez, E. Semenza, A. Zanferrari, and C. Zanolla, (1987). Modello sismotettonico dell'Italia nord-orientale. CNR GNDT Rendiconto 1, Ricci, Trieste, 82 pp.

Transalp Working Group (2002). First deep seismic reflection images of the Eastern Alps reveal giant crustal wedges and transcrustal ramps. *Geophys. Res. Let.* 29 (10), 10.1029/ 2002GL014911, 92-1 - 92-4.

Williams S.D.P. (2003). Offsets in Global Positioning System time series, *Journal of Geophysical Research*, 108, No. B6, 2310.

Zadro M. and C. Braitenberg (1999) Measurements and interpretations of tilt-strain gauges in seismically active areas. *Earth Science Reviews*, 47, 151-187.

Zanferrari A., Bollettinari G., Carobene L., Carton A., Carulli G.B., Castaldini D., Cavallin A., Panizza M., Pellegrini G.B., Pianetti F., Sauro U. (1982). Evoluzione neotettonica dell'Italia nord-orientale, *Mem. Sci. Geol.* 35, 355-376.

Zanferrari A., S.Rogledi, and M.E Poli (2000). Il fronte della Catena Sudalpina orientale nelle Prealpi Carniche centro-orientali (Friuli). In: *Riassunti 80° Riunione Estiva Società Geologica Italiana,* EUT, Trieste, 2000, 467-468.

Zerbini S., F. Matonti, F. Raicich, B. Richter, and T. Van Dam (2004). Observing and assessing nontidal ocean loading using ocean, continous GPS and gravity data in the Adriatic area. *Geophysical Research Letters*, vol. 31, L23609.

Deformation Monitoring in Northern Israel between the Years 1996 and 2002

L. Shahar, G. Even-Tzur
Department of Civil and Environmental Engineering,
Technion – Israel Institute of Technology, Haifa 32000, Israel

Abstract. The geology of Israel is dominated by the Dead Sea Rift separating the Arabian plate from the African plate. Its influence on the geological structure in Northern Israel is strongly evident in the secondary faulting running in a general northwesterly direction. Northern Israel is enriched with seismological activity which causes significant horizontal and vertical deformations. In Israel there is an infrastructure of 16 permanent GPS stations scattered around the country. This infrastructure enables sequential monitoring of main faults, but its relatively low density does not enable monitoring of secondary faults. During the beginning of the Nineties the geodetic-geodynamic network in Israel was designed and constructed by the Survey of Israel and the Geological Survey of Israel. It contains about 160 points that homogeneously cover the state of Israel, when about third of them are located in Northern Israel. The location of the points was designed mainly according to geological considerations so that it would have the potential of monitoring deformations in primary and secondary known faults over Israel. The network was measured twice in 1996 and in 2002. The data analysis enables the study of movements in northern Israel with high resolution. The GPS campaigns were planned and carried out in a way that ensures high precision and reliability. Each campaign was processed using scientific software. In order to obtain the velocity field of the network points in Northern Israel, the monitoring analysis was carried out by the two-step analysis method of dynamical networks. The analysis enables, for the first time, to map in detail the active faults in the region.

Keywords. crustal movement, deformations, velocity field, two-step analysis, G1 Network, the Golan Heights, Carmel Mountain, Northern Israel

1 Introduction

1.1 Geophysical background

The Syrian-African Fault constitutes a fault line between two tectonic plates, the Arabian plate in the east and the African plate in the west. The majority of Israel is located on the Sinai sub plate, a part of the African plate.

The fault section that passes Israel called the Dead-Sea Rift (see Fig 1). This rift is tectonically active as a result of relative movement between the plates in its surroundings. The two plates move to the north but with different velocities. Research shows that over time, the Arabian plate moves faster than Sinai sub plate. Due to the velocity differences, the rift defines with a left horizontal movement which creates secondary fault systems on its two sides. Northern Israel is enriched with these types of systems that take an important part in the designing of the geological and geomorphologic landscape of its surroundings.

Most of the area of research is mountainous and defined with a lot of exposed rock. The valleys of the area are often defined with components of clay in the soil that reduces the stability of ground level. A possibility to deal with this problem is to position points on bed rock.

In Israel there are two main underground water reservoirs, the coast aquifer and the mountain aquifer. The southern part of the research area mixes with the northern parts of the two aquifers and some of the points located above them, especially above the mountain aquifer.

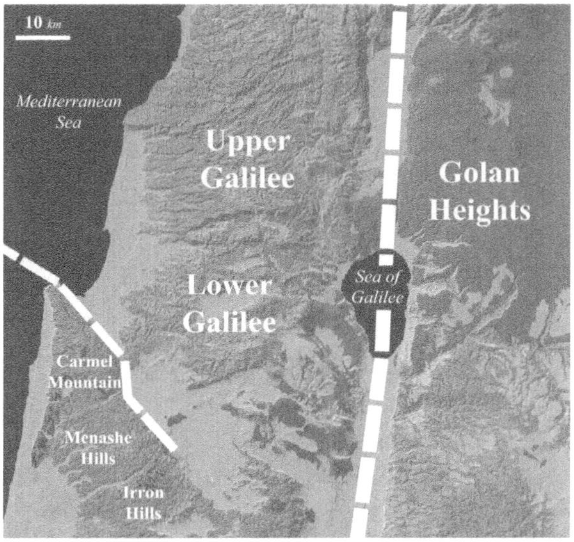

Fig 1. Geological Relief Map of Northern Israel. The two white broken lines represent faults. The Eastern thick one is a part of the Dead Sea Rift that constitutes a fault line between the Arabian plate in the east and Sinai sub plate in the west. The western thin line is the Carmel Fault

1.2 Geodetic background

The G1 Network was designed and constructed during the beginning of the Nineties by the Survey of Israel (SOI) and the Geological Survey of Israel (GSI). The network serves as a geodetic-geodynamic network in Israel and contains about 160 points that are homogeneously scattered around the country. About a third of the G1 points are located in the northern part of Israel. The location of the points was planned mainly according to geological considerations, to enable optimum monitoring of deformations in primary and secondary known faults over Israel. The network points were built according to high technical specifications to ensure their high geotechnical stability (Karcz, 1994).
The network was planned to be measured about twice a decade. Until now it was measured in 1996 and in 2002. In both the measuring campaigns four GPS dual frequency receivers were used. In 1996 the network was measured with Trimble SSE and SSI receivers and in 2002 with Trimble 5700 receivers with Zephyr Geodetic antennas.
The 1996 campaign includes two measurements of each point in two independent sessions of about 24 hours (Ostrovsky, 2001). In 2002 each point was measured independently at least in three sessions of about 8 hours (Even-Tzur et al., 2004).
GPS permanent sites that are located in northern Israel were in operation and were used as part of this research, one of them measured in 1996 and five in 2002.

2. The data Processing

2.1 The two-step analysis of dynamical networks

Most of the research for deformation analysis uses simultaneous methods in which all the measured GPS vectors constitute a single base for the adjustment model of the movement of the network points.
This research makes use of the two-step analysis of dynamical networks (Papo and Perelmuter, 1993). This method is based on a first step of independently adjusting every measurement campaign for a set of parameters. In the second step, we use the adjusted parameters from the first step as pseudo-observations for the adjustment of a set of deformation parameters. The reliability of the results at the second step depends on the suitability of the chosen movement model to the real movement which in fact is not known from the outset. Therefore, the reliability of the deformation analysis depends a great deal upon the chosen model. The method has several clear advantages over the simultaneous method:
1. At the end of the first step adjusted coordinates and their covariance matrices are formed for each measurement campaign. These factors assist in the formation of a quality control for each measurement campaign and a general assumption about the network and its ability to detect deformation. This method of control allows matching of the movement model. The residuals at the end of the second step can prove the model's suitability.
2. Modularity as opposed to the simultaneous model helps in future monitoring. While in the simultaneous method there is a need to store the measured data for each measurement campaign, in the two-step method it is sufficient to store the results of the first step only, as in, the adjusted network components and their covariance matrix of each measurement campaign.
3. The method is very convenient in discovering gross errors, difficult monitoring or problematic points.

2.2 The vector processing

The measurements were processed with RTD software, semi-scientific software for processing GPS vectors, was developed by Geodetics (http://www.geodetics.com). The software can solve the measured vectors of the same session simultaneously and calculates its full covariance matrix.

2.3 The network adjustment

The network solution for each of the measurement campaigns will be done by the least squares method (Koch, 1999)

$$v + L = Ax \qquad (1)$$

where x is an unknown vector which includes the adjusted coordinates of the network according to the measurement of the solved campaign. A is the design matrix. L is the observation matrix which is built of the observation vectors of the measurement campaign and v is the residuals vector. The full covariance matrix of the observations is to be constructed for every measurement campaign. Therefore, it is a symmetric block diagonal matrix which looks as follows

$$\Sigma_{L_k} = \begin{bmatrix} \Sigma_{L_1} & 0 & \cdots & 0 & \cdots \\ 0 & \Sigma_{L_2} & & 0 & \cdots \\ \vdots & & \ddots & & \\ 0 & 0 & & \Sigma_{L_i} & \\ \vdots & \vdots & & & \ddots \end{bmatrix} \qquad (2)$$

when Σ_{L_i} is the observation covariance matrix of session i.

3. Reliability of the control network

An important part in the data processing stage is the controlled reliability of the control network. The adjusted results, according to the least squares method, are based on several assumptions which give anchor to the reliability of the statistical tests. When several of such assumptions are not fulfilled in reality the solution is defined as biased. Gross errors are the main reason for dismissal of the assumptions (Even-Tzur 1997).

The residual vector is assumed to be normally distributed $v \sim N(\mu, \sigma^2)$ with mean μ and variance σ^2. When gross errors are removed, the mean value of the vector components is accomplishing the condition $\mu = 0$, then, an unbiased solution is achieved. Gross errors bring the accomplishment of the condition $\mu \neq 0$ and therefore the solution is biased. It is not possible to detect this problem at the stage of the control network adjustment and there is a need to test the measured vectors during the carrying out of the first step of the two-step method. In order to improve the network reliability there is a need to confine the mistaken vectors and to neutralize them in advance.

Gross error detection in GPS measurements can be achieved by two statistical tests a general hypothesis test and a partial hypothesis test. The general test examines the relationship between the posterior variance of unit weight to the a prior value (Cooper, 1997

$$\begin{aligned} H_0 &: \hat{\sigma}_0^2 = \sigma_0^2 \\ H_1 &: \hat{\sigma}_0^2 \neq \sigma_0^2 \end{aligned} \qquad (3)$$

We reject H_0 at a significance level of α if

$$\frac{\hat{\sigma}_0^2}{\sigma_0^2} > F_{r,\infty,\alpha} \qquad (4)$$

when r is the degree of freedom.

Gross errors are usually the main reason for rejection H_0. To detect GPS vectors which contain errors we can continue to the partial test. It is based on the quantity (Chen et al., 1987)

$$\Delta R = v^T PE(E^T P\Sigma_v PE)^{-1} E^T Pv \qquad (5)$$

when E, for GPS measurements, is a three column matrix as are the GPS vector components being tested. In the matrix, sub matrices will be aligned, these matrices are 3×3 unit matrices, placed with relation according to the serial position of the tested vectors in the observations vector L and the rest will be zeros. Σ_v is the covariance matrix of the residuals v and it is calculated as

$$\Sigma_v = \Sigma_L - A\Sigma_x A^T \qquad (6)$$

when A is the design matrix and Σ_x is the covariance matrix of the solution vector x.

In the literature it is possible to find three main tests for partial gross error detection (Chen et al., 1987), w test when

$$w = n_2 \sigma_0^2 F_{n_2,\infty,\alpha} \qquad (7)$$

t test when

$$t = \frac{n_2 F_{n_2,r-n_2,\alpha}}{n_2(F_{n_2,r-n_2,\alpha} - 1) + r} v^T Pv \qquad (8)$$

And τ test when

$$\tau = \frac{n_2 F_{n_2,r-n_2,\alpha}}{F_{n_2,r-n_2,\alpha} - n_2 + r} v^T Pv \qquad (9)$$

when n_2 is the number of suspicious observations.

In the case of w test, the tested hypotheses are

$$\begin{aligned} H_0 &: \Delta R \leq w \\ H_1 &: \Delta R > w \end{aligned} \qquad (10)$$

when the null hypothesis H_0 will represent the absence of gross errors in the measured vectors. In the same way, it is possible to do the hypothesis test using the t test or the τ test.

It is best to eliminate one vector only which has the highest value of ΔR. In that case, the procedure will prevent a possibility of eliminating useful vectors.

After having sorted out gross errors, completion of both tests and network adjustment for each one of the measurements campaign, the first step of the two-step method has been completed for the k different campaigns.

4. The mathematical model

The quality of the result of the second step of the two-step method is not absolute and it depends on matching the mathematical model to the geophysical definitions of the area surveyed. The instability of the model is directly expressed to its noise.

Although the relative advantages of the dynamic model in estimating the geophysical parameters such a model has not been chosen in this stage of the research. The small number of measurement campaigns and the large surveyed area which is mostly expressed by one type of tectonic movement bring about the choosing of the kinematical model of velocities field. Two measurement campaigns bring us to use the linear model which solves the location of the points and its velocities without degree of freedom. Therefore, the model which we use is

$$x_i = x_0 + \dot{x}(t_i - t_o) \tag{11}$$

when x_i is the vector of the coordinates of the network points at the time t_i. x_0 contains the coordinates for the reference epoch t_0 and $\dot{x}$ contains velocities.

The adjustment of the first step looks like

$$L^{\mathrm{I}} = A^{\mathrm{I}} x^{\mathrm{I}} - v^{\mathrm{I}} \tag{12}$$

And in specific writing, the matrices which build up the final solution for the first step will be created by lining the i sub matrix, in a following order for k measurement campaigns in the following way

$$\begin{bmatrix} L_1^{\mathrm{I}} \\ \vdots \\ L_i^{\mathrm{I}} \\ \vdots \\ L_k^{\mathrm{I}} \end{bmatrix} = \begin{bmatrix} A_1^{\mathrm{I}} & & 0 & \cdots & 0 \\ & \ddots & & & \vdots \\ 0 & & A_i^{\mathrm{I}} & & 0 \\ \vdots & & & \ddots & \\ 0 & \cdots & 0 & & A_k^{\mathrm{I}} \end{bmatrix} \begin{bmatrix} x_1^{\mathrm{I}} \\ \vdots \\ x_i^{\mathrm{I}} \\ \vdots \\ x_k^{\mathrm{I}} \end{bmatrix} + \begin{bmatrix} v_1^{\mathrm{I}} \\ \vdots \\ v_i^{\mathrm{I}} \\ \vdots \\ v_k^{\mathrm{I}} \end{bmatrix}$$

The result of the first step are used as pseudo measurements for the second step and therefore

$$L^{\mathrm{II}} = x^{\mathrm{I}} \tag{13}$$

And therefore, similar to the first step

$$x^{\mathrm{I}} = A^{\mathrm{II}} x^{\mathrm{II}} - V^{\mathrm{II}} \tag{14}$$

where I and II indexes represent matrices from the first and the second step, respectively.

5. The datum defect and its solution

The lack of reference system for the location and for the velocity in the adjustment system of the second step causes a datum defect of $d = 6$. In order to overcome this defect, it is necessary to decide upon one of the points which is involved in all the adjustments as a reference point and solves the defect of the adjustment system. Similarly, we will arbitrarily decide that this point is motionless over time with relation to itself and so we will solve the velocity defect (Agmon, 2001).

Solving the datum defect in the mentioned way is problematic because there is no evidence that the point chosen as a datum has not changed its place over time and so the solution is biased and the sum square of the velocities is not a minimum. This problem can be solved with projecting the system into a free solution. For the projection, we use the matrix

$$J = I - H(H^T P_x H)^{-1} H^T P_x \tag{15}$$

when H is geometrical constraint matrix of Helmert and P_x is a weight matrix. The weighting is noticed diagonally on the matrix, when the chosen point as a datum will be represented with ones on the matrix diagonal.

A free solution is one where all the network points are chosen as a datum and therefore in this case $P_x = I$ and the free network will be calculated as,

$$x^f = J x^{\mathrm{II}} \tag{16}$$

According to the law of covariance propagation

$$\Sigma_{x^f} = J \Sigma_{x^{\mathrm{II}}} J^T . \tag{17}$$

6. Selection of datum and conversion to a free weighted solution

In deformation analysis, we usually completely separated between the horizontal and vertical sub networks. However, often they have a characteristic dependence on the type of tectonic action that is taking place in the research area.

This separation that takes place in the determination of datum, that is to say, the geographical area

chosen to use the datum is likely to contain different points for each one of the sub networks.
Considerations of choosing the datum are the inner stability and the geophysical logic. The meaning of inner stability is the permanent differences in coordinates, over time, between all datum points, that is to say, the absence of relative movement between the points, horizontal movement for horizontal datum and vertical movement for vertical datum.
The stability is checked with statistic tools explained throughout the paper. The geophysical consideration is made up of geological, geomorphologic and hydrological logic whereby its absence is likely to affect the inner stability over time. It is required that every datum points will be located in the area which have the most homogenous geological characteristics as possible. It is necessary for the geomorphologic examination to take place even before arranging the position of the points in the field so that it will not be carried out in an area where the level of ground is likely to change.
Hydrological examination should apply to areas where aquifer with changing in its volume over time that cause changes in the ground level. The application of these rules is likely to minimize the noise of the measurements in a way that determines the deformations.
In a similar way, to continue to a free solution this procedure from the free solution to the weighted free solution where weight constraints are put only on specific chosen datum

$$x^w = Jx^f \tag{18}$$

$$\Sigma_{x^w} = J\Sigma_{x^f} J^T . \tag{19}$$

7. Research results

7.1 The solution of the first step

The network includes about 60 G1 Network points and permanent GPS stations. 51 of those points were measured in the first campaign and 33 in the second campaign. The degree of freedom of the network at the first step, in the first and the second campaigns was 63 so that the degree of freedom f in the second step was 126.
The arbitrary single datum point which was chosen to solve the datum defect was KATZ that located in the Golan Heights.
In Fig 2 we can see a systematic weakness of the horizontal network in 1996 in the directions of north-west and south-east.

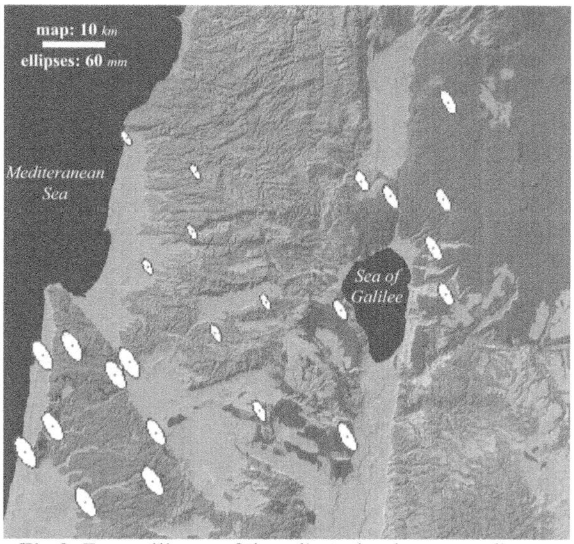

Fig 2. Error ellipses of the adjusted points according to 1996 campaign. The significance level α is 0.05.

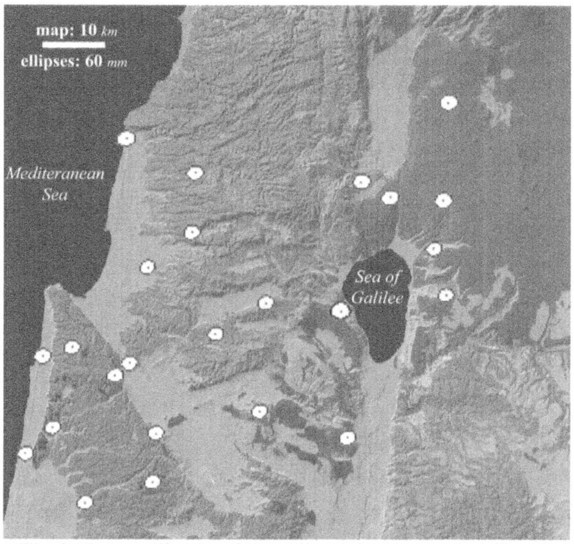

Fig 3. Error ellipses of the adjusted points according to 2002 campaign. The significance level α is 0.05.

Planning the measurement of the second campaign brought about changes. Observation in Fig 3 shows that there was no real flatting in the error ellipses of the measured points in 2002 and it meant that the network was defined with homogenous sensitivity in all directions. The examination of accuracy of this step allows the prediction of measurement noise expected in the results of the second step.

7.2 Second step solution

In the second step, 25 points took part which were used in both measurement campaigns. Since we are only analyzing two campaigns, it is possible to calculate and to introduce the results as displacement vector d between the two measurement campaigns

$$d = x_2 - x_1 \quad (20)$$

$$\Sigma_d = \Sigma_{x_2} + \Sigma_{x_1} \quad (21)$$

x is the coordinates of the network points for each campaign and Σ_d is the displacement covariance matrix.

The results of the general hypothesis test for the two sub networks are introduced in Table 1. In both cases we reject H_0 and conclude that there are deformations in the network.

components	statistic	$F_{b,f,0.05}$	H_0
2d (n, e)	2.3	1.7	rejected
1d (up)	6.7	1.6	rejected

Table 1. Results of the general hypothesis tests for the two sub-systems.

Analysis of the results is exemplified in this paper twice via the choosing of two different geographic areas as datums. The two specific areas are from two edges of the network the Golan Heights in the east and the Carmel Mountain and its coast in the west.

These datum show in the best way the movement of the network that because the primary fault of the Dead Sea and all secondary faults that separate between them including the Carmel-Tirza fault system that is known with tectonic activity.

In the all cases, we use significance level of $\alpha = 0.05$.

In Table 2 it is possible to see the results of the partial hypothesis tests for the two datums that were chosen. In these two cases a stable datum was found for the horizontal sub network and an unstable one for the vertical sub network. The reason of the instability of the datum in the case of the Carmel is the point ATLT that is located on the Carmel coast. Previous monitoring shows that the Carmel Mountain is raised relative to the coastal plane (Shahar and Even-Tzur, 2004, Even-Tzur and Agmon, 2005) and it is possible to use the Carmel as a vertical datum if it does not include ATLT point.

In regards to the datum of the Golan Heights, we examined the possibilities to reduce instabilities points (SHAL and KATZ) and we found it as a weak datum.

Horizontal network

datum	statistic	$F_{b,f,0.05}$	H_0
Carmel	0.8	2.0	accepted
Golan	0.4	2.0	accepted

Vertical network

datum	statistic	$F_{b,f,0.05}$	H_0
Carmel	18.3	2.5	rejected
Golan	2.9	2.5	rejected
Carmel (3 points)	0.8	2.7	accepted

Table 2. The results of the datum hypothesis tests.

The result analyzes was done independently for each sub system, horizontal or vertical.

It is convenient to analyze the results using local horizontal system. Therefore, if this transfer is not done at an earlier stage it has to be done now. After the convertion, it is necessary to divide the matrices of results and the covariance for the two sub networks.

Statistic analysis of the deformations starts regarding the general hypothesis test for the free network results and partly tests for the weighted solution of the sub network.

The statistic of the test will be computed,

$$\lambda = \frac{x^T \Sigma_x^+ x}{b} \quad (22)$$

where b is the number of the sub network components. The hypotheses will be

$$H_0 : \lambda = 0 \qquad H_1 : \lambda \neq 0 \quad (23)$$

We reject H_0 with a significance level of α if

$$\lambda > F_{b,f,\alpha}. \quad (24)$$

For the general test the meaning of accepting H_0 hypothesis is an indication for absence of significant deformations. Rejecting H_0 allows the using of partial hypothesis test in the same way to examine the sub networks weighted with a datum. Rejecting H_0 in this test shows existence of deformations.

The weighted free solution with the datum of the Carmel according to the points SFIA, MRKA, KRML and ATLT for the horizontal sub system are introduced in Fig 4.

The error ellipses are with significance level of $\alpha = 0.05$ and movement and its direction displayed as a line, when movement outside of the ellipse area is a significant movement.

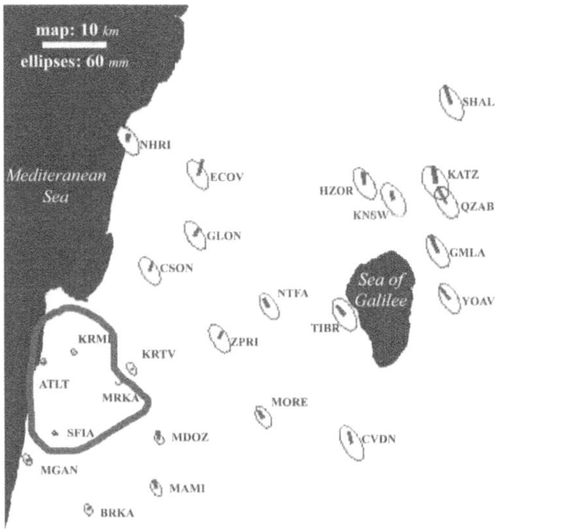

Fig 4. The error ellipses of the points, according to the Carmel datum, that bound with the thick closed line, $\alpha = 0.05$. The lines start in the ellipses center sign the displacements.

In Table 3, points and displacements were introduced which have significant movement in relation to the Carmel datum.

point	n	e
MAMI	**14**	**-3**
MDOZ	**14**	**-1**
CSON	**16**	**4**
CVDN	**20**	**-6**
YOAV	**21**	**-18**
ECOV	**26**	**8**
GLON	**14**	**8**
HZOR	**21**	**-1**
KATZ	**26**	**-2**
GMLA	**30**	**-12**
QZAB	**26**	**-13**
SHAL	**29**	**-11**

Table 3. The significant ($\alpha = 0.05$) horizontal displacements according to the Carmel datum in mm.

Analysis of the results show that even in the cases without significant movement the direction is north relation to the datum. The main regional significant movement is in the Golan Heights with a north-north-west movement relation to the Carmel.

Fig 5 shows vertical movement of the points relative to the Carmel datum. The rectangles boundaries express significance level of 0.05. Movement outside of that is significant and there are four points with significant vertical movement that introduced in Table 4 with their values. In this case, the datum does not include the point ATLT that is located on the Carmel coast. This point sinks relative to the Carmel and the fact is familiar from a leveling network monitoring (Shahar and Even-Tzur, 2004).

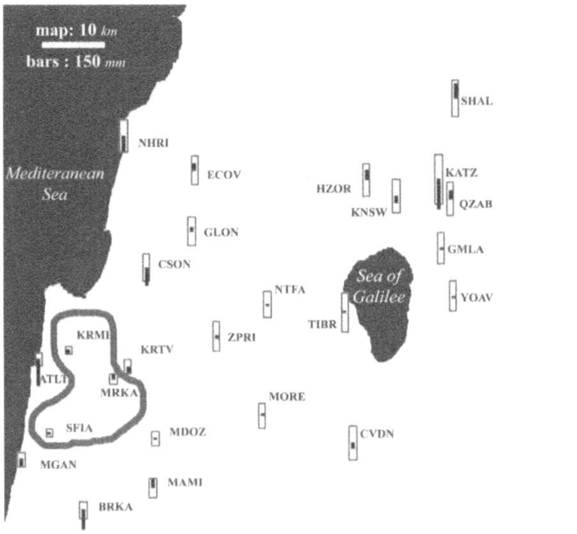

Fig 5. The vertical displacements, according to the reduced Carmel datum that bounds with the thick closed line. The top and the bottom of the rectangles refer to 1.96σ that means significant level of $\alpha = 0.05$. The lines start in the rectangles center sign the displacements.

point	u
ATLT	**-72**
BRKA	**-53**
CSON	**-50**
KATZ	**-84**

Table 4. The significant $\alpha = 0.05$ vertical displacements according to the reduced Carmel datum in mm.

In the case of the results according to the Golan Heights datum the examination shows a significant movement of the points introduced in Table 5, relative to the Golan Heights datum, usually in the direction of north-north-west. It is possible to visibly see Fig 6. The significant displacements are shown in Table 5. The Golan Heights is located on the Arabian Plate and this result supports the fact that was mentioned in the introduction about a northern movement of the Arabian Plate in relation to the Sinai sub plate.

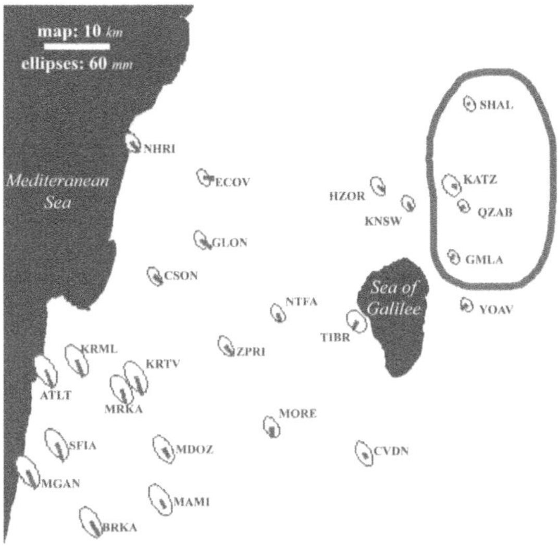

Fig 6. The error ellipses of the points, according to the Golan Heights datum, that bound with the thick closed line, $\alpha = 0.05$. The lines start in the ellipses center sign the displacements.

point	n	e
MRKA	**-22**	**6**
KRML	**-29**	**10**
SFIA	**-30**	**11**
ATLT	**-31**	**11**
BRKA	**-23**	**14**
MGAN	**-30**	**12**
CSON	**-12**	**13**
KRTV	**-30**	**9**
MORE	**-16**	**1**
NTFA	**-13**	**2**
ZPRI	**-17**	**12**
YOAV	**-6**	**-8**
ECOV	**-2**	**17**
GLON	**-14**	**17**
KNSW	**-13**	**6**
NHRI	**-16**	**11**

Table 5. The significant ($\alpha = 0.05$) horizontal displacements according to the Golan datum in mm.

8. Summary and Conclusions

The monitoring includes two campaigns of measurement of the network and consists of 60 points of which 25 took part in the two measurement campaigns. In the examination of the monitoring we used the two-step method for dynamical networks. The movement model which was used was the linear model of velocities.

The network proved productive to find significant velocities of 3 mm per year.

The results exemplified here through the use of two datums support the assumptions known from geophysical and geological research. The vertical results are similar to previous monitoring results via the use of leveling network and recent publication of its results (Shahar and Even-Tzur, 2004).

References

Agmon, E. (2001). *Algorithm for the Analysis of Deformations Monitoring Networks*. M.Sc. Thesis, Technion-Israel Institute of Technology, Haifa, Israel (in Hebrew).

Chen, Y.Q., Kavouuras M., Chrzanowski A., (1987). "A Strategy for Detection of Outlying Observations in Measurements of High Precision". *The Canadian Surveyor*, 41(4), pp. 529-540.

Cooper, M. A. R., (1987). *Control Surveys in Civil Engineering*. Collins, London.

Even-Tzur, G., Salmon E., Kozakov M., Rosenblum M. (2004). "Designing a Geodetic-Geodynamic network: A Comparative Study of Data Processing tools". *GPS Solutions*, 8(1), pp. 30-35.

Even-Tzur, G., Agmon, E. (2005). "Monitoring Vertical Movements in Mount Carmel by Means of GPS and Precise Leveling". *Survey Review*, 38(296), pp. 146-157.

Even-Tzur, G. (1997). *Optimal Design of GPS Based Geodetic Networks for Monitoring Deformation*. D.Sc. Thesis, Technion-Israel Institute of Technology, Haifa, Israel (in Hebrew).

Karcz, I. (1994). "Geological Considerations in Design of the Seminal Dead Sea Rift Network". Perelmuter Workshop on Dynamic Deformation Models, Haifa, Israel.

Koch, K. R. (1999). *Parameter Estimation and Hypothesis Testing in Linear Models*. Springer Verlag, New-York.

Ostrovsky, E. (2001). "The G1 GPS Geodetic-Geodynamic Reference Network: Final Processing Results". *Israel Journal of Earth Science*, Vol. 50, No. 1, pp. 29-37.

Papo, H. B., Perelmuter A. (1993). "Two steps Analysis of Dynamical Networks". *Manuscripta Geodetica*, Vol. 18, No. 6, pp. 422-430.

Shahar, L., Even-Tzur G. (2004). "Vertical Movements in the Carmel Mountain". FIG Working week 2004, Athens, Greece.

Designing the Configuration of the Geodetic-Geodynamic Network in Israel

G. Even-Tzur
Department of Civil and Environmental Engineering,
Technion – Israel Institute of Technology, Haifa 32000, Israel

Abstract. The Geodetic-Geodynamic network (G1) was established with the intention of monitoring recent crustal deformations in Israel and serving as a major control network. In 1996 the network was measured for the first time using GPS. The goal of the proposed design is to improve the ability of the network to detect and measure movements and deformations, based on the geological phenomena to be monitored, while satisfying the demand for high accuracy and reliability. This paper presents the network configuration design of the second GPS measurement campaign of the G1 network held during 2002. Sensitivity analysis of the network, as dependent upon a postulated velocity field, was used as the basis for the network configuration design.

Keywords. Network design, sensitivity, velocity field, Geodetic-Geodynamic network

1 Introduction

The Geodetic-Geodynamic network in Israel (G1) was designed and constructed during the beginning of the nineties by the Survey of Israel (SOI) and the Geological Survey of Israel (GSOI). It contains about 160 points that homogeneously cover the state of Israel (Fig. 1). The location of the points was designed mainly according to geological considerations, so that it would have the potential of monitoring deformations in primary and secondary known faults over Israel. Additionally, the G1 network was planned to serve as the major geodetic control network of Israel. The points were built according to very high technical specifications to ensure their geotechnical stability (Karcz, 1994).

During 1996 and the beginning of 1997 the G1 network was measured for the first time. The network was measured by four Trimble (SSE and SSI) GPS dual frequency receivers with a session duration of 24 hours. Most network points were measured in two independent sessions (Ostrovsky, 2001). The GP data processing was carried out by Survey of Isra using the BERNESE software (Rothacher an Mervart, 1996). The second GPS measureme campaign of the G1 network was held in 2002.

The most important elements of a monitoring networ are its accuracy, reliability, and sensitivity (Coope 1987). Accuracy refers to the quality of the network i terms of random errors. Reliability is defined as th ability of the network to sense and identify gros errors in the measurements, as the undetecte influence of gross errors in a monitoring network ca be easily interpreted as a deformation, thus creatin an undesirable situation. The sensitivity of monitoring network refers to its capacity to detect an measure movements and deformations in the are covered by the network. A network is defined a sensitive when the geological phenomena monitore can be detected with specified probabilities α and β (being the level of significance, β being the te power). Thus, it is most reasonable to use sensitivit as the main criteria in the design of monitorin networks.

The performance of geodetic monitoring networks heavily influenced by the configuration of th measured GPS vectors. An effective design of th GPS measurements will decrease the campaign cos and increase the accuracy and reliability of th network. As the main task of the G1 network is t monitor deformations, it is most reasonable to ado the concept of sensitivity as the main criterion for th design of the G1 network (Even-Tzur, 1999). Th goal of this design is to give the network the ability t detect and measure movements and deformation based on the geological phenomena to be monitored. This paper presents the method used for the GP vector configuration design, as well as the actu configuration design used in the second measureme campaign of the G1 network and its results.

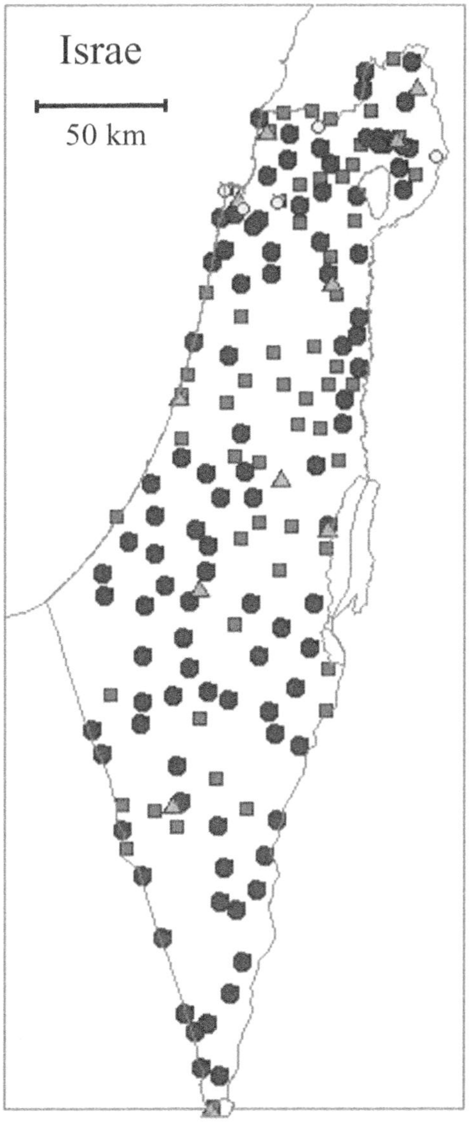

Figure 1 - The G1 network. Triangles – the permanent GPS stations in Israel. Circles – points that were measured in 1996 and 2002. White circles – new points measured in 2002. Squares – points measured only in 1996.

2 The Sensitivity Criteria

For two epochs i=1,2 there are two sets of estimation coordinates and a singular cofactor matrices: $\hat{\mathbf{x}}_1, \mathbf{Q}_{\hat{x}_1}$ and $\hat{\mathbf{x}}_2, \mathbf{Q}_{\hat{x}_2}$. The reference system definition for the two sets must be identical.

Let the time interval between two epochs of observations be Δt. The velocity vector

$$\hat{\dot{\mathbf{x}}} = (\hat{\mathbf{x}}_2 - \hat{\mathbf{x}}_1)/\Delta t \qquad (1)$$

represents the velocity of the network points, assuming velocities are constant. The cofactor matrix of the velocity vector $\hat{\dot{\mathbf{x}}}$ is:

$$\mathbf{Q}_{\hat{\dot{x}}} = \frac{\mathbf{Q}_{\hat{x}_1} + \mathbf{Q}_{\hat{x}_2}}{(\Delta t)^2}. \qquad (2)$$

The null hypothesis

$$H_0 : \quad \hat{\dot{\mathbf{x}}} = 0$$

can be tested against the alternative hypothesis

$$H_1 : \quad \hat{\dot{\mathbf{x}}} \neq 0.$$

Acceptance of the null hypothesis signifies no movement of points in the network. The cofactor matrix $\mathbf{Q}_{\hat{\dot{x}}}$ has rank r and d degrees of freedom in the estimation of $\hat{\dot{\mathbf{x}}}$, and the a-posteriori variance factor is $\hat{\sigma}_0^2$. The test statistic, including all velocities of the network points, is:

$$t = \frac{\hat{\dot{\mathbf{x}}}^T \mathbf{Q}_{\hat{\dot{x}}}^+ \hat{\dot{\mathbf{x}}}}{r\hat{\sigma}_0^{\,2}} \sim F_{r,d} \qquad (3)$$

For three-dimensional networks with u points r=3u-7, for two-dimensional networks r=2u-4, and for single-dimension single networks r=u-1.

If $t > F_{r,d,\alpha}$, the null hypothesis is rejected with probability α, indicating that at least one point has moved significantly.

The parameters $\hat{\dot{\mathbf{x}}}$, $\hat{\sigma}_0^2$, and $\mathbf{Q}_{\hat{\dot{x}}}$ are unknown, but a-priori values can be assigned to them.

Let the a-priori variance factor be $\sigma_0^2 = 1$, the weight matrix of the measurements set as **P**, and the rank deficient design matrix set as **A**, then:

$$\mathbf{Q}_x = \mathbf{N}^+ = \left(\mathbf{A}^T\mathbf{P}\mathbf{A}\right)^+ \qquad (4)$$

Assuming the measurements for the two epochs are identical, the covariance matrix of the velocity vector $\dot{\mathbf{x}}$ is:

$$\mathbf{Q}_{\dot{x}} = \frac{2\mathbf{Q}_x}{(\Delta t)^2} \qquad (5)$$

The pseudo-inverse $\mathbf{Q}_{\dot{x}}^+$ is:

$$\mathbf{Q}_{\dot{x}}^+ = \left(\frac{2(\mathbf{N})^+}{(\Delta t)^2}\right)^+ = \frac{(\Delta t)^2}{2}\mathbf{N} = \frac{(\Delta t)^2}{2}(\mathbf{A}^T\mathbf{P}\mathbf{A}) \qquad (6)$$

Based on an assumed geological model a-priori velocity values ($\dot{\mathbf{x}}$) can be calculate. Therefore, the value of the test statistic t can now be computed as:

$$t = \frac{\dot{\mathbf{x}}^T \mathbf{Q}_{\dot{x}}^+ \dot{\mathbf{x}}}{r\sigma_0^{\,2}} \sim F_{r,\infty} \qquad (7)$$

Since the test statistic is calculated using estimated values, the F distribution is calculated under the assumption of infinite degrees of freedom ($d \rightarrow \infty$). If H_1 is accepted ($t > F_{r,\infty}$) the test statistic has a non-central F distribution with a non-centrality parameter λ given by

$$\lambda = \frac{\dot{\mathbf{x}}^T \mathbf{Q}_{\dot{x}}^+ \dot{\mathbf{x}}}{\sigma_0^2} \quad (8)$$

We can define λ_0 as the boundary value of λ, which will cause the rejection of the null hypothesis at probability levels α and β. The value λ_0 is an implicit function. Values of λ_0 for $\beta = 0.7$, 0.8 and 0.9 were given by Baarda (1968) for different values of α and the degrees of freedom in the estimation. As stated before, the main aim is establishing a monitoring network, which will enable the rejection of the null hypothesis H_0 and the acceptance of the alternative hypothesis H_1, thus enabling the definition of the network as sensitive for monitoring the expected movements and deformation.

3 The Sensitivity of the Northern Part of the G1 Network

The geology of Israel is dominated by the Dead Sea rift, separating the Arabian plate from the African plate. The relative movement between the two sides of the Dead Sea rift could amount to a few millimeters per year (Wdowinski et al., 2004, Pe'eri et al., 2002). The G1 network is located along the Dead Sea rift, where in the north of Israel both sides of the rift are accessible (see Fig. 2). For this reason the sensitivity analysis is performed only for the northern part of the network.

To evaluate the sensitivity of the network, the analysis was performed based on the first measurement campaign. This sensitivity analysis was a necessary step before launching the second measurement campaign. The analysis results served in the design of the second campaign.

The sensitivity analysis was preformed for two possible deformation models, the Simple Transform Fault and the Locked Fault models. The northern part constellation, the coordinates and their variance-covariance matrix, based on the GPS surveying campaigns measured during the fall of 1996 and processed by the BERNESE software, served in the analysis. The sensitivity analysis pointed to poor network sensitivity. The value λ, calculated by usin equation (8), was very small relative to λ_0 fc $\Delta t = 1$ (Even-Tzur, 2001). The network proved to b more sensitive for displacement in the East-We direction than in the North-South directio contradictory to the expected displacement in th North-South direction along the Dead Sea rift. I general, adding several remote control points to th east or the west of the fault proved to increase th sensitivity of the network. As there are geographic limitations (the Mediterranean Sea limits the G network to the west) and political limitations (th Syrian border limits the network to the east), the onl possible addition of two points to the east side of th network proved to be of great contribution to th sensitivity of the network (see Fig. 1) (Even-Tzu 2001).

4 The Network Configuration

A network of 11 continuous permanent GPS statior is located across Israel (Fig. 1). These points, togeth with the G1 network, are used as the mai deformation monitoring network of Israel. Financi considerations lead to the decision that only two thir of the G1 network points that were measured durin the first campaign in 1996 will be measured in th second campaign. A survey held during the beginnin of 2001 found that about 10% of the original poin were destroyed during passing years. 100 poin covering the state of Israel homogenously we chosen (Fig. 1), and five new points were establishe (with identical technical specifications) in th northern part of the network. Two points, on permanent GPS station (ELRO) and one new G1 poi were established in the North-East part of the netwo for increasing the sensitivity of the network (a suggested in the previous section). Another 4 poin were established to replace destroyed points and wit the intention of strengthening the network. In tota 105 G1 points were measured in the secon campaign.

5 The Method of GPS Vecto Configuration Design

The performance of geodetic monitoring networks i heavily influenced by the configuration of th measured GPS vectors. An effective design of th GPS measurements will decrease campaign costs an increase the accuracy and reliability of the network. When $\lambda < \lambda_0$ we are able to increase the networ sensitivity by adding a new GPS vector **a** with corresponding weight **p**, where **a** contains three rows, one per each vector component and the dimension of the weight matrix **p** is 3 by 3. The new value $\bar{\lambda}$, describing the network sensitivity including the additional vector, is $\bar{\lambda} = \lambda + \Delta\lambda$ (Even-Tzur, 2002), while

$$\Delta\lambda = \frac{(\Delta t)^2}{2\sigma_0^2}[\dot{\mathbf{x}}^T(\mathbf{a}^T\mathbf{pa})\dot{\mathbf{x}}] \tag{9}$$

For the purpose of achieving the largest possible improvement in the sensitivity of the network, one should seek a GPS vector that will maximize $\Delta\lambda$. Based on a set of possible vectors, one can identify the contribution of each vector to the λ value. Naturally, the preferred addition would be of the vector with the maximal $\Delta\lambda$. If the new value $\bar{\lambda}$ is still lower than λ_0, a further addition of the next vector with the highest value of $\Delta\lambda$ should be made. This process should be continued until $\bar{\lambda} \geq \lambda_0$. This process is carried under the assumption that the measurement vectors for the two epochs were identical. In a situation where a first GPS campaign has already been carried out, the aim is to identify a set of GPS vectors that should be measured in a second campaign, in order to improve the sensitivity of the network.

The pseudo-inverse $\mathbf{Q}_{\dot{x}}^+$ which is used in equation (8) is now

$$\mathbf{Q}_{\dot{x}}^+ = \left(\frac{\mathbf{Q}_{x_1} + \mathbf{Q}_{x_2}}{(\Delta t)^2}\right)^+ = (\Delta t)^2\left(\mathbf{N}_1^+ + \mathbf{N}_2^+\right)^+ \tag{10}$$

The following expression $\left(\mathbf{C}^+ + \mathbf{D}^+\right)^+$, referred to in literature as the parallel sum of matrices, was defined by (Rao and Mitra, 1971):

$$\left(\mathbf{C}^+ + \mathbf{D}^+\right)^+ = \mathbf{C}(\mathbf{C} + \mathbf{D})^+\mathbf{D} \tag{11}$$

Therefore:

$$\mathbf{Q}_{\dot{x}}^+ = (\Delta t)^2\left[\mathbf{N}_1(\mathbf{N}_1 + \mathbf{N}_2)^+\mathbf{N}_2\right] \tag{12}$$

The non-centrality parameter λ is:

$$\lambda = \frac{(\Delta t)^2}{\sigma_0^2}\dot{\mathbf{x}}^T\left[\mathbf{N}_1(\mathbf{N}_1 + \mathbf{N}_2)^+\mathbf{N}_2\right]\dot{x} \tag{13}$$

Following the first campaign a normal matrix $\mathbf{N}_1$ has been defined. From the set of possible GPS vectors one can create the normal matrix $\mathbf{N}_2$ for each vector and calculate the respective λ value. As stated before, the preferred GPS vector for the first measurement would be the vector with the highest value of λ and the optimal contribution to the network sensitivity. After adding the first vector, one should repeat the process using the normal matrix $\mathbf{N}_2$, already containing the additional vector, and another vector from the possible GPS vector list. Following the calculation of the new λ value for each vector, the preferred GPS vector for the second measurement would be the vector with the highest value of λ. The process is continued until $\lambda > \lambda_0$. The additional vectors will ensure the ability of the network to monitor possible deformations with probabilities α and β thus enabling the improvement of the network sensitivity in the planned campaign.

GPS measurements are usually carried out in several separate sessions when more than one vector is measured at a time. Based on the number of GPS receivers participating in the measurements a set of possible GPS sessions can be planned. The ordered list of vectors with the highest contribution to the λ value could be used in the planning of these possible sessions. In that case all the vectors that are measured in each session are represented in matrix $\mathbf{N}_2$.

6 Vector Configuration Design of the G1 Network

The proposed method was applied to the design of the GPS sessions planned for the Geodetic-Geodynamic network in Israel. The northern part of the G1 points constellation, the coordinates and their variance-covariance matrix, based on the GPS surveying campaign measured during 1996 and processed by the BERNESE software, served in designing the vector configuration for the second campaign. The configuration design process was confined to the horizontal plain, since the expected vertical motion was relatively small and the accuracy of GPS measurements in the vertical direction is poor. The basic assumption governing this design was that the only reason for movements and deformations in the area covered by the network was the motion of the Dead Sea Rift. The variance in meters of vector ℓ_{ij} between points i and j was given by $\sigma_{\ell_i} = 0.003 + \ell_{ij} \times 0.1\text{ppm}$, and a correlation of 7% was assumed between any two vector components. The correlation between the two vector components was derived from the actual variance-covariance matrix of the first GPS campaign. The variance of unit weight is $\sigma_0^2 = 1$.

The design was based on an assumed Locked Fault model in the Dead Sea Rift, since evidence showed no recent movements along the rift. According to this model the fault is locked from the surface down to depth *D*, and slips freely below this depth by *V* millimeters per year. While y is the horizontal distance of a point perpendicular to the fault its velocity parallel to the fault, v_x, is (Weertman and Weertman, 1964):

$$v_x = \frac{V}{\pi} tan^{-1}\left[\frac{y}{D}\right] \qquad (14)$$

When setting $V = 5_{mm/yr}$ and $D = 10_{km}$, the velocity field ($\dot{\mathbf{x}}$) of the G1 network points can be computed. For the detection of a Locked Fault model, 13 vectors provided a sufficiently sensitive network when measuring vectors shorter than 40_{km}. For vectors shorter than 20_{km}, 25 vectors provided a sufficiently sensitive network.

The network was designed for measurements using four receivers, and previous experience proved that the use of four receivers was logistically optimal. Based on the results of the vectors with the highest contribution to the λ value a set of 25 possible sessions was created. By using the proposed method a set of GPS sessions that provide a sufficiently sensitive network can be presented, thus, for the detection of a Locked Fault model 8 sessions will provide a sufficiently sensitive network (see Fig.2).
The design pointed to a greater contribution to the sensitivity of the network by the vectors in the east-west direction than the contribution of the vectors in the North-South direction, and that measurements of longer vectors could further increase the contribution to the sensitivity of the network. These conclusions assisted in the configuration design of the GPS sessions in the southern part of the network, where only one side of the Dead Sea rift is covered by the G1 points.
To increase the reliability of the network each point was measured in three independent sessions, in order to detect and identify gross errors resulting from the positioning of the GPS antenna at the marker points, identify alignment errors, and measure the antenna height. The duration of each session was planned for 8 hours.
The results of the vector configuration design, incorporating the demand for three independent sessions in each network point, yielded 92 sessions for 105 of the network points. For illustration purposes, the designed sessions in the northern part of the network are presented in Fig. 2.

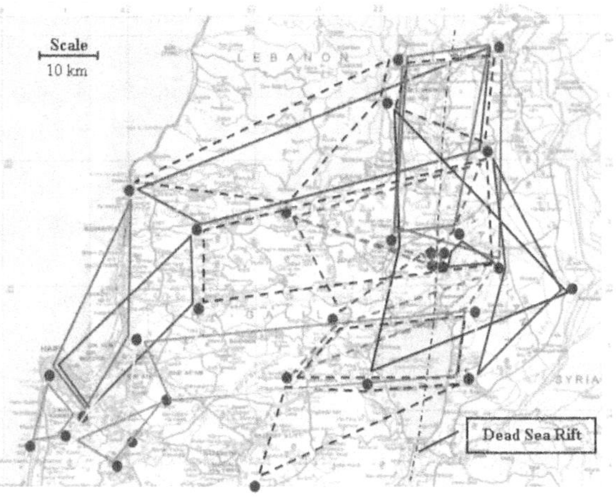

Figure 2 - The northern part of the G1 network an the designed sessions carried out with four GP revisers. The eight sessions that provide sufficientl sensitive network are marked with broken lines.

7 Field Data Collection and Dat Processing

Four 5700 Trimble dual frequency geodetic GP receivers with Zephyr Geodetic antennas were use for the G1 network measurements. A GPS antenn was set up at least three times at each point. Th observations were carried out in 92 sessions, whic were measured between day 343 of 2001 and day 13 of 2002. Twelve additional sessions were measure between days 282 and 331 of 2002. The sessions wei measured with a 30-second sampling rate.
The GPS raw data was transformed from th TRIMBLE receiver format into RINEX and examine by a quality check (QC) software developed b UNAVCO (http://www.unavco.org). All the dat coming from the G1 network measurements was (sufficiently good quality.
The processing was performed in a session mod with all the available permanent GPS sites in a ionosphere free linear combination solution. Th ADDNEQ program was used to compute a mult session solution based on all 104 single-sessio solutions.
The typical precision of the network points wa estimated to be 2-3 millimeters for the horizont component and 4-7 millimeters for the heigl component. The error ellipses of the G1 networ points were mostly circles, indicating a homogenou accuracy.
The precision of the network points in the fir campaign was estimated to be 3-4 millimeters for th horizontal component (Ostrovsky, 2001). Therefor we can roughly estimate the ability of the network t

sense horizontal velocity along the Dead Sea rift for a velocity of two millimeters per year.

8 Summary and Conclusions

The Geodetic-Geodynamic network (G1) was established with the intention of monitoring recent crustal deformations in Israel, and was therefore designed based on the geological phenomena monitored. The sensitivity of a monitoring network is defined as its capacity to detect and measure movements and deformations in the area it covers. The purpose of the deformation survey is to determine whether the monitored object is being deformed or not. It therefore seems reasonable to adopt the concept of sensitivity as the main criterion in the design of a monitoring network.

Based on the surveying campaign measured during the fall of 1996, a sensitivity analysis of the G1 network in the north of Israel was performed for two possible deformation models, the Simple Transform Fault and the Locked Fault. This analysis pointed to a low sensitivity of the G1 network, indicating that an identical second measurement campaign will not enable detecting possible movements and deformations, thus requiring a new network design. Though there was no need for a major redesign of the network points' location, two new points were added in the North-East part of the G1 network for increasing the network sensitivity.

This paper presented a method of GPS vector configuration design and its use in the design of the GPS sessions for the second measurement campaign of the G1 network. The designed configuration contained the sessions with the most effective contribution to the sensitivity of the network. As a further step, a deformation analysis should be performed to define the actual movements, based on the two GPS measurement campaigns.

9 References

Baarda, W. (1968). A Testing Procedure for Use in Geodetic Networks. Netherlands Geodetic Commission, Publications on Geodesy, New Series 2, No. 5.

Cooper, M. A. R. (1987). *Control Surveys in Civil Engineering*. Collins, London.

Even-Tzur, G. (2002). "GPS Vector Configuration Design for Monitoring Deformation Networks." *Journal of Geodesy*, 76(8), 455-461.

Even-Tzur G. (2001). "Sensitivity Analysis of Monitoring Deformation Networks in The North of Israel." *Israel Journal of Earth Sciences*, 50(1): 23-27.

Even-Tzur, G. (1999). "Sensitivity Design for Monitoring Deformation Networks." *Bollettino Di Geodesia E Scienze Affini*,. 58(4): 314-324.

Karcz, I. (1994). "Geological Considerations in Design of the Seminal Dead Sea Rift Network." Perelmuter Workshop on Dynamic Deformation Models, Haifa, Israel.

Ostrovsky, E. (2001). "The G1 GPS Geodetic-Geodynamic Reference Network: Final Processing Results." *Israel Journal of Earth Science*,. 50(1): 29-37.

Pe'eri, S., Wdowinski, S., Shtibelman, A. and Bechor, N. (2002). "Current Plate Motion Across the Dead Sea Fault from Three Years of Continues GPS Monitoring." *Geophysical Research Letters*,. 29(14): 42-1 – 42-4.

Rao C. R., Mitra S. K. (1971). *Generalized Inverse of Matrices and its Application*. John Wiley and Sons, New York.

Rothacher, M. and L. Mervart (Eds.) (1996). *Bernese GPS Software. Ver. 4.0.* Astronomical Institute, University of Berne.

Wdowinski, S., Y. Bock, G. Baer, L. Prawirodirdjo, N. Bechor, S. Naaman, R. Knafo, Y. Forrai, and Y. Melzer (2004). "GPS measurements of current crustal movements along the Dead Sea Fault." J. Geophys. Res., 109(B5)

Weertman, J. and Weertman, J.R. (1964). *Elementary Dislocation Theory*. Macmillan, New-York.

Monitoring of Deformations Along Fethiye-Burdur Fault Zone (Turkey) with GPS

S. Erdogan[3], M. Sahin[1], H. Yavasoglu[1], H. Karaman[1], T. Erden[1], S. Bilgi[1], G. Ruzgar[1], E. Tari[1], Z. Coskun[1], O. Tuysuz[2], M. Gullu[3], T. Baybura[3], I. Tiryakioglu[3], O. Erdogan[3], F. Taktak[3], E. Gokalp[4], Y. Boz[4].

[1] Istanbul Technical University (ITU), Department of Geodesy & Photogrammetry, Maslak, Istanbul, Turkey.

[2] ITU Eurasian Institute, Istanbul, Turkey.

[3] Afyon Kocatepe University, Department of Geodesy & Photogrammetry, Afyonkarahisar, Turkey.

[4] Karadeniz Technical University, Department of Geodesy & Photogrammetry, Trabzon, Turkey.

Abstract. Afyonkarahisar, Burdur and Fethiye region, southwest of Turkey is located on a tectonically active area. According to the historical recordings, the region was destructed by earthquakes many times. Since the beginning of the 20th century, more than 35 earthquakes with magnitudes bigger than 4 occurred in Afyonkarahisar region. Seven earthquakes recorded over the last 90 years in the region between Burdur and Fethiye whose magnitudes are between 5.0 and 7.0. This new project "Monitoring of Deformations Along Fethiye-Burdur Fault Zone With GPS", is founded by The Scientific and Technical Research Council of Turkey (TUBITAK), Istanbul Technical University (ITU), Karadeniz Technical University (KTU) and Afyon Kocatepe University (AKU). The aim of the project is to determine strain accumulation of the study area.

To determine tectonic characteristic of the region, a network having ten sites is established. First and second campaigns were carried out with in August 2003 and March 2004. The duration of measurements in each day was about 12 hours. After first two periods, our network is extended with six new points. Then, the third campaign was performed with 16 sites at two stages in 6 days. In this period the duration of measurements in each day was about 8 hours. The GPS measurements for three campaigns are processed by using GAMIT/GLOBK software package. This paper examines the results from the three GPS campaigns.

The velocity vectors between the SRKK and SLVR points show an opening which is caused by the activeness of the movements between the area of Çay Tatarlı in the South of Sultandağı fault. The GPS velocities of SLVR and ISRT points show that the horizontal velocities of the area are in the same direction showing that there isn't active movement in this area. The velocity vectors of GKPN and KYBS stations are interpreted that constriction is dominant in the area and it is because of the left lateral oblique faults in the region. Examining SIRA and KZLR stations confirm that the area is active and that there are left lateral faults in the region.

Keywords. Tectonic, GPS, Turkey, Fethiye Burdur fault-zone.

1 Introduction

Fethiye Burdur fault zone is approximately 300 km long left oblique normal fault which lies in the southwest of Turkey from the southern city of Fethiye to the northern part of Çay-Şarkikaraağaç region. This fault is not a single structural line but consists of developed intermittent segments parallel to each other. Fethiye Burdur fault zone intersects in the North in the Çay area, with the faults of Akşehir, Sultandağı and Simav. In the South, outside the open seas of Rhodes, it forms a bend with Cyprus and Hellenic Arcs. The area, restrict the geological structure known as Isparta Angle with Akşehir and Simav faults in the northeast. In the South west, the fault zone also form the South border of the Aegean Graben System. (Figure 1)

The locations and the magnitudes of the earthquakes in the last century on Fethiye Burdur fault zones are shown in Figure 3. In the last century there has been around 90 earthquakes with

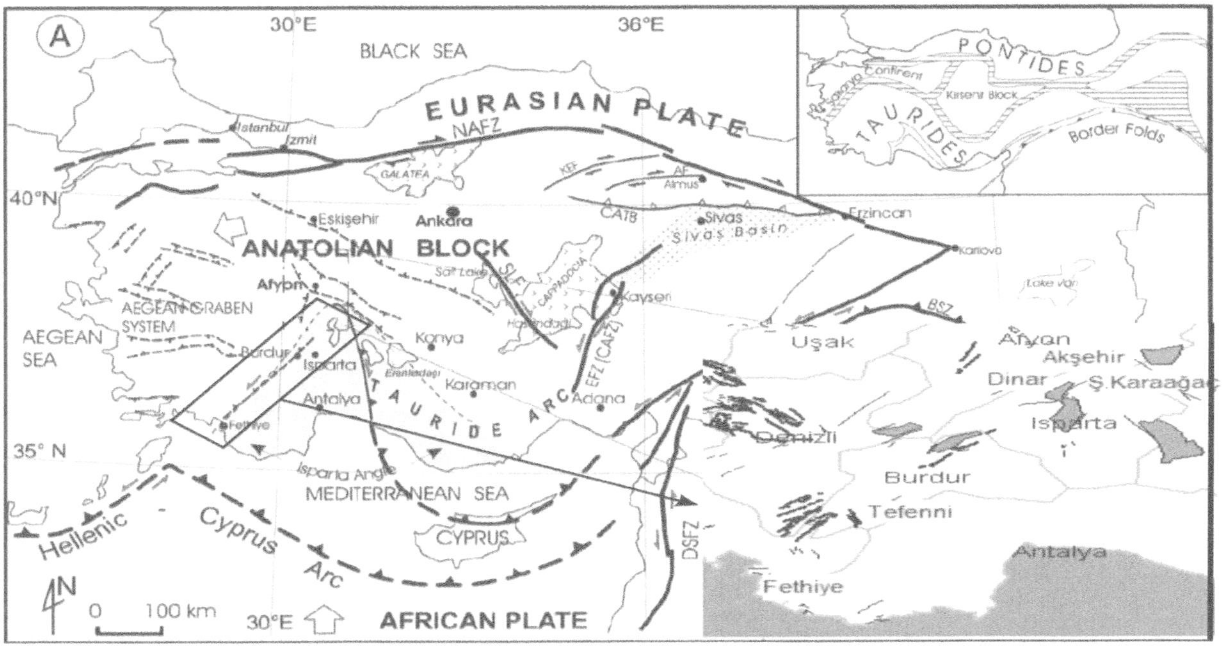

Fig. 1 The location of Fethiye Burdur Fault Zone according to the neighbor plates

the magnitude of 5.0 and over in the area between Rhodes and Afyonkarahisar. Among these were that affected Burdur 3 October 1914 (Ms 7.0), 12 May 1971 (Ms 6.2), that affected Dinar, 7 August 1925 (Ms. 6.0), 1 October 1995 (Ms 6.1), 15 December in Akşehir (Ms 5.9), 3 February 2002 in Çay (Ms 6.0) and that affected Fethiye, 24 April 1957 (Ms 7.1) are important signs of the seismic activities in the last century.

The earthquakes which occurred after the Burdur earthquake, in 1914 in the North of Burdur, were more effective in the south of Burdur between 1922 and 1963. During this period the earthquakes began in 1922 in Karpathas (Ms 6.7) and in time moved towards Southwest and occurred in orderly Rhodes in 1926, Fethiye in 1957, Tefenni in 1963 and once again Burdur (Ms 6.2) in 1971 (Taymaz and Price, 1992). The earthquakes that occurred in Karpathas, Rhodes, Fethiye, Tefenni and Burdur during the 40 years period of 1957-1992 importantly indicate that the fault has progressed from the South to the North. This progress shows that the fault segments triggered each other and transferred tension in the Fethiye Burdur fault zone (Taymaz and Price, 1992, Yağmurlu, 1997) (Figure 2). Compared to the two well known main active fault systems of North and east Anatolia faults or the West Anatolian Graben system in Turkey, the Fethiye-Burdur fault zone is not a well identified fault zone. Earlier studies and mapping works in the region regarding the Fethiye-Burdur fault zone have focused on the paleotectonic structure (Oral, 1994).

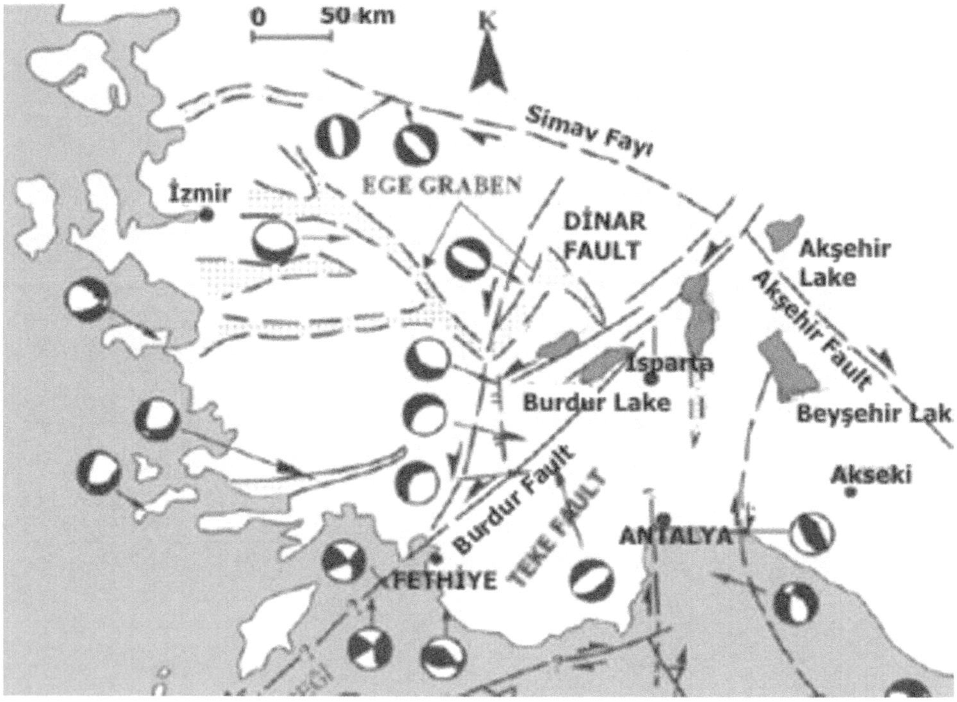

Fig. 2 Fault plane solutions in the region (Changed from Mc Kenzie 1978)

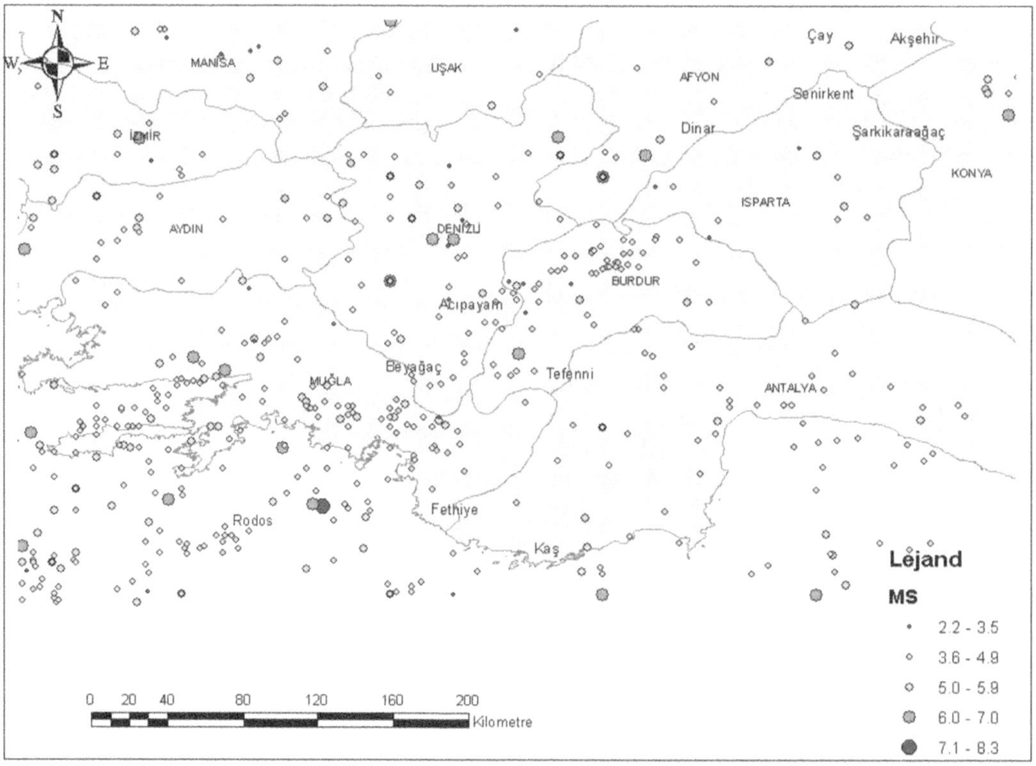

Fig. 3 The earthquakes that occurred in the area in the last century

2 GPS Campaigns in the Area

To determine the tectonic movements on the Fethiye Burdur fault zone the 10 points which represent the area's characteristics were determined (Table 1), (Figure 4). Six of the 10 points determined were pillar shaped and the remaining 4 were found in the rocks as bronze shaped. The first, the second and the third period studies were performed in August 2003, March 2004 and September 2004, respectively. In the last period of the measurements, 6 new points have been added to the network. The first 2 periodic measurements were performed in 12 hourly measurements and the third periodic measurements were repeated 3 times in 8 hourly measurements.

Table 1 Fethiye Burdur GPS Stations

Station Name	Region	Station Name	Region
SLVR	Selevir	CLTK	Çeltikçi
SRKK	Şarkikaraağaç	GKPN	Acıpayam
TKIN	Tekin	KYBS	Kayabaşı
ISRT	Isparta	KZLR	Kozlar
YSLV	Yeşilova	SIRA	Fethiye

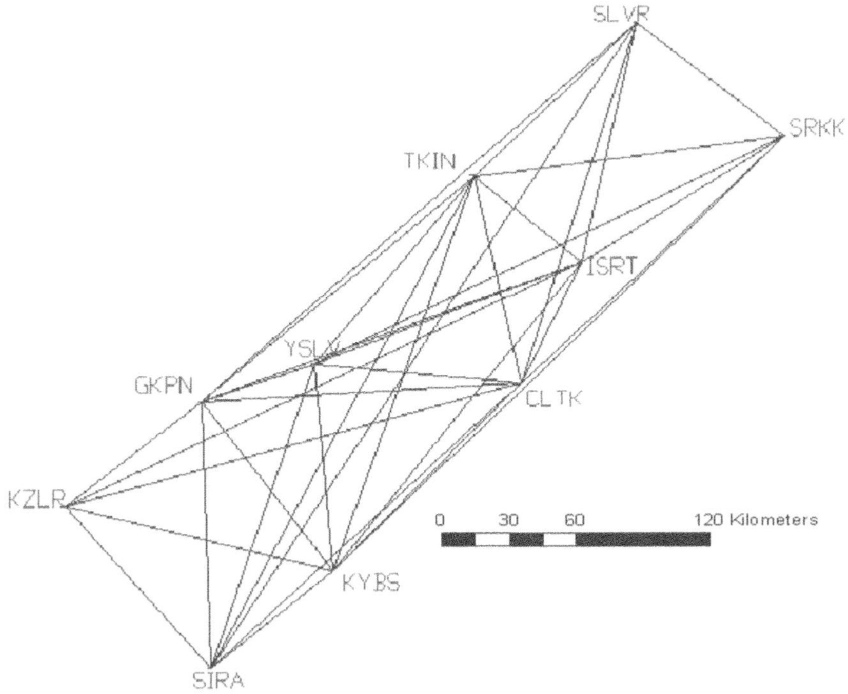

Fig. 4 GPS network and baselines

In the first and the second campaign 6 Z Xtreme GPS receivers with dual frequency are used belong to Afyon Kocatepe University and Istanbul Technical University. In the third campaign 3 more Z Xtreme GPS receivers are provided to the campaign from Karadeniz Technical University. The measurements were done in 15 second time intervals and the evaluations were based in 30 second time intervals. In the working area, the longest baseline was SIRA-SLVR being 240 km and the shortest was TKIN-ISRT being 50 km.

3 The Analysis of the Data

The GPS data of the campaigns were evaluated in a three step approach by using the GAMIT-GLOBK software package (Feigl et al. 1993). In the first step the station coordinates, zenith delays, atmosphere gradients and earth's rotation parameters were achieved by double differences GPS phase measurements by using GAMIT software. In this phase, 8-10 IGS stations were added to the evaluation in every period. These were ISTA, TUBI, WTZR, ZECK, MATE, NICO, SOFI, GRAZ, MERS and KIT3 stations. By doing so it was possible to make a connection between regional network and global IGS network. In this phase, loose a priori constraints were applied to all parameters in session and station tables in GAMIT (Table 2). In the second step, in order to get the received station, satellite coordinates, covariance matrixes, definite velocity and coordinate measurements of each and every day were processed by the Kalman filter. The weights for each data set of quasi-observations are determined by averaging the increments in chi-square per degree of freedom from a forward and backward filtering of data. In this phase, the daily regional analyses were evaluated together with the global daily IGS analyses which are being published by SOPAC. With this approach for the analysis with global data included is allows us to define the reference frame by minimizing the departure from a priori values of the positions and velocities of a set of well determined stations, usually globally distributed but for some applications concentrated within the region of interest. In the third phase, Eurasian reference frame for the estimation of velocity were defined. In this phase in order to define the reference frame different point combinations have been used. Out of these considered as the most suitable points, (MATE, SOFI, ZECK, TROM, KIRU, ONSA, TROM, CAGL, VILL, POTZ, GRAZ, LAMA, WTZR) the Eurasian frame is defined by minimizing the horizontal velocities received by the priory coordinates given by ITRF00 (Boucher et al., 1998; McClusky et al., 2000), (Figure 5). According to the Eurasian frame the station velocities are shown in Figure 6 and Table 3.

Table 2 Constrains used in processing

Stations	**Coordinates' Constrains**					
	Lat. (m)	Long. (m)	Radius(m)			
MATE	0.050	0.050	0.200			
GRAZ	0.050	0.050	0.200			
WTZR	0.050	0.050	0.200			
SOFI	0.050	0.050	0.200			
KIRU	0.100	0.100	0.200			
ZECK	0.050	0.050	0.200			
ONSA	0.050	0.050	0.200			
VILL	0.050	0.050	0.200			
All other	100	100	100			
Semiaxis	**Eccentricity**	**Inclination**	**Ascension**	**Arg. of per.**	**Mean Anomaly**	**Rad1..9**
0.01	0.01	0.01	0.01	0.01	0.01	0.01

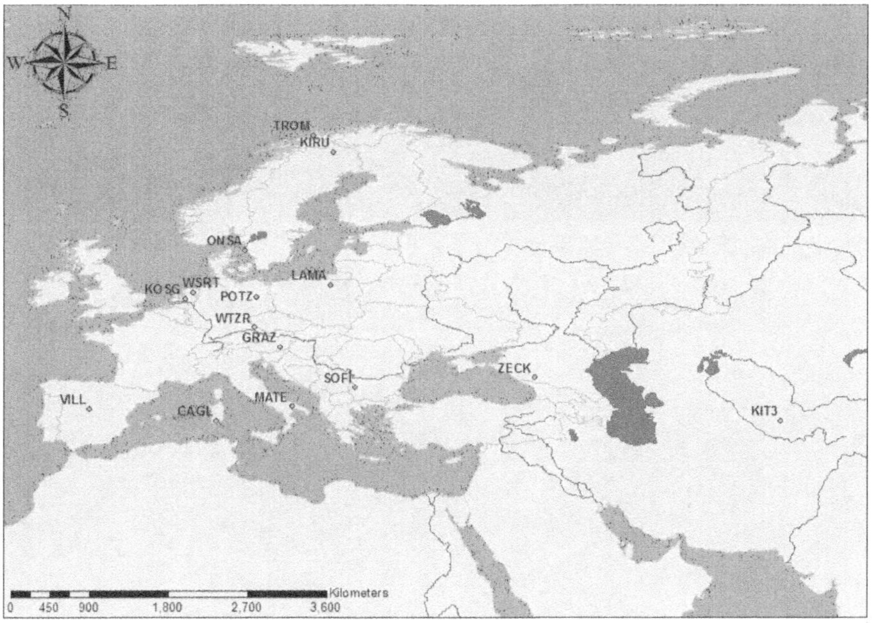

Fig 5 IGS Stations used in defining the Eurasian Frame

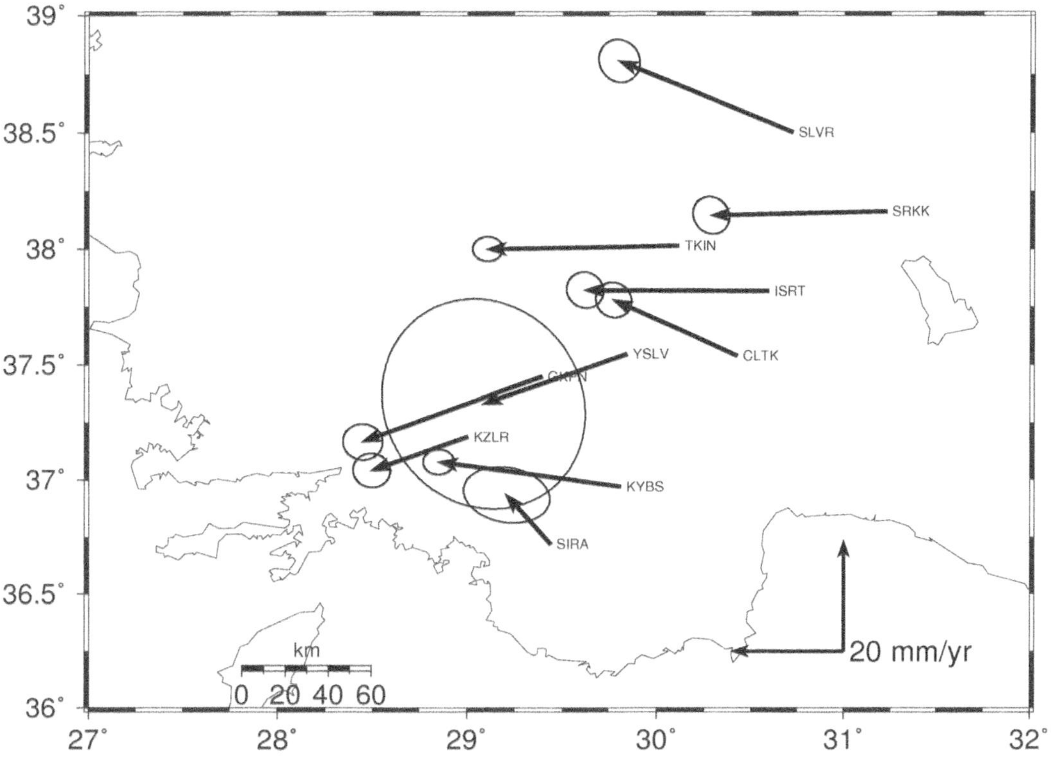

Fig 6 GPS velocity vectors with their %95 confidence ellipses at the points on Fethiye Burdur Fault zone according to the Eurasian Frame

Fig 7 GPS velocity vectors, faults in the region and new GPS Sites

Table 3 GPS Velocity Values

GPS Sites	V_E (mm)	V_N (mm)
SRKK	-27.95	-4.74
SLVR	-27.81	10.90
TKIN	-31.18	-3.30
ISRT	-29.86	-1.08
CLTK	-23.12	6.45
YSLV	-23.99	4.81
GKPN	-31.40	-12.70
KYBS	-30.86	1.64
KZLR	-19.48	-8.34
SIRA	-27.28	16.49

4 Conclusions

As we can see in Figure 6 according to the Eurasian reference system the Aegean plate shows a movement towards the South west with a yearly speed of 30-35 mm. If we check the velocity vectors at the station points, it can be seen an opening between the SRKK and SLVR points. It is thought that this opening is caused by the activeness of the movements in between the area of Çay Tatarlı in the South of Sultandağı fault. Once again, when we examine the GPS velocities of TKIN and ISRT points we see that the horizontal velocities of the area are in the same direction. Also, these 2 points do not nearly have a relative horizontal velocity. In connection to this, it is thought that there isn't active movement in this area. Since the gained fault ellipses of the velocity vectors at the YSLV point were bigger than the planned the accuracy, the explanation of the movements in the area will have to wait for the next campaign measurements. When we examine the GKPN and KYBS point's velocity vectors it is thought that constriction is dominant in the area and this is caused by the left lateral oblique faults in the area. Based on the examination of the SIRA and KZLR points it can be seen that the area is active and that there are left lateral faults operating in the area (Tüysüz, O., 2005). Because of the complex structure of the zone, it is very difficult to understand the tectonic movements. Meanwhile, while analyzing velocity vectors, block movements in the zone must be taken into consideration. So, additional studies with the new GPS sites will provide more reliable information about the tectonic structure of the region.

References

Boucher, C., Z. Altamimi, and P. Sillard, Results and Analysis of the ITRF96, IERS Technical Note 24, 166 pp., Cebtral Beureau of IERSObservatoire de Paris, 61 Avenue de l'Observatoire, F-75014, Paris, France, 1998.

Feigl, K. L, D. C. Agnew, Y. Bock, D.-N. Dong, A. Donnellan, B. H. Hager, T. A. Herring, D. D. Jackson, R.W. King, S. K. Larsen, K. M. Larson, M. H. Murray, and Z.-K. Shen, Measurement of the velocity field in central and southern California, J. Geophys. Res., 98, 21667–21712, 1993.

McClusky, S., Balassanian, S., Barka, A., Demir, C., Ergintav, S., Georgiev, I., Gurkan. O., Hamburger, M., Hurst, K., Kahle, H., Kastens, K., Kekelidze, G., King, R., Kotzev, V., Lenk, O., Mahmoud, S., Mishin, A., Nadriya, M., Ouzounis, A., Paradissis, D., Peter, Y., Prilepin, M., Reilinger, R., Sanli, I., Seeger, H., Tealeb, A., Toksoz, M. N. And Veis, G., 2000. Global Positioning System Constraints On Plate Kinematics And Dynamics In The Eastern Mediterranean and Caucaus, Journal of Geophysical Research, 105, 5695-5719.

McKenzie, D., 1978. Active Tectonics of the Alpine-Himalayan belt: the Aegean Sea and surrounding regions: Geophys. Jour. Royal Astro. Soc., 55, 217-254 s.

Oral, B., 1994. GPS Measurement in Turkey (1988-1992) Kinematics of the Africa-Arabia Eurasia Plata Collision Zone, PhD Thesis, Massachusetts Institute of Technology, ABD.

Tari, E., Şahin, M., Barka, A., Reilinger, R., King, R., McClusky, S. and Prilepin, M., 2000. Active tectonics of the Black Sea with GPS, Earth Planets Space, 52, 747–751.

Taymaz, T., Jackson, J.A. and McKenzie, D., 1991. Active Tectonics of the North and Central Aegean Sea, Geophys. J. Int., 106, 433-490.

Taymaz, T., Price, S., 1992. The 1971, Burdur Earthquake Sequence, SW Turkey: A Synthesis Of Seismological And Geological Observations, Geophys. J. Int. 108, 589–603.

Tüysüz, O., 2005, Personal Communication.

Yağmurlu, F., Savaşçın, Y. ve Ergün, M., 1997, Relation of Alkaline Volcanism and Active Tectonism within the Evolution of the Isparta Angle, SW Turkey: J. Geol. 105, 717-728.

Yağmurlu, F., 2000, Burdur Fayının Sismotektonik Özellikleri, Batı Anadolu'nun Depremselliği Sempozyumu, İzmir, 143-151.

Monitoring Crustal Movements and Sea Level in Lanzarote

L. García-Cañada, M.J. Sevilla
Instituto de Astronomía y Geodesia (CSIC-UCM),
Facultad de Ciencias Matemáticas, Universidad Complutense de Madrid, 28040 Madrid, Spain.
laura_garcia@mat.ucm.es

Abstract. The Institute of Astronomy and Geodesy is measuring sea level variations in Lanzarote Island by two automated tide gauges of precision since 1993. In order to obtain "real" sea level variations a permanent GPS station has been installed near the tide gauges. The goal of this GPS station is to measurement vertical crustal movements in order to obtain absolute sea level variations removing these movements from tide gauge data.

A vertical tie with high accuracy between the reference point of the pillar where the GPS antenna is installed and the tide gauges bench marks is absolutely necessary. We have carried out this altimetric link yearly since year 2000. Methods of repeated geometric and trigonometric levelling of very short tracts have been used due to the great level difference existent.

The control of the local stability of the pillar where the GPS antenna has been established, is carried out by a micro-geodetic control network building around it. This network of 13 benchmarks has been regularly observed by classic and GPS geodetic techniques in years 2000, 2001, 2002, 2003 and 2004. The data of these campaigns have been processed by different types of adjustments in the same reference system. The precision of the measurements and the reliability of the networks have been calculated.

In this work the results obtained, the evaluation of the altimetric links and levelling campaigns, and the comparison of the levels of different measurements are presented.

Keywords. Sea level, crustal movements, geodetic networks and GPS.

1 Introduction

Changes in sea level are usually recorder by tide gauges, but their data corrupted by vertical land movements and they measure relative sea level. These vertical land movements at tide gauges have to be measure in order to obtain absolute sea level and separate crustal movements and sea level variations (Zerbini et al., 1996).

The aim of this work is monitoring the area where the tide gauges are installed using GPS and levelling techniques in order to provide absolute sea level instead "land relative" sea level. That is why continuous geodetic positioning of tide gauge bench marks (TGBMs) is required (IOC, 2000).

This allow to remove vertical motions of the land where tide gauge is attached from the sea level data and studying the correlation between relative sea level and crustal vertical movements.

2 Tide gauges and GPS sites

Lanzarote is the most Northeast island of the Canarian Archipelago, it is 100 km off the African coast and 1000 km off Iberian Peninsula. "Jameos del Agua" is a place in the north coast where the Institute of Astronomy and Geodesy has a geodynamic laboratory (Vieira, 1994).

There, the Corona volcano lava tube intersects with the ocean forming two lakes. It is an ideal place for very sensitive instruments to be hosted and a good area to sea level researches because there are not perturbations like winds and waves.

Two automated tide gauges of precision (Druck pressure sensors) are located in the two lakes that measure raw sea level in a relative reference system since 1993. Really near each tide gauge there is an altimetric bench mark.

A permanent GPS station has been installed to detect vertical velocity at tide gauges. It is located on the top of "La Casa de los Volcanes" building, which is a good place for a CGPS because it has a clear view of sky in all directions, area is free of interference sources and it is near tide gauges (Figure 1).

Fig. 1 Antenna of the CGPS Lanzarote station.

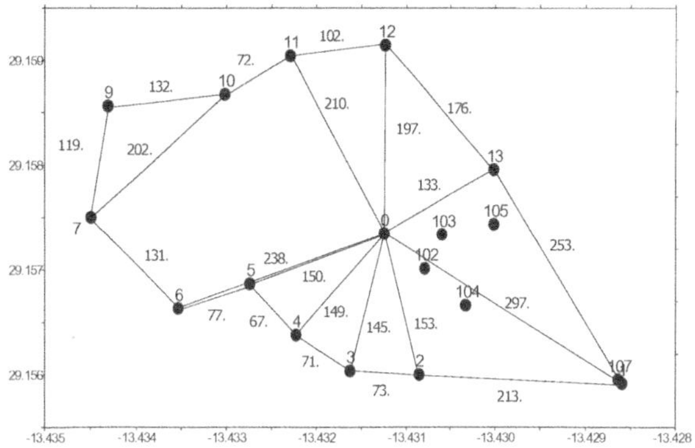

Fig. 2 Micro geodetic control network installed around CGPS (point number 0).

The instrumentation installed is a dual frequency code and phase measuring Ashtech Z-Surveyor receiver and Dorne Margolin choke ring ASH701945B_M antenna. It follows the guidelines of the IGS.

In addition tide gauges and a permanent GPS station there are other sensors to measure water and air temperature, pressure, humidity and gravity.

3 Geodesic control

Top of a building is not the best place to have a CGPS, that is why a micro geodetic control network has been designed and installed around it in order to monitor the stability of the monument and the building.

The network has 13 bench marks and the distance between these points is from 65 to 300 m (Figure 2). It has been regularly observed every year by classical geodetic techniques and GPS since year 2000.

But if a somewhat remote CGPS station is moving by an unknown amount relative to its tide gauge, then the absolute position of the tide gauge is being determined only as frequently as the levelling tie is being made (Figure 3).

The methodology to measure in the campaigns is different depending on the topography of the land.

Classical precise spirit levelling is use when height difference between bench marks is not so great. But when it is more important, trigonometric levelling with reciprocal and simultaneous vertical angles and geometric distance measure is used.

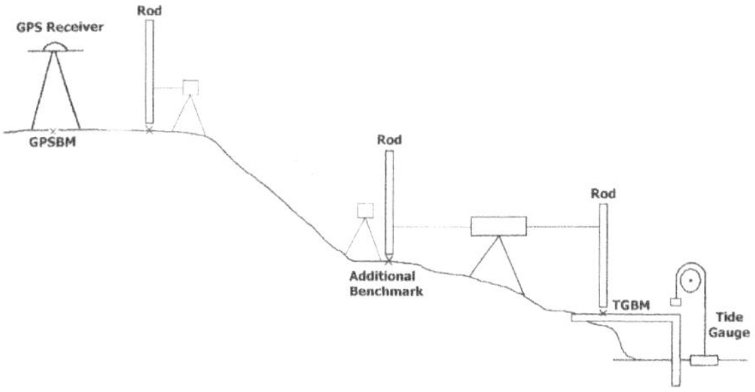

Fig. 3 Schematic of levelling tie between TGBM and CGPS.

In trigonometric method (Figure 4) the height difference between two points, *A* and *B*, is calculated using (Sevilla and Romero, 1991):

$$\Delta h = hi_2 - hi_1 = (R + hi_1)\frac{2 \cdot sin\left(\frac{z_1 - z_2}{2}\right) \cdot sin\frac{\omega}{2}}{\cos\left(\frac{z_2 - z_1}{2} + \frac{\omega}{2}\right)} \quad (1)$$

where z_1 y z_2 are vertical angles measured, hi_1 y hi_2 are instruments heights, *R* is terrestrial mean radio and ω is the angle that form the vertical of points *A* and *B* in the centre of the Earth.

The angle ω is calculated by:

$$\omega = \frac{L}{R} \quad (2)$$

where *L* is the distance on the cord of ellipsoid,

$$L^2 = \frac{d_{12}^2 - \Delta H_{12}^2}{\left(1 + \frac{h_1}{R}\right)\left(1 + \frac{h_2}{2}\right)} \qquad (3)$$

Δh is obtained by iteration method, using a approximated L in the first iteration.

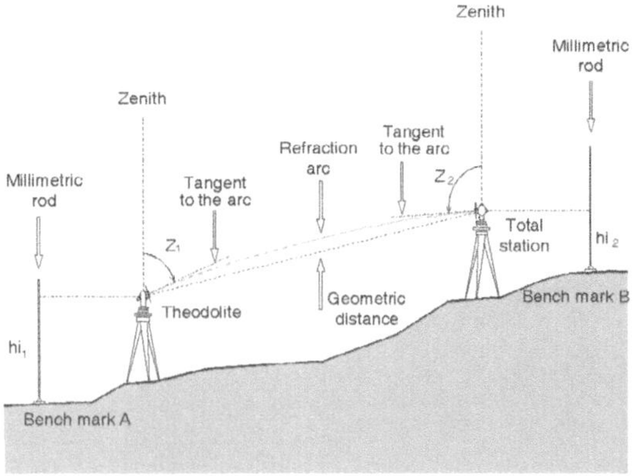

Fig. 4 Schematic of reciprocal and simultaneous trigonometric levelling.

The height difference using geometric levelling is calculated by:

$$\Delta Z_{AB} = L_A - L_B \qquad (4)$$

where L_A and L_B are the readings in the rods located in the points A and B respectively (Figure 5).

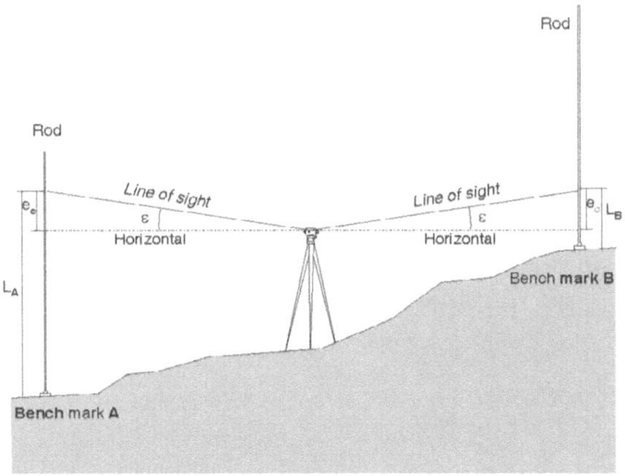

Fig. 5 Schematic of spirit levelling.

4 Levelling tie CGPS-tide gauge results

The trigonometric methodology has been used in the levelling tie between TGMBs and CGPS since 2000. It is carried out in four tracts using three auxiliary points in year 2000. In 2001 some additional bench marks had installed in order to have a better monitoring in the area in case any deformation were detected.

The results of altimetric profile in 5-year campaigns (Figure 6) show there is not significant differences in consecutive years, less than 3mm; this suggest that the tide gauge did not experience any uplift or subsidence relative to CGPS.

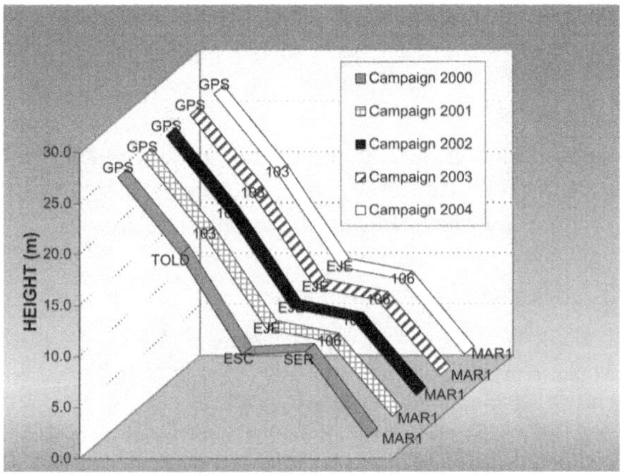

Fig. 6 Altimetric profiles GPS-MAR1 in 2000-2004 campaigns, with precision trigonometric levelling technique.

5 Control network results

The micro-geodesic control network (Figure 2) has been measurement yearly since 2000 too. Reciprocal and simultaneous trigonometric levelling is use to tie CGPS with network bench marks when there is visibility. Spirit levelling is carried out in the surveys for the external ring.

The least-square adjustment of the campaign data has been carried out with the method of variation coordinates using four different models:

1. Free network and equal weights (M1).
2. CGPS fixed and equal weights (M2).
3. Free network and weights inversely proportional to the distance (M3).

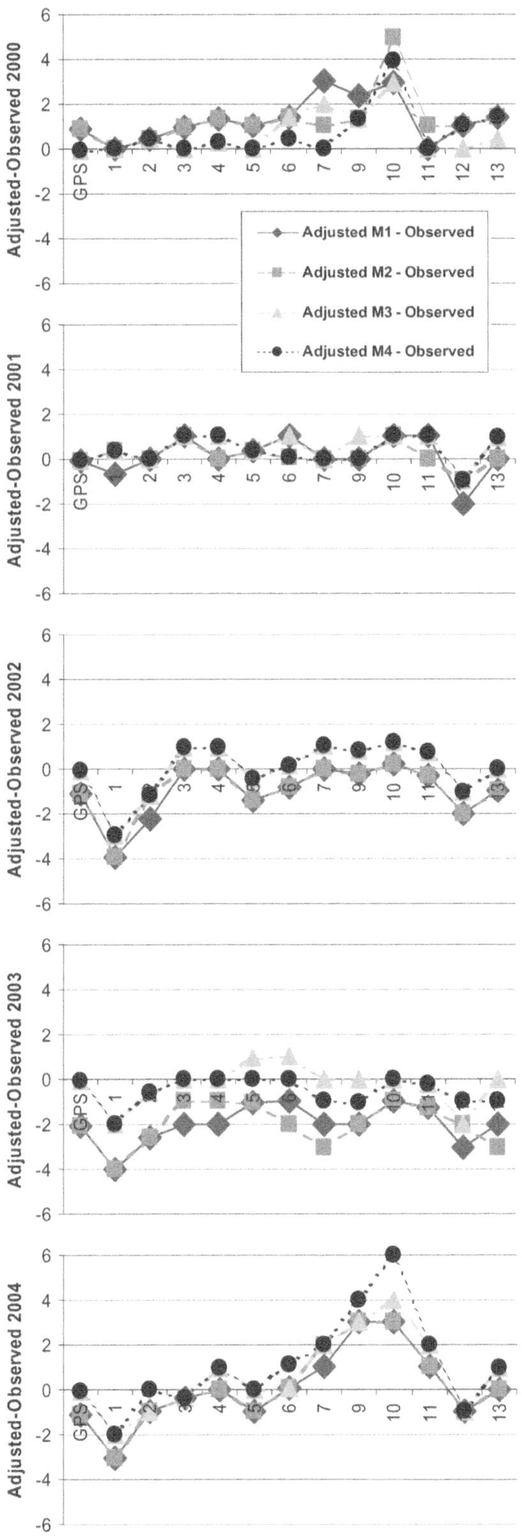

Fig. 7 Adjusted network results with four models in the five surveys.

4. CGPS fixed and weights inversely proportional to the distance (M4).

In Figure 7 we can see the difference between the difference height observed and adjusted with the four models in the 5 years and there is no significant difference in the results using one method or another.

If the results in the 5 years with one model are compared (Figure 8), a large subsidence at points 6 (13 mm) and 13 (10 mm) has been detected in the period 2000-2001. The reason was that these bench marks had been moved by a man-effect and they had to be repaired and the problem was solved since 2001.

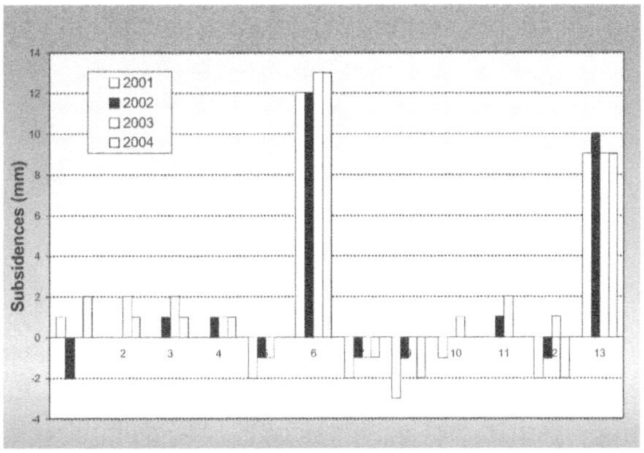

Fig. 8 Subsidences (in mm) detected in years 2001, 2002, 2003 and 2004 in the control network points refer to 2000 campaign.

The results of the repeated precise levelling surveys showed no significant changes in height within the control network over the period from 2000 to 2004. This confirms the stability of the monument and building on which the antenna GPS is located relative to the area.

6 GPS time series

CGPS data has been processed in a regional network of permanent stations formed by four IGS stations, MAS1, MADR. RABT and SFER, four EUREF stations, ALAC, CASC, CEUT and LPAL, and the Lanzarote station (LACV) (Figure 9).

The network is processed with Bernese v.4.2 software using IGS final precise orbits and Quasi Ionosphere Free (QIF) strategy for the ambiguity resolution (Hugentobler et al., 2001). Elevation cut off angle of 10°, elevation dependent observation weighting and the dry-Niell mapping function (Niell, 1996) has been used.

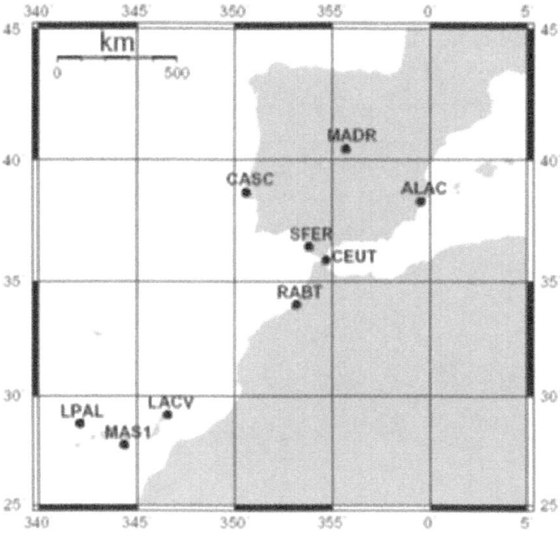

Fig. 9 Map of the CGPS regional network.

The Saastamoinen global tropospheric delay model is used to generate a priori values for the tropospheric delay but total troposphere zenith delay has been estimated for all the stations every hour. The ionosphere effect is eliminated introducing the ionosphere free linear combination of the observables.

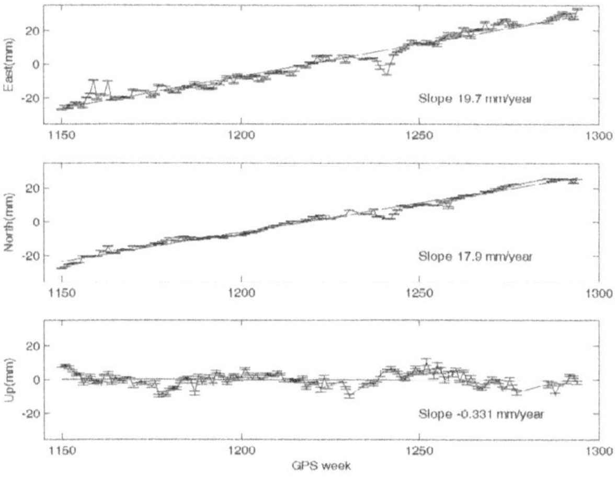

Fig. 10 LACV weekly time series.

The ocean tide loading effect has been taken into account according to the GOT00.2 model. Amplitudes and phases of this model have been obtained from the H.G. Scherneck site (http://www.oso.chalmers.se/~loading/).

Sessions for 24 hours have been defined and for each session the network has been adjusted in Cartesian threedimensional coordinates in the ITRF2000 reference frame. The IGS stations MAS1, MADR and SFER have been considered fiducial stations constrain their ITRF2000 coordinates.

Daily coordinates and normal equations are calculated in this way and they are combined by ADDNEQ subroutine (Hugentobler et al., 2001) to obtain weekly solution.

Although time span is too short and there are some outlier period, preliminary results of the weekly time series (Figure 10) show there is no vertical rate and horizontal velocity in the same order of other stations in the same tectonic plate.

7 Conclusions

After measure sea level by tide gauges in Jameos del Agua (Lanzarote Island) since 1993 we adverted that geodetic techniques for monitoring rates of vertical land movements were needed if we want to obtain sea level change without land movements at tide gauges.

That is why a CGPS stations has been installed near tide gauges, a micro-geodesic control network has been established around it and it is surveyed regularly; levelling tie between CGPS and TGBM is carried out every year since 2000.

The geodesic control network results over a period of four years show the monument of the GPS antenna and the building where the CGPS is, are stable and there is not any deformation between the pillar in the top of the building and the land is this area around it.

Due to the great height difference, the tie CGPS-TGBMs is measure using precise trigonometric levelling techniques. The results of these campaigns confirm that the CGPS do not show any vertical deformation respect to TGBMs.

This ensures that the relative sea level measured by the tide gauges is not affected by land movements in the area, so it is representative of the surrounding area.

About time series of CGPS, preliminary results based on 2.5 years time span, suggest that the vertical component does not present any rate, but observations spanning a longer time period will be necessary to get more reliable values.

Acknowledgement

By the support coming from research project REN2001-2271, "Geodesic networks, space techniques and gravimetry applied to the study of the structure and dynamic of the crust in Canary" by Science and Technology Spanish Ministry.

References

Hugentobler, U., S. Schaer and P. Fridez (Edts.), 2001. Bernese GPS Software Version 4.2. Astronomical Institute. University of Berne.

IOC, 2000. Manual on sea-level measurement and interpretation. Volume 3 - Reappraisals and Recommendations as of the year 2000. Intergovernmental Oceanographic Commission. Manuals and Guides No. 14. IOC, Paris, 52pp.

Niell, A.E., 1996. Global mapping functions for the atmosphere delay at radio wavelengths. Journal of Geophysical Research, Vol. 101, No. B2, pp. 3227-3246.

Sevilla M. J. and P. Romero, 1991. Ground deformation control by statistical analysis of a Geodetic network in the caldera of Teide. Journal of Volcanology and Geothermal Research, Vol. 47 pp. 65-74. Elsevier Sc. Pub. Amsterdam.

Vieira, R., 1994. La estación geodinámica de Lanzarote. Serie Casa de los Volcanes. Monografía nº 3, pp. 31-40.

Zerbini, S., H.P. Plag, T. Baker, M. Becker, H. Billiris, B. Bürki, H.G. Kahle, I. Marson, L. Pezzoli, B. Richter, C. Romagnoli, M. Sztobryn, P. Tomasi, M. Tsimplis, G. Veis, and G. Verrone, 1996. Sea level in the Mediterranean: a first step towards separating crustal movements and absolute sea level variations. Global and Planetary Change, 14, pp. 1-48.

GPS Measurements along the North Anatolian Fault Zone on the Mid-Anatolia Segment

H. Yavasoglu, E. Tari, H. Karaman, M. Sahin, O. Baykal, T. Erden, S. Bilgi, G. Ruzgar, C. D. Ince
ITU Faculty Of Civil Engineering, Department Of Geodesy & Photogrammetry
Maslak-Istanbul, Turkey
Tel: +90 212 285 6653, Fax: +90 212 285 6587

S. Ergintav, R. Çakmak
TUBITAK, Marmara Research Center, Earth & Marine Sciences Research Institute
Gebze-Kacaeli, Turkey

U. Tari, O. Tuysuz
ITU Eurasian Institute Of Earth Sciences
Maslak Istanbul, Turkey

Abstract. The North Anatolian Fault (NAF) is one of the longest active strike slip systems. Slip rate of the NAF was estimated from the GPS data as 20-24mm/yr approximately from previous studies. One of the important features of the NAF is seen in the central part. Here NAF consists of southward splitting concave branches. These splines generally have right-lateral slip and can be compared with Riedel fractures. The important splays are Sungurlu, Merzifon and Lacin faults. A new project "Determination of Kinematics along the North Anatolian Fault Branch between Ladik and Ilgaz with GPS Measurements", founded by The Scientific and Technical Research Council of Turkey (TUBITAK) and Istanbul Technical University (ITU) Research Fund has been carried out in order to determine the magnitude and direction of the block movements in the region by using GPS since 2001. The knowledge about the neotectonics of the region with the contributions of geology and seismology after the GPS campaigns is expected to provide further information on the assessment of the earthquake potential.

To determine tectonic characteristic of the region, a network having sixteen sites was established. Measurements with three day pairs were performed in six days at two stages. The duration of measurement in each day was about 8 hours with an interval of 30 seconds. All stations were observed for at least three days. The GPS measurements for four campaigns were processed using GAMIT/GLOBK software package. This paper examines the results from the four GPS campaigns.

According to the obtained GPS velocity vectors, the movement of the region has the same direction with Anatolian plate escaping from East to West.

All stations on the north of the NAF (KRGI (Kargı), IHGZ (Ihsangazi), KVAK (Kavak) and GOL1 (Gol)) have smaller vectors than the stations on the south. The magnitude and direction differences between velocity vectors present the anticlockwise movement of the Anatolian plate with the information abut the movements of the blocks in the region.

Keywords : Tectonics, GPS, Turkey, Deformation, North Anatolian Fault.

1 Introduction

The tectonic framework of the Eastern Mediterranean is dominated by the collision of the Arabian and African plates with the Eurasia (Hubert-Ferrari et al. (2002)). This collision has created wide variety of tectonic processes such as folds and thrust belts, major continental strike-slip faults, opening of pull-apart basins etc. All these tectonic activities have caused long-term destructive earthquakes in Anatolia (Hubert-Ferrari et al. (2002)) (Stein et al. (1997)).

The Last earthquakes occurring at the end of the 20th Century, on August 17 and November 12, 1999 in Golcuk and Duzce, also drew the attention of international scientific community over the tectonics and kinematics of the NAF. A westward migrating earthquake series starting from 1939 Erzincan earthquake, produced more than 1000 kilometers of ruptures between Erzincan and Sea of Marmara (Figure 1) (Barka A. et al. (1988) (1996)).

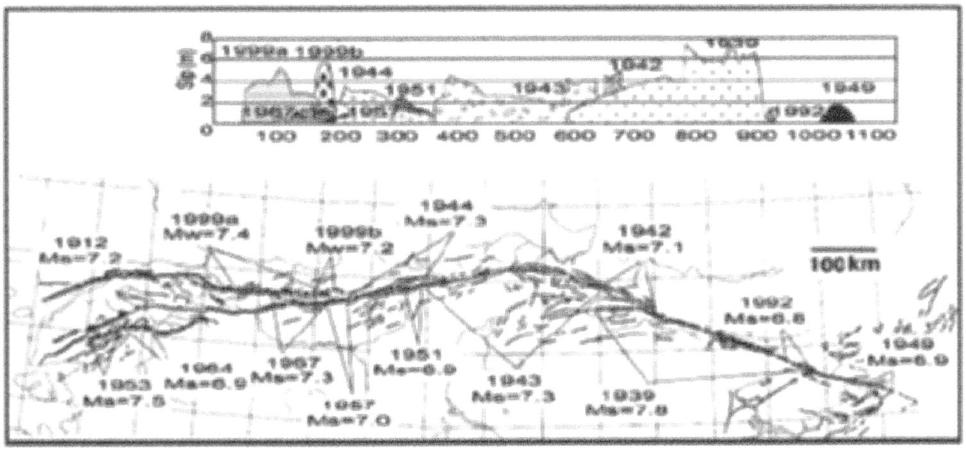

Fig. 1. Westward Migrating Earthquake of Turkey (Şengör et al. (2005)).

2 Geological Features of NAF

The North Anatolian Fault (NAF) is one of the largest currently active strike-slip faults in the world (Dewey and Şengör (1979); Jackson and McKenzie (1984) Şengör et al. (2005)). The collision of the African and Arabian plates with Eurasia effected the Anatolian plate. The Arabian plate moves in the north-northwest direction relative to the Eurasia at a rate of about 18-25 mm/yr (McClusky et al (2000)). The African plate also moves at a rate of about 10 mm/yr in north direction relative to the Eurasia. The Anatolian plate escapes from eastward to westward and rotating anticlockwise. Slip rate of the NAF was estimated from the GPS data as approximately 24mm/yr (McClusky et al (2000)) (Meade et al (2002)).

The right-lateral NAF extends for 1000 km from Eastern Turkey to the Aegean Sea in an arc parallel to, and 80–90 km from, the Black Sea coast (Ambraseys (1970)), (McKenzie (1972)), (Barka (1992)). Our study area (LVKI) is between Ladik and Ilgaz with a total length of about 200 km. The project area is 16500 km^2.

One of the important features of the NAF is seen in the central part (Koçyiğit et al. (1998)). Here NAF consists of southward splitting concave branches. These splines have generally right-lateral slip and can be compared with the Riedel fractures. One of the biggest splays is known as Sungurlu fault. The other important splays are Merzifon and Lacin faults. In the twentoeth century, the faults forming the central part of the NAF and its splays produced different earthquakes greater than magnitude 5 (Canıtez and Üçer (1967)) (Eyidoğan, et al (1991)) (Blumenthal, (1945)) (Blumenthal (1948)) (Harvard GMT) (Taymaz (2000)) (Table 1).

Table 1. Major Earthquakes on The Mid-Anatolia Segment of The NAF in The Last Century

DATE	SETTLEMENT	MAGNITUDE
9 March 1902	Çankırı	5.5
25 June 1910	Osmancık	6.1
21 November 1942	Osmancık	5.6
2 December 1942	Osmancık	5.9
20 December 1942	Erbaa	7.1
27 November 1943	Ladik	7.2
13 August 1951	Kurşunlu	6.9
7 Semtenber 1953	Kurşunlu	6.0
14 August 1996	Mecitözü	5.6
6 June 2000	Orta	6.1

In contrast to the other parts of the NAF, the central part has not been studied using GPS and other geodetic methods in detail yet (McClusky et al (2000)). The data, which will be produced in this project, is expected to add an important contribution to the present knowledge on the NAF.

3 The GPS Measurements

3.1 The Design of Mid-NAF GPS Network

The estimated lateral movement on the LVKI segment of NAF is approximately 24mm per year (McClusky et al (2000)) (Meade et al (2002)). In order to determine movements at the level millimeters, point marks in the network are planned to be built with forced centering instruments (pillars or steel rods etc.) (Oral et al (1988)).

The first work was carried out in the study area to find out convenient old pillars. As a result of this study, 25 existing pillar points were determined in the region. The network was formed by 16 stations (Table 2) because of financial limitations. The stations were named with the four letter abbreviations of the nearest settlement. The stations were chosen so that they could represent the motions of the blocks in the region.

Table 2. Mid-NAF GPS Network Sites

Site ID	Site Name
ALAC	Alaca
CNKR	Çankırı
DDRG	Dodurga
GBAG	Göllübağlar
GHAC	Gümüşhacıköy
GKCB	Gökçebel
GOL1	Göl
GYNC	Göynücek
HMMZ	Hamamözü
HVZA	Havza
IHGZ	İhsangazi
KRGI	Kargı
KVAK	Kavak
ORTC	Ortalıca
OSMC	Osmancık
SNGR	Sungurlu

The GPS sites were mainly chosen as representative of the fault-bounded continental blocks. Although there are lots of faults in the area, active and recently earthquake producing faults and continental blocks that are bounded by these faults were taken into consideration.

3.2 GPS Measurements

The campaigns were carried out in or around the month of August between 2001 and 2004. Each station was observed for at least three days in each campaign. SNGR (Sungurlu) and IHGZ (Ihsangazi) stations for the first, third and fourth campaigns, and IHGZ (Ihsangazi) and ALAC (Alaca) stations for the second campaign were selected as continuous stations to control the network against any error and to connect the measurements that observed at different times. The duration of the measurement in each day was about 8 hours with the interval of 30 seconds. Throughout the campaigns, dual frequency 5 Ashtech and 5 Trimble GPS equipments and 1 Leica GPS equipment were used.

3.3 GPS Analysis

GPS measurements were analyzed using GAMIT/GLOBK software package (Herring, (1998)), (King and Bock (1998)) in three-step approach described before (Feigl et al. (1993)), and (Dong et al. (1998)). Firstly, doubly differenced GPS phase measurements from each day were used to estimate station coordinates EOP (Earth Orientation Parameters) and zenith delay. Loose a priori constraints were applied to all parameters (Table 3).

Table 3. Constraints of Parameters Used in GAMIT Process

Stations	Coordinate Constrains Latitude (m)	Longitude (m)	Radius (m)
MATE	0.050	0.050	0.100
KIT3	0.050	0.050	0.100
WTZR	0.050	0.050	0.050
SOFI	1.000	1.000	1.000
NICO	1.000	1.000	1.000
ZECK	1.000	1.000	1.000
TUBI	1.000	1.000	1.000
others	100	100	100

Semi axis	Ecc.	Incl..	Asc.	Arg. Of Per.	Mean Ano.	Rad 1..9
0.01	0.01	0.01	0.01	0.01	0.01	0.01

Furthermore, in the GPS campaign analysis sites from IGS as global sites were included in the analysis to link IGS network and our regional network (Table 4).

Table 4. Global GPS Sites Used in Analysis

First Campaign		Other Campaigns	
Station	**Location**	**Station**	**Location**
ANKR	Ankara (Turkey)	ANKR	Ankara (Turkey)
BUCU	Bucuresti (Romania)	GRAZ	Graz (Austria)
ISTA	Istanbul (Turkey)	ISTA	Istanbul (Turkey)
MATE	Matera (Italy)	MATE	Matera (Italy)
NICO	Nicosia (South Cyprus)	NICO	Nicosia (Souht Cyprus)
TELA	Tel Aviv (Israel)	MERS	Mersin (Turkey)
TUBI	Tubitak (Turkey)	TUBI	Tubitak (Turkey)
		ONSA	Onsala (Sweden)
		SOFI	Sofia (Bulgaria)
		WTZR	Wettzell (Germany)
		ZECK	Zelenchukskaya (Russia)

Secondly, loosely constrained estimation of station coordinates, orbits and EOP and their covariance from each day as quasi-observations were utilized to estimate consistent set of coordinates and velocities with using Kalman filter. Quasi-observations are estimated parameters and associated covariance matrices for station coordinates, earth-rotation parameters and sources positions generated from analyses of the daily

GAMIT solution (Herring, (1998)). In addition to this, quasi-observations from each day were combined with IGS quasi-observation for same day performed by SOPAC. After this, all quasi-observations from our process and IGS were combined into single quasi-observation. We determined weight from IGS and our process solutions which were combined separately. In addition, repeatabilities of measured days and their nrms (normalized root mean square) and wrms (weighted root mean square) were used to check measurements accuracy (Table 5).

Table 5. General Results of Four Campaigns

	2001 (First Campaign General Results)					
	nrms (NEU)			wrms (NEU) (mm)		
Average	1,28	1,55	1,62	1,66	1,87	7,69
Max	2,21	3,24	2,96	2,90	6,05	17,00
Min	0,54	0,59	0,81	0,55	0,60	3,50

	2002 (Second Campaign General Results)					
	nrms (NEU)			wrms (NEU) (mm)		
Average	1,21	1,24	1,20	2,37	3,43	12,43
Max	2,07	2,19	2,21	4,15	6,90	23,50
Min	0,36	0,42	0,56	0,90	0,70	5,20

	2003 (Third Campaign General Results)					
	nrms (NEU)			wrms (NEU) (mm)		
Average	0,85	1,10	0,63	1,69	3,05	6,20
Max	1,29	1,91	0,90	2,75	5,95	8,65
Min	0,31	0,33	0,41	0,75	0,75	4,05

	2004 (Fourth Campaign General Results)					
	nrms (NEU)			wrms (NEU) (mm)		
Average	0,55	0,88	0,55	1,50	2,82	6,26
Max	1,12	1,43	0,91	2,75	4,80	10,90
Min	0,27	0,50	0,26	0,90	1,35	2,50

In the third step, reference frame for our velocity estimates were defined by using generalized constraints for transformation parameters. (McClusky et al (2000). Eurasian plate was defined by minimizing horizontal velocities of sites (Table 6). After the transformation the root mean square (rms) of station was 0.7 mm/yr. A lot of alternatives were tested to attain the current solution.

Table 6. GPS Sites Used for Definition of Eurasia Plate

Site ID	Location
YAKT	Yakutsk, , Russian Federation
IRKT	Irkutsk , Russia
KSTU	Kransnoyarsk, Russia
ARTU	Arti , Russian Federation
ZWEN	Zwenigorod, Russia
GLSV	Kiev/Golosiiv , Ukraine
GRAZ	Graz-Lustbuehel, Austria
WSRT	Westerbork, The Netherlands
POTS	Potsdam, Germany
WTZR	Wettzell, Germany
KOSG	Kootwijk, The Netherlands
CAGL	Cagliari, Italy
NRIL	Norilsk, Russian Federation
NVSK	Novosibirsk, Russia
VILL	Villafranca, Madrid, Spain

After the fourth campaigns process, our velocities belonging to IGS stations which were close our study area were compared with IGS solutions (Table 7). The velocities in the region following the analysis of the fourth campaign were computed (Figure2).

Table 7. Velocity Length Comparison

STATIONS	The velocity vector's lengths of our process	The velocity vector's lengths processed by IGS
ANKR	26.00 mm/yr	24.76 mm/yr
ISTA	3.50 mm/yr	3.34 mm/yr
ZECK	1.50 mm/yr	0.46 mm/yr
NICO	8.00 mm/yr	7.99 mm/yr

4. Conclusion

Velocity vector directions and magnitudes are in harmony with the escape of Anatolian Plate towards the west along the North and Eastern Anatolian Faults.

According to GPS velocity vectors, there is no movement on Sungurlu fault. For this reason, the fault is seemed as locked or it has lost activity. As there is no record of a big earthquake on this fault in the historical and instrumental period on this fault, the second possibility is most favorable.

When GYNC (Göynücek) and GBAG (Göllübağlar) vectors are examined, there are differences due to their directions. These differences show that Merzifon Plain has continued to improve as a pull-apart basin.

The stations in the middle of the project region (GHAC, HMMZ, OSMC, DDRG) show that Merzifon-Hamamözü-Osmancık fault zone is active, but has been moving slowly.

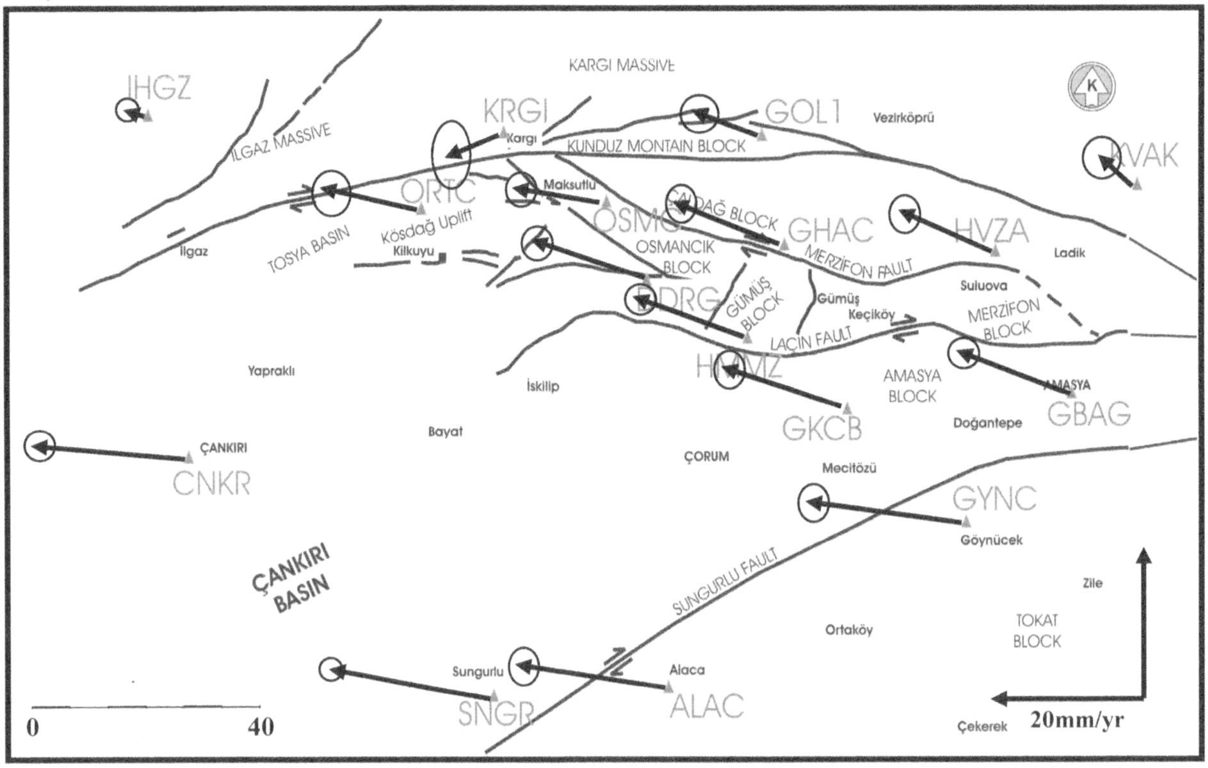

Fig. 2. According to Eurasia Plate, MID-NAF GPS Network Velocity Vectors in ITRF00

Acknowledgment

The project, "Determination of Kinematics along the North Anatolian Fault Branch between Ladik and Ilgaz with GPS Measurements", has been founded by The Scientific and Technical Research Council of Turkey (TUBITAK) and Istanbul Technical University (ITU) Research Fund under the grants of No.101Y035 and No.1636, respectively.

References

Ambraseys, N.N., 1970. Some characteristic features of the North Anatolian Fault Zone. Tectonophysics 9, 143-165.

Barka A. 1996. Slip distribution along the North Anatolian Fault associated with the large earthquakes of the period 1939 to 1967. *Bull. Seismol. Soc. Am.* 86:1238–54

Barka A. 1992. The North Anatolian Fault zone.*Ann. Tecton.* 6:164–95

Barka, A. A., and K. Kadinsky-Cade, Strike-slip fault geometry in Turkey and its influence on earthquake activity, Tectonophysics, 7, 663 – 684, 1988.

Blumenthal, M., La ligne sismique de Lâdik: Maden Tetkik ve Arama Mecmuası, no. 1/33, pp. 162-174, 1945.

Blumenthal, M., Bolu Civarı ile Aşağı Kızılırmak Mecrası Arasındaki Kuzey Anadolu Silsilelerinin Jeolojisi: Maden Tetkik ve Arama Enstitüsü Yayınlarından, Seri B, no. 13, Ankara, 265 pp.+54 photographic plates+numerous unnumbered foldout plates, 1948.

Canıtez N. and Üçer, S.B., A Catalogue of Focal Mechanism Diagrams for Turkey and Adjoining Areas: İ.T.Ü. Maden Fakültesi, Arz Fiziği Enstitüsü Yayınları, no.25, 111 pp, 1967.

Dewey, J.F. and Şengör, A.M.C., Aegean and surrounding regions: complex multiplate and continuum tectonics in a convergent zone: Geological Society of America Bulletin , Part I, v. 90, p. 84-92, 1979.

Dong, D.-N., T. A. Herring, and R. W. King, Estimating regional deformation from a combination of space and terrestrial geodetic data, *J. Geod.*, 72, 200–214, 1998.

Eyidoğan, H., Utku, Z., Güçlü, U. and Değirmenci, E., Türkiye Büyük Depremleri Makro-Sismik Rehberi (1900-1988): İstanbul Teknik Üniversitesi, Maden Fakültesi, Jeofizik Mühendisliği Bölümü, [II]+198 pp, 1991.

Feigl, K. L, D. C. Agnew, Y. Bock, D.-N. Dong, A. Donnellan, B. H. Hager, T. A. Herring, D. D. Jackson, R.W. King, S. K. Larsen, K. M. Larson, M. H. Murray, and Z.-K. Shen, Measurement of the velocity field in central and southern California, *J. Geophys. Res.*, 98, 21667–21712, 1993.

Herring, T. A., GLOBK: Global Kalman filter VLBI and GPS analysis program Version 4.1 Internal Memorandum, Massachusetts Institute of Technology, Cambridge, 1998.

Hubert-Ferrari A, Armijo R, King GCP, Meyer B, Barka A. 2002. Morphology, displacement, and slip rates along the North Anatolian Fault, Turkey. *J. Geophys. Res.* 107: 2235

Jackson, J. and McKenzie, D.P, Active tectonics of Alpine-Himalayan Belt between western Turkey and Pakistan: Geophysical Journal of the Royal Astronomical Society, v.77, pp. 185-264, 1984.

King, R. W. and Y. Bock, Documentation for the GAMIT analysis software, release 9.7, Massachusetts Institute of Technology, Cambridge, MA, 1998.

Kocyiğit A, Beyhan A. 1998. A new intracontinental transcurrent structure: The Central Anatolian Fault Zone, Turkey. *Tectonophysics* 284:317–36

Meade BJ, Hager BH, McClusky SC, Reilinger RE, Ergintav S, et al. 2002. Estimates of seismic potential in the Marmara Sea region from block models of secular deformation constrained by Global Positioning System measurements. *Bull. Seismol. Soc. Am.* 92:208–15

McClusky S, Balassanian S, Barka A, Demir C, Ergintav S, et al. 2000. Global Positioning System constraints on plate kinematics and dynamics in the eastern mediterranean and Caucasus. *J. Geophys. Res.* 105:5695–719

McKenzie, D., 1972. Active tectonics of the Mediterranean region. Geophysical Journal of the Royal Astronomical Society 30, 109-185.

Oral, M.B., Robert, E.R., Toksoz, N.M., Barka, A.A., Kinik, I., 1993. Preliminary results of 1988 and 1990 GPS measurements in western Turkey and their tectonic implications. Crustal Geodynamics 23, 407-416.

Stein, R.S., Barka, A.A., Dieterich, J.H., 1997. Progressive failure on the North Anatolian fault since 1939 by earthquake stress triggering. Geophysical Journal International 128, 594-604.

Şengör, A.M.C., Tüysüz, O., İmren, C., Sakınç, M., Eyidoğan, H., Görür, N., Le Pichon, X.,Claude Rangin, C., 2005, The North Anatolian Fault. A new look. Ann. Rev. Earth Planet. Sci., 33,1–75

Taymaz, T, 2000. 6 June 2000 Orta-Çankırı Earthquake-Mw=6.0 –and seismicity of surrounding area: Earthquake potentiel of the North Anatolian fault and fault plane solutions. Aktif Tektonik Araştırma Grubu Dördüncü Toplantısı, Eskişehir, Abstracts, p.7 [in Turkish], 16-17 Kasım 2000

Assessment of Coulomb Stress Changes Associated with the 1995 Aigion Earthquake in the Gulf of Corinth (Greece)

C. Mitsakaki, M. Sakellariou, D. Tsinas, A. Marinou
Centre for the Assessment of Natural Hazards and Proactive Planning
School of Rural and Surveying Engineering, National Technical University of Athens, Athens, Greece

Abstract The Gulf of Corinth has long been recognised as one of the most active rifts in the Aegean. Although, recent studies indicate that the rift is bordered both north and south by active faults, the fault line along the coastline of north Peloponnesos has been more rigorously studied so far, since it was associated with the 1995 Aigion earthquake, which was a well documented event resulting from normal faulting.

Since 1990 an extensive geodetic network has been installed in the area in a piecewise manner, covering, by now, the central and eastern parts of the gulf rather densely. Several campaigns of GPS (Global Positioning System) observations were carried out. So far, the analysis of these data has provided estimates of the secular motion of the free surface as well as, in combination with tectonic and seismological data, new insights in the discontinuous (co-seismic) deformation in the area.

In the present work an attempt is made to assess the Coulomb stress change (ΔCFF) associated with the 1995 Aigion earthquake, assuming the earthquake can be modeled as a static dislocation in an elastic half-space. The stress changes are estimated for normal faulting and for the appropriate parameters of the Aigion event. The free surface displacement field is compared with the observed one as derived from the GPS observations.

Keywords. Coulomb failure stress change (ΔCFS), Coulomb failure stress function (ΔCFF), earthquake stress drop, displacement fields.

---

1 Introduction

A region, where intensive and repeated geological, geophysical and geotechnical studies have been carried out, is eastern Mediterranean. The complicated tectonics of this region are dominated by the collision of the Arabian and African plates with Eurasia [*Reilinger R.E., et al, 1997*]. This collision, in the Caucasus region and eastern Turkey, is probably the reason for the westward motion of Turkey with respect to Eurasia.

The regions of western Turkey, eastern and northern Greece and the north Aegean Sea are seismically active, dominated by extension, as the southern Aegean Sea moves in a southwest direction with respect to Eurasia. This westward motion is mostly accommodated on the North and East Anatolian strike-slip fault systems. However, the Aegean Sea extension cannot be wholly attributed to the buoyancy forces of eastern Turkey. It is thought that it is mostly driven by the "roll-back" of the subducting slab beneath the south Aegean as it sinks into the mantle [*Taymaz T., et al, 1991*], [*Le Pichon X., et al, 1995*], [*Kahle H. G., et al, 2000*].

The mechanical interpretations of the relations between the driving forces and the resultant relative motions, in the region, differ to a significant extend. According to several researchers [*Le Pichon X., et al, 1995*], [*Reilinger R.E., et al, 1997*], [*Nyst M., Thatcher W., 2004*] deformation in the region occurs in the form of motion between rigid microplates with zones of diffused deformation in the interior of the plates.

Alternatively, the continental lithosphere is generally considered weak enough to deform in a distributed fashion under the forces acting on it. The relatively thin (~10km depth) brittle crust is considered as moving, on the large scale, in a pattern reflecting the spatially smooth deformation of the underlying ductile parts of the lithosphere [*Flerit F., et al, 2004*]. However, *Taymaz T, et al,* [*1991*] suggest that, although the distributed character of the deformation in the Aegean may be due to the ductile deformation of the lower lithosphere, the details of the surface kinematics may be governed by interactions of crustal blocks that perhaps reflect pre-existing structures of the upper crust.

While the convergence rate between Eurasia and Africa is of the order of 1cm/yr, the total extension rate across the Aegean, estimated from satellite geodetic observations [*Reilinger R.E., et al, 1997*], is about 4-5cm/yr. Most of the Aegean, from the north Aegean trough to the southwest Hellenic trench, is involved, while an intricate pattern of strike-slip and normal faults accommodate this extensional deformation [*Hatzfeld D., et al, 2000*].

For actively extending regions on the continents the normal faulting is organized into sub-parallel systems distributed over areas of tens or even hundred kilometres wide. One of the most prominent and active features of such a system of faults is the rift along the Gulf of Corinth.

The Gulf of Corinth is the most rapidly extending rift system in Greece with about 120km length and 30km width and a WNW-ESE trend. It is believed to be active at the present rates since the last 5Myrs [*Taymaz T., et al, 1991*], [*Armijo R., et al, 1996*].

The rift is asymmetric and bounded along its south coast, on Peloponnesos, by north-dipping faults of en échelon pattern with maximum segment lengths of 15-25km, while some south-dipping faults appear on its northern side (Sterea Hellas)[*Armijo R., et al, 1996*]. Seismic, geomorphologic and geological observations suggest that the northern coastal and offshore normal faults on the Peloponnesos side of the Gulf appear to be the most active today, while no such strong indications exist for the faults in the northern mainland of Peloponnesos.

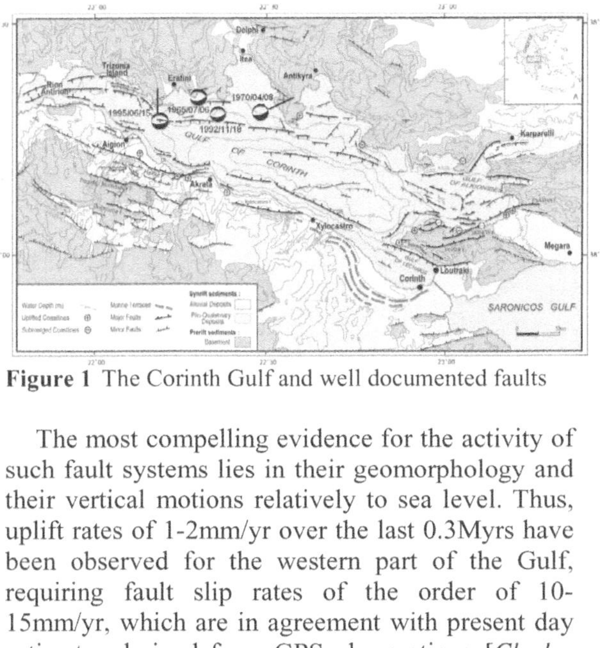

Figure 1 The Corinth Gulf and well documented faults

The most compelling evidence for the activity of such fault systems lies in their geomorphology and their vertical motions relatively to sea level. Thus, uplift rates of 1-2mm/yr over the last 0.3Myrs have been observed for the western part of the Gulf, requiring fault slip rates of the order of 10-15mm/yr, which are in agreement with present day estimates derived from GPS observations [*Clarke, P.J., et al, 1997*], [*Avallone A., et al, 2004*]. Present day extension rates, estimated from GPS observations, for the eastern part of the Gulf are significantly lower (~5 -6mm / yr) than the western ones [*Briole P., et al, 2000*], [*Clarke P.J., et al, 1997*] (*Figure 1*).

One of the differences, between the western (near Patra) and the eastern (near Corinth) provinces of the Gulf, is that in the western part the faults dip at a shallow angle of 30°N, while at the eastern part they have a steeper dip (40°-45°) [*Hatzfeld D., et al, 2000*].

Another significant observation is that the main activity in the Gulf appears to have shifted from one large system of faults to another in the last 1Myrs. Since such changes of spatial and temporal fault activity have been observed in other regions of Greece and over the world, they may represent general features of extending or shortening regions. They might, also, help improving the simplistic models so far used for describing the behaviour of blocks of faults, like the simple domino model of extension [*Jackson J., 1999*].

2 Stress Change Due to an Earthquake

Although geometric information regarding an earthquake and the fault that is ruptured by it may be estimated from seismological, geodetic and geological data, the complete picture of earthquake mechanics remains elusive.

However, calculations of the static stress changes due to a main shock affecting subsequent earthquake locations were presented as soon as the early seventies, but were not adopted by the scientific community in earthquake hazard assessment. Such stress changes, of the order of 0.1MPa (≈1 bar) which are only a fraction of the stress drop during an earthquake, were regarded insignificant, while the results at the time were predominantly qualitative.

In the last 15-20 years a lot of studies have been carried out and a large amount of static stress changes data have been gathered all over the world.

A widely used model to describe fault interactions is the *Coulomb failure stress change* (ΔCFS) or *Coulomb failure stress function* (ΔCFF).

When a fault ruptures producing an earthquake,, it, also, induces a change in the stress field on the nearby faults. To estimate this change, the so-called Coulomb stress is calculated using elastic dislocations on rectangular planes in a homogeneous and isotropic half space [*Okada, Y., 1992*].

Assuming a cohesionless material behaviour for the earth crust, the strength envelope is represented by a straight line passing through the origin of the axes. So, a simple formula expresses the change of the difference of the absolute value of the shear stress developed in a particular direction minus the shear strength along it, the so-called "Coulomb stress", as (*Figure 2*):

$$\Delta CFS = |\Delta\tau_{slip}| - \mu'\cdot\Delta(\sigma_n - p) \quad (1)$$

where:

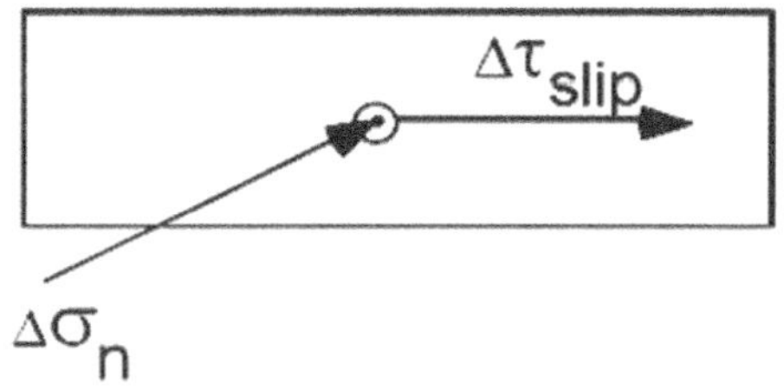

Figure 2 Coulomb failure stress change

$\Delta\tau_{slip}$ is the change in shear stress due to the first fault rupture, resolved in the slip direction of a second fault,

σ_n is the normal stress (*positive if compressive*) due to the first rupture, resolved in the direction orthogonal to a second fault plane,

μ' is the effective coefficient of friction, and

p is the pore fluid pressure.

ΔCFS is resolved onto the fault plane and in the slip direction of the second "receiver" fault

If ΔCFS > 0 then the first fault rupture may bring the second fault closer to failure; if ΔCFS < 0, then the first event may send the second event farther away from failure, and into a so-called stress shadow.

The stress shadow lasts as long as it takes the second fault plane to recover from the stress decrement. One manner of recovery is through long-term tectonic process.

3 Analysis of Data

3.1 Geodetic Data

Since 1990 the Higher Geodesy Laboratory of the National Technical University of Athens participates in a European multi-disciplinary research programme for monitoring the tectonic behaviour of the Corinth rift region. A GPS network, part of a much larger network established in Greece by several research teams [*Briole P, et al, 2000*], [*Clarke P.J. et al, 1997*], includes about 50 first order points (measured at least three times in a given campaign). In addition, about 150 second order points were observed one or two times during at least one campaign. It should be mentioned that approximately 140 of these GPS network points are pillars of the *Hellenic triangulation network*. Since the area under consideration is about 80x100km^2, the resulting average density is of the order of~1point / 5km^2 (*Figures 3* and *4*) providing a reasonable sampling of the main well documented active faults.

The main objectives were:

a. To measure the *inter-seismic strain* across the Gulf of Corinth using both the dense geodetic network, parts of which were observed periodically, as well as a recently established smaller network with five permanent stations for continuous monitoring (red rectangles in *Figure 6*)

b. To characterise *displacement fields* associated with earthquakes (*Figure 5*)

c. To model the observed deformation (*Figure 6*).

One of the large earthquakes in the region (M_S ~6.2) took place on 15 June 1995, close to the town of Aigion on the northern coast of Peloponnesos.

Time series of the geodetic coordinates for points located close to the influence of the 1995 earthquake (e.g. Points C, E in *Figure 5*) in combination with geotectonic and seismic data helped to estimate *a "co-seismic" offset,* as well as to infer the location and the characteristics of the fault for the 1995 main event.

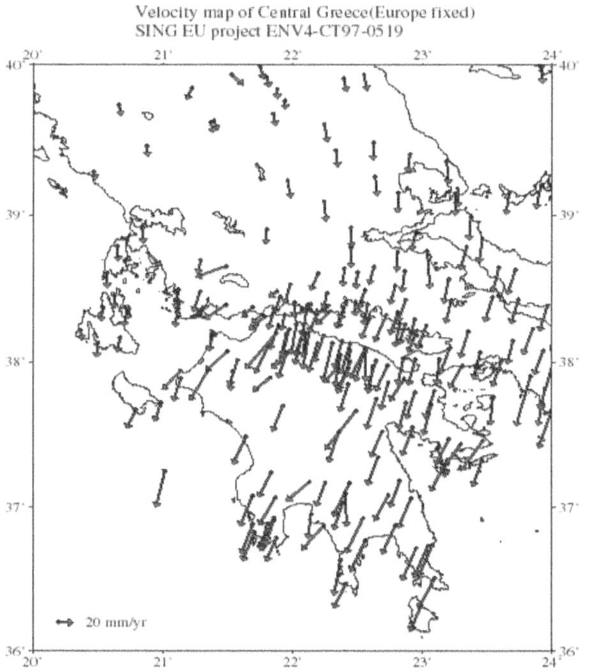

Figure 3 Velocity of Central Greece (Europe fixed) from GPS observations between 1990 and 2000 [*Briole P., et al., 2001*]

A dislocation model [*Okada Y., 1992*] which best satisfied the geodetic data was then applied [*Bernard P., et al, 1997*]. The inversion of the geodetic data gave as the best fitting model a low angle north-dipping fault *(~35°)*,of *~10km* length, located in the central part of the rift, with depth between *4.5km* and *9.5km* (*Figure 6*). The choice for the upper depth of the fault was constrained by the moderate *"co-seismic" slips* observed at several network points (e.g. points A and G) (Figure *5*).

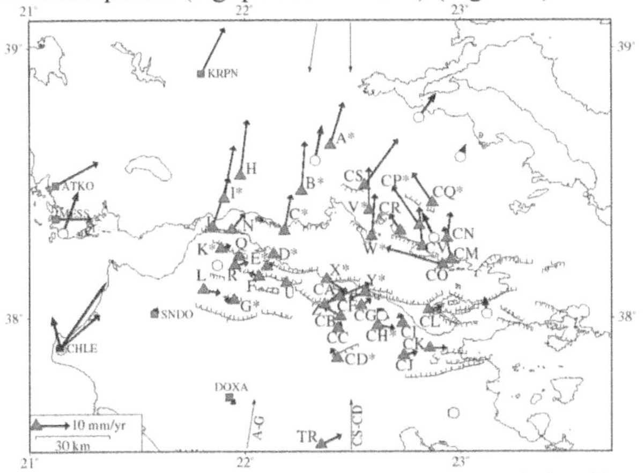

Figure 4 Extension across the Corinth rift expressed in a local reference system (Peloponnisos fixed) [*Corinth Rift Lab., 2001*]

After removing the 1995 effect for the sites where this "co-seismic" offset was significant (>5mm), there still remained what might be called a *"long-term" inter-seismic displacement field* (*Figure 3*). This residual displacement field of the free surface of the earth may be considered as what is left after subtracting the episodic seismic response. It may be attributed to secular motions of the region such as the lithospheric plates' motion. It may still be modified by small episodic motions due to micro-seismicity. However, the GPS displacement field, discrete in both space and time, cannot discern such effects.

The GPS observations accumulated so far, spanning only a little more than a ten years interval, provide a rather short time window with respect to the recurrence time of large events. However, when in previous work [*Agatza-Balodimou A.M, et al, 2002*] GPS data were combined with old geodetic ones the displacement field remained consistent.

Furthermore, the geodetic coordinate time series slopes prior and after the 1995 event *remain steady*. The assumption of a steady rate velocity field seems acceptable. Although, if the velocities are expressed in a *"fixed Europe" reference frame* (*Figure 3*), a rotation of the northern part of the gulf is indicated, the *extension across* the gulf is the main feature of this velocity field. This picture is more pronounced if the velocity field refers to a *local reference frame* (*Figure 4*).

It appears that the *extension of the rift* is contained almost entirely offshore, in a narrow zone of no more than 10km close to Aigion, and increasing from east to west. This rate, estimated from eleven years of GPS observations is *~11mm/yr* in the central part and *~16mm/yr* in the vicinity of

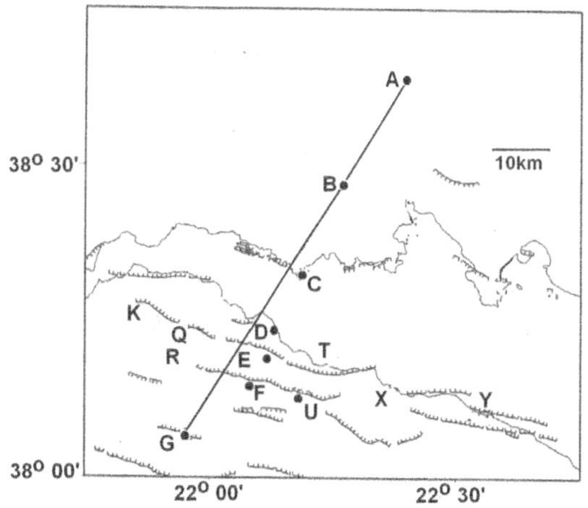

Aigion on an N185°E direction [*Briole P., et al, 2000*].

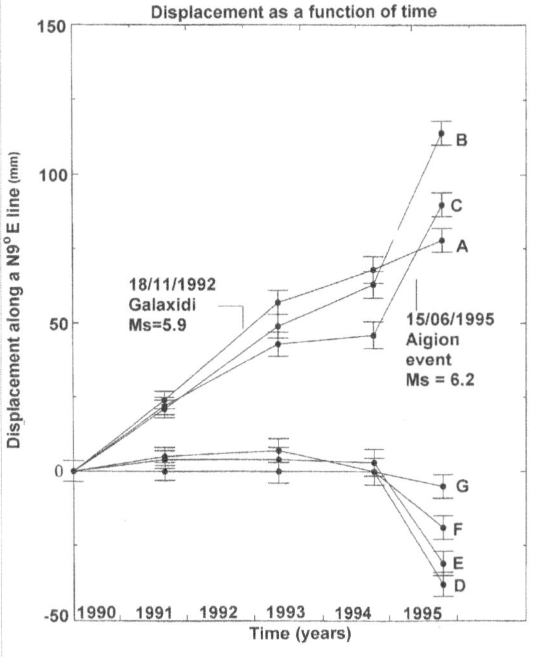

Figure 5 Temporal evolution of geodetic positions for the line A-G between 1990 and 2000 (Points A, G located away from influence of 1995 earthquake)

The fact that the western part of the rift opens more rapidly than the eastern one indicates that the east tends to follow the SE motion of Euboea and South Peloponnesos, while the NW side of the rift is attached to the Apoulia-Adriatic micro-plate and N Greece [*Avallone A., et al, 2004*].

Since the major faults of the southern part of the Gulf appear to deform in a slow rate, the smaller structures in the inner part of the rift (like the 1995 event) accommodate most of the extension with rather frequent earthquakes of medium magnitude (M_S~5.5-6.5). Thus, large events on faults of the mainland (M_S~6.5-7) may have long recurrence times (> 500 years) [*Avallone A., et al, 2004*].

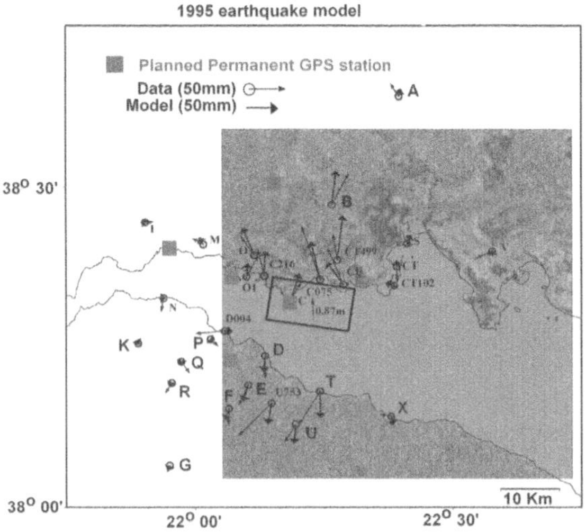

Figure 6 GPS deformation associated with the 15/6/1995 Aigion earthquake together with the planned location of permanent GPS sites. The dislocation model is also depicted.

3.2 Coulomb Stress Change

In the present work a first assessment of the Coulomb stress change (ΔCFF) associated with the 1995 Aigion earthquake and its offshore ruptured fault has been carried out.

The software used in this analysis was the Coulomb software, version 2.6, [*Toda S., et al, 1998*]. It calculates, on any surface and at any depth, static displacements, strains and stresses caused by fault slip, point sources of inflation /deflation, and dike expansion /contraction.

All calculations are made in a half-space with uniform isotropic elastic properties. The software implements elastic dislocation formulae [*Okada Y., 1992*] and boundary element formulae [*Crouch S.L. & Starfield A.M, 1983*].

The elastic parameters necessary for the model were either taken from the USGS (NEIC) data or were computed (*Table 1*). Thus the shear modulus (or modulus of rigidity) was computed as:

$$\mu = \frac{E}{2\cdot(1+\nu)} \qquad (2)$$

where E is the Young modulus and ν is the Poisson ratio.

To estimate the stress drop (Δσ in bars), the following expression from *Bayrak Y. & Yilmazturk A., [1998]* was used:

$$\log(\Delta\sigma / \mu) = 1.485 M_S - \log Mo + 12.35 \qquad (3)$$

where μ is the modulus of rigidity and Mo is the seismic moment expressed in (dyn·cm).

The stress changes were estimated for normal faulting and for the appropriate parameters of the geometry and slip direction of the 1995 Aigion earthquake rupture. The fault parameters (*Table **2***) were chosen from *Bernard P. et al, [1997]* where the geodetic observations were used in order to solve for the Okada dislocation model, as mentioned above. Some additional information was gathered from USGS (NEIC) data files. It should be mentioned that, for stress change estimations, the exact details of the geometry and slip of a main event become less important the farther one goes from the rupture.

The choice of the appropriate "receiver" faults was based on studies carried out on recent micro-seismicity (2000-2001) and other geophysical data for the western part of the rift. The active well documented faults in the region appear to be in the late part of their seismic cycle. The probability of an earthquake of M_S*~6.0-6.5* is very high for the next few decades, while even the possibility of cascading events should not be entirely discarded [*Bernard P., et al, 2005, in preparation*].

Coulomb stress changes on planes with optimum (meaning critical) orientation can also be calculated for a comparison with the actual specified faults. It is presumed that a sufficient number of fractures (small faults) exist having all possible orientations and that the faults optimally oriented for failure will be the most likely to slip in small earthquakes.

After a fault rupture occurs, the stress changes on the specified fault, as well the optimum direction are determined as function of the direction and magnitude of the fault slip and the regional in situ stress field.

Among these active faults, which are almost parallel (*Figure 7*), those that might be affected by the stress change due to the 1995 event and with rake direction and fault plane orientation known were chosen as "receiver" faults (*Table **3***).

Model Parameters				
Young modulus *(GPa)*	Shear modulus *(GPa)*	Poisson's ratio	Coefficient of friction	Calculation depth *(km)*
73.2	29	0.28	0.6	6

Table 1 Model elastic parameters

The geometry parameters were taken from: *http://www.ingv.it*. In this case the stress change calculations are dependent on knowledge of the prevailing regional stresses or any pre-existing stress field from other events.

Thus, some knowledge of this stress field is required, although simplifying assumptions can be made.

1995 Earthquake					
Date	**Ms**	**Mo** *(Nm)*	**Lat** *(⁰)*	**Long** *(⁰)*	**Length** *(km)*
15/6/1995	6.2	$3.9\ E10^{18}$	38.401N	22.283E	15

1995 Earthquake (*continued*)					
Strike	**Dip**	**Rake**	**Top/Bottom** *(km)*	**Slip** *(m)*	**Stress Drop** *(bar)*
275°	35°	-83°	4.5/9.7	0.87	27

Table 2 Parameters for the offshore fault of the 1995 earthquake

Fault Parameters							
Fault	**Lat** *(⁰)*	**Long** *(⁰)*	**Length** *(km)*	**Strike** *(⁰)*	**Dip** *(⁰)*	**Rake** *(⁰)*	**Top / Bottom depth** *(km)*
East Heliki	38.193N	22.150E	16.6	279°	50°	270°	0.2/7.5
West Heliki	38.231N	22.030E	12	283°	50°	270°	0.2/7.5
Aigion	38.265N	22.035E	10	277°	50°	270°	0.2/7.5

Table 3 Fault geometry for "receiver" faults

The principal stresses of the "regional" stress field were chosen as: $\sigma_1 = 27$bars, equal to the estimated stress drop due to the 1995 event, $\sigma_3=0$ and the intermediate $\sigma_2 = 13.5$bars (*Table 4*). .

Principal Axes					
σ_1		σ_2		σ_3	
Azimuth *(⁰)*	**Plunge** *(⁰)*	**Azimuth** *(⁰)*	**Plunge** *(⁰)*	**Azimuth** *(⁰)*	**Plunge** *(⁰)*
174°	12°	82°	9°	316°	75°

Table 4 "Regional" in-situ stress field orientation (***Source***: USGS (NEIC)

It is important to emphasize that the change of Coulomb stress function is calculated on the specified orientations, as well as, on planes of optimal (critical) orientation with respect to the regional stress field. [*King G. C.P, et al, 1994*].

In the following figures (*Figures 7-9*) the ΔCFS for both specified and optimal orientations are depicted.

One of the specified orientations chosen is the strike of the fault that gave the 1995 earthquake, in order to examine whether most of the aftershocks fell inside the positive ΔCFS change area (*Figure 7a, b*).

All calculations of ΔCFS refer to a depth of 6km. This depth was chosen as being halfway between the top and bottom depths of the fault planes, as well as the middle of the seismogenic zone in the area.

Cross-sections are depicted to a maximum depth of 40km below the free surface.

The average orientation of the most active well documented faults on the north coast of Peloponnessos (*Table 3*) was also chosen as a specified orientation (*Figure 8 a, b, c*).

Finally, the map view and the relative cross section for the optimally oriented faults were calculated and the results are also depicted. (*Figure 9 a, b*).

4 Discussion – Suggestions

The present work is one of the few that normal faulting, as the one that gave the 15 June 1995 event has been studied with respect to static stress changes (Coulomb fracture criterion).

The geodetic GPS network established in the Gulf of Corinth and re-observed for more than a decade now, is the only source of the displacement field on the free surface for the region. The inversion of the geodetic data helped in acquiring a reliable estimation of the geometry and the slip of the offshore normal fault, which gave the 1995 event [*Bernard P., et al, 1995*].

Furthermore, the same displacement field after the removal of the 1995 "co-seismic" effect may provide information for the so-called secular (inter-seismic) motion of the area.

Thus, it appears that geodetic observations are indispensable in evaluating, in combination with seismo-tectonic information, the fault plane geometry and the size of slip on it for a seismic event as well as the secular velocity field for a region.

With respect to the Coulomb stress change for the 1995 Aigion event it appears that, as in the case of strike-slip faulting, the spatial aftershock pattern correlate well with regions *of positive Coulomb stress changes* (*Figure 7**a**, **b***).

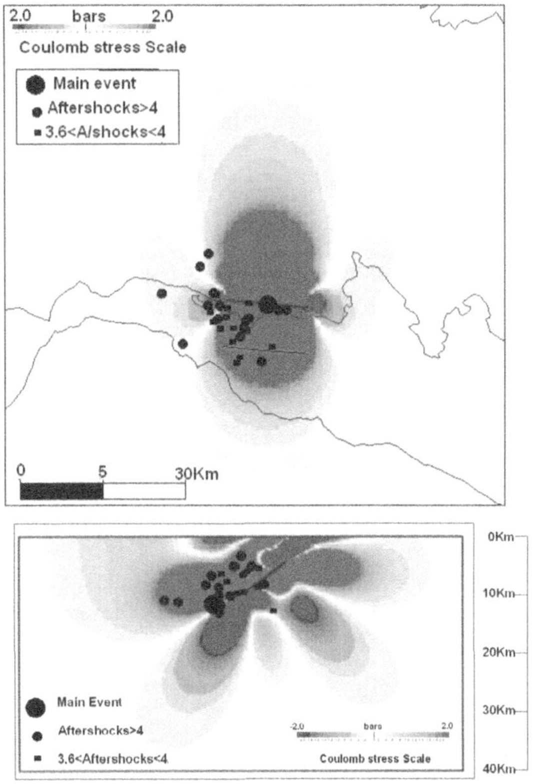

Figure 7 a. Map view for the 1995 event with specified direction and aftershocks locations **b.** The relative cross section with the locations of the aftershocks at depth

Also, the *East Heliki fault,* which is one of the active faults in the area, appears to be *inside the positive stress change (~2bars)* and is a probable candidate for rupture in the next few years (*Figure **8 a, b***). This estimation is corroborated by the micro-seismicity in the region for the years 2000-2001 [*Bernard P., et al, 2005*].

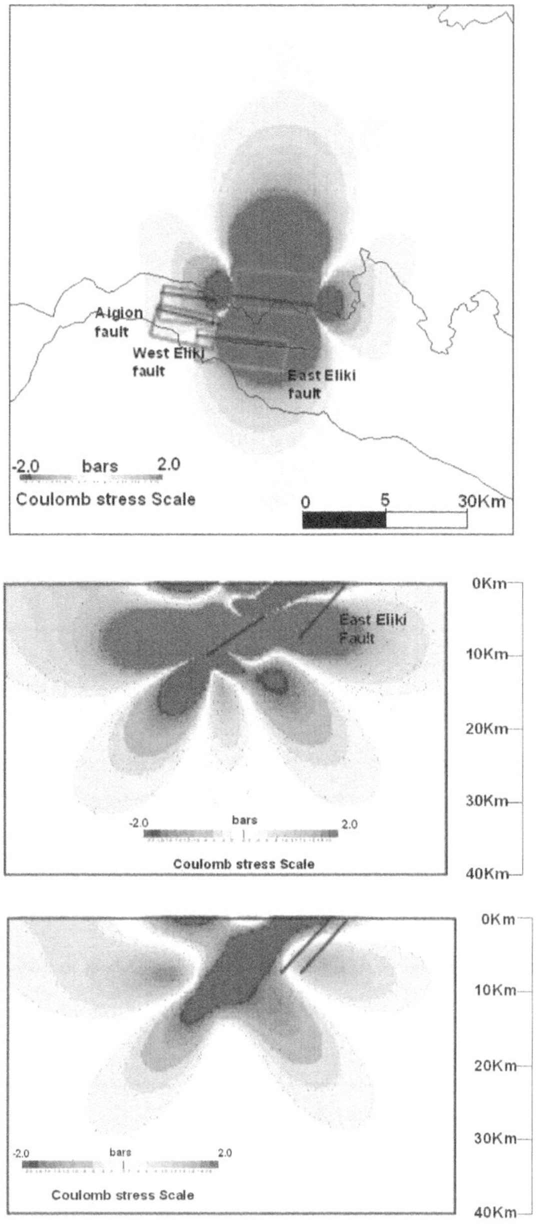

Figure 8 a. Map view of the 1995 event with all nearby active and well documented faults included **b.** Cross section at the centre of the fault of the 1995 event and East Heliki fault **c.** Cross section at the western tip of the fault of the 1995 event and both East and West Heliki faults.

The cross section at the extreme western tip of the offshore fault, that gave the 1995 event, indicates that *even the West Heliki fault is also affected, but to a lower degree (less than 1bar stress rise*) (*Figure **8 a, c***).

The Aigion fault appears to be *in the stress shadow zone* of the 1995 event.

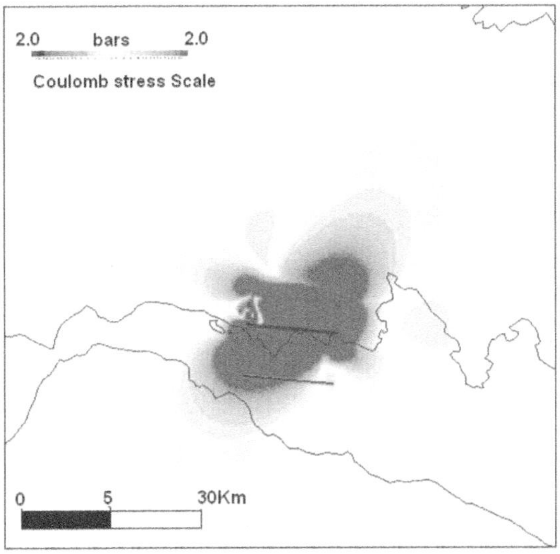

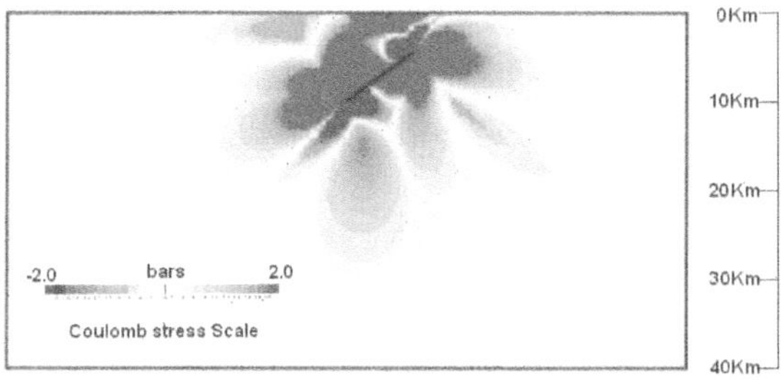

Figure 9 a. Map view of the 1995 event for optimally (critically) oriented faults **b.** Cross section of the above

Future analysis should include the major aftershocks in the time sequence of static stress change in the region, as well as the only significant event prior to the 1995 one, the 1992 Galaxidi earthquake of M_S~5.8.

Since the region has a significant (~15mm/yr) *inter-seismic displacement field*, well documented from more than 10 years GPS observations, it would be interesting to calculate the respective accumulated static stress change.

References

Agatza-Balodimou A.M., Briole P., Mitsakaki C., Papazissi K., Crustal Deformations From Sparse Geodetic Data, *Survey Review, No 283, Vol. 36,* pp. Jan. 2002.

Armijo R., Meyer B., King G., Rigo A., Papanastassiou D., (1996). Quaternary evolution of the Corinth Rift and its implications for the Late Cenozoic evolution of the Aegean, *Geophys. J. Int., 126*, 11–53.

Avallone A., Briole P., Agatza-Balodimou A.M., Billiris H., Charade O., Mitsakaki C., Nercessian A., Papazissi K., Paradissis D., Veis G. (2004) Analysis of eleven years of deformation measured by GPS in the Corinth Rift Laboratory area, *Comptes Rendus Geoscience 336, 4-5,* 301–311.

Bernard P., Lyon-Caen H., Briole B., Deschamps A., Pitilakis K., Manakou M., Boudin F., Berge C., Makropoulos K., Diagourtas D., Papadimitriou P., Lemeille F., Patau G., Billiris H., Castarède H., Charade O., Nercessian A., Avallone A., Zahradnik J., Sacks S., Linde A., (2005). Seismicity, deformation and seismic hazard in the western rift of Corinth: New insights from the Corinth Rift Laboratory (CRL). *In preparation.*

Bernard P., Briole R., Meyer B., Lyon-Caen H., Gomez J.-M., Tiberi C., Berge C., Cattin R., Hatzfeld D., Lachet C., Lebrun B., Deschamps A., Courboulex F., Larroque C., Rigo A., Massonet D., Papadimitriou R., Kassaras J., Diagourtas D., Makropoulos K., Veis G., Papazissi E., Mitsakaki C., Karakostas V., Papadimitriou E., Papanastassiou D., Chouliaras G., Stavrakakis G., (1997). The Ms. 6.2, June 15, 1995 Aigion earthquake (Greece): evidence for low angle normal faulting in the Corinth rift. *Journal of Seismology 1,* 131-150.

Briole P. Rigo A., Lyon-Caen H., Ruegg J.C., Papazissi K., Mitsakaki C., Agatza-Balodimou A.M., Veis G., Hatzfeld D., Deschamps A., (2000). Active Deformation of the Gulf of Korinthos, Greece: Results From Repeated GPS Surveys Between 1990 and 1995. *JGR, Solid Earth, Vol.105, No B11,* 25605-25625

Briole, P.; Billiris, H.; Felikis, S.; Papazissi, K.; Paradissis, D.; Veis, G.; Avallone, A.; Charade, O.; Nercessian, A. (2001). The Gulf of Corinth CORSEIS" permanent GPS network: framework and perspectives. Presented.at *EGS XXVI General Assembly, Nice, France, March 2001.*

Clarke P.J., Davies R.R., England P.C., Parsons B.E., Billiris H., Paradissis D., Veis G., Denys P.H., Cross P.A., Ashkenazi V., Bingley R., (1997). Geodetic estimate of seismic hazard in the Gulf of Korinthos. *Geophysical Research Letters 24,* 1303-1306.

Crouch, S.L., and Starfield A.M., (1983) *Boundary Element Methods in Solid Mechanics*, 322 pp., Allen Unwin, London.

Flerit F., Armijo R., King G., Meyer B., (2004). The mechanical interaction between the propagating North Anatolian Fault and the back-arc extension in the Aegean. *Earth and Planetary Science Letters 224,* 347–362.

Harris R. A. (2000) Earthquake stress triggers, stress shadows, and seismic hazard, *Current Science, Vol. 79, No. 9, Special Section: Seismology 2000.*

Hatzfeld D., Karakostas V., Ziazia M., Kassaras I., Papadimitriou E., Makropoulos K., Voulgaris N., Papaioannou C., (2000). Microseismicity and faulting geometry in the Gulf of Corinth (Greece). *Geophysical Journal International 141*, 438-456.

Jackson, J., (1999). Fault death: a perspective from actively deforming regions, *J. struct. geol.,* 21, 1003–1010.

Kahle H.-G., Cocard M., Peter Y., Geiger A., Reilinger R., Barka A., Veis G. (2000). GPS derived strain rate field within the boundary zones of the Eurasian, African, and Arabian Plates. *J. Geophys.Res.*, *105*, 23.353–23.370.

King G. C.P, Stein R. S., and Lin J. (1994) Static Stress Changes and the Triggering of Earth-quakes, *Bull. Seismol. Soc. Am. 84,* 935-953.

Le Pichon X., Chamot-Rooke N., Lallemant S., Noomen R., Veis G., (1995). Geodetic determination of the kinematics of central Greece with respect to Europe, *J. Geophys. Res., 100,* 12.675– 12.690.

Nyst M., Thatcher W., (2004). New constraints on the active tectonic deformation of the Aegean, *J.G.R.., 109*, B11406.

Okada, Y., Internal deformation due to shear and tensile faults in a half-space, (1992) *Bull. Seismol. Soc. Amer.*, *82* (2), 1018-1040.

Reilinger R. E., McClusky S. C., Oral M. B., King R. W., Toksoz M. N., Barka A. A., Kinik I., Lenk O., Sanli I., (1997). Global Positioning System measurements of present-day crustal movements in the Arabia-Africa-Eurasia plate collision zone. *J. Geophys. Res., 102*, 9983-9999.

Taymaz T., Jackson J.A., McKenzie D.P. (1991). Active tectonics of the north and central Aegean Sea, *Geophys. J. Int. 106*, 433– 490.

Toda, S., Stein R.S., Reasenberg P.A., Dieterich J.H., (1998)Stress transferred by the Mw=6.5 Kobe, Japan, shock: Effect on aftershocks and future earthquake probabilities, *J. Geophys. Res.*, *103*, 24.543-24.565.

Studies on Crustal Structure and Gravity in the Eastern Alps

E. Brückl, U. Mitterbauer, M. Behm
Institute of Geodesy and Geophysics
Vienna University of Technology, Gusshausstrasse 27-29, 1040 Vienna, Austria

CELEBRATION 2000 and ALP 2002 Working Groups
http://paces.geo.utep.edu/celebration_web/celebration.shtml; http://www.alp2002.info

Abstract. This study concentrates on the Eastern Alps and their transition into the surrounding Bohemian Massif, the Pannonian Basin, the Carpathians and the Dinarides. The geodynamic setting is characterized by the ~N-S directed head-on collision between the European and Adriatic-Apulian plates in the central part of the Eastern Alps, leading to the E-directed lateral extrusion of the Eastern Alps into the Pannonian Basin, and the transition of the Eastern Alps to the Dinarides. New seismic data about the lithosphere in this area has been derived from the wide-angle reflection and refraction experiments CELEBRATION 2000 and ALP 2002. A 3D model of the P-wave crustal velocity has been generated by tomographic methods. Further, a map of the Moho discontinuity has been constructed. Both stacking techniques and travel time inversion have been applied. Interactive modeling by 2D ray tracing along selected lines has been used to supplement the 3D evaluation. The tomographic model of the crust supplies continuous information about the P-wave velocity only in the upper crust. Bouguer gravity data has been implemented to better constrain the velocity of the lower crust by the use of a velocity-density relation. For the uppermost 10 km, the density has been derived from the seismic model. For the lower crust, a linear velocity-depth function has been deduced removing the gravity effect of the Moho topography from the Bouguer gravity. The residual gravity shows a significant regional pattern that can be related to geologic provinces (e.g. Bohemian Massif, Molasse, Pannonian Basin, and Vienna Basin) and the geodynamic situation (existence of a tectonic block forming a triple junction with the European and Adriatic-Apulian plates). An integrated model has been constructed, which fits well seismic and gravimetric data and is very close to Airy-isostatic equilibrium.

Keywords. Eastern Alps, crust, Moho, reflection and refraction seismology, gravity, isostasy.

1 Introduction

The target of this study is the lithosphere of the Eastern Alps and their transition into the Bohemian Massif, the Pannonian Basin, the Carpathians and the Dinarides (Figure 1). The geodynamic setting is characterized by the approximately N-S directed head-on collision between the European and Adriatic-Apulian plates in the central part of the Eastern Alps. This collision started in the Oligocene. Continued convergence through the Tertiary led to the present orogenic pattern. The continent-continent collision is not the only imprinting tectonic process. In the late Oligocene and Miocene, lateral escape took place from the compressed Central Alps to the unconstrained margin of the Pannonian Basin. The significant eastward thinning of the crust corresponds to the extrusion and tensional processes (Ratschbacher et al., 1991). The present day seismicity in the area of our investigation follows a pattern compatible with the escape tectonic model (Aric et al., 1987).

Seismic exploration of the lithosphere of the Eastern Alps has a long tradition. Refraction and wide-angle reflection experiments started around the Eschenlohe quarry 40 years ago. Seismic lines spread out from this shot point over the Alps to the shot points Lago Lagorai and Trieste (Giese et al., 1976). The projects ALP 75, ALP 77, and ALP 78 supplied further refraction and wide-angle reflection lines (Yan and Mechie 1989; Scarascia and Cassinis 1997). Deep reflection seismic profiling started in former Czechoslovakia in 1980. The profiles 3T and 8HR (Tomek 1993a, 1993b) focused on the transition from the Bohemian Massif to the Carpathians and the Pannonian Basin. In Hungary, the Pannonian Basin was explored by deep reflection seismic profiling during the occasion of the Pannonian Geotraverse (Posgay et al., 1996). In Austria, deep reflection seismic profiling was carried out in the area of the Rechnitz Penninic window in NE Styria (Weber et al. 1996; Graßl et al. 2004). An important deep reflection line bounding the investigation area of this study to the W is TRANSALP (Transalp Working Group 2002), which crosses the Eastern Alps from Munich to Verona. The most recent results on the seismic structure derive from the 3D refraction experiments CELEBRATION 2000 (Guterch et al. 2003) and ALP 2002 (Brückl et al. 2003a). Preliminary 2D- and 3D-models of the crust and Moho are available from these two experiments.

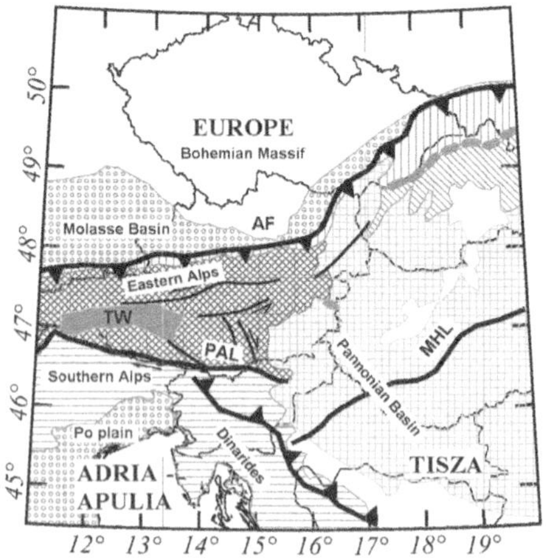

Fig. 1 Tectonic setting of the investigation area
PAL: Periadriatic Lineament
AF: Alpine Front
MHL: Mid-Hungarian Line
TW: Tauern Window

Gravimetric investigations have also supplied valuable information on the structure of the Eastern Alpine crust. Before 1973 gravity measurements in the Eastern Alps were performed mainly along the benchmarks of leveling lines. A Bouguer gravity map of Austria was derived from approximately 2000 stations by Senftl (1965). In 1973 a densification of the gravity net and modeling started in Austria with the Gravimetric Alpine Traverse (Meurers et al. 1987). Another study by Meurers (1993) covered the plutonic and metamorphic part of the Bohemian Massif in Austria. Lillie et al. (1994) focused on regional studies on gravity and isostasy, extending from the Eastern Alps to the Carpahtians and Pannonian Basin. One profile, crossing the Alps from N to S at 13°E longitude and another, W-E oriented at 47° latitude cover the investigation area of this study. The Eschenlohe-Trieste (NW-SE) and Eschenlohe-Vicenza (N-S) seismic profiles were used as a basis for a density model and as constraints for a gravity inversion by Braitenberg et al. (1997) and Dal Moro et al. (1998). The 3D density model along the TRANSALP profile was developed by Ebbing et al. (2001), Ebbing (2002), and Ebbing (2004). The model was then used for calculating internal crustal loads and evaluate the elastic thickness variations and the isostatic equilibrium in the frame of the flexural isostatic model in Braitenberg et al. (2002).

In this paper we present preliminary seismic models of the crust and the Moho derived from CELEBRATION 2000 and ALP 2002 data. Information on the P-wave velocity of the lower crust is incomplete or not very reliable. Gravity data will be used as a constraint on the velocity of the lower crust. The implementation of the gravimetric data into the seismic model will be described and a correlation of the main structural features with the regional geology and geodynamic situation will be attempted.

2 Preliminary Seismic Model of the Eastern Alps Derived from CELEBRATION 2000 and ALP 2002 Data

In this study we consider a merged data set from the 3rd deployment of CELEBRATION 2000 (55 shots, 844 receivers, total profile length 2 800 km) and ALP 2002 (39 shots, 947 receivers, total profile length 4 300 km) (Figure 2). This data set comprises more than 78,000 traces.

2D-interpretations by interactive ray tracing modeling (Červený, V. and I. Psencik 1984) have been published for the line CEL09 (Hrubcová et al, 2005). Preliminary interpretations of Alp01 and Alp02 have been presented as a poster (Bleibinhaus et al., 2004).

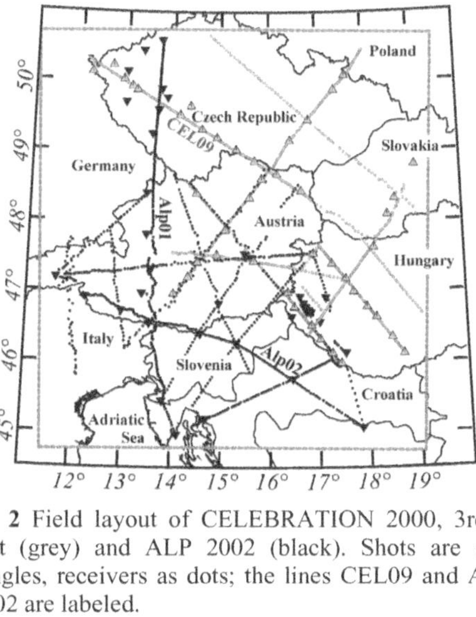

Fig. 2 Field layout of CELEBRATION 2000, 3rd deployment (grey) and ALP 2002 (black). Shots are shown as triangles, receivers as dots; the lines CEL09 and Alp01 and Alp02 are labeled.

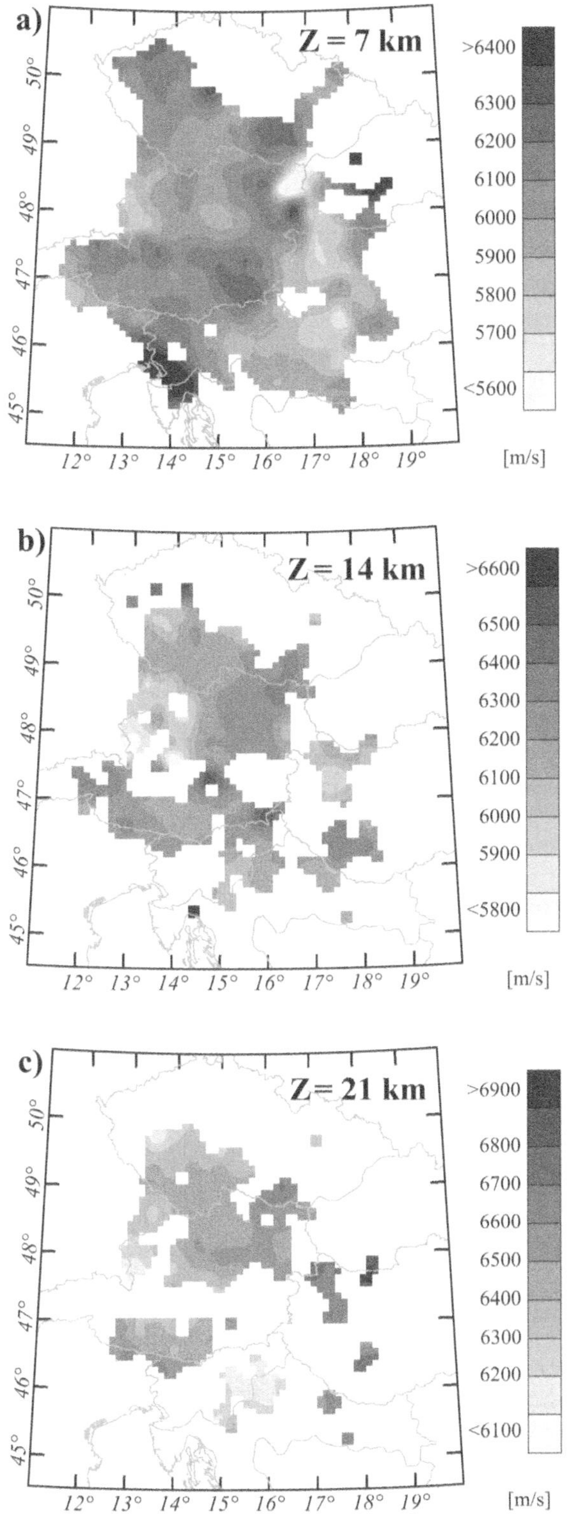

Fig.3 Depth slices through the 3D P-wave velocity model of the crust in (a) 7 km, (b) 14 km, and (c) 21 km depth.

A 3D P-wave velocity model of the crust has been derived from Pg-diving wave tomography as well as a map of the Moho discontinuity from Pn delay time analysis and PmP wide-angle reflections. Both stacking techniques to improve signal to noise ratio and travel time inversions were applied. The methods used were presented at the AGU fall meeting (Brückl et al. 2003b) and are described in detail in Behm (2006). Horizontal slices through the P-wave velocity model of the crust at 7, 14 and 21 km depth are shown in Figures 3a-c.

Significant velocity structures can be recognized, which correlate well with the regional tectonic setting. Low velocities down to about 7 km depth are found in the Pannonian Basin. Other velocity lows corresponding to the Molasse Basin, the Vienna Basin and the granite part of the Bohemian Massif reach below 10 km. The most prominent velocity high is in and north of Istria; others are the metamorphic part of the Bohemian Massif and an E-W stretching zone N to the central range of the Eastern Alps. These velocity highs reach below 10 km. The transition from the Diarides in N to the Tisza unit in S (Mid Hungarian Line) can be identified by an increase of velocities at depths around 10 km.

The analysis of the Pn delay times and the wide-angle PmP reflections have been referenced to the depth datum Z=10 km. The average and standard deviation of the Pn velocity is 7990 +/- 90 ms^{-1}. Figure 4 shows a map of the Pn-velocity. Low velocities can be found in an E-W oriented belt at the northern front of the Eastern Alps. Another area of low Pn velocities is the Croatian part of the Pannonian Basin. The highest Pn velocities are below Istria (Adriatic-Apulian plate) and the central part of the Bohemian Massif.

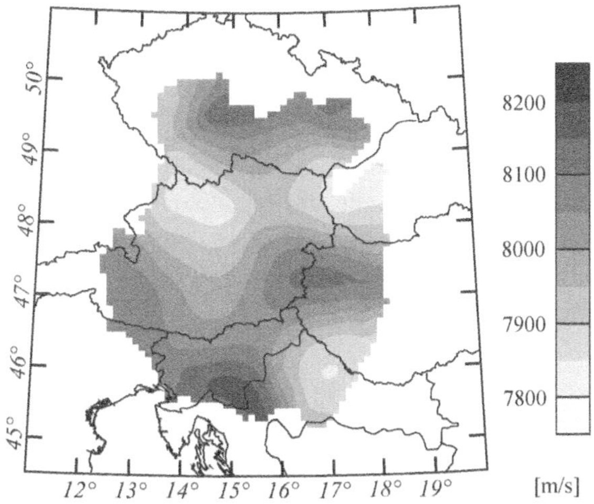

Fig. 4 Upper mantle velocity derived from a delay time analysis of the Pn-phase.

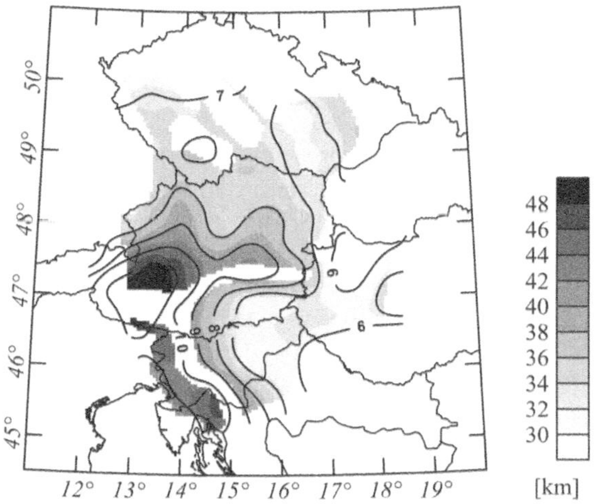

Fig. 5 Moho depth map derived from Pn delay time analysis and PmP wide-angle reflection processing (grey); iso-lines refer to vertical (migrated) 2-way Moho reflection travel time T0, reference datum Z = 10 km.

Figure 5 shows a map of the Moho discontinuity. The Moho depth is visualized by a grey code. Additionally, iso-lines of the vertical (or migrated) 2-way Moho reflection travel time, referenced to the datum Z = 10 km are superimposed. This transformation to the time domain makes the Moho map insensitive against errors in the velocities of the lower crust and will serve as "hard" information on the Moho in our later considerations.

A preliminary interpretation of the Moho structure is an S-dipping European Moho, underthrusting the Adriatic Moho from the western boundary of the investigation area to about 14°E. Further to the E we introduced a tectonic block or microplate, which is underthrusted by the European plate from the N and the Adriatic-Apulian plate from the SSW. This tectonic block shall be called "Pannonia" in the following. Scarascia and Cassinis (1997) separated this fragment as "Thin Adriatic Moho" from the surrounding Moho parts. A maximum Moho depth of about 50 km is found at the triple junction between the European and Adriatic-Apulian plate and "Pannonia", a minimum Moho depth of 27 km is found in the southern Pannonian Basin.

Support for the suggested 3D Moho model comes also from the 2D-interpretations by interactive ray tracing along ALP75 (Yan and Mechie, 1989), CEL09 (Hrubcová et al. 2005) and the preliminary results from Alp01 and Alp02 (Bleibinhaus et al. 2004). The difference in Moho depth between the independently derived 2D and 3D evaluations is about 1.0 +/- 2.0 km. The high crustal velocities of the Adriatic-Apulian plate were also verified by the 2D interpretations. Teleseismic tomography in this area (Lippitsch 2002, Lippitsch et al. 2003; Schmid et al. 2004) indicates a change in the direction of the subduction. According to these results and interpretations W of TRANSALP the European plate is subducted below the Adriatic-Apulian plate. Roughly at TRANSALP the direction of subduction changes and E of TRANSALP the Adriatic-Apulian plate is subducted below Europe. At present we cannot conclude whether a tectonic model in agreement with Moho structure shown in Figure 5 and the teleseismic model can be constructed.

3 Constraining the P-Wave Velocity of the Lower Crust by Gravity

The tomographic model of the crust supplies nearly continuous information about the P-wave velocity only in the uppermost 10 km of the crust. In most areas the diving wave tomography does not penetrate down to the Moho (Figure 6). A preliminary model of the average P-wave velocity of the crust between Z = 10 km and the Moho (Figure 7) has been constructed by estimating the P-wave velocity between the maximum penetration depth and the Moho from RMS-velocities (V_{RMS}) of PmP phases and constraints from global data (Christensen and Mooney, 1995). The map of the Moho in time and depth domain (Figure 5) has been constructed by the use of this preliminary velocity model.

An additional constraint on the P-wave velocity of the lower crust (below the penetration depth of the diving wave tomography) can be derived from gravity assuming a relation between P-wave velocity and density. The Christensen and Mooney (1995) velocity-density relationship is based on a world wide collection of refraction and wide-angle reflection data. Using this relation we do not consider the influence of the geothermal gradient or the mineralogy. Sobolev and Babeyko (1994) derived a relation between P-wave velocity and density that considers the thermal gradient and the mineral transformations within the crust with changing PT conditions. Ebbing (2002) used this relation together with the regional heat-flow data supplied by Cermak et al. (1992) for the 3D modeling of the gravity in the area TRANSALP (Eastern Alps between 10° - 14° E). However, only few observations on the geothermal gradient are available within the Eastern Alps (Sachsenhofer 2001). The additional use of S-wave velocity or Poisson's ratio could resolve to some degree the influence of mineral composition (Christensen 1996). The evaluation of S-wave velocities and Poisson's ratios will be possible from the CELEBRATION 2000 and ALP 2002 data in the future. The implementation of the gravity constraints presented in this paper has the character of a pilot study. Therefore we decided to choose the Christensen and Mooney (1995) velocity-density relationship because only P-wave velocity data are needed to estimate crustal densities. These simplifications must be kept in our mind in our later interpretations.

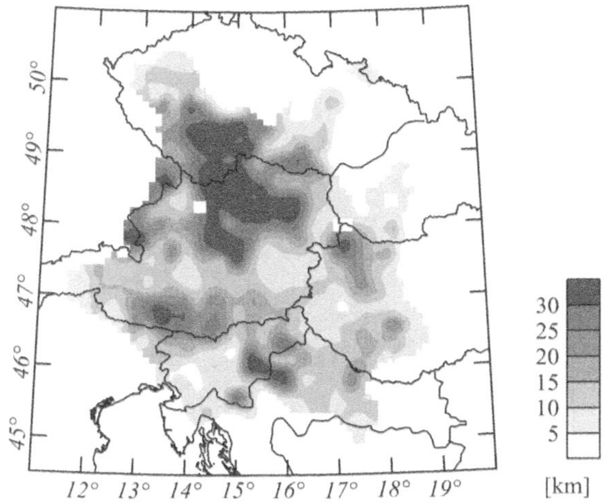

Fig. 6 Penetration depth of Pg-diving wave tomography.

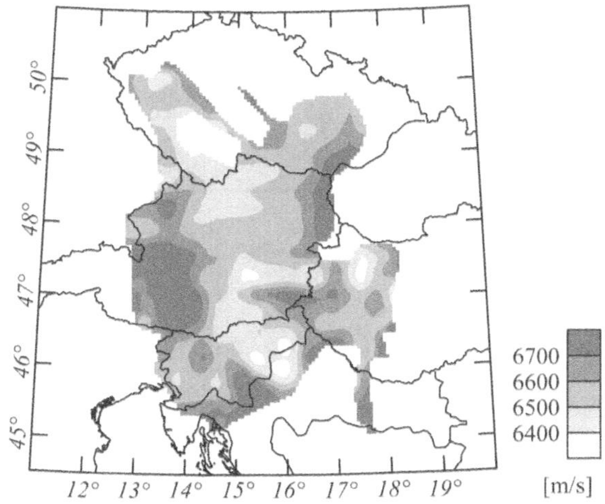

Fig. 7 Preliminary model of the average P-wave velocity of the crust between Z = 10 km and the Moho.

The gravity data used in this study (Figure 8) is the regional Bouguer gravity map of the West East European Gravity Project (http://www.getech.com). Within Austria this data has been replaced by the "New Austrian Bouguer Map" (Kraiger and Kühtreiber 1992). We call the Bouguer gravity derived from this data dgB0. The density for reduction of masses above the Ellipsoid is d = 2670 kg/m^3.

The following steps will be carried out in order to constrain the P-wave velocity of the lower crust by gravity:

· Stripping of the uppermost 10 km of the crust yielding the Bouguer gravity residuum dgB1.

· Subtracting the gravity effect of the lower crust yielding the residuum dgB2; an average linear velocity-depth function of the lower crust and corresponding densities are determined by the condition such that the correlation coefficient between dgB2 and T0 (Moho depth in time domain) is zero.

· Modeling the Bouguer gravity residuum (dgB2) by masses in the lower crust and derivation of corresponding variations of the P-wave velocity.

· Combination of the interval velocities of the lower crust (between Z = 10 km and the Moho) derived from seismic and gravimetric data.

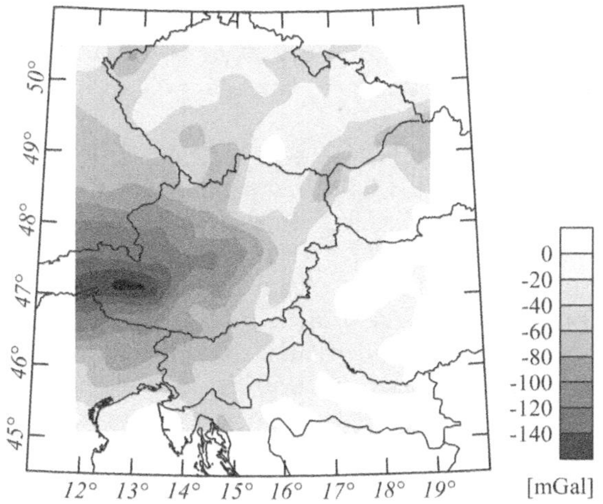

Fig. 8 Bouguer gravity map of the investigation area compiled from data of the West East European Gravity Project and the "New Austrian Bouguer Map" (for references see text).

Gravity modeling will be done by a superposition of the gravity effects of prismatic bodies after Nagy (1966). Densities relative to a reference model will be used in order to avoid edge effects. The steps for the implementation of the constraints from gravity in order to resolve the P-wave velocity of the lower crust will be described in detail in the following chapters.

4 Stripping the Uppermost 10 km of the Crust

Between the surface and the level Z = 10 km the 3D P-wave velocity model covers almost the whole area. Therefore, the densities are well constrained by the P-wave velocities. The calculation of the gravity effect is straight forward (Figure 9) without the need of assumptions. The reference density for the uppermost 10 km of the crust is 2670 kg/m^3. Subtraction of this gravity effect from the Bouguer gravity dgB0 yields a residual gravity dgB1 (Figure 10) that still shows the effect of the varying depth of the Moho, but also additional masses below sedimentary basins, compensating the mass deficit of the sediments. Such a gravity signal can be detected below the Vienna and Pannonian Basins.

The implementation of the detailed surface density map of Austria (Steinhauser et al. 1984) into the stripping of the uppermost 10 km of the crust could improve the result. However, proper weighting of densities derived from near surface samples and from P-wave velocity is a task beyond the scope of this study.

5 Average Linear Velocity Function of the Lower Crust

Moho depths have been calculated by the assumption of the following parameters of a linear velocity-depth function, valid below Z = 10 km. P-wave velocities at Z = 10 km (V_{10km}): 5900, 6000, 6100, 6200, 6300 ms^{-1}; vertical velocity gradients (k): 0.10, 0.15, 0.20, 0.25, 0.30 s^{-1}.

For these 25 velocity-models of the lower crust we calculated also the densities and the gravity effect of the masses between Z = 10 km and the Moho assuming two values for the upper mantle density, d_M=3250 kg m^{-3} and d_M=3300 kg m^{-3}. Subtraction of these gravity effects from dgB1 yields the residual Bouguer gravity dgB2. A lower crust density of 2900 kg m^{-3} and a Moho depth of 33 km were used as reference. The reference density of the upper mantle was always the density d_M of the model.

We assume that the best choice for an average velocity-depth function of the lower crust yields minimum correlations between dgB2 and T0, the time information on the Moho depth (migrated 2-way reflection travel time). The signature of the varying crustal thickness will be removed by this choice. The assumption has been made, that there are no deeper (upper mantle) density structures that produce a correlation between dgB2 and T0.

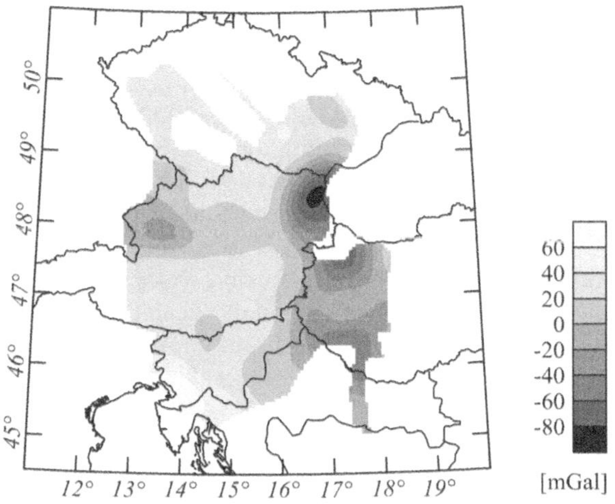

Fig. 9 Gravity effect of the upper crust (Z <= 10 km).

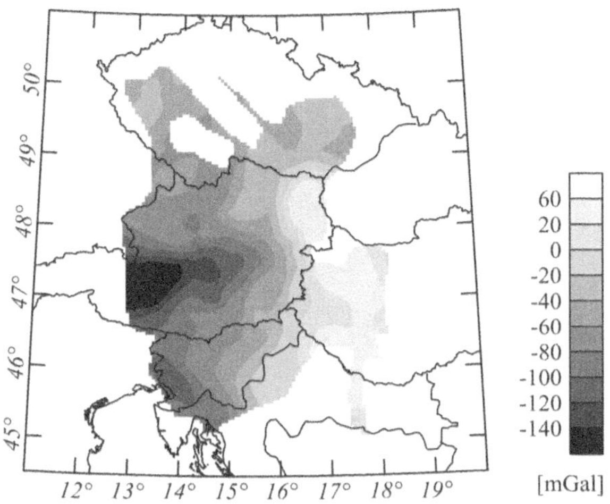

Fig. 10 Residual gravity dgB1 after stripping the upper crust (Z <= 10 km).

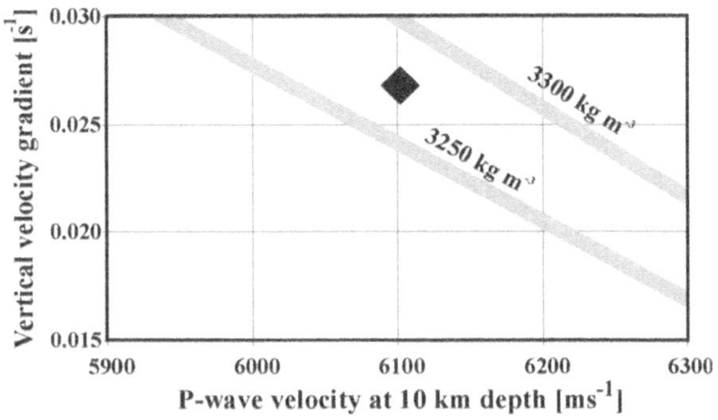

Fig. 11 Parameter combinations (grey lines) of linear velocity-depth functions for the lower crust (between 10 km depth and the Moho) that yield zero correlation between T0 (two way vertical travel time to Moho, datum Z = 10 km) and dgB2 (Bouguer gravity residuum after correction for the crust) for two mantle densities 3250 and 3300 kg m^3; the diamond represents the world wide average for orogens (Christensen and Mooney 1995).

The idea of our method is similar to Nettleton's method to determine the density of topographic masses by minimizing the correlation of the Bouguer anomaly with the terrain elevation. Figure 11 shows the pairs of V_{10km} and k that result in zero correlation between dgB2 and T0. We see there is an ambiguity in a sense that either lower V_{10km} and higher k, or higher V_{10km} and lower k result in zero correlation. In order to get additional constraints on V_{10km} and k we approximated the world wide average velocities for orogens (Christensen and Mooney 1995) by a linear velocity-depth function and plotted the corresponding values for V_{10km} = 6102 ms^{-1} and k = 0.0268 s^{-1} in the same diagram. With this velocity function for the crust below 10 km depth and an upper mantle density d_M = 3270 kgm^3 we also yield a zero correlation of dgB2 with T0. Therefore we take this parameter triple as our optimum solution. The average velocity of our 3D seismic model at Z = 10 km is 6090 ms^{-1}, which is in good agreement with this choice. The residual gravity dgB2 for these optimum parameters is shown in Figure 12.

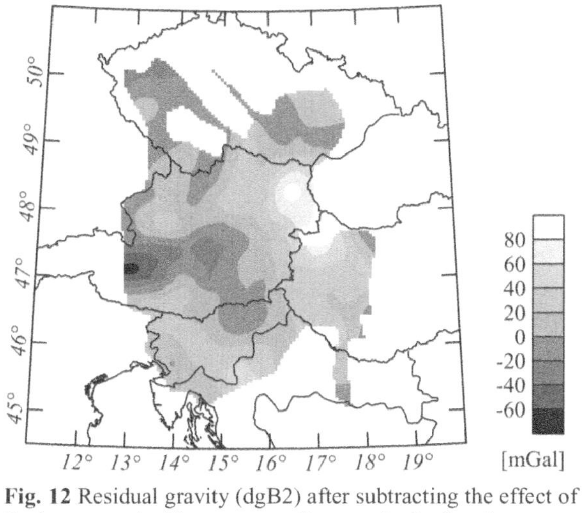

Fig. 12 Residual gravity (dgB2) after subtracting the effect of the lower crust with an average linear velocity function.

6 Modeling of the Bouguer Gravity Residuum

The Bouguer gravity residuum (dgB2) can be related to tectonic structures. An anomaly low correlates with the South Bohemian Pluton that has significantly lower densities than the metamorphic part of the Bohemian Massif (Meurers 1993). Corresponding variations of the P-wave velocity can be traced below 10 km depth in our 3D seismic model. Gravimetric modeling (Meurers 1993) also confirms that the sources of this anomaly are distributed over a depth range of at least 10 km. Another well-known anomaly is the gneiss kernel of the Tauern Window. Quantitative modeling of the low density body of the Tauern Window (Ebbing 2000) limit these low densities to the uppermost 10 km of the crust. The 3D seismic model does not show a corresponding low in the velocities of the upper crust.

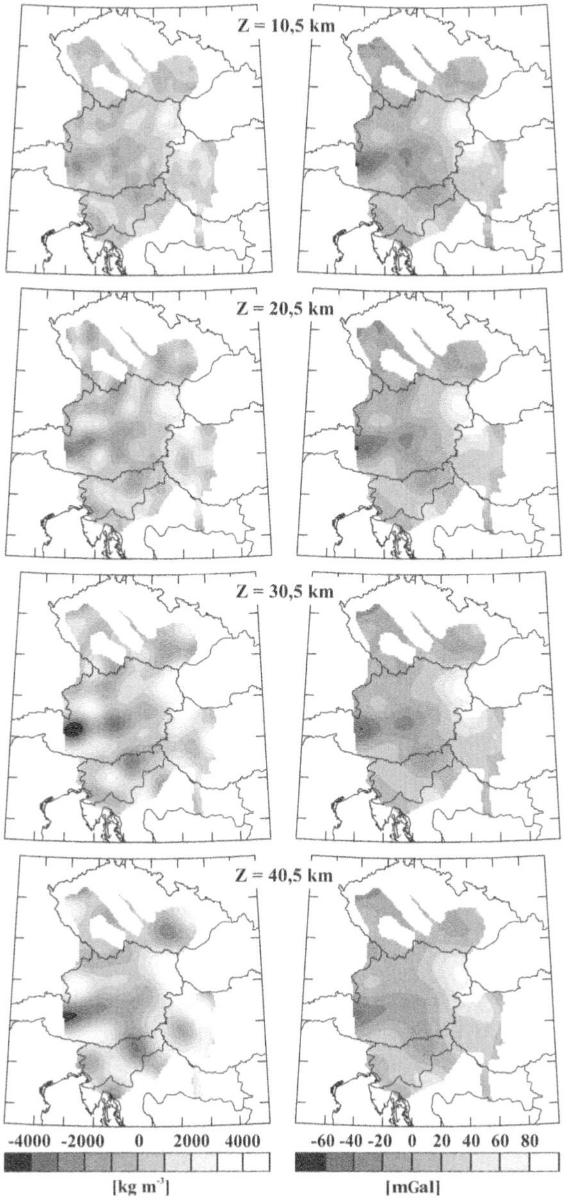

Fig. 13 Modeling of residual gravity by sources in 1 km thick layers at 4 depth levels between 10.5 and 40.5 km; left column shows the model densities of the 1 km thick layers, right column the calculated gravity effect of these layers.

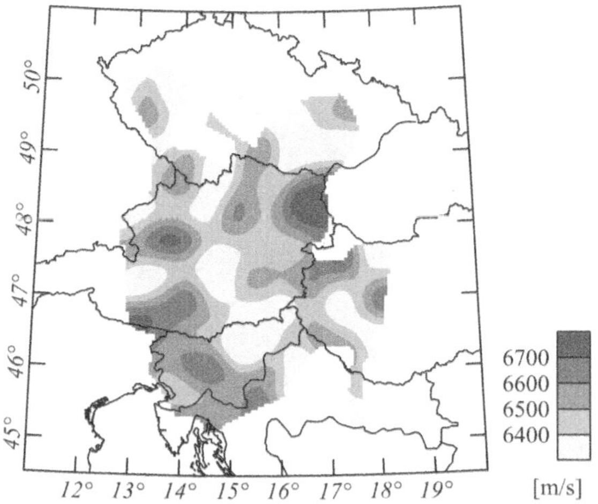

Fig. 14 Average P-wave velocity V_{grav} of the crust between Z = 10 km and the Moho derived from gravimetric data.

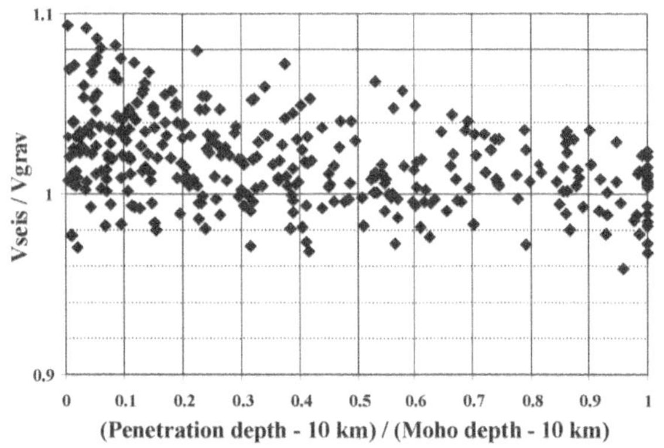

Fig. 15 Ratio R of V_{seis} (seismically determined average velocity of the lower crust) to V_{grav} (gravimetrically determined average velocity of the lower crust) versus the ratio of penetration depth to crustal thickness.

Evidence for an Adriatic-Apulian plate dipping below the assumed tectonic block "Pannonia" and the Dinarides is indicated by positive anomalies. Strong positive anomalies are also found in the transition from the Northern Pannonian domain and the Vienna Basin to the European platform, where also high velocities in the lower crust are obtained from the seismic model.

In order to get an estimate of the geologically reasonable depth range of the sources, we modeled the residual Bouguer gravity (dgB2) by masses concentrated in one layer of 1 km thickness at the depth levels of Z = 10.5, 20.5, 30.5, and 40,5 km (Figure 13).

As to be expected, the shallowest distribution of sources models the residual gravity best, but also down to Z = 30.5 km the calculated distribution of masses are geologically reasonable and fit the data well. At Z = 40.5 km the distribution of masses becomes less geologically reasonable. Thus we conclude that it is geologically reasonable to distribute the sources of the residual Bouguer gravity (dgB2) evenly over the depth range of the lower crust. Conversion of the resulting density changes to velocity changes and adding these improvements to the average linear P-wave velocity function of the lower crust results in V_{grav}, the gravimetrically determined interval velocity between the datum Z = 10 km and the Moho (Figure 14). The values of V_{grav} at the location of the Tauern Window will be excluded from our further analysis, because the source of dgB2 is the gneiss kernel in the uppermost 10 km of the crust and must not be projected to the lower crust.

7 Combination of Seismic and Gravimetric Velocity Information on the Lower Crust

Figure 15 shows the ratio V_{seis}/V_{grav} versus the ratio R of penetration depth below Z = 10 km to thickness of the lower crust (Between Z = 10 km and Moho). It is evident, the greater the penetration depth of the diving wave tomography, the less differs the ratio V_{seis}/V_{grav} from unity. The ratio V_{seis}/V_{grav} is 1.003 +/- 0.018 for R>0.85. We therefore construct a velocity solution from a weighted mean of V_{seis} and V_{grav} seismic solution (Figure 16): $V_{seis+grav} = R \cdot V_{seis} + (1-R) \cdot V_{grav}$. Because of the weighting scheme, the seismic information about the lower crust is not changed where information from diving wave tomography is available down to the Moho.

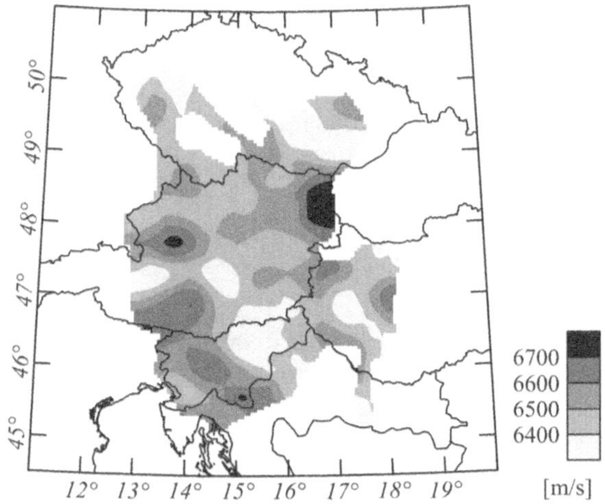

Fig. 16 Average P-wave velocity between Z = 10 km and Moho derived from seismic and gravimetric data.

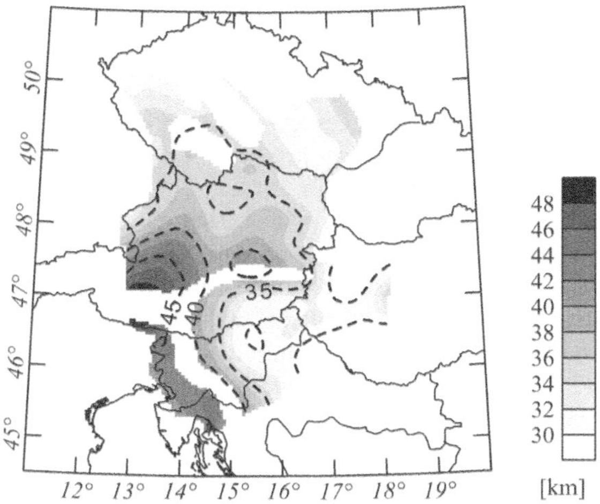

Fig. 17 Moho depth map calculated by the use of the combined seismic-gravimetric velocity model of lower crust.

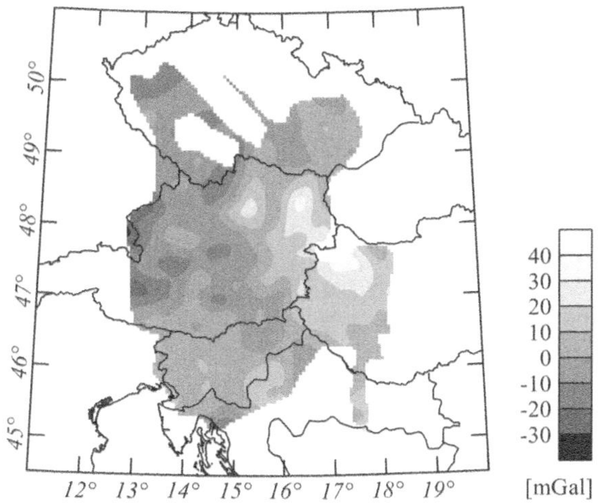

Fig. 18 Bouguer gravity residuum (dgB3) after removing the effect of the whole crust.

Figure 17 shows the depth map of the Moho calculated from the time map (Figure 5) and the interval velocities shown in Figure 16.

Finally, we calculated the gravity effect of the lower crust between Z = 10 km and the Moho on the basis of the improved P-wave velocities and their transformation to densities. The final gravity residuum (dgB3) after removing the effect of the whole crust is shown in Figure 18. The standard deviation of dgB3 is 11 mGal, the standard deviation of the Bouguer anomaly without any corrections for the crust (dgB0) has been 30 mGal. The residuum dgB3 does not show any more the signature of the topography or the Moho structure.

8 Isostasy

A density model of the lithosphere should also represent a reasonable solution for isostasy. A map of the isostatic anomaly of the Eastern Alps was calculated by Wagini et al. (1988). The residuals vary between -25 mGal and +25 mGal with an average value of +8 mGal. Meurers (1996) points out that density variations in the upper crust may significantly change the isostatic residual and he concludes that the Eastern Alps are at least near an isostatic equilibrium. Braitenberg et al. (2002) and Ebbing (2004) come to the conclusion that the internal crustal loads contribute to the isostatic equilibrium in the same order of magnitude as the topographic loads in the area of TRANSALP. Again the Airy-isostatic residual shows a strong correlation with tectonics and the near surface density distribution. Furthermore, calculations of the flexural rigidity result in relatively low values.

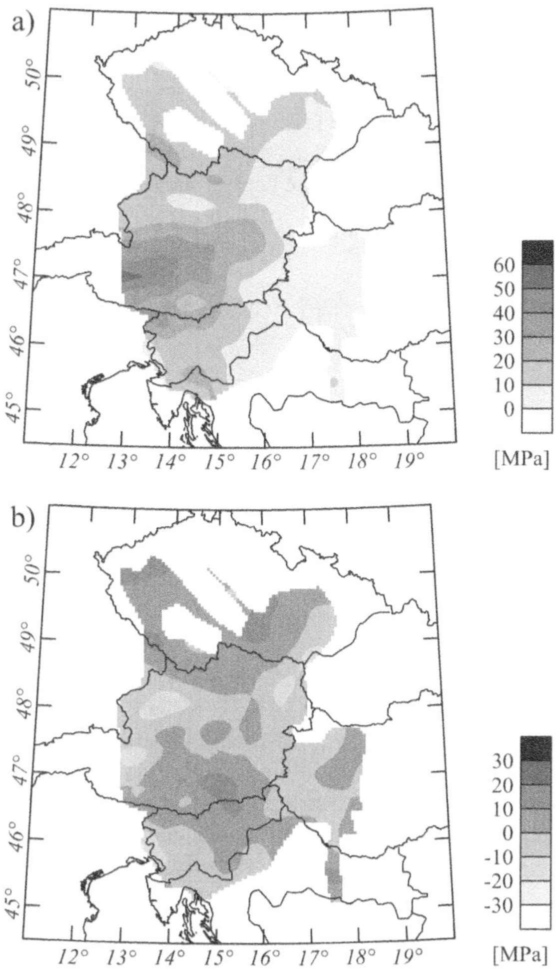

Fig. 19 Vertical loading stress, derived from the loads given by our density model at **(a)** Z = 0 km and **(b)** Z = 50 km. The values at Z = 50 km are relative to the reference lithosphere.

The study of Lillie et al. (1994) in the Eastern Alpine - Western Carpathian - Pannonian Basin region demonstrate that the lithosphere - asthenosphere boundary must be included into the gravimetric modeling to achieve isostatic equilibrium.

From these findings we conclude, the vertical loading stress at Z=50 km (a level below the orogen roots), derived from the loads given by our density model should be near an Airy-isostatic equilibrium. Long wave length deviations from equilibrium should find their compensation in the lower lithosphere and asthenosphere (Lillie et al. 1994).

Figure 19 shows the vertical loading stresses at Z=0 km (a) and Z=50 km (b). The load of the masses above Z=0 km were calculated from a smoothed topography and a constant density of 2670 kg m^{-3}. The values of the vertical loading stresses at Z=50 km are calculated from the loads given by our 3D density model relative to the reference lithosphere defined in paragraphs 4 and 5. The standard deviation from isostatic equilibrium at Z=50 km is +/- 6 MPa. Ebbing et al. (2005) derived deviations from isostatic equilibrium in the same order of magnitude in the central part of the Eastern Alps (TRANSALP) at depth levels 60 - 300 km from a high resolution 3D density model.

9 Conclusions and Future Plans

The presented work is based on a preliminary 3D seismic model and Bouguer gravity map of the Eastern Alpine crust. It has therefore the character of pilot study. The P-wave velocity of the crust was mainly determined by diving wave tomography. In a considerable part of the investigation area the penetration depth of this method was not deep enough to achieve sufficient information on the velocity of the lower crust.

The implementation of gravity data resulted in a well-constrained solution for the P-wave velocity of the lower crust in the interval between Z=10 km and the Moho. The variations of the seismic velocities of the crust, the Moho topography and the Bouguer gravity residual after stripping the uppermost 10 km of the crust and removing the influence of the Moho topography (dgB2) can be related to tectonic provinces and geodynamic processes to some extent.

Conversion of the seismic model constrained by gravity to a density model using the relations from Christensen and Mooney (1995) and subtraction of the gravity effect of this model from the observed Bouguer gravity (dgB0) results in a residuum (dgB3) that reduces the variance of dgB0 by 86% (Figure 18). The vertical loading stress at the depth level Z=50 km corresponding to this density model (Figure 19b) shows only little variations. Adjustments of the average densities of the 50 km lithospheric columns by less than 1% could restore Airy-isostatic equilibrium.

Despite the good fit of our velocity and density model the final residuals of gravity and the deviation from isostatic equilibrium show patterns, which again could be related to tectonic structures and processes. However, before we consider these relations we plan to improve our results by the use of an upgraded gravity map, which will be available for our investigation area in the near future (Meurers, personal communication). Furthermore, we will implement the surface density map (Steinhauser et al. 1984). The evaluation of S-wave velocities will provide better constraints on the mineralogy. Consideration of thermal flux and temperature as well as other velocity density relations (f.e. Sobolev and Babeyko 1994) will be necessary too. Also teleseismic models of the lower lithosphere and asthenosphere (f.e. Lippitsch et al. 2003) should be incorporated to model the deep structures.

Acknowledgements

This work is part of the ALP 2002 and ALPASS projects and has been funded by the Austrian Academy of Sciences and the Austrian Science Fund, project P15576-N06. We say our sincere thanks to C. Braitenberg and M.G. Doufexopoulou for their thorough reviews that helped to improve the paper and to A.J. Gil and F. Sansó for their efforts as editors.

Members of the CELEBRATION 2000 and ALP 2002 Working Groups

S. Acevedo, K. Aric, M. Behm, A. Belinsky, F. Bleibinhaus, T. Bodoky, E. Brückl, R. Clowes, W. Chwatal, W. Czuba, E. Gaczyński, H. Gebrande, A. Gosar, M. Grad, H. Grassl, A. Guterch, Z. Hajnal, S. Harder, E. Hegedüs, S. Hock, V. Hoeck, P.Hrubcová, T. Janik, G. Jentzsch, P. Joergensen, A. Kabas, G. Kaip, K. Komminhaho, G.R. Keller, F. Kohlbeck, S. Kostiuchenko, D. Kracke, K.C. Miller, A. Morozov, J. Oresko, K. Posgay, E.-M. Rumpfhuber, C. Schmid, R. Schmöller, O. Selvi, C. Snelson, A. Špicák, P. Środa, F. Sumanovac, E. Takács, H. Thybo, T. Tiira, C. Tomek, C. Ullrich, J. Vozár, F. Weber, M. Wilde-Piórko, J. Yliniemi.

References

Aric, K., et al. (1987). Seismological studies in the Eastern Alps. In: H.W. Flügel, P. Faupl (Ed.): Geodynamics of the Eastern Alps, Franz Deuticke Verlag.

Behm, M. (2006). Accuracy and resolution of a 3D seismic model of the Eastern Alps. Ph.D.-Theses, Vienna University of Technology, Vienna.

Behm, M., E. Brückl, W. Chwatal, H. Thybo and CELEBRATION 2000 Working Group, ALP 2002 Working Group (2004). Seismic structure of the Eastern Alps - evidence for a "Pannonian" microplate. Poster Presentation at IGC Florence, Italy, August 20-28, 2004.

Bleibinhaus, F., E. Brückl, A. Gosar, M. Grad, E. Hegedüs, P. Hrubcová, G.R. Keller, F. Sumanovac, J. Yliniemi and ALP 2002 Working Group (2004). Alp 2002 Experiment - 2D Raytracing modelling and seismic tomography of selected profiles. Poster Presentation at IGC Florence, Italy, August 20-28, 2004.

Braitenberg C., J. Ebbing, H.-J. Götze (2002). Inverse modelling of elastic thickness by convolution method - the Eastern Alps as a case example. Earth and Planetary Science Letters, 2002, 387-404.

Braitenberg C., F. Pettenati, M. Zadro (1997). Spectal and classical methods in the evaluation of Moho undulations from gravity data: the NE-Italian Alps and isostasy. Journal of Geodynamics, 23, 5-22.

Brückl, E., T. Bodoky, A. Gosar, M. Grad, A. Guterch, Z. Hajnal, E. Hegedüs, P. Hrubcová, G.R. Keller, A. Špičák, F. Sumanovac, H. Thybo, F. Weber and ALP 2002 Working Group (2003a). ALP 2002 seismic experiments. Stud. Geoph. Geod., 47 (2003), 671-679.

Brückl E., M. Behm, W. Chwatal (2003b). The application of signal detection and stacking techniques to refraction seismic data. Oral Presentation at AGU, San Francisco, 08-12 December 2003.

Čermak, V., Balling, N., Della Vedova, B., Lucazeau, F., Pasquale, V., Pellis, G., Schulz, R., and Verdoya, M. (1992). Heat-flow density. In D. Blundell, R. Freeman, and S. Mueller, editors, A Continent Revealed: The European Geotraverse. Cambridge University Press.

Červený, V. and I. Psencik, (1984). Documentation of Earthquake Algorithms. SEIS83 - Numerical modeling of seismic wave fields in 2-D laterally varying layered structures by the ray method. E. R. Engdahl edit., Report SE-35, Boulder, 36-40.

Christensen, N.I. and W.D. Mooney (1995). Seismic velocity structure and composition of the continental crust: A global view, J. Geophys. Res., 100, 97619788.

Christensen, N. (1996). Poisson ratio and crustal seismology, J. Geophys. Res 101, 3139-3156.

Dal Moro G., C. Braitenberg, and M .Zadro (1998). Geometry and mechanical and crustal properties in NE Italy based on seismic and gravity data. B olletino di Geofisica Teorica ed Applicata,Vol. 39, N.1, 37-46.

Ebbing, J. (2002). 3-D Dichteverteilung und isostatisches Verhalten der Lithosphäre in den Ostalpen. Thesis, Freie Universität Berlin, 143.

Ebbing, J. (2004). The crustal structure of the Eastern Alps from a combination of 3D gravity modelling and isostatic investigations. Tectonophysics, 380/1-2, 80-104.

Ebbing J., C. Braitenberg, H.-J. Götze (2001). Forward and inverse modell ing of gravity revealing insight into crustal structures of the Eastern Alps. Tectonophysics, 337/3-4, 191-208.

Ebbing J., C. Braitenberg, H.-J. Götze (2005). The lithospheric density structure of the Eastern Alps, Tectonophysics, in press (Transalp special volume).

Giese, P., et al. (1976). Explosion Seismology in Central Europe. Eds. P. Giese, C. Prodehl, A. Stein. Springer Verlag Berlin, Heidelberg, New York.

Graßl, H. (1999). NESTMK - Ein Tiefenseismikprofil in der Nordoststeiermark. Dissertation, Institut für Geophysik, Montanuniversität Leoben, Mai 1999.

Graßl, H., F. Neubauer, K. Millhan, F. Weber, (2004). Seismic image of the deep crust at the eastern margin of the Alps (Austria): indications for crustal extension in a convergent origin. Tectonophysics 380, 105-122.

Guterch, A., M. Grad, G.R. Keller, K. Posgay, J. Vozár, A. Špičák, E. Brueckl, Z. Hajnal, H. Thybo, O. Selvi, and CELEBRATION 2000 Experiment Team (2003). CELEBRATION 2000 Seismic Experiment, *Stud. Geoph. Geod.*, 47, 659-669.

Hrubcová P, P. Sroda, A. Špičák and CELEBRATION 2000 Working Group (2005). Crustal and uppermost mantle structure of the Bohemian Massif based on CELEBRATION 2000 data. J. Geophys. Res, in press.

Kraiger, G. and N. Kühtreiber (1992). Preliminary results of a new Bouguer Map of Austria. Geodesy and Physics of the Earth: Geodetic Contributions to Geodynamics, 7 th Symposium Nr. 112, 5-10 October, Potsdam. Eds. H. Montag and C. Reigber: Springer-Verlag, p.133

Lillie, R.J., M. Bielik, V. Babuska, and J. Plomerova (1994). Gravity modelling of the lithosphere in the Eastern Alpine - Western Carpathian - Pannonian Basin region. Tectonophysics, 231, 215-235.

Lippitsch, R. (2002). Lithosphere and Upper Mantle P-Wave Velocity Structure Beneath the Alps by High-Resolution Teleseismic Tomography. Ph.D.-Thesis, Swiss Federate Institute of Technology, Zürich.

Lippitsch, R., E. Kissling, J. Ansorge, (2003). Upper mantle structure beneath the Alpine orogen from high-resolution teleseismic tomography, Journal of Geophysical Research, 108, B8, 2376, doi: 10.1029/2002JB002016

Meurers, B. (1996). Investigation of the isostatic anomaly of the Eastern Alps. Acta Geod. Geophys. Hung., 31 (3-4), 389-403

Meurers, B. (1993). Die Böhmische Masse Österreichs im Schwerefeld. 6. Int. Alpengrav. Koll., Leoben 1993, Österr. Beitr. Met. Geoph., 8, 69-81.

Meurers, B., D. Ruess, P. Steinhauser, (1987). The Gravimetric Alpine Traverse. In: Flügel, H., and Faupl, P. (Editors): Geodynamics of the Eastern Alps. Deuticke, Vienna, 334-344.

Nagy, D. (1996). The gravitational attraction of right angular prism. Geophysics 31, pp. 362371, 1966

Posgay, K., et al. (1996). International deep reflection survey along the Hungarian Geotraverse, Geoph. Trans. 40 (1-2), 1-44.

Ratschbacher, L., Frisch, W., Linzer, H.-G., and Merle, O. (1991). Lateral extrusion in the Eastern Alps, Part II: Structural analysis. Tectonics, 10, 256-272.

Sachsenhofer, R. (2001). Syn- and post-collisional heat flow in the Cenozoic Eastern Alps. Int. J. Earth Sciences (Geol. Rundsch.), 90, 579-592.

Scarascia, S. and R. Cassinis, (1997). Crustal structures in the central-eastern Alpine sector: a revision of the available DSS data. Tectonophysics 271, 157-188.

Schmid, S., Fügenschuh, B., Kissling, E., and Schuster, R. (2004). Tectonic map and overall achitecture of the Alpine orogen. Eclogae Geologicar Helvetiae, Swiss Journal of Geosciences, 97/1, 93-117.

Senftl, E. (1965). Schcrckarte von Österreich, Bouguer-Isanomalen, 1:1 Mill., Bundesamt für Eich- u. Vermessungswesen, Wien.

Sobolev, S.V and A.Y. Babeyko, (1994). Modelling of mineralogical composition, density and elastic wave velocities in anhydrous magamatic rocks. Surveys in Geophysics 15, 515-544.

Steinhauser, P., D. Ruess, D. Zych, H. Haitzmann, G. Walach (1984). The geoid in Austria: Digital models of mean topographic heights and rock densities. Proc. 18th Gen. Ass. IUGG, IAG, Vol. 1, 322-338.

Tomek, C. (1993a). Deep crustal structure beneath the central and inner West Carpathians. Tectonophysics 226, p. 417-431.

Tomek, C. (1993b). Subducted continental margin imaged in the Carpathians of Czechoslovakia. Geology, v.21, p. 535-538.

Transalp Working Group (2002). First deep seismic reflection images of the Eastern Alps reveal giant crustal wedges and transcrustal ramps. Geophys. Res. Let. 29 (10), 10.1029/2002GL014911, 92-1 - 92-4.

Wagini, A., Steinhauser, P., Meurers, B. (1988). Isostatic residual gravity map of Austria. USGSA, Open file report 87-402.

Weber. F., R. Schmöller, R.K. Frühwirth (1996). Results of a deep reflection seismic measurement south of Rechnitz/Burgenland/Austria. Geophys. Trans., 40, 79-93.

Yan, Q.Z., J. Mechie, (1989). A fine section through the crust and lower lithosphere along the axial region of the Alps. Geophysical Journal 98, 465-488.

Observing Fennoscandian Gravity Change by Absolute Gravimetry

L. Timmen, O. Gitlein, J. Müller, H. Denker
Institut für Erdmessung (IfE), University of Hannover, Am Schneiderberg 50, D-30167 Hannover, Germany

J. Mäkinen, M. Bilker
Finnish Geodetic Institute (FGI), Geodeetinrinne 2, FIN-02430 Masala, Finland

B.R. Pettersen, D.I. Lysaker, O.C.D. Omang, J.G.G. Svendsen
Department of Mathematical Sciences and Technology, UMB, P.O. Box 5003, N-1432 Ås, Norway

H. Wilmes, R. Falk, A. Reinhold, W. Hoppe
Bundesamt für Kartographie und Geodäsie (BKG), Richard-Strauss-Allee 11, D-60598 Frankfurt, Germany

H.-G. Scherneck
Chalmers Univ. of Technology, Onsala Space Observatory, SE-439 92 Onsala, Sweden

B. Engen, B. G. Harsson
Statens Kartverk (Norwegian Mapping Authority), Kartverksveien 21, 3511 Hønefoss, Norway

A. Engfeldt, M. Lilje
Lantmäteriet (National Landsurvey of Sweden), Lantmäterigatan 2, S-80182 Gävle, Sweden

G. Strykowski, R. Forsberg
KMS (National Survey and Cadastre), Rentemestervej 8, DK-2400 Copenhagen NV, Denmark

Abstract. The Nordic countries Norway, Sweden, Denmark and Finland are a key study region for the research of glacial isostasy, and, in addition, it offers a unique opportunity for validating and testing the results of the GRACE experiment. Over a period of five years, the expected life time of GRACE, a temporal geoid variation of 3.0 mm is expected in the centre of the Fennoscandian land uplift area, corresponding to a gravity change of about 100 nm/s². This is expected to be within the detection capabilities of GRACE. With terrestrial absolute gravimetry, the gravity change due to the land uplift can be observed with an accuracy of ±10 to 20 nm/s² for a 5-year period. Thus, the terrestrial in-situ observations (ground-truth) may be used to validate and test the GRACE results.

Since 2003, absolute gravity measurements have been performed in Fennoscandia at about 30 stations covering Norway, Sweden, Finland and Denmark. Four groups with FG5 absolute gravimeters (BKG, FGI, IfE, UMB) are engaged to survey the uplift network annually by a mutually controlled procedure. Nearly all absolute stations are co-located with permanent GPS stations. From the 2003 and 2004 comparisons between the instruments, an overall accuracy of ±30 nm/s² is indicated for a single absolute gravimeter and a single station determination. This is in full agreement with the project goal.

Keywords. Absolute gravimetry, Fennoscandian land uplift, geoid change, glacial isostatic adjustment (GIA), postglacial rebound (PGR), GRACE validation

1 Determination of the Fennoscandian Land Uplift

In Fennoscandia, the Earth's crust is rising continuously since the last glacial maximum due to the deloading of the ice. The region is dominated by the Precambrian basement rocks of the Baltic Shield comprising mainland Norway, Sweden, Finland, the Kola Peninsula, and Russian Karelia. The maximum spatial extension is about 2000 km in northeast-southwest direction; see Fig. 1 for the approximate shape and location (after Ekman and Mäkinen (1996)). Geophysical approaches to study

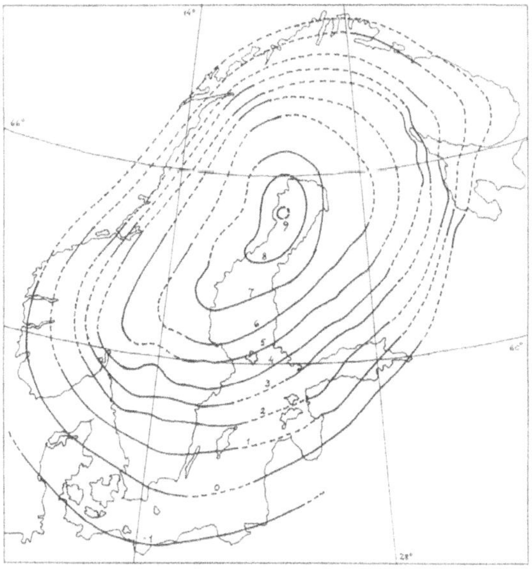

Fig. 1 Map of the postglacial uplift of Fennoscandia in accordance with Ekman and Mäkinen (1996).

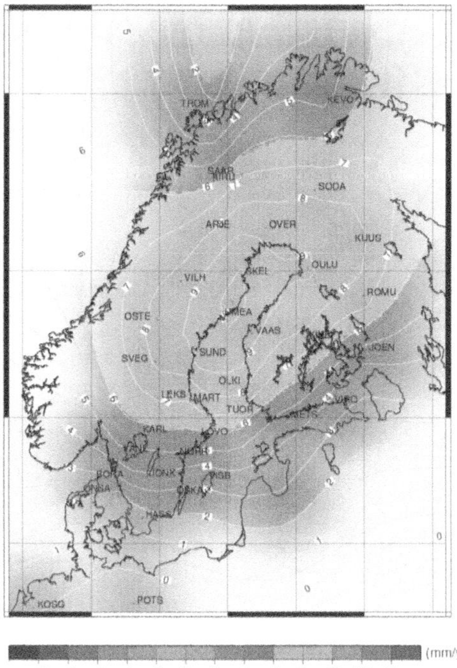

Fig. 2 Measured rates at BIFROST permanent GPS stations after Scherneck et al. (2003).

the postglacial rebound are associated with the evidence for past sea levels, the knowledge or assumptions about the geometry of the ice sheets (thickness, position), and some Earth model parameters (lithosphere thickness, mantle viscosity). After Lambeck et al. (1998), the inverse solution for the sea level data includes both ice and Earth model parameters as unknowns. Despite the recent progress in understanding the underlying models, definite models for the isostatic rebound do not yet exist. Lateral rheological variations have to be taken into account to obtain a more realistic glacially induced uplift model (Kaufmann et al. (2000)).

To monitor and investigate the recent land uplift in Fennoscandia, various measurements were collected since 1892: mareograph records, geodetic levellings, and relative gravity measurements since 1966. With these observations, the capability of terrestrial point measurement techniques to determine the land uplift was proven along east-west profiles. According to Ekman (1996), these observations reveal a maximum orthometric height change of 1 cm/a in the northern part of the Bothnian Gulf (see Fig. 1) and show symmetry around the maximum, closely correlated to the former Late Pleistocene Fennoscandian Ice Sheet. The height change in the centre is associated with a maximum gravity change of −20 nm/s^2 per year. Based on these numbers, a geoid change of 0.6 mm/a has been derived for the central area (Ekman and Mäkinen (1996)). An eustatic sea level rise of 1.2 mm/a has been deduced from the tide gauge observations (Nakiboglu and Lambeck (1991)).

Since 1993, permanent GPS stations were established in Fennoscandia to implement a further geodetic method with several advantages compared to the classical techniques (continuous data acquisition, homogeneous point distribution, large extension of the measurement area, low cost, three-dimensional survey). In this respect, the project BIFROST (Baseline Inferences for Rebound Observations, Sea Level, and Tectonics) was based on the GPS technique and geophysical modelling, and has delivered a maximum height variation (with respect to a geocentric reference ellipsoid) of more than 11 mm/a (see Fig. 2, cf. Milne et al. (2001), Johansson et al. (2002), Scherneck et al. (2003)). The location of the centre and the geometrical structure of the uplift process differ from the previous model, with no clear zero line and more regional structures being visible.

During the mission duration of GRACE (about five years), a temporal geoid change of 3.0 mm can be expected in the centre of the Fennoscandian land uplift area, corresponding to a gravity change of about 100 nm/s^2 (≡ 10 μGal). This is a clear secular gravity change of regional scale, and it might be a challenging task to detect this signal from satellite gravity data, in particular GRACE. After Wahr and

Velicogna (2003), the land uplift causes a measurable signal in the GRACE observations. Early results from Tapley et al. (2004) confirm that this satellite mission is able to resolve geoid variations for a range of spatial scales down to 400 km for particular regions with large signals. They found that the error level in the 2003 solutions was in the order of 2 to 3 mm for spatial features of about 600 km. Considering the large extension of the land uplift area, Fennoscandia is a suitable application region for GRACE. Vice versa, the temporal gravity field change can be used for the validation of the GRACE results. Because the observation of the rebound signal is interfered with mass variations due to oceanographic, land hydrology and atmospheric processes, these effects have to be accounted for in the GRACE data analysis with appropriate mathematical approaches (e.g. Wiehl et al. (2005)). Hence, the combination with other geological and geodetic measurements is inevitable.

2 Absolute Gravimetry

Besides the geometrical methods, terrestrial absolute gravimetry is a further geodetic technique to study the land uplift. In addition and complementary to the other geodetic measurements, absolute gravimetry has the following positive characteristics:

- accuracy ±20 to 30 nm/s^2 (10 $nm/s^2 \equiv 1\ \mu Gal$) for a single station determination,
- absolute monitoring of mass movements and vertical displacements (no problems with calibration and datum level),
- accuracy of absolute gravity net is independent of geographical extension,
- independent validation method for GPS, VLBI, SLR, and superconducting gravimetry,
- combined with geometrical methods, vertical surface deformations and subsurface mass movements can be separated.

The benefit of absolute gravimetry has already been exploited in different research projects, covering areas of global or regional extensions, and monitoring variations caused by mass movements. The International Absolute Gravity Basestation Network (IAGBN) serves, among other purposes, to determine large scale tectonic plate movements (Boedecker and Fritzer (1986)). The recommendations of the Interunion Commission of the Lithosphere on Mean Sea Level and Tides propose the regular implementation of absolute gravity measurements at coastal points, 1 to 10 km away from tide gauges (Carter et al. (1989)). The height differences between gravity points and tide gauges have to be controlled by levelling or GPS. In Great Britain, the main tide gauges are controlled by repeated absolute gravity determinations in combination with episodic or continuous GPS measurements (Williams et al. (2001)).

To determine crustal deformations by absolute gravimetry, secular gravity changes should be measured with a precision of about ±5 nm/s^2 per year. This can be achieved by annual measurements over five years. To exploit absolute gravimetry in combination with GPS for a "pointwise" validation of the GRACE results or to support the GRACE data evaluation by terrestrial gravimetry, the temporal variations of gravity disturbances (or gravity anomalies) are needed in accordance with the resolution of the GRACE data. Because of the long-wavelength nature of the GRACE results, the terrestrial point results derived from absolute gravimetry have to be reduced for all local effects changing gravity with time. In this connection, a severe problem is the subsurface water mass movement (change of groundwater table, temporary water storage in clefts and crevasses). Such impacts are partly considered by the station selection. Moreover, by measuring every year the absolute gravity value at a station over a 5-year period, the impact of ground water variations is averaged out to a large extent within the computation of the linear gravity rate. In addition, observations of the ground water table in boreholes and in nearby wells during the absolute gravity surveys are taken and used to assess the disturbing impact. Furthermore, a second favourable averaging effect arises from deriving a spatial mean over a larger area with a few hundred of kilometres extension. For that reason, and to allow the determination of an asymmetric uplift model with possible regional structures, a rather large number of stations is occupied every year over the whole Fennoscandian area.

3 The Conversion of Measured Land Uplift to Geoid Change

Proceeding with an appropriate data set of terrestrial geodetic observations acquired over a period of at least five years, a model of temporal geoid changes in Fennoscandia can be derived, which is suitable for direct comparisons with the GRACE results. The calculation is based on a formula found by Hotine (1969), which is extended here for the conversion of temporal gravity and height changes to geoid changes:

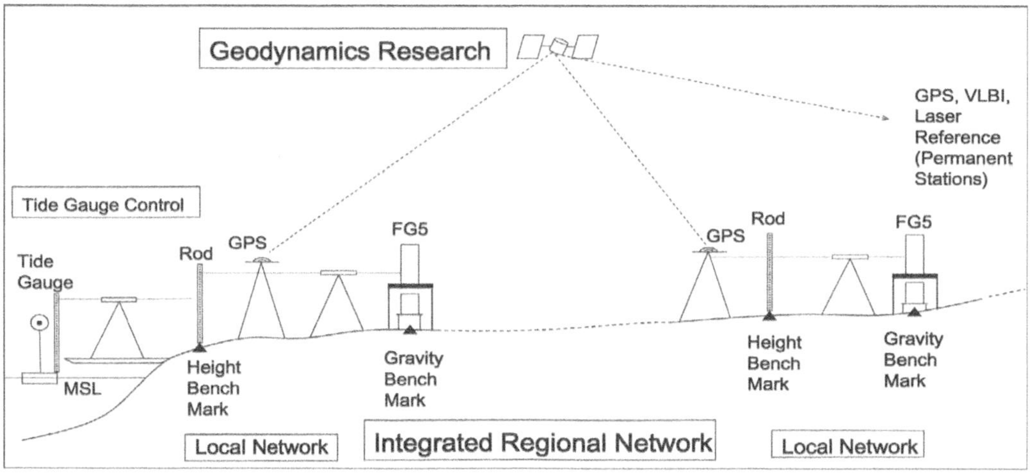

Fig. 3 Integration of different geodetic techniques to survey the temporal gravity and geoid variations of the Fennoscandian land uplift area.

$$\dot{N} = \frac{R}{4\pi\gamma}\iint_{\sigma} \mathrm{H}(\Psi)\left(\dot{g} + \frac{2\gamma}{r}\dot{h}\right)\mathrm{d}\sigma, \quad (1)$$

$$\textit{with}\quad \mathrm{H}(\Psi) = \left(\sin\frac{\Psi}{2}\right)^{-1} - \ln\left(1+\left(\sin\frac{\Psi}{2}\right)^{-1}\right).$$

$\dot{N}, \dot{h}, \dot{g}$ are the temporal changes of geoid heights, ellipsoidal heights, and gravity; γ is the normal gravity; R is the mean Earth radius; r is the radius of computation point P, and Ψ is the spherical distance. This equation shows that the measured gravity changes from absolute gravimetry have to be combined with the ellipsoidal height changes from GPS to obtain geoid height changes. The accuracy of the observed gravity and height changes should be in accordance to each other (e.g., 20 nm/s^2 and 10 mm).

4 Project Realisation

In 2002, the Institut für Erdmessung (IfE) of the University of Hannover has received a new FG5 absolute gravity meter from Micro-g Solutions, Inc. (Erie, Colorado), which is a "state-of-the-art" instrument (Niebauer et al. 1995). This version replaced the older JILAG3 system at IfE, which was successfully employed in South America, Europe and China since 1986. In the meantime, a joint project for the survey of the land uplift in Fennoscandia was established. The Working Group for Geodynamics of the Nordic Geodetic Commission (NKG) serves as a platform to organise the project. Besides the IfE from Hannover, the following institutions are participating in the project: National Survey and Cadastre (KMS, Copenhagen/Denmark), Finnish Geodetic Institute (FGI, Masala/Finland), Bundesamt für Kartographie und Geodäsie (BKG, Frankfurt/Germany), Department of Mathematical Sciences and Technology, Norwegian University of Life Sciences (UMB, Ås), Statens Kartverk (SK, Hønefoss/Norway), Onsala Space Observatory (Chalmers University of Technology, Onsala/Sweden), Lantmäteriet (Gävle/Sweden). The FGI procured a new FG5 at the beginning of 2003, and the UMB in spring 2004. The project realisation and strategy may be summarised as follows:

- absolute gravity determinations at 22 stations in 2003 with 3 gravimeters employed: BKG (FG5-301), FGI (FG5-221), IfE (FG5-220),
- absolute gravity determinations at 24 stations in 2004 with 3 gravimeters employed: FGI (FG5-221), IfE (FG5-220), UMB (FG5-226),
- repetition measurements in the years 2005, 2006 and 2007,
- simultaneous GPS measurements (gravity stations are normally co-located with permanent GPS stations),
- geometrical connections between gravity stations and tide gauges (mainly by GPS),
- auxiliary levelling measurements to eccentres (control of local variations, ties between gravity and GPS stations),
- elaboration of reduction models (air mass movements, ocean tides, etc.),
- integration of already existing geodetic data sets (e.g., FG5 surveys by Wilmes et al. (2004)),
- utilization of the project products (temporal changes of gravity disturbances, two-dimensional model describing the geoid change) for GRACE validation.

Fig. 3 demonstrates the principles to combine different geodetic methods. Nearly all absolute

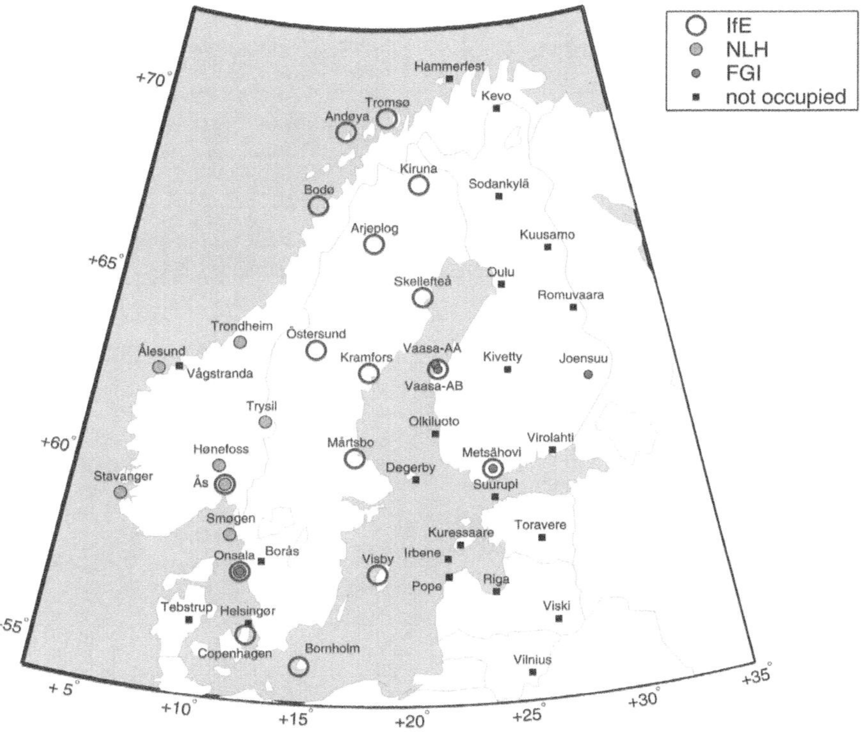

Fig. 4 Observed absolute gravity stations in 2004 occupied by the absolute gravimeters FG5-220 (IfE), FG5-221 (FGI), FG5-226 (UMB).

gravity sites are co-located with permanent GPS stations, and also tide gauges are in the vicinity of coastal stations. The gravity points, GPS stations and tide gauges are connected locally, using terrestrial surveying techniques, like levelling, to monitor the local vertical variations. Fig. 4 shows the stations occupied in 2004. The employment of more than one absolute gravimeter allows simultaneous (parallel) observations in stations with two sites close to each other and control measurements on identical sites. This strategy increases the network reliability and accuracy because it helps to identify possible offsets of the instruments. To exclude uncertainties introduced by relative gravimetry (e.g. via the measured vertical gravity gradient) into the absolute gravimetric results, the final absolute values are all related to a common height at 1.200 m above the reference mark at floor level, cf. Timmen (2003). Figures 5 and 6 show typical conditions for the absolute gravimetric field work in Fennoscandia.

Fig. 5 Station Skellefteå (Sweden) with an absolute gravity pier inside and permanent GPS antenna outside.

Fig. 6 Absolute gravimeter FG5-220 of IfE installed at station Östersund.

5 Measurement Accuracy of Absolute Gravimetry

Table 1 shows the results of the comparisons of the IfE absolute gravimeter FG5-220 with the FG5-221 of FGI, FG5-301 of BKG, and FG5-226 of UMB carried out during the absolute gravity campaigns in 2003 and 2004. The comparisons were made by simultaneous parallel registrations at stations Bad Homburg, Metsähovi (see Fig. 7), Vaasa, Onsala or double occupations (Onsala, Skellefteå, Trondheim, Trysil, Hønefoss, Ås). The decline of the observed g-values at the IfE reference station in Clausthal (Harz mountains) in 2003 can be explained by the decrease of the subsurface water content from spring to autumn 2003 due to a very dry season. The re-occupation of 4 sites by FG5-220 in August/September, observed already by FG5-301 in July, shows the most deviating result for Hønefoss. In March 2004 the IfE group (FG5-220) visited the new gravity laboratory of UMB in Ås to provide a reference for the new FG5-226. The result obtained with the new instrument one month later differs by 22 nm/s^2. A nearly simultaneous comparison of the FG5-220 to FG5-221 (FGI) in Vaasa in May 2004 yielded a difference of 14 nm/s^2. Comparisons at Onsala (after the campaigns in 2004) in October with three absolute gravimeters agreed within 20 nm/s^2.

The overall discrepancy (rms) of the comparisons of about ±20 to 30 nm/s^2 proves the high accuracy of the absolute gravimetric survey of Fennoscandia. The mean values show that no significant offset of the IfE instrument can be assumed compared with the other three absolute gravimeters.

6 Summary and Conclusions

As a short status report, the ongoing project may be described as follows:

Fig. 7 Parallel measurements with FG5-220 (IfE) and FG5-221 (FGI) at station Metsähovi in Finland.

Table 1. Comparisons of the IfE absolute gravimeter FG5-220 with FG5-221 (FGI), FG5-226 (UMB), and FG5-301 (BKG), and with Clausthal reference value (mean in 2003). The difference is calulated as g(FG5-220) minus g(i).

Date	Station	FG5-220 compared with	Diff. [nm/s^2]	Remark
Jan 03	Clausthal	Mean 2003	27	
Feb 03	Bad Hb. AA	FG5-301	25	Simult.
Feb 03	Bad Hb. BA	FG5-301	-4	Simult.
May 03	Clausthal	Mean 2003	8	
Jun+Jul 03	Onsala	FG5-301	11	
Jun 03	Clausthal	Mean 2003	-9	
Aug+Jul 03	Skellefteå	FG5-301	-30	
Sep+Jun 03	Trondheim	FG5-301	-42	
Sep+Jun 03	Trysil	FG5-301	-2	
Sep+Jun 03	Hønefoss	FG5-301	-48	
Aug 03	Metsäh. AB	FG5-221	5	Simult.
Aug 03	Metsäh. AC	FG5-221	-37	Simult.
Aug 03	Vaasa AA	FG5-221	-32	Simult.
Aug 03	Vaasa AB	FG5-221	-32	Simult.
Oct 03	Clausthal	Mean 2003	-26	
Nov 03	Bad Hb. AA	FG5-301	38	Simult.
Nov 03	Bad Hb. BA	FG5-301	36	Simult.
		2003 Mean	-7	
		rms	±28	
Mar+Apr04	Ås	FG5-226	22	
May 04	Vaasa AB	FG5-221	14	
May 04	Metsäh. AB	FG5-221	-3	Simult.
May 04	Metsäh. AC	FG5-221	44	Simult.
Oct 04	Onsala AS	FG5-226	-23	Simult.
Oct 04	Onsala AS	FG5-221	3	Simult.
Oct 04	Onsala AN	FG5-226	-14	Simult.
Oct 04	Onsala AN	FG5-221	15	Simult.
		2004 Mean	7	
		rms	±21	
		Total Mean	-2	
		rms	±26	

- The whole uplift network includes more than 30 absolute gravity stations located beside or in the vicinity of permanent GPS points. More than 20 stations will be occupied annually over a period of five years. In the first two years, absolute gravity determinations were performed at 22 stations in 2003 and 24 stations in 2004, employing 4 FG5 absolute gravimeters.
- Results derived from surveys with the IfE absolute gravimeter FG5-220 have been compared with results from the other participating instruments and with results from repetition measurements at the IfE reference station in Clausthal. The comparison leads to an rms discrepancy of approx. ±20 to 30 nm/s^2.
- The check of the instruments by parallel and reference measurements is essential. Especially for geodynamics projects more than one absolute gravimeter should be employed to increase the reliability of the results and to detect instrumental offsets. This procedure improves the absolute accuracy of the whole network.

- An overall accuracy of better than ±30 nm/s^2 is indicated for a single absolute gravimeter and a single station determination. This agrees well with the project aim to provide observed gravity changes in Fennoscandia with a sufficient accuracy to support the GRACE data evaluation.

Acknowledgements

The research is supported by Deutsche Forschungsgemeinschaft (DFG, German Research Foundation), through the research grant Mu 1141/3-1 and 3-2.

References

Boedecker, G., Fritzer, Th. (1986): International Absolute Gravity Basestation Network. Veröff. Bayer. Komm. für die Internat. Erdmessung der Bayer. Akad. d. Wissensch., Astron.-Geod. Arb. 47, München.

Carter, W.E., Aubrey, D.G., Baker, T. et al.(1989): Geodetic fixing of tide gauge bench marks. Woods Hole Oceanographic Institution Report WHOI-89-31/CRC-89-5, Woods Hole, Mass., U.S.A.

Ekman, M. (1996): A consistent map of the postglacial uplift of Fennoscandia. Terra Nova 8, 158-165.

Ekman, M., Mäkinen, J. (1996): Recent postglacial rebound, gravity change and mantle flow in Fennoscandia. Geophys. J. Int. 126: 229-234.

Hotine, M.(1969): Mathematical Geodesy. ESSA Monograph 2, U.S. Dept. of Commerce, Washington, D.C.

Johansson, J.M., Davis, J.L., Scherneck, H.-G., Milne, G.A., Vermeer, M., Mitrovica, J.X., Bennett, R.A., Jonsson, B., Elgered, G., Elósegui, P., Koivula, H., Poutanen, M., Rönnäng, B.O., Shapiro, I.I. (2002): Continuous GPS measurements of postglacial adjustment in Fennoscandia, 1. geodetic results. J. Geophys. Res. 107, B8, ETG 3: 1-27.

Kaufmann, G., Wu, P. Li, G. (2000): Glacial isostatic adjustment in Fennoscandia for a laterally heterogeneous Earth. Geophys. J. Int. 143: 262-273.

Lambeck, K., Smither, C., Johnston, P. (1998): Sea level-change, glacial rebound and mantle viscosity for northern Europe. Geophys. J. Int. 134: 102-144.

Milne, G.A., Davis, J.L., Mitrovica, J.X., Scherneck, H.-G., Johannson, J.M., Vermeer, M., Koivuly, H. (2001): Space-geodetic constraints on glacial isostatic adjustment in Fennoscandia. Science 291: 2381-2385.

Nakiboglu, S.M., Lambeck, K. (1991): Secular sea-level change. In: Sabadini, R. et al. (eds.) Glacial isostasy, sea level and mantle rheology. Kluwer Acad. Publ, pp 237-258.

Niebauer T.M., Sasagava G.S., Faller J.E., Hilt R., Klopping F. (1995): A new generation of absolute gravimeters. Metrologia 32: 159-180.

Scherneck, H.-G., Johansson, J., Koivula, H., van Dam, T., Davis, J. (2003): Vertical crustal motion observed in the BIFROST project. J. of Geodyn. 35: 425-441.

Tapley, B.D., Bettadpur, S., Ries, J.C., Thompson, P.F., Watkins, M.M. (2004): GRACE measurements of mass variability in the Earth system. Science 305: 503-505.

Timmen, L. (2003): Precise definition of the effective measurement height of free-fall absolute gravimeters. Metrologia 40: 62-65.

Wahr, J., Velicogna, I. (2003): What might GRACE contribute to studies of post glacial rebound. Space Science Reviews 108: 319-330.

Wiehl, M., Dietrich, R., Lehmann, A. (2005): How Baltic sea water mass variations mask the postglacial rebound signal in CHAMP and GRACE gravity field solutions. In: Reigber, Ch., Lühr, H., Schwintzer, P., Wickert, J. (eds.) Earth observation with CHAMP. Springer-Verlag Berlin Heidelberg, pp 181-186.

Williams, S.D.P., Baker. T.F., Jeffries, G. (2001): Absolute gravity measurements at UK tide gauges. Geophys. Res. Letters 28: 2317-2329.

Wilmes, H., Falk, R., Klopping, F., Roland, E., Lothhammer, A., Reinhold, A., Richter, B., Plag, H.-P., Mäkinen, J. (2004): Long-term gravity variations in Scandinavia from repeated absolute gravity measurements in the period 1991 to 2003, proceedings (CD publ.) of the IAG symposium Gravity, Geoid and Space Missions, Porto.

A Model of Plate Motions

F. Riguzzi
Istituto Nazionale di Geofisica e Vulcanologia, Rome, Italy
University *La Sapienza*, Rome, Italy

M. Crespi, M. Cuffaro, C. Doglioni, F. Giannone
University *La Sapienza*, Rome, Italy

Abstract. The wide use of space geodesy techniques devoted to geophysical and geodynamical purposes has recently evidenced some limitations due to the intrinsic Terrestrial Reference Frame (TRF) definition. Current TRFs are defined under hypotheses suited to overcome the rank deficiency of the observations with respect to the parameters that have to be estimated, i.e. coordinates and velocities (Dermanis, 2001; Dermanis, 2002).

From a geodetic point of view, one possibility implies the application of the no-net-rotation condition (NNR). One of the main geophysical consequences due to the application of this condition is that it allows only accurate estimations of relative motions, whilst other motions of geodynamical interest, for instance with respect to the inner layers of the Earth body, are not determinable.

The main purpose of this paper is to propose a unified way to describe plate motions, overcoming the problems introduced by the NNR condition, in order to establish a new reference frame useful for geodynamical applications too.

Since we believe relevant the role played by global tectonics inferences, we introduce the concept of the *main tectonic sinusoid* to propose an analytical description of the plate motions flow, which is polarized to the "west" in the hotspot reference frame.

Keywords. Terrestrial Reference Frame, *main tectonic sinusoid*, westward drift, plate kinematics

1 Introduction

The most updated information on present plate motions is based on space geodesy data (Heflin et al., 2004), where motions are essentially estimated from GPS continuous observations in a no-net-rotation frame (NNR), as assumed by the International Terrestrial Reference Frame (Altamimi et al., 2002 a, b).

It is useful to recall that the TRF origin may be sensed by geodetic techniques; this is realized in the geocenter, being well defined by SLR.

The scale is metric, but depends on the speed of light, because the observing sites and the targets in the space are linked by electromagnetic signals. However, TRF orientation cannot be sensed by any geodetic technique, so that it is conventionally defined at a starting epoch and its time evolution is ensured by imposing the NNR condition over the whole Earth.

This condition is currently applied by aligning the TRF (Altamimi et al., 2002) to the NNR-NUVEL-1A model (De Mets et al., 1990; Argus and Gordon, 1991; De Mets et al., 1994), to try to guarantee its co-rotation with the Earth surface.

The practical TRF realization consists of a set of coordinates and velocities of the observing sites at a given epoch.

Figure 1 shows the current ITRF2000 velocities provided by JPL (Heflin et al., 2004), according to which, for what stated above, only accurate relative plate motions are defined.

However, current realization of this condition involves some problems:

- it theoretically requires a whole integral over the Earth, the so called Tisserand condition, but space geodesy observed sites are discrete and quite far from optimally distributed,
- it prevents from actually describing the Earth plate motions, for instance w.r.t the underlying mantle, that may be considered a fundamental geodynamical task

The aim of this paper is to give a first attempt to define an alternative plate motion model, useful for geodynamical tasks, on the basis of the main global tectonic features.

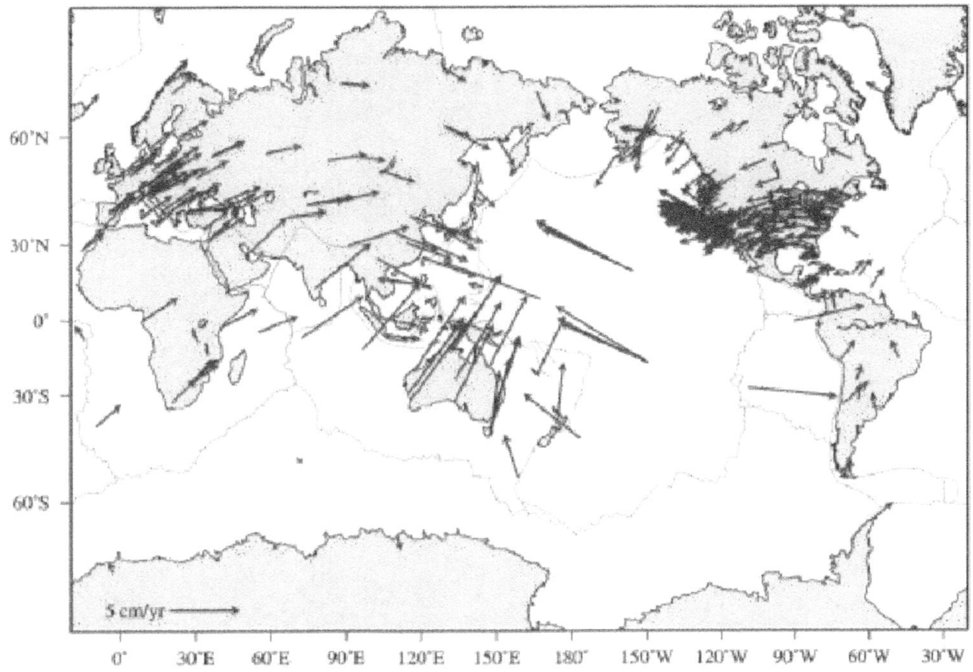

Fig. 1 Current ITRF2000 velocities (Heflin et al., 2004)

2 Tectonic mainstream

In order to establish the geological constraints for the definition of the analytical model, let us consider the first order tectonic structures along the boundaries of six large plates of Earth (Pacific, Nazca, South America, Africa, Arabia-India, and Eurasia), reported in Figure 2: the East Pacific Rise (1), the Mid Atlantic ridge (2), and the Red Sea - Indian ridge (3), for extensional margins, and the western Pacific subduction zones (4), the western northern and southern Americas Cordilleras (5), and the Alpine-Himalayas system (6) for convergent margins.

In the extensional tectonic settings, we assume that transform faults are parallel to the relative plate motions, whereas in convergent settings, the relative plate motions are constrained by the dominant trend of folds and thrusts, where no significant transpressive tectonics occurs. Analyzing the relative motions across these tectonic structures crossing the whole lithosphere, it appears that all the lithospheric plates do not move randomly, but follow a global mainstream, with a sinusoidal shape.

The tectonic mainstream (Doglioni, 1990; Doglioni, 1993) can be described here as an imaginary line named the *main tectonic sinusoid*, with a great undulation from east Africa to the western Pacific.

There are independent evidences of a "westward" drift of the lithosphere with respect to the mantle based on geophysical and geological evidences (Bostrom, 1971; Ricard et al., 1991; O'Connell et al., 1991; Doglioni et al., 1999; 2003).

The westward drift is then polarizing the tectonic mainstream, i.e., plates move, although at different velocities, toward the "west" with respect to the underlying mantle.

The tectonic mainstream may be described as a series of flow lines representing the main plate motion trajectories (Figure 3). The *main tectonic sinusoid* is the line roughly in the middle of the flow where the velocity toward the "west" is maximum within the plates crossed by the sinusoid.

The aim of this work is to give an analytical representation of the proposed *main tectonic sinusoid* useful to describe the plate motions with respect to the mantle.

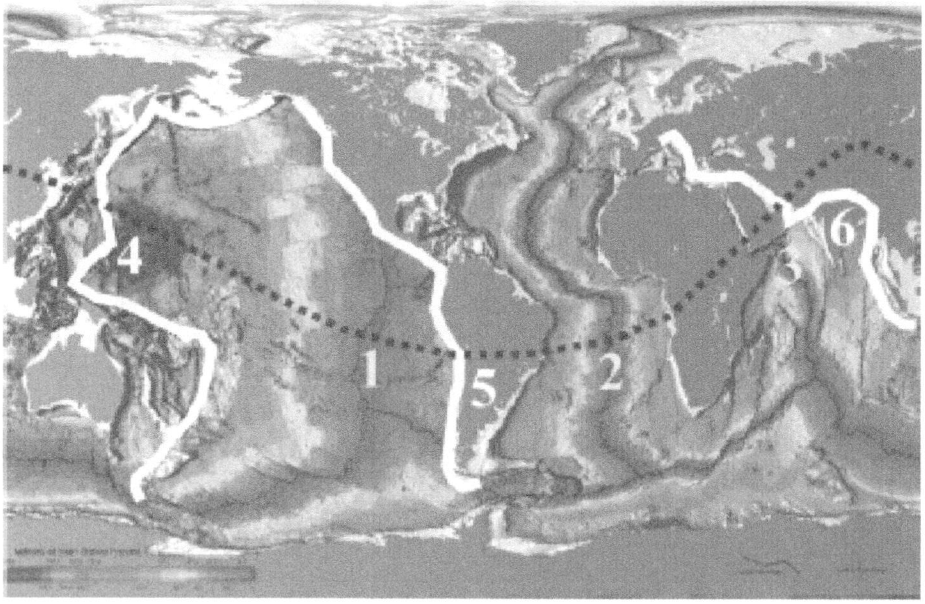

Fig. 2 Construction of a *main tectonic sinusoid*, starting from the Pacific motion direction and linking all the other relative motions in a global circuit using first order tectonic features such as the East Pacific Rise (1), the Atlantic rift (2), the Red Sea, the Indian Ocean rift (3) for the rift zones, and the west Pacific subduction (4), the Andean subduction (5), and the Zagros-Himalayas subduction (6) for convergent margins; base map, Age of the Ocean Floor, World Data Center-A for Marine Geology and Geophysics Report MGG-12, 1996, National Geophysical Data Center, US.

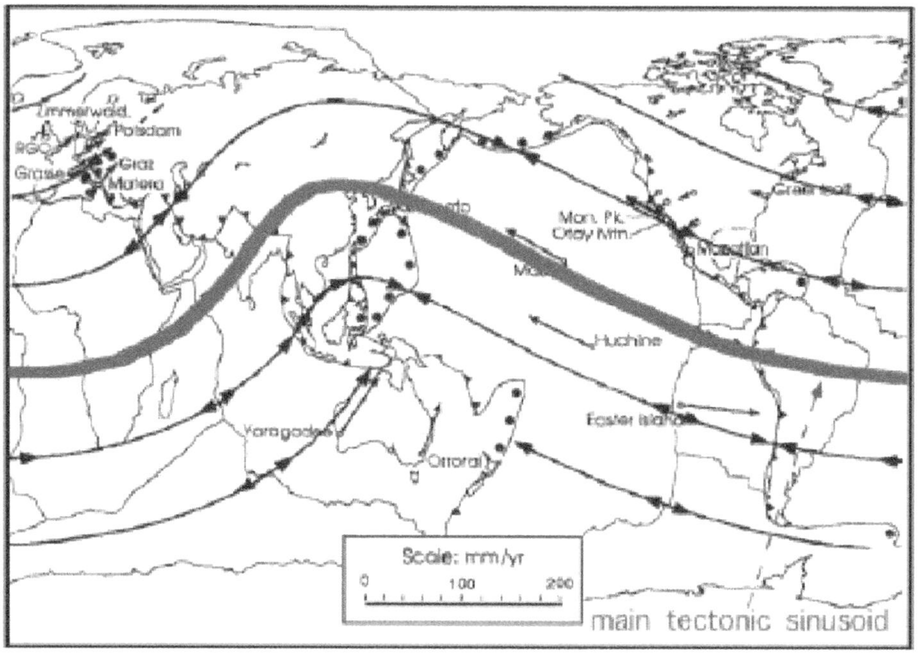

Fig. 3 The flow lines represent the mainstream of plate motions. Since the lithosphere has a net "westward" rotation, the underlying mantle is relatively moving "eastward" .

This can be done starting from a 3rd order Fourier series in geographic coordinates (φ, λ), whose 7 coefficients have to be estimated taking into account the aforementioned geological evidences

$$\varphi(\lambda) = \frac{a_0}{2} + \sum_{i=1}^{3}(a_i \cos(i\lambda) + b_i \sin(i\lambda))$$

The parameters (unknowns) of this equation are: the seven coefficients of the Fourier series (global parameters) and the ratios between the rotation components for each plate (local parameters).

The basic assumptions for the estimation are the following: spherical approximation, plate motions modeled as 3D rotations; moreover for each plate it is required:

- orthogonality between the plane of the eulerian equator and the rotation axes, where the eulerian equator is the mean plane of the main tectonic sinusoid within each plate crossed by the main tectonic sinusoid itself
- velocity horizontal with maximum intensity along the eulerian equator (Figure 4).

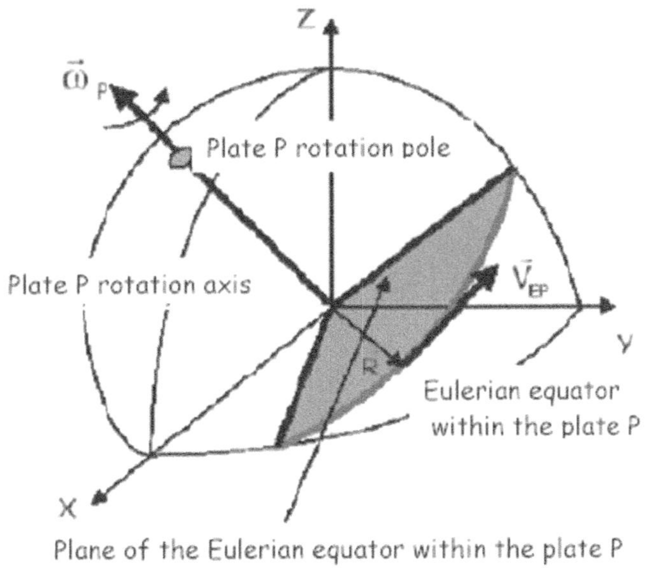

Fig. 4 The kinematic condition under the spherical approximation

This implies that

$$\omega_{XPL} X + \omega_{YPL} Y + \omega_{ZPL} Z = 0$$

The requested orthogonality between the rotation axis and the eulerian equator (kinematic condition) leads to the first equation of the *main tectonic sinusoid* useful for its analytical representation. This is obtained by equating the kinematic condition to the analytical representation:

$$\frac{a_0}{2} + \sum_{i=1}^{3}(a_i \cos(i\lambda) + b_i \sin(i\lambda)) = -\arctan\frac{\omega_{XPL}\cos\lambda + \omega_{YPL}\sin\lambda}{\omega_{ZPL}}$$

The second equation is derived from a geological condition: the direction of the tectonic line must be equal to the mean azimuth α of the direction of motion across the largest tectonic features (Searle, 1986; Gordon, 1995).

This implies to equate the first derivative of the analytical expression of the *tectonic mainstream* to a quantity that depends, on the tangent of the mean azimuth.

$$\sum_{i=1}^{3}[-ia_i \sin(i\lambda) + ib_i \cos(i\lambda)] = \frac{\cos\varphi}{\tan\alpha}$$

For this purpose, we selected the directions (Table1) of some principal tectonic structures (Figure 5), used for the analytical representation of the global tectonic pattern: the Mid Atlantic Ridge (MAR), the Red Sea and East Africa Rift (RSEAR), the Japan Subduction (JS), the Hawaiian sea-mount chain (HH), the East Pacific Rise (EPR) and the Andean Subduction (AS).

Table 1 Azimuth of the selected tectonic features

	λ (rad)	φ (rad)	α (deg)
1 MAR	6.021	-0.262	80 ± 3
2 RSEAR	1.047	0.524	36 ± 3
3 JS	2.583	0.785	301 ± 5
4 HH	3.578	0.349	293 ± 5
5 EPR	4.538	0.000	279 ± 3
6 AS	4.800	0.000	90 ± 5

We introduce these geological conditions into the equation system.

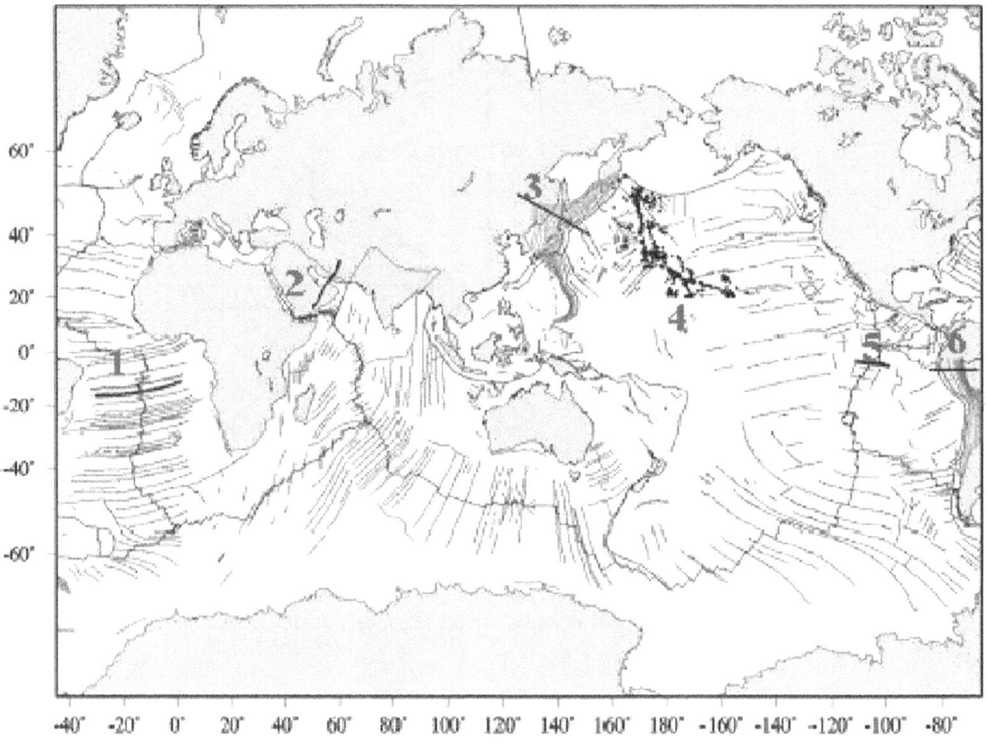

Fig. 5
Map of the main tectonic features selected to introduce azimuthal constraints into the *main tectonic sinusoid* estimation
1=MAR; 2=RSEAR; 3=JS ; 4=HH ; 5=EPR ; 6= AS

The over-determined system is composed by

• a first block of equations, written for points located along the tectonic mainstream, suitably chosen in order to guarantee parameters estimability and reliability for each plate crossed by the main tectonic sinusoid

• a second block of equations, written for the aforementioned tectonic features

The parameters are estimated according to the least squares principle, in an iterative fashion, due to the non-linearity of the equation system, starting from their approximated values.

The final precision of the *main tectonic sinusoid* after the least squares estimation is $\hat{\sigma}_0 = 0.2\ rad$, which corresponds to about 1000 km of uncertainty along the N-S direction.

3 The Hot Spot Reference Frame (HSRF)

The estimated *main tectonic sinusoid*, defined by the coefficients a_i and b_i, is now suited to establish a new reference frame, according to which it is possible to represent plate motions.

The kinematics of the plates crossed by the *main tectonic sinusoid* can be defined by assigning a starting value of tangential velocity along a sector of the line.

In particular, we focus our attention on the Pacific plate, the fastest plate displaying HS stability with respect to the mantle (Gripp and Gordon, 2002) .

Fixing the Hawaiian HS velocity with respect to the mantle under two different hypotheses and introducing the information on relative plate motions, it is possible to define the rotation components of each plate crossed by the *main tectonic sinusoid* and the velocities of sites located on each of the plates.

This is done by assigning

- a mean velocity $V_{eq,PA}$ to the Pacific plate
- a mean relative velocity between the crossed plates $V_{eq,rel}$

We computed the relative velocities $V_{eq,rel}$ between plates along the *main tectonic sinusoid* from the APKIM2000.0 model, the most recent plate kinematic model incorporating space geodesy observations (Drewes and Meisel, 2003).

In this way, the motions of the plates crossed by the *main tectonic sinusoid* can be defined with respect to the mantle.

In Figure 6, we represent the Pacific velocity under four different reference frame choices. In all the scenarios the relative velocity across the EPR has the same value of 11 cm/yr.

In the first and second case we have the plate motions represented in relative fashion. More in detail, the first is with respect to the Pacific plate, the second attains the well-known NNR solution. In the last two cases, on the contrary, we represent the motion with respect to the HS reference frame.

The idea is based on the fact that the Hawaiian volcanic track indicates that there is a decoupling between the magma source and the lithosphere, which is moving WNW.

We consider two different scenarios for the Pacific HS: the first hypothesizes a deep source located in the mantle, so that the track records the entire shear between lithosphere and mantle and, according to Gripp and Gordon (2002), it reaches the value of 10 cm/yr.

If the source is shallower, for instance located in the middle asthenosphere, the track cannot account for the entire shear between lithosphere and mantle, so there is a missing part of motion.

In this case, considering only the Hawaiian HS, according to Doglioni et al. (2005), the entire motion could reach 20 cm/yr.

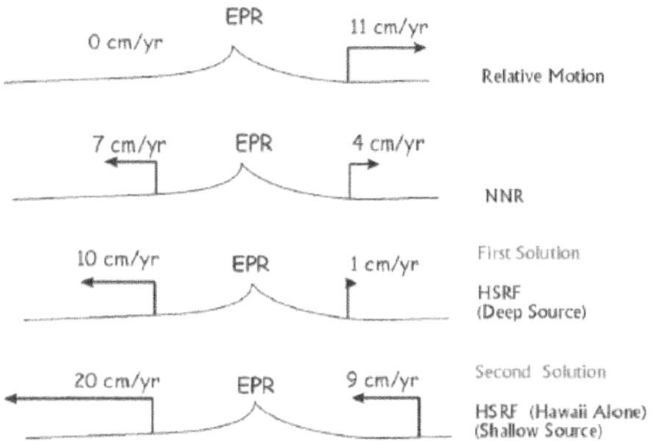

Fig. 6 Pacific plate motion and situation across the EPR. From the top to the bottom: Pacific plate fixed; classical ITRF solution; HSRF deep source; HSRF shallow source

4 Preliminary kinematic model

Under these hypotheses about the Pacific plate velocity, we jointly estimated the rotation poles (Table 2) and the angular velocities (Tables 3 and 4) of the six plates (Pacific-PA, Nazca-NZ, South America- SA, Africa-AF, Arabia-AR, Eurasia-EU) crossed by the *main tectonic sinusoid*, without the corresponding uncertainties, as preliminary proof.

The least square estimation was performed starting from some approximated values of the parameters, i.e. the rotation components from APKIM2000 model (Drewes and Meisel, 2003) and the *main tectonic sinusoid* coefficients from the *twelve ordinates* method, selecting an appropriate number of points in agreement with the geological evidences (Von Karman and Biot, 1951), since the equation system is not linear.

Table 2 Estimated rotation poles

PLATE	φ °N	λ °E
PA	-58.3	149.8
NZ	-77.4	168.7
SA	-75.1	172.3
AF	-67.0	110.4
AR	-43.8	118.0
EU	-37.3	124.7

Table 3 Estimated angular velocities if $V_{eq,PA}$ =10 cm/yr

PLATE	ω °Myr^{-1}	ω_x °Myr^{-1}	ω_y ° Myr^{-1}	ω_z ° Myr^{-1}
PA	0.93	-0.42	0.24	-0.79
NZ	-0.14	0.03	-0.01	0.14
SA	0.38	-0.10	0.01	-0.36
AF	0.19	-0.03	0.07	-0.18
AR	0.02	-0.01	0.01	-0.02
EU	0.16	-0.07	0.10	-0.10

Table 4 Estimated angular velocities if $V_{eq,PA}$ =20 cm/yr

PLATE	ω °Myr^{-1}	ω_x °Myr^{-1}	ω_y ° Myr^{-1}	ω_z ° Myr^{-1}
PA	1.80	-0.82	0.47	-1.53
NZ	0.73	-0.16	0.03	-0.71
SA	1.25	-0.32	0.04	-1.21
AF	1.06	-0.14	0.39	-0.98
AR	0.89	-0.30	0.57	-0.62
EU	1.03	-0.47	0.67	-0.63

We applied the estimated kinematic model to some GPS sites located on the plates crossed by the *main tectonic sinusoid*, to present the estimated global pattern.

Figures 7 and 8 show their velocities with respect to the underlying mantle, under the two different HSRF hypotheses.

Both the solutions confirm the presence of a global mainstream, a coherent undulated flow toward the "west", along which plates move at different velocities. It has to be noted that only under the first hypothesis the Nazca plate remains counterflow.

Concerning the relative motions between plates resulting after the least square estimation of the new plate kinematic parameters, they remain in agreement with their initial approximated values, given by the APKIM2000 model (Drewes and Meisel, 2003), if estimation errors are taken into account.

5 Conclusions and future perspectives

On the basis of geological evidences the concept of *main tectonic sinusoid* is, for the first time, introduced in order to describe the plate motions with respect to the mantle.

Two basic equations, which must be satisfied by the *main tectonic sinusoid* are derived and two preliminar solutions based on different kinematic hypotheses are computed.

The final precision in the *main tectonic sinusoid* definition is $\hat{\sigma}_0 = 0.2\ rad$, which corresponds to about 1000 km of uncertainty in N-S direction.

For the future, the motions of the plates not crossed by the *main tectonic sinusoid* must be computed too, in order to derive a complete description of plate motions.

As a final speculation, the *main tectonic sinusoid* is tilted 25-30° with respect to the Earth's equator, but close to the ecliptic plane of the Earth's revolution plane, and within the band of oscillation of the Moon's revolution (Fig. 9). This evidence might support an astronomical origin of the global ordered tectonic pattern observed on Earth.

Acknowledgements

Discussions with D. Boccaletti, E. Bonatti, M. Caputo, E. Carminati, F. Innocenti and B. Scoppola were very fruitful. Critical reading by M. Bevis was very constructive.

Maps were created by the Generic Mapping Tool (Wessel & Smith, 1995).

A special thank is due to J. Garate and G. Panza which reviewed this manuscript and to A. J. Gil Cruz and colleagues to organize this special issue.

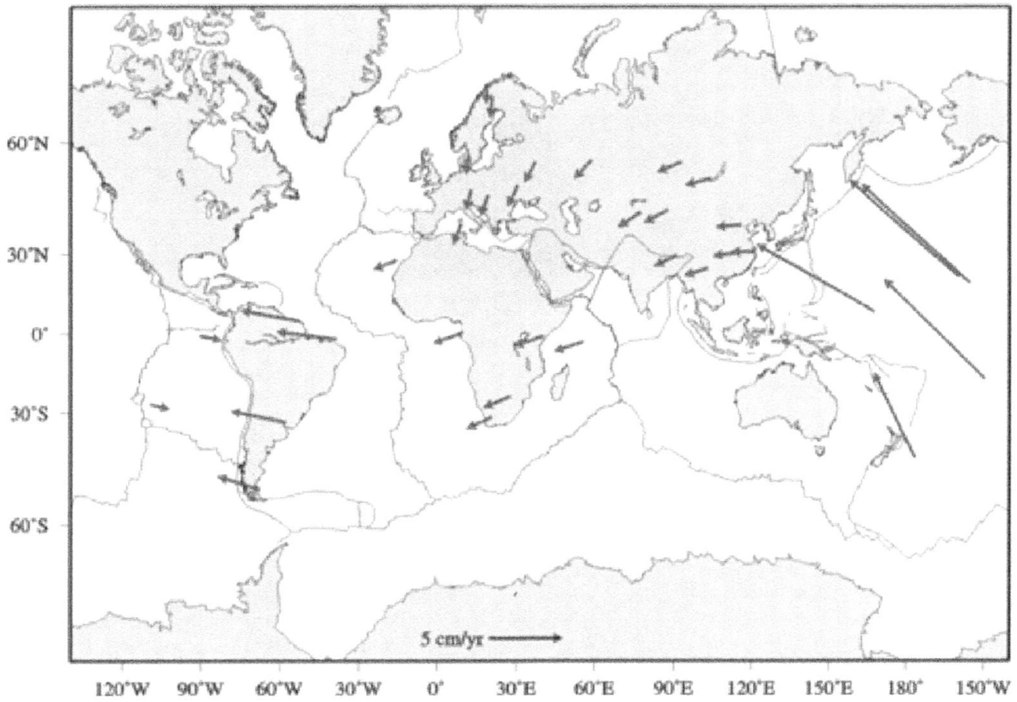

Fig. 7 Plate motion under the first HSRF hypothesis

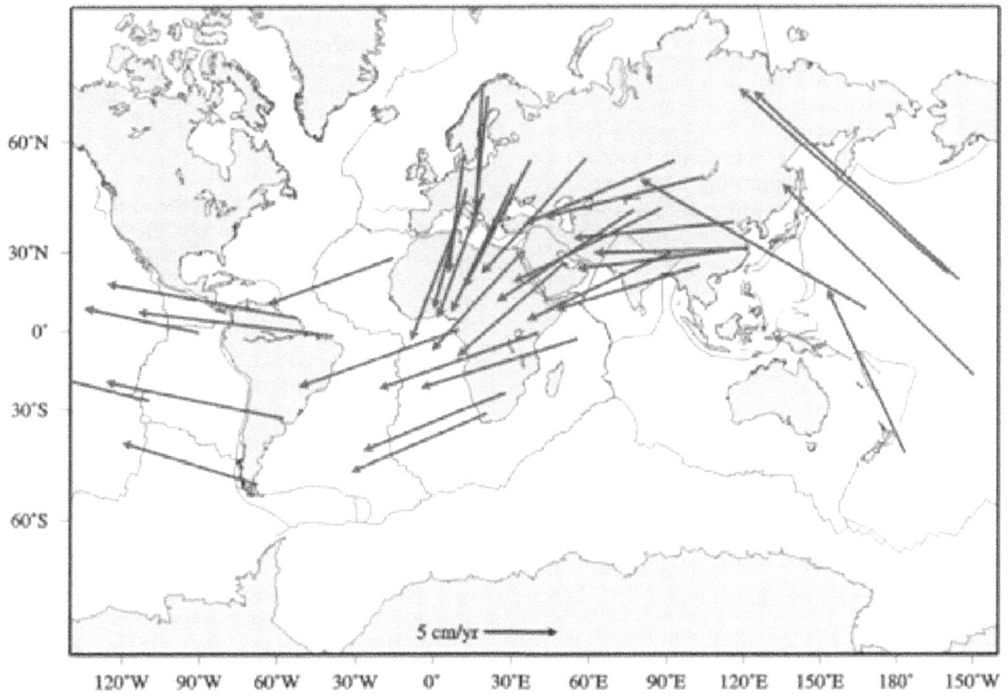

Fig. 8 Plate motion under the second HSRF hypothesis

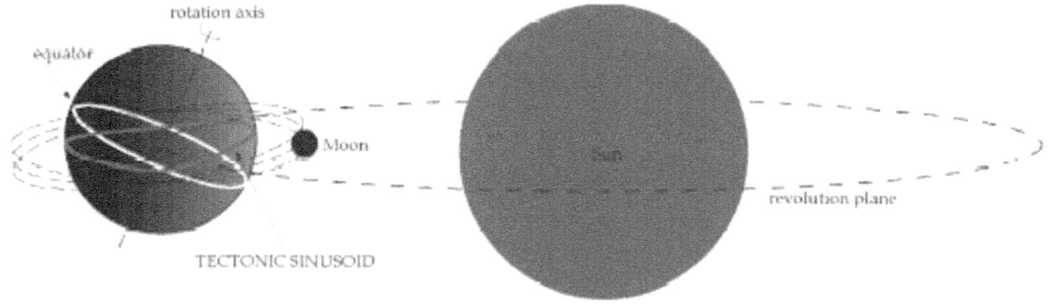

Fig. 9 Cartoon, not to scale, showing how the main tectonic sinusoid falls close to the ecliptic plane, and within the band of oscillation of the Moon's revolution, suggesting a rotational origin of the tectonic flow pattern.

References

Altamimi Z., Sillard, P., Boucher, C. (2002) ITRF2000: A new release of the International Terrestrial Reference Frame for earth science applications, J. Geophys. Res., 107, B10, 2214, doi:10.1029/2001JB000561.

Altamimi Z., Sillard, P., Boucher, C. (2002) The impact of No-Net-Rotation Condition on ITRF2000, Geophys. Res. Lett., 30, 2, 1064, doi:10.1029/2002GL016279.

Bostrom R.C. (1971) Westward displacement of the lithosphere, Nature, 234, 356-538.

Bostrom R.C. (2000) Tectonic consequences of the Earth's rotation, Oxford University Press.

DeMets C., Gordon, R.G., Argus, F., Stein, S. (1990) Current plate motions, Geophys. J. Int., 101, 425-478.

Argus D.F, R.G. Gordon (1991) No-net-rotation model of current plate velocities incorporating plate motion model NUVEL-1 Geophys. Res. Lett., 18, 2039-2042.

DeMets C., R.G. Gordon, D.F. Argus, S. Stein (1994) Effect of recent revisions to the geomagnetic reversal time scale on estimates of current plate motions, Geophys. Res. Letters, 21, 2191-2194.

Dermanis A. (2001) Global Reference Frames: Connecting Observation to Theory and Geodesy to Geophysics, IAG 2001 Scientific Assembly "*Vistas for Geodesy in the New Millenium*" 2-8 Sept. 2001, Budapest, Hungary. Budapest.

Dermanis A. (2002) The rank deficiency in estimation theory and the definition of reference frames, Proc. of V Hotine-Marussi Symposium on Mathematical Geodesy, F. Sanso` Edt., IAG Symposia, vol. 127, 145-156, Springer.

Doglioni C. (1990) The global tectonic pattern, J. Geodyn., 12, 21-38.

Doglioni C. (1993) Geological evidence for a global tectonic polarity, Journal of the Geological Society, London, 150, 991-1002.

Doglioni C., Harabaglia P., Merlini S., Mongelli F., Peccerillo A. Piromallo C. (1999) Orogens and slabs vs their direction of subduction, Earth Sci. Reviews, 45, 167-208

Doglioni C., Carminati E., Bonatti E. (2003) Rift asymmetry and continental uplift, Tectonics, 22, 3, 1024, doi:10.1029/2002TC001459.

Doglioni C., Green D., Mongelli F. (2005) On the shallow origin of hotspots and the westward drift of the lithosphere: in Plates, Plumes and Paradigms, G.R. Foulger, J.H. Natland, D.C. Presnall, and D.L. Anderson (Eds), GSA Sp. Paper 388, in press.

Drewes H., Meisel B. (2003) An Actual Plate Motion and Deformation Model as a Kinematic Terrestrial Reference System, Geotechnologien Science Report No. 3, 40-43, Potsdam.

Gordon R.G. (1995) Present plate motion and plate boundaries, Glob. Earth Phys., AGU Ref. S., 1, 66-87.

Gripp A.E., Gordon R.G. (2002) Young tracks of hotspots and current plate velocities, Geophys. J. Int., 150, 321-361

Heflin M. et al. (2004) http://sideshow.jpl.nasa.gov/mbh/series.html

O'Connell R., Gable C.G., Hager B. (1991) Toroidal-poloidal partitioning of lithospheric plate motions. In: Glacial Isostasy, Sea-Level and Mantle Rheology (R. Sabadini et al. Eds.), Kluwer Ac. Publ., 334, 535-551.

Ricard Y., Doglioni C., Sabadini R. (1991) Differential rotation between lithosphere and mantle: a consequence of lateral viscosity variations, J. Geophys. Res., 96, 8407-8415.

Searle R.C. (1986) GLORIA investigations of oceanic fracture zones: comparative study of the transform fault zone, J. Geol. Soc., London, 143, 743-756.

Von Kármán, T., and Biot , M. A., Metodi matematici nell'ingegneria, Einaudi edition, 1951, in Italian.

Wessel P., Smith W. H. F. (1995) *The Generic Mapping Tools (GMT) version 3.0.* Technical Reference & Cookbook,SOEST/NOAA.

Geodetic Control of the Present Tectonic Deformation of the Betic Cordillera (Spain)

P. Alfaro, A. Estévez
Dpto. Ciencias de la Tierra y Medio Ambiente, Universidad de Alicante, Apdo.99, San Vicente del Raspeig, Alicante, Spain

E.B. Blázquez
Dpto. Expresión Gráfica, Diseño y Proyectos, E.U.P., Universidad de Málaga. Campus El Ejido, s/n, 29071, Málaga, Spain

M.J. Borque, M.S. Garrido, A.J. Gil, M.C. Lacy, A.M. Ruiz
Dpto. de Ingeniería Cartográfica, Geodésica y Fotogrametría, Escuela Politécnica Superior de Jaén, Campus Las Lagunillas, s/n, Edif. A3, Universidad de Jaén, 23071 Jaén, Spain

J. Giménez
Servei d'Estudis i Planificació, D.G. Recursos Hídrics, Conselleria de Medi Ambient, Govern de les Illes Balears, Palma de Mallorca, Spain

S. Molina
Dpto. Ciencias del Mar y Biología Aplicada
Universidad de Alicante, Apdo 99, San Vicente del Raspeig, Alicante, Spain

G. Rodríguez-Caderot
Dpto. Astronomía y Geodesia, Facultad de Ciencias Matemáticas, Plz. de Ciencias, 3. Universidad Complutense de Madrid, Spain

M. Ruiz-Morales
Instituto Geográfico Nacional, c/. Divina Pastora, 7, 9, 18012 Granada, Spain

C. Sanz de Galdeano
Instituto Andaluz de Ciencias de la Tierra, CSIC-Universidad de Granada, 18071 Granada, Spain

Abstract. The Betic Cordillera, located in the south of the Iberian Peninsula, is currently subjected to NW-SE compression and NE-SW extension. Until a few years ago, there were hardly any geodetic studies to quantify the tectonic deformation taking place in the Betic Cordillera. In 1999, in the framework of an interdisciplinary research project to monitor the deformation, several geodetic studies were initiated by the Universities of Jaén, Granada and Alicante. These studies were centred on the basins of Granada and the Bajo Segura (Alicante), sectors in which the last two destructive earthquakes in the Iberian Peninsula took place. This paper describes the geodynamic context of these study areas and the main features of the geodetic studies, focusing on the deformation analysis of the three observation surveys carried out in 1999, 2000 and 2001 over a non-permanent GPS network established in the Granada Basin. The results agree with the low rates of deformation previously calculated from geological data. To detect such small rates, observation over a longer time span will be needed.

Keywords. Active Tectonic, GPS, levelling, crustal deformation monitoring, Granada Basin, Betic Cordillera.

1 Introduction

The Betic Cordillera, located to the south of the Iberian Peninsula, stands in an area of convergence where the African and Eurasian plates are approaching each other at a rate of 5 mm/year (Argus et al. 1989). The African-Eurasian plate boundary is a complex deformation zone featuring continental collision, subduction, volcanism and wide-spread earthquake activity. Fernandes et al. (2003) studied the relative motion between Africa and Eurasia. From a geodynamic point of view, mention must be made of the fact that the referred 5 mm/year convergence rate is accommodated in a broad deformation area that includes the North of Africa (Rif, Tell and Kabilias), the Alboran Sea and the Betic Cordillera. The present-day crustal deformation of the Iberian Peninsula, the Maghreb, and the adjacent offshore regions in the Mediterranean and the Atlantic is driven mainly by the NW-SE directed convergence along this part of the African-Eurasian plate boundary (Jiménez-Munt et al. 2001). Interference from the Iberian microplate as well as the coexistence of compressional and extensional tectonics complicate the understanding of the collision, and many aspects of the tetonic structure and development of the region remain a mater of debate (Stich et al. 2003).

Until a few years ago, there were hardly any geodetic studies to quantify the tectonic deformation taking place in the Betic Cordillera. In 1984, the Instituto Geográfico Nacional set up the "Red Geodinámica del Estrecho de Gibraltar". In 1997, the Instituto Cartográfico de Cataluña and the University of Barcelona installed the CuaTeNeo network (Castellote et al. 2000; Gárate et al. 2003). The purpose of this network was to monitor the deformation of the Eastern Betic Shear Zone in the provinces of Almería and Murcia. In 2001, the University of Barcelona made an estimation of the vertical movements of the eastern sector of the Cordillera using high-precision levelling profiles (Giménez et al. 2000; Giménez 2001; Giménez et al. 2004). Recently, radar interferometry (InSAR) has also been applied to the Mula earthquake in 1999, with an Mb magnitude of 4.8 (Martínez Díaz et al. 2002).

In 1999, several collaboration initiatives were implemented between the Universities of Jaén, Granada and Alicante, with the aim of carrying out several studies to monitor the deformation in the basins of Granada and the Bajo Segura (Alicante) (Figure 1), sectors in which the last two destructive earthquakes in the Iberian Peninsula were located: The Torrevieja earthquake of 1829 and the earthquake of Andalusia of 1884. Gil et al. (2002) and Ruiz et al. (2003) include preliminary results of the geodetic studies in the Granada Basin, such as Alfaro et al. (2000) and Giménez et al. (2004) do in the Bajo Segura Basin.

This contribution offers a general description of the referred geodetic studies carried out to quantify the tectonic deformation in the Betic Cordillera. In particular, this paper focuses on the analysis of geometrical detection of spatial deformation derived from three observation campaigns of a non-permanent GPS network established in the Granada Basin.

In section 2, a geodynamical description of the area is presented. The methodology followed to design the non-permanent GPS network as well as the GPS data processing are shown in section 3. Section 4 deals with the statistical testing used to compare the coordinate differences. Finally, the conclusions are presented.

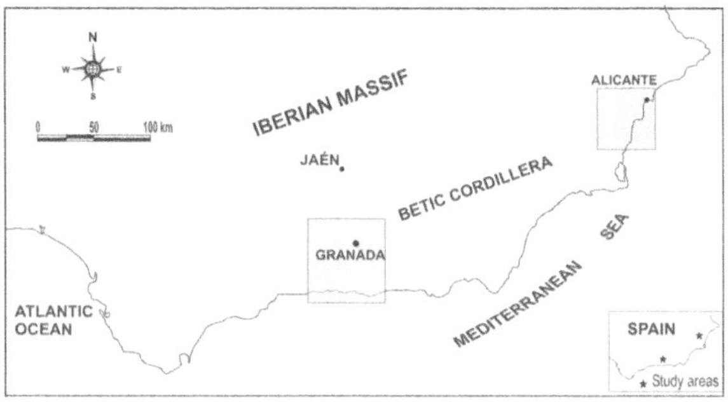

Fig. 1 Regional map of the South Iberian Peninsula showing the location of the study areas.

2 Geodynamic context of the study areas

Regional studies of paleostress indicate that the Betic Cordillera has undergone an almost N-S compression from the Late Miocene period to the Present (Montenat and Ott d'Estevou 1990; Sanz de Galdeano 1990; Galindo-Zaldívar et al. 1993; Herraiz et al. 2000). This compression has been accompanied by a major ENE-WSW extension (Galindo-Zaldívar et al. 1999; Sanz de Galdeano and López-Garrido 2000), mainly in the central sector of the Betic Cordillera.

2.1 Granada Basin

This basin is currently subject to NW-SE compression and a linked ENE-WSW extension. The NW-SE compression has produced a shortening of the basement and the Upper Miocene-Quaternary sedimentary cover. The E-W Sierra Nevada antiform is one of the most noticeable structures in which this shortening takes place.

On the other hand, the NE-SW extension has been accommodated by normal faults which are mainly concentrated on the NE part of the basin. Gil et al. (2002), using geodetic techniques where several NW-SE active normal faults crop out, calculated a minimum extensional rate of 0.15-0.30 mm/year.

Moreover, the most outstanding vertical movements related to active folding and normal faults are located in this central sector of the Betic Cordillera (Sanz de Galdeano and Alfaro 2004). At present, the Granada Basin and surrounding areas are rising at different rates, which vary between 0.4 and 0.6 mm/year.

In order to control the NW-SE shortening, the ENE-WSW extension and the relative uplift of the basin and surrounding areas, a GPS non permanent network, comprising 15 control sites distributed throughout the whole basin (Figure 2), was installed in the Granada basin. Three observation campaigns were carried out in 1999, 2000 and 2001.

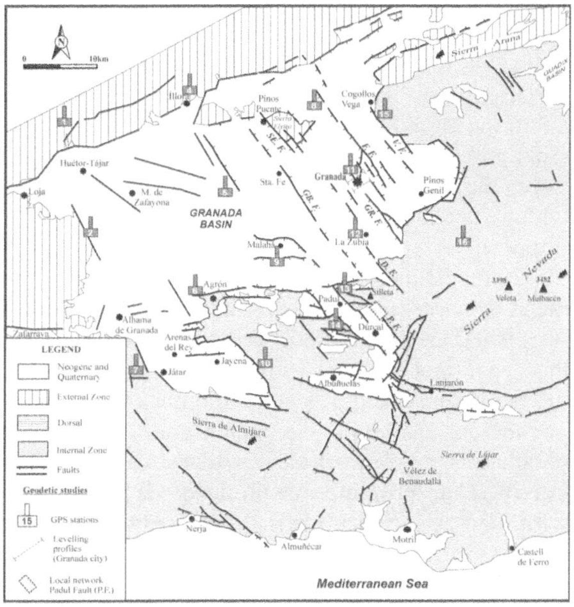

Fig. 2 Geological map of the Granada Basin showing location of the geodetic studies: GPS control points, triangulation-trilateration network of the Padul Fault and two levelling profiles in the city of Granada.

In addition, detailed geodetic studies have been performed in order to quantify the present deformation of the Padul and Granada faults (Ruiz et al. 2003). Along the NW-SE Padul fault, which shows widespread geomorphological evidence of recent activity in the area, an eight-station Triangulation-Trilateration network was established, and two levelling profiles, approximately two kilometres long each, were developed in order to measure vertical movements in the Granada fault. One of them (Genil Profile) has fourteen benchmarks, and the other (Víznar Profile) sixteen (Fig. 2).

2.2 Bajo Segura Basin

The Bajo Segura Basin is located in the northern terminal splay of the Eastern Betic Shear Zone (Silva et al. 1993), also known as the Betic segment of the Trans-Alboran Shear Zone (De Larouzière et al. 1988). This area has functioned as a transpressive zone since, at least, Plio-Pleistocene times (Silva et al. 1993), under a NW-SE/N-S compression (Montenat et al. 1990; Alfaro 1995).

This compression produces a shortening in the sedimentary cover and its basement which shows active folding and faulting. The active folds, which accommodate this NW-SE/N-S shortening, have an average E-W strike. In addition, this shortening is also accommodated by ENE-WSW reverse faults (Crevillente and Bajo Segura), as well as NW-SE right-lateral faults (Guardamar, Torrevieja and San Miguel de Salinas).

At present, vertical movements also occur in the study area related to growth anticlines and synclines. Zazo et al. (1993) established mean rates of uplift and subsidence for the last 100 kyr varying between +5.2 and 3.0 cm/kyr.

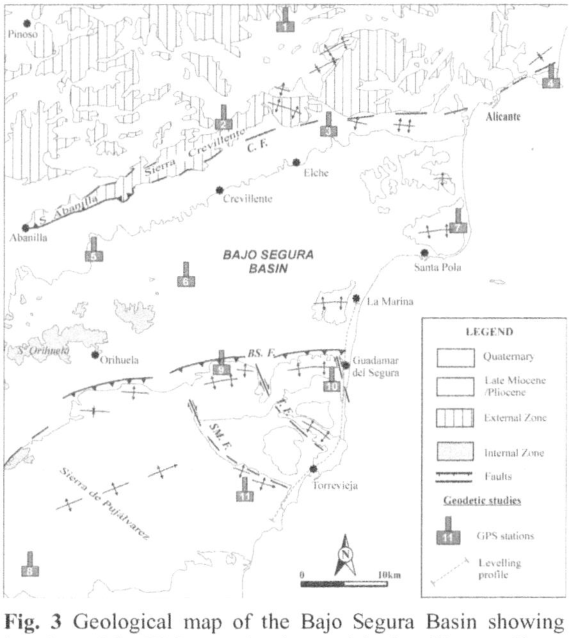

Fig. 3 Geological map of the Bajo Segura Basin showing location of the GPS control points and the levelling profile.

In order to quantify the shortening of the Bajo Segura Basin and the associated regional vertical movements, a GPS network with 11 control sites was installed in the basin and surrounding areas (Fig. 3). This was measured in 1999, 2000 and 2001. Furthermore, in 2003 a measurement campaign was carried out along a 30 km high-precision levelling line in order to quantify vertical movements related to active folding in the active ENE-WSW Bajo Segura fault and several other active faults such as the NW-SE Guardamar, Torrevieja and San Miguel de Salinas. The data collected during this campaign have been compared with those of 1976 and 1997. A result of this comparison is that the inferred vertical movements do not exceed the rate of 0.2 mm/year in the last 25 year interval (Giménez et al. 2004).

3 GPS monitoring of the Granada Basin

With the aim of quantifying the deformation that is currently occurring in the Granada Basin, the first non-permanent GPS network of this area was outlined (Fig. 2). The design of the network as well as the individual site selection required consideration of geoscientific, logistical and observational aspects (Klotz and Lelgemann 1999). Since a geodynamic network was considered as a very long-term project the monumentation should guarantee reoccupation after a period of time. This network was made up of fifteen reinforced concrete pillars anchored to rock. Moreover, to assure that the antennas would be placed exactly at the same position in different reoccupations, the pillars had an embedded forced centring system. Nine of the fifteen monitoring points were located over the basement rocks in the External and Internal Zones of the Cordillera that surround the Granada Basin. The remaining six sites were located within the basin, over rocks of various ages. Figure 2 shows the geological aspects of the Granada Basin and GPS sites. The sites #5, #6, #9, #11 and #12 were located in the centre of the basin, over the "relative" subsidence sectors of the Granada Basin. The others were distributed over the uplifted relief which surrounds the basin. Sites #13 and #14 were closer in order to establish a local control of the Padul normal fault, one of the most active and spectacular faults of the basin. Site #16, the highest control point, was located in Sierra Nevada, over the Internal Zone basement in the area with apparent maximum uplift. All sites met the following requirements: no obstruction above 15 degrees; no high power lines nearby and easy access. New sites #22 and #25 were built in a restricted area for the second observation campaign after sites #12 and #15 were vandalised destroyed. The first survey was carried out from February 27th to March 7th, 1999, the second from June 18th to June 25th, 2000 and the third from June 30th to July 7th, 2001. The precise IGS orbits were in ITRF96 during the first GPS survey and in ITRF97 during the other two observation periods. The GPS constellation was tracked for eight-hour sessions over baseline lengths ranging from 10 km to 55 km. The equipment used was three Leica SR399 and two Leica SR9500 dual frequency carrier phase GPS receivers which belong to the University of Jaén. The network was link to the permanent sites of Villafranca (Madrid), Alicante and San Fernando (Cádiz), from which the coordinates of the network's central point were computed. GPS data processing was performed using Bernese 4.2 software (Beutler et al. 2000) computing single sessions in multibaseline mode. The final solution for each session was obtained using the iono-free observable with precise ephemeris (http://igscb.jpl.nasa.gov/) and antenna phase centre variation files. The coordinates of the stations, apart from the fixed point (#55) and relative integer ambiguities, were estimated in three campaigns using the QIF method. The Saastamoinen model, with a standard atmosphere, was used to model the troposphere effect. The troposphere parameters were not estimated because in this case, the root mean square errors of coordinates and residuals were smaller than those of a data process in which these parameters were estimated. This means that the final adjustment procedure did not contain clock parameters, orbital elements and atmosphere parameters. From these results we used an intermediate program to produce GPS baselines with their covariance matrices. Using the NETGPS program (Crespi 1996) that performs the adjustment of GPS baselines accounting for their full covariance matrices, the minimal-constrained network adjustment was performed. This adjustment suddenly raised the problem of the incorrect stochastic model and showed the necessity of its reestimation, taking into account the reliability of the network. It is well known that the longer the baselines estimated from GPS observation are, the greater the estimate biases due to some possible mismodelling effects affecting these observations (e.g. residual troposphere effect, residual ionosphere effect). Therefore, since GPS phase model may not be completely correct, the covariance matrix of GPS baselines is often underestimated. Therefore, this covariance matrix cannot be a correct stochastic model for the GPS network adjustment and it needs to be reestimated a posteriori. The appropriate software was subsequently developed. The reestimation procedure caused some changes both in coordinates and in their covariance matrix, whose estimates are biased when wrong stochastic models are used. The discrepancies in the covariance matrix are particularly relevant with respect to the subsequent significance analysis of coordinate differences. Moreover, internal and external reliability are remarkably underestimated. Table 1 shows some statistical data of the minimum constrained network adjustment in the 1999, 2000 and 2001 campaigns. Tables 2, 3 and 4 show the geodetic coordinates from these adjustments. From these tables it is clear that the mean precision in height is remarkably worse than in horizontal position. This difference is due to the weakness of the GPS height determination.

4 Testing the coordinates differences

Usually the measurement errors are assumed to be normal (Gaussian) and statistically uncorrelated from one another (white noise). However several geodetic data sets have provided evidence for error sources that introduce temporal correlations into the data. This is why some authors use as a power-law process as a model for describing many types of geophysical signal (which may contribute to the noise), (Williams, 2003).

In our case, the definition of the statistical tests to compare two GPS networks is based on the assumption that the observations, cartesian components of the GPS baselines, the coordinate and their differences are normally distributed The significance of the coordinate differences is tested by the well-known statistic:

$$\frac{\delta\hat{x}^T Q_{\delta\delta}^{-1} \delta\hat{x}}{m\hat{\sigma}_o^2} = F_{sp}, \tag{1}$$

where $F_{sp} \approx F_{m,(r_1+r_2),\alpha}$, if $H_0 : \hat{x}_1 = \hat{x}_2$ is true; m being the number of tested parameters and r_1, r_2 the redundancies of the first and second adjustment, respectively. If

$$F_{sp} \leq F_{m,(r_1+r_2),\alpha} \tag{2}$$

the null hypothesis H_0 is accepted and the coordinates show no significant changes at significance level α. On the contrary, if

$$F_{sp} > F_{m,(r_1+r_2),\alpha} \tag{3}$$

the null hypothesis H_0 is rejected and it is possible to conclude that the coordinates have changed significantly. To identify a point whose coordinates have changed it is necessary to test each point separately (Crespi 1996). To establish the final datum, i.e. to find the stable points, an iterative procedure is used. In the first step, each point is tested to define a starting datum including only the point i for which the statistic (1) is minimum. In the second step, the differences of the coordinates and their covariance matrix are calculated in the current datum. Finally, a new test on each point is carried out to choose another point j to be included in the datum. The procedure continues anew in the second step and stops when the datum cannot be enlarged.

It is important to stress that a fixed common point in the campaigns is needed in order to compare the point network. In this case the point called #55 was the common point. It is necessary to map the station coordinates in the same reference frame, so a previous transformation is required. There are different strategies to achieve this purpose. We have chosen to estimate the reference frame translations and rotations in an LS adjustment before comparing the station coordinates. A translation between both adjusted networks to be checked was performed by estimating the translation vector between the respective coordinates of the point #55. After that, a spatial rotation between the two sets of coordinates taking into account their full covariance matrices was estimated to reduce the coordinates to the same reference frame before testing the statistical significance of the coordinate differences. The results are shown in Table 5. It is clear that no rotation is significant.

Table 1. Minimum constrained adjustment parameters (1999, 2000 and 2001 surveys). Ses: sessions, Eq: Number of equations, Unk: unknown parameters, Red: redundancy, σ_0^2: unit weight variance; χ^2_{exp}: experimental χ^2 with "redundancy" degrees of freedom; χ^2_{teo} theoretical χ^2 with "redundancy" degrees of freedom at the 99% confidence level; rms: average SQM values in mm, *Smaj*: semimajor axis of the 99% confidence ellipse in mm. *Smin*: semiminor axis at the 99% confidence level in mm; CI: 99% confidence height interval in mm.

Ses	Eq	Unk	Red	σ_0^2	χ^2_{exp}	χ^2_{teo}	rms X	rms Y	rms Z	S_{min}	S_{maj}	*CI*
8	78	36	42	1.25	65.9	66.21	7	3	5	10	17	20
8	93	42	42	1.02	52.76	77.39	6	3	4	9	20	18
8	84	36	48	1.23	72.86	73.68	8	4	6	11	23	23

Table 2. Cartesian Coordinates at epoch 1999.2 in ITRF96 [m].

Point	Latitude [° ′ ″]	rms [m]	Longitude [° ′ ″]	rms [m]	Height [m]	rms [m]
1	37 14 07.747643	0.003	-4 02 14.455796	0.002	734.503	0.006
2	37 05 55.47988	0.003	-4 00 56.155030	0.002	989.949	0.006
4	37 16 48.357057	0.003	-3 50 13.165428	0.002	964.831	0.006
6	37 01 29.647364	0.003	-3 52 19.545908	0.002	1221.186	0.006
7	36 56 08.961273	0.006	-3 56 14.032563	0.004	1135.688	0.011
8	37 15 28.572444	0.002	-3 37 16.447923	0.002	946.578	0.004
9	37 04 10.342209	0.004	-3 43 47.698904	0.003	908.391	0.008
10	36 56 44.209388	0.005	-3 41 50.498579	0.005	1553.464	0.010
11	37 11 36.740577	0.002	-3 35 17.182073	0.001	988.398	0.004
13	37 02 12.639935	0.005	-3 36 54.928532	0.004	1286.037	0.011
14	36 59 42.819484	0.005	-3 37 12.613944	0.004	943.273	0.010
16	37 07 07.488081	0.006	-3 25 35.963844	0.004	2253.056	0.011
55	37 09 26.861668	fixed	-3 50 16.088967		724.671	

Table 3. Cartesian Coordinates at epoch 2000.5 in ITRF97 [m].

Point	Latitude [° ′ ″]	rms [m]	Longitude [° ′ ″]	rms [m]	Height [m]	rms [m]
1	37 14 7.747325	0.004	- 4 2 14.454385	0.004	734.573	0.010
2	37 5 55.479859	0.002	- 4 0 56.154400	0.002	990.029	0.005
4	37 16 48.357497	0.002	-3 50 13.164522	0.001	964.921	0.003
6	37 1 29.647386	0.005	-3 52 19.545301	0.004	1221.247	0.010
7	36 56 8.961053	0.002	-3 56 14.031921	0.002	1135.777	0.005
8	37 15 28.572787	0.001	-3 37 16.447245	0.001	946.676	0.003
9	37 4 10.342449	0.001	-3 43 47.697877	0.001	908.467	0.003
10	36 56 44.209533	0.002	-3 41 50.498492	0.002	1553.549	0.005
11	37 11 36.740691	0.001	-3 35 17.181612	0.001	988.499	0.003
13	37 2 12.640021	0.007	-3 36 54.928003	0.006	1286.142	0.014
14	36 59 42.819609	0.002	-3 37 12.613264	0.002	943.392	0.005
16	37 7 7.488083	0.007	-3 25 35.962967	0.006	2253.190	0.014
22	37 0 45.497060	0.003	-3 33 52.017316	0.003	910.297	0.007
25	37 14 30.103725	0.004	-3 31 20.931064	0.003	1325.979	0.008
55	37 9 26.861663	fixed	-3 50 16.088394		724.751	

Table 4. Cartesian Coordinates at epoch 2001.5 in ITRF97 [m].

Point	Latitude [° ′ ″]	rms [m]	Longitude [° ′ ″]	rms [m]	Height [m]	rms [m]
1	37 14 7.748523	0.004	- 4 2 14.454543	0.003	734.627	0.008
2	37 5 55.480380	0.003	- 4 0 56.153672	0.003	990.031	0.007
4	37 16 48.357855	0.003	-3 50 13.163595	0.004	964.940	0.008
6	37 1 29.648028	0.003	-3 52 19.544409	0.003	1221.241	0.007
7	36 56 8.961727	0.004	-3 56 14.031410	0.004	1135.771	0.009
8	37 15 28.573236	0.003	-3 37 16.446399	0.003	946.692	0.007
9	37 4 10.342916	0.003	-3 43 47.697123	0.003	908.491	0.007
10	36 56 44.209889	0.005	-3 41 50.497844	0.005	1553.597	0.012
11	37 11 36.741258	0.003	-3 35 17.180654	0.003	988.495	0.006
13	37 2 12.640461	0.008	-3 36 54.927093	0.007	1286.155	0.018
14	36 59 42.819965	0.005	-3 37 12.612741	0.005	943.385	0.012
16	37 7 7.488678	0.008	-3 25 35.962755	0.008	2253.219	0.020
22	37 0 45.497931	0.004	-3 33 52.016455	0.004	910.258	0.010
25	37 14 30.104135	0.004	-3 31 20.930175	0.004	1325.993	0.010
55	37 9 26.862055	fixed	-3 50 16.087679		724.766	

Table 5. Rotation of the second network around the first one. ω: rotation angle around the axis *0X*; φ: rotation angle around the axis *0Y*; κ: rotation angle around the axis *0Z*. All are measured in clockwise direction.

Network	ω (rad)	rms (rad)	φ (rad)	rms (rad)	κ (rad)	rms (rad)
99-00	$-0.563 \cdot 10^{-6}$	$0.185 \cdot 10^{-6}$	$-0.226 \cdot 10^{-6}$	$0.381 \cdot 10^{-6}$	$0.621 \cdot 10^{-6}$	$0.242 \cdot 10^{-6}$
00-01	$0.168 \cdot 10^{-6}$	$0.212 \cdot 10^{-6}$	$-0.527 \cdot 10^{-6}$	$0.379 \cdot 10^{-6}$	$-0.209 \cdot 10^{-6}$	$0.277 \cdot 10^{-6}$
99-01	$-0.211 \cdot 10^{-6}$	$0.222 \cdot 10^{-6}$	$-0.123 \cdot 10^{-5}$	$0.503 \cdot 10^{-6}$	$0.121 \cdot 10^{-7}$	$0.301 \cdot 10^{-6}$

The DENETGPS program (Crespi 1996) was used to evaluate the significance of the coordinate differences from Tables 2 and 3, Tables 3 and 4 and Tables 2 and 4. No initial hypothesis on the points behaviour or any geophysical constraint was assumed. This assumption is equivalent to fixing the centroid of the whole network.

Horizontal displacement vectors of all common points belonging to the 1999 and 2000 GPS networks, 2000 and 2001 GPS networks and 1999 and 2001 GPS networks are shown in Figures 4, 5 and 6, respectively.

5 Conclusions

Geological and seismological data indicate that, at present, the Betic Cordillera is subjected to NW-SE shortening and a linked ENE-WSW extension. These studies also suggest low rates of deformation broadly distributed throughout the whole Cordillera. In order to quantify the current deformation of the Betic Cordillera some geodetic studies in the Bajo Segura and Granada Basin were established in 1999. The combination of these geodetic and geologic data will be essential in order to obtain the first short-term deformation rates in these areas of the Betic Cordllera. This paper focuses on the GPS works in the Granada Basin. From Figure 6, all the points show no significant coordinates changes at the 1% significance level in the period 1999-2001. This result agrees with the low rate of extension calculated from geological data.

In the Granada Basin, enough time has not yet elapsed for the detection of movements greater than the measurement errors. In order to detect such small rates, observations over a longer time span are needed, probably a decade. The present and future results will be very significant for a better understanding of the recent movements and in order to obtain slip rates of the main faults of the Granada Basin.

Acknowledgements. The authors gratefully acknowledge the support given by *Dirección General de Enseñanza Superior e Investigación Científica (Spain)* (Project PB97-1267-C03). This study was also partly financed by the Projects BTE2000-3337-E, BTE2003-1113 and by the Generalitat Valenciana (GRUPOS03/085, OCYT).

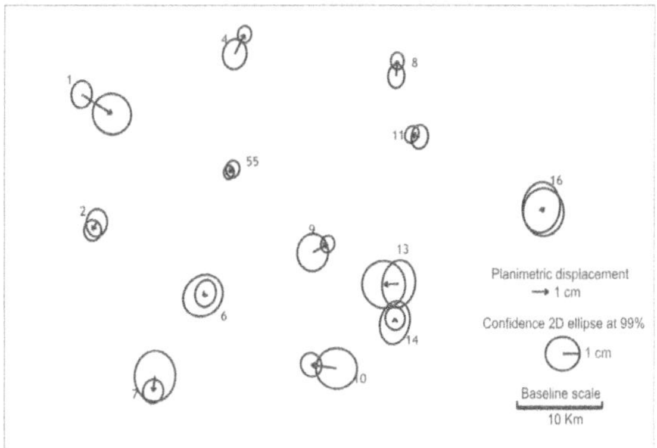

Fig. 4 Displacement vectors of all common points belonging to the 1999 and 2000 surveys in the frame of final datum and 99% reliability.

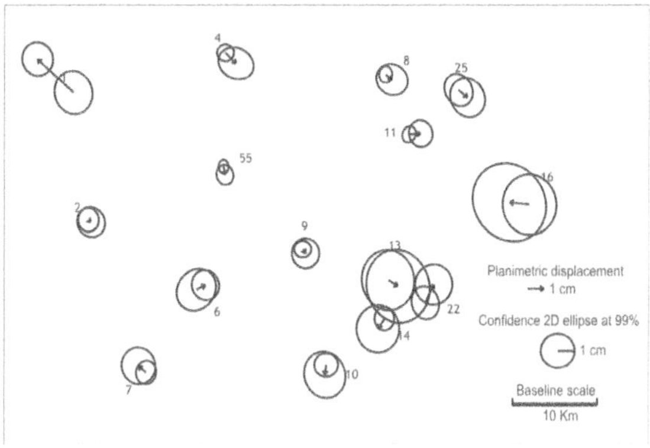

Fig. 5 Displacement vectors of all common points belonging to the 2000 and 2001 surveys in the frame of final datum and 99% reliability.

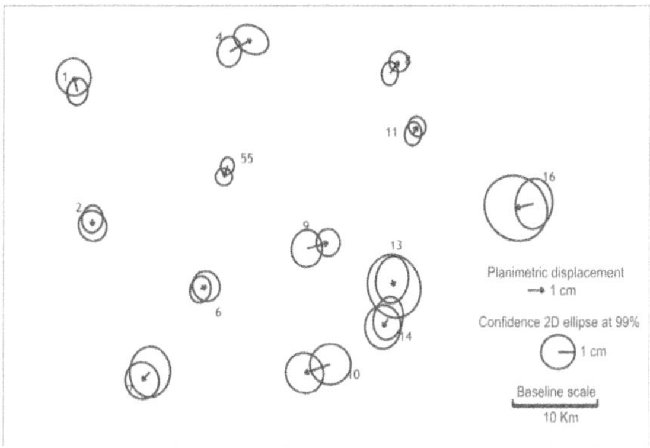

Fig. 6 Displacement vectors of all common points belonging to the 1999 and 2001 surveys in the frame of final datum and 99% reliability.

References

Alfaro, P. (1995). Neotectónica de la Cuenca del Bajo Segura (Cordillera Bética oriental). Ph.D. Thesis, University of Alicante, Spain, 211 pp.

Alfaro P., M.J. Borque, A. Estévez, A.J. Gil and S. Molina (2000). Red de control GPS de la Cuenca del Bajo Segura (Cordillera Bética oriental). *Geotemas*, 1, 4, 245-248.

Argus, D.F., R.G. Gordon, C. De Mets and S. Stein (1989). Closure of the Africa-Eurasia-North America plate motion circuit and tectonics of the Gloria Fault. *J. Geophys. Res.*, 94, 5585–5602.

Beutler, G., E. Brockmann, R. Dach, P. Fridez, W. Gurtner, U. Hugentobler, J. Johnson, L. Mervant, M. Rothacher, S. Schaer, T. Springer & R. Weber, 2000. *BERNESE GPS software version 4.2.* Astronomical Institute University of Berna.

Castellote, M., X. Goula, E. Suriñach, J. Talaya and A. Térmens (2000). Red GPS CuaTeNeo en el SE de la Península Ibérica con fines geodinámicos. In: *Proc. of 2ª Asamblea Hispano-Portuguesa de Geodesia y Geofísica*. S01-24, 63-64. Lagos (Portugal).

Crespi, M. (1996). Software Package for the Adjustment and the Analysis of GPS Control Networks. In: M. Unguendoli (ed.), *Reports on Survey and Geodesy in memoria of Proff. A. Gubellini and G. Folloni*. Edizione Nautilus, Bologna, 237–264.

De Larouzière, F.D., Bolze, J. Bordet, P., Hernández, J., Montenat, C. and P. Ott d'Estevou (1988). The Betic segment of the lithospheric trans-Alboran shear zone during the late Miocene. *Tectonophysics*, 152, 41-52.

Fernandes, R.M.S., B.A.C. Ambrosius, R. Noomen, L. Bastos, M.J.R. Wortel, W. Spakman and R, Govers (2003) The relative motion between Africa and Eurasia as derived from ITRF2000 and GPS data. *Geophysical Research Letters*, 30, 16, 1828, SDE 1-1 - 1-5.

Galindo-Zaldívar, J., F. González Lodeiro and A. Jabaloy (1993). Stress and palaeostress in the Betic-Rif Cordilleras (Miocene to Present). *Tectonophysics*, 227, 105-126.

Galindo-Zaldívar J., A. Jabaloy, I. Serrano, J. Morales, F. González-Lodeiro and F. Torcal (1999). Recent and present-day stresses in the Granada Basin (Betic Cordilleras): Example of a late Miocene-present-day extensional Basin in a convergent plate boundary. *Tectonics*, 18, 686–702.

Gárate, J., J. Martín, E. Suriñach, M. Berrocoso, A. Pérez-Peña and J. Talaya (2003). Cuateneo network: preliminary results alter first reobservation. *Geophysical Research Abstract*, 5, 09549. European Geophysical Society.

Gil, A.J., G. Rodríguez-Caderot, M.C. Lacy, A.M. Ruiz, C. Sanz de Galdeano and P. Alfaro (2002). Establishment of a non-permanent GPS network to monitor the recent NE-SW deformation in the Granada Basin (Betic Cordillera, Southern Spain). *Studia Geophysica et Geodaetica*, 46, 395-410.

Giménez, J. (2001). Cuantificación de deformaciones recientes en el este de la Península Ibérica mediante la comparación de datos de precisión. *Acta Geológica Hispánica* 36, 1-2, 21-51.

Giménez, J., M.J. Borque, A. Gil, P. Alfaro, A. Estévez and E. Suriñach (2004). Nivelación de alta precisión a lo largo de la costa entre La Marina y Torrevieja (Alicante) para controlar movimientos tectónicos verticales. *Geo-Temas*, 7, 137-141.

Giménez, J., E. Suriñach and X. Goula (2000). Quantification of vertical movements in the eastern Betics (Spain) by comparing levelling data. *Tectonophysics*, 317, 3-4, 237-258.

Herraiz, M., G. De Vicente, R. Lindo-Ñaupari, J. Giner, J.L. Simón, J.M. González-Casado, O. Vadillo, M.A. Rodríguez-Pascua, J.I. Cicuéndez, A. Casas, L. Cabañas, P. Rincón, A.L. Cortés, M. Ramírez and M. Lucini (2000). The recent (Upper Miocene to Quaternary) and present tectonic stress distributions in the Iberian Peninsula. *Tectonics*, 19, 4, 762-786.

Jiménez-Munt, I. P. Bird and M. Fernández (2001) Thin-shell modeling of neotectonics in the Azores-Gibraltar region. *Geophysical Research Letters*, 28(6), 1083-1086.

Klotz, J. and D. Lelgemann (1999). Present State of the Central Anden GPS-Traverse ANSA. GPS Techniques Applied to Geodesy and Surveying. *Lectures Notes in Earth Sciences*, 19, 427–436.

Martínez-Díaz, J.J., A. Rigo, L. Louis, R. Capote, J.L. Hernández-Enrile, E. Carreño and M. Tsige (2002). Caracterización geológica y sismotectónica del terremoto de Mula (febrero de 1999, MB: 4,8) mediante la utilización de datos geológicos, sismológicos y de interferometría de RADAR (INSAR). *Boletín Geológico y Minero*, 113, 1, 23-33.

Montenat, C. and P. Ott d'Estevou (eds.). (1990). Les bassins néogènes du domaine bétique oriental (Espagne). *Documents et Travaux de l'IGAL*, Paris, 12-13, 392 p.

Montenat, C., P. Ott D'Estevou and G. Coppier (1990). Les bassins néogènes entre Alicante et Cartagena. In: *Les basins néogènes du domaines Bétique orientel (Espagne).* Doc. et Trav. IGAL, 12-13, 313-368.

Ruiz, A.M., G. Ferhat, P. Alfaro, C. Sanz de Galdeano, M.C. De Lacy, G. Rodríguez-Caderot and A.J. Gil (2003). Geodetic measurements of crustal deformation on NW-SE faults of the Betic Cordillera, Southern Spain, 1999-2001. *Journal of Geodynamics*, 35, 259-272.

Sanz de Galdeano, C. (1990). Geologic evolution of the Betic Cordillera in the Western Mediterranean, Miocene to Present. *Tectonophysics*, 172, 107-119.

Sanz de Galdeano, C. and P. Alfaro (2004). Tectonic significance of the present relief of the Betic Cordillera. *Geomorphology*, 63, 175-190.

Sanz de Galdeano, C. and A.C. López-Garrido (2000). Las fallas tortonienses a cuaternarias entre Granada y la costa: el límite occidental del nevado-filábride y de las unidades alpujárrides inferiores. *Rev. Soc. Geol. España*, 13(3-4), 519–528 (in Spanish).

Silva, P.G., J.L. Goy, L. Somoza, C. Zazo and T. Bardají (1993). Landscape response to strike-slip faulting linked to collisional settings: Quaternary tectonics and basin formation in the Eastern Betics, southern Spain. *Tectonophysics*, 224, 289-303.

Stich, D., C.J. Ammon and J. Morales (2003) Moment tensor solutions for small and moderate earthquakes in the Ibero-Magreb region. *Geophysical Research Letters*, 108, B3, 2148, ESE 7-1 – 7-20.

Williams, S.D.P. (2003) The effect of coloured noise on the uncertainties of rates estimated from geodetic time series. *Journal of Geodesy*, 76, 483-494.

Zazo, C., J.L. Goy, C.J. Dabrio, T. Bardají, L. Somoza, and P.G. Silva (1993). The Last Interglacial in the Mediterranean as a model for the present interglacial. *Global and Planetary Change*, 7, 109-117.

Horizontal Deformation Models for Deception Island (South Shetland Islands, Antarctica)

M. Berrocoso, M. E. Ramírez, A. Fernández
Laboratorio de Astronomía, Geodesia y Cartografía, Dep. Matemáticas, Universidad de Cádiz.
Campus Universitario Río San Pedro, s/n. 11510, Puerto Real, Cádiz (Spain)

Abstract. Deception Island (62.93°S, 60.57°W) is one of the few active volcanoes in the Antarctica, whose most recent eruptions took place in 1842, 1967, 1969 and 1970.

During the continuous Spanish campaigns in the Antarctica, several scientific groups have developed different investigation projects in order to control the deformation the island suffers as a result of its volcanic activity. With this purpose, a geodetic network has been designed and improved.
Nowadays, the network consists of twelve stations around Port Foster which are provided with WGS-84 geodetic coordinates with respect to the ITRF2000, and another station at the Spanish Base Juan Carlos I in Livingston Island. Time analysis of these coordinates will lead us to get the horizontal deformation model. On the other hand, a levelling network has been designed to obtain the vertical deformation model. This network is denser in those areas where the volcanic activity is stronger, as the Fumarole Bay and the Hill of Obsidians.

GPS, levelling and gravimetric measurements have been also collected in secondary points to obtain an experimental geoid model which make possible an adequate reference frame for physical applications.

Keywords: GPS surveying, GPS data processing, superficial deformation models.

1 Geographical and Geological Situation

The South Shetland Islands and the Antarctic Peninsula constitute two of the few active areas of volcanic and seismic activity of the Antarctica. Deception Island is situated at the spreading centre of the Bransfield Strait back-arc on the south western side of the Scotia Sea region. (Fig. 1) It is a young strato-volcano 25 km. large in its submerged basal diameter and 15 km. of diameter in its emerged part (Ortiz et al. 1992, Ortiz et al. 1997).

From 1987-88 several Spanish Groups have carried out different researches projects to study the volcanic activity in the island and its geodynamic consequences. After the increase of the seismic activity detected during 1998-99 austral summer (Ibañez et al. 2000), with more than 2000 earthquakes in two months, a cross-disciplinary study was carried out to evaluate the volcanological status in the island. (García A. and DECVOL working group 2002).

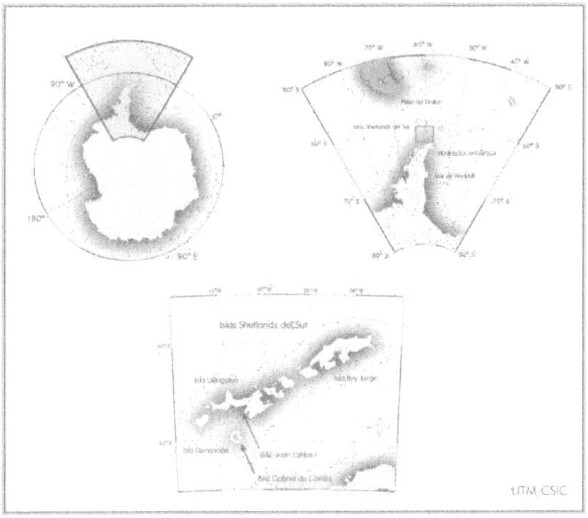

Fig. 1 Geographical situation of Deception Island.

This study was part of the DECVOL project, which took place in December 1999, with the aim of carrying out a reobservation of the geodetic network; the seismic register, the analysis of fumarolic gasses (Caselli et al. 2002), the acquisition of geomagnetic and gravimetric data, and thermometric measurements in hot areas. The volcanic crisis were modelled and monitored by means of geodetic and geophysical techniques.

The main goals of the geodetic researches which have been done in the island are the establishment of a local reference frame to get an accurate position of scientific data, the determination of a deformation model for the island and the detection and monitoring of the volcanic activity (Berrocoso et al. 2002).

In order to obtain a consistent reference frame, which makes possible the accurate position of the scientific data, as well as its deformation models from the detection and monitoring of its volcanic activity, it has been designed and established a geodetic network (REGID) and a levelling network (RENID). The horizontal deformation models are obtained from the processing of the data after the surveying of the geodetic network, while the vertical deformation models are provided by the observation of the levelling lines by means of geodetic and trigonometric levelling. The distribution of the geodetic stations and the levelling benchmarks in the island are shown in the figure 2.

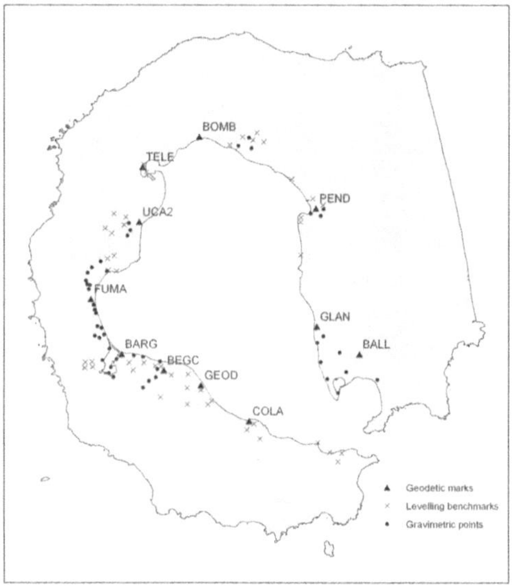

Fig. 2 REGID Geodetic Network. Leveling marks and points with gravimetric measurements are also shown.

On the other hand, GPS and levelling measurements were also collected in secondary points, as well as gravimetric measurements to obtain an experimental geoid (Berrooso et al. 2004), although we do not go deeply into it since that is not the scope of the present paper.

2 REGID Geodetic Network

Nowadays, REGID geodetic network consists of twelve stations in Deception Island and another station in Livingston Island, 20 km away. First surveying were carried out during the austral summer of 1988-89.

In a first stage the geodetic network consisted of four stations at Deception Island, in the Argentinian Base (BARG), in Fumarole Bay (FUMA); Pendulum Cove (PEND) and in Whalers' Bay (BALL), and a fifth one in Livingston Island, in the Spanish Antarctic Base Juan Carlos I. Some of these stations were built from existing structures, as the ones in Pendulum Cove, Whalers' Bay and the one at the Argentinian Antarctic Base.

Until 1995-96, the reference point of the network was the one at the Argentinian Base (BARG). During this campaign a new station was built near the Spanish Antarctic Base Gabriel de Castilla, which was set as the main station of the network (BEGC). By this time, the network consisted of five stations, which observed along the successive campaigns in 1995-96, 1998-99, and 1999-00.

These five stations provided a global vision of the deformation in the island, but the lack of stations in the North and South area made the models insufficient.

Seven more stations were built during the 2001-02 campaign to make the network denser, completing the number of stations up to the twelve current ones: COLA, GEODEC, UCA1, TELE, BOMB, CR70, and GLAN. The situation of these stations is shown in figure 2.

The new geodetic marks were well-rooted in the permafrost, with steel bars to fix them and a low height from the ground, making them more stable and involving a poor environmental impact. A standard screw was fixed in one of the corners of the post to let the use of geodetic instrumentation. The structure of the new vertices is shown in figure 3.

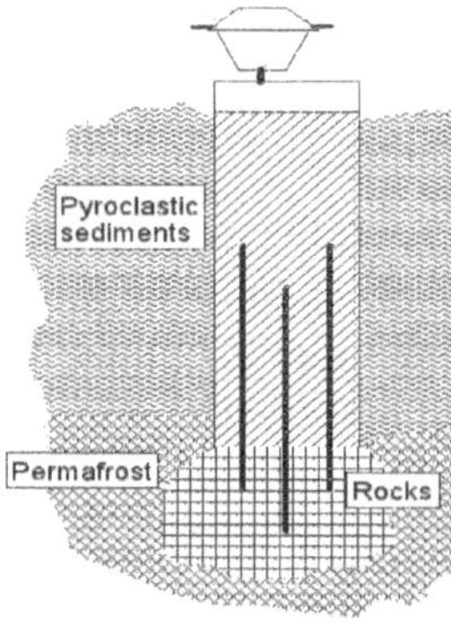

Fig. 3: Geodetic station monumentation.

Finally, in the 2003-04 campaign a station at Caleta Cierva (Antarctic Peninsula) was built with the aim of differentiating the deformation due to the volcanic and the tectonic activity.

The most commonly used instrumentation for the GPS surveying is constituted by TRIMBLE receivers, from its series 4000 and 5000, although ASHTECH and LEICA products have been also used. Data were processed with the GPSurvey software in a first stage and now they are being reprocessed using the BERNESE v4.2 software (Beutler et al. 2001), with accuracy of the order of the millimetre.

3 Obtaining the Horizontal Deformation Models

As it was mentioned above, available data from 1989 were processed in a first stage using the GPSurvey software. They have been reprocessed using the BERNESE V4.2 software.

The processing of GPS data requires available accurate orbits, good satellite geometry and suitable models for tropospheric and ionospheric effects, as well as the possible interference that can affect the GPS signal. Precise satellite orbits produced by the International GNSS Service were used, since they have the sufficient accuracy for the processing of a local network with the dimensions of the one in Deception Island. The corresponding Earth rotation parameter is also required.

Data were processed on a daily basis solution using the BERNESE GPS Software 4.2, which uses a double-differences processing strategy. Baseline or single-difference files were computed from a predefined baseline file, were the reference station is set to be the fixed one of every baseline. Ambiguities were resolved using the quasi-ionosphere-free QIF strategy.

The sampling rate of the data was 30 s., and observations below an elevation mask of 10° were rejected. An elevation-dependent weighting was also taken into account.

One tropospheric parameter per one and station is estimated, as the BERNESE GPS software recommends. Dry-Niell mapping function was used.

After the processing of the data a set of daily coordinates were got. Just the reference station was constrained. A coordinate per station and campaign was obtained after the adjustment of the daily coordinates.

Since medium distances between stations are up to 10 km, except OHIG station, in the Antarctic Peninsula (100 km away), and the station BEJC, in Livingston Island (20 km away), we have chosen the processing options suggested for the adjustment of local networks by Beutler et al. (2001). Precise ephemerides were used for the processing of the 1991/92 campaign (referred to the ITRF91), 95/96 (ITRF93), 99/00 (ITRF94), and 01/02 campaign (ITRF97).

To obtain the horizontal deformation models, the adjusted coordinates from the processing of the data of successive campaigns have been transformed to the ITRF00 reference system, according to the following expression

$$\begin{bmatrix} X_{yy} \\ Y_{yy} \\ Z_{yy} \end{bmatrix} = \begin{bmatrix} X \\ Y \\ Z \end{bmatrix} + \begin{bmatrix} T_1 \\ T_2 \\ T_3 \end{bmatrix} + \begin{bmatrix} D & -R_3 & R_2 \\ R_3 & D & -R_1 \\ -R_2 & R_1 & D \end{bmatrix} \cdot \begin{bmatrix} X \\ Y \\ Z \end{bmatrix}$$

where $[X, Y, Z]^t$, $[X_{yy}, Y_{yy}, Z_{yy}]^t$ and the station coordinates in the ITRF00 reference system (Altamimi et al. 2002) and in the other frame, respectively. The required parameters for the transformations were obtained from the ITRF web site (http://itrf.ensg.ign.fr/).

BARG station was considered as the reference station of REGID network and also the center of the local reference system to obtain the coordinates variations in the NS and EW directions.

To eliminate the local displacements it was also processed the baseline BARG-BEJC, which links the geodetic stations in Deception Island with Livingston Island. After the adjustment of the network with the ADDNEQ module that BERNESE GPS software provides, the coordinates of the remaining stations have been recalculated.

From the 2001/02 campaign, the reference station BARG was replaced by BEGC station, situated in the surroundings of the Spanish Base Gabriel de Castilla, a less active area in the island.

The displacement models were obtained by the comparison of the coordinates along the successive campaigns. Velocity fields have been also obtained from these results. To model the total area in the island an interpolation method was applied, based on a regular grid (10 x 10 km) covering the island. The existing deformation is constrained only to inner effects in the island, and no displacement is considered outside.

The interpolation criterion is based on a weight which depends on the distance from the grid point to the geodetic stations. For the points inner to the network, we consider the two nearest geodetic stations, whereas for outer points we constrain with two points of null deformation in its surroundings.

The increase in the volcanic activity during the 1998/99 campaign can be interpreted from the deformation model in figure 4b and its comparison to the one in figure 4a, which shows the deformation model in the island from 1991/92 to 1995/96 campaign.

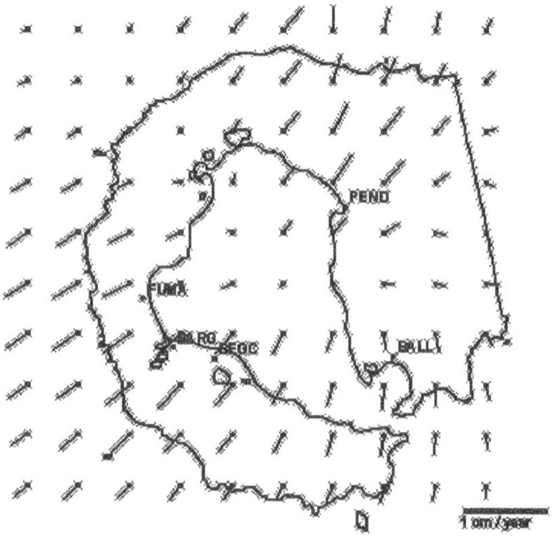

a

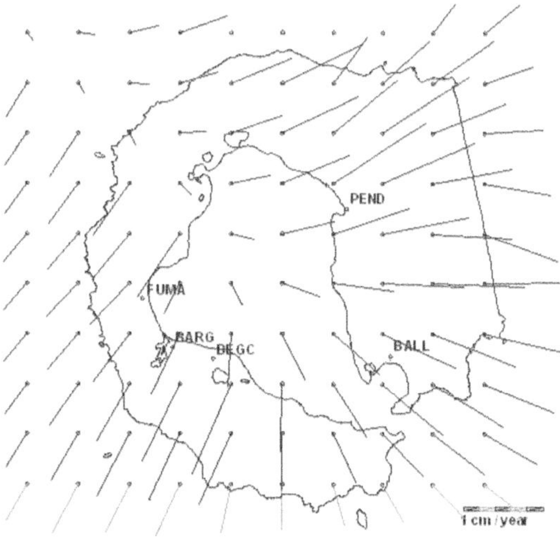

b

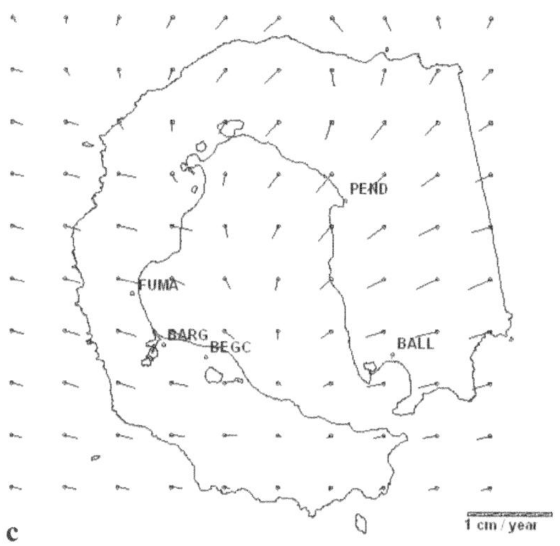

c

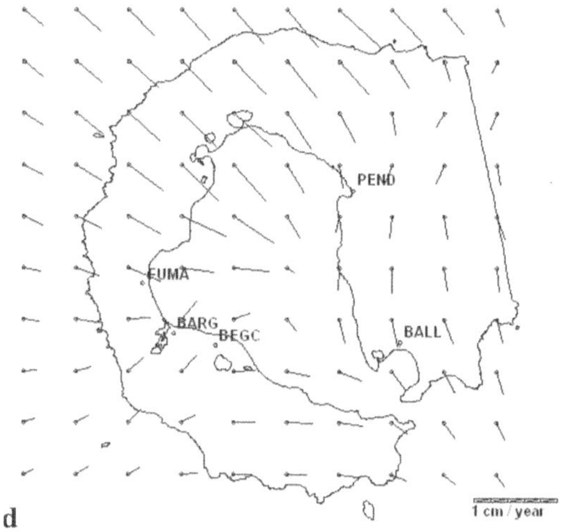

d

Fig. 4 Horizontal Displacement Models for Deception Island: a. from 1991/92 to 1995/96 campaign; b. from 1995/96 to 1999/00 campaign; c. from 1999/00 to 2001/02 campaign; d. from 2001/02 to 2002/03 campaign.

4 Future Works

We are now considering a new approach to the analysis of the GPS data: all the available data have been reprocessed using 30 min length sessions with the aim of detecting sudden changes in the signal within a campaign and the presence of short period variation.

Wavelets techniques are now being used to denoise the data. Times series will be analysed both with statistical classical methods and with concepts related to wavelet theory.

Velocity fields will be also determined to compare with the ones we have obtained by usual strategies. The analysis of these time series will provide not only the deformation models due to the tectonic and volcanic activity in the island but also the kind of noise and seasonal fluctuations presented in the data due to the subdaily processing strategy we have applied.

4 References

Altamimi Z, Sillard P, Boucher C (2002) ITRF2000: a new releases of the International Terrerstrial Frame for earth science applications. J. Geophys. Res., 107, 2214.

Berrocoso M, Martín-Dávila J, Astiz M, Catalán- Morollón M, García A (2002) El proyecto GEODEC: un estudio multidisciplinar de la actividad geodinámica de la Isla Decepción (Islas Shetalnd del Sur, Antártida). Proceedings de la 3ª Asamblea Luso-española de Geodesia y Geofisica. 2, 766-769.

Berrocoso M., Enríquez de Salamanca J. M., Jiménez Y., Fernández- Ros A., Torrecillas C., Ramírez M. E., González Fuentes M. J., Pérez- Peña A., Páez R., Tárraga M.., García García A., García García F., Soto R. (2004) Determination of an experimental geoid for Deception Island (Antarctica). IAG International Symposium GGSM 2004, Porto (Portugal).

Beutler G., Bock H., Brockmann E., Dach R., Fridez P., Gurtner W., Hugentobler U., Ineichen D., Johnson J., Meindl M., Mervart L., Rothacher M., Schaer S., Springer S., Weber R. (2001) BERNESE GPS Software Version v4.2. Ed. By U. Hugentobler, S. Schaer, P. Fridez. Astronomical Institute University of Berne.

Caselli A., M. Dos Santos A, Risso C, García A, Ortiz R. (2002) Caracterización geoquímica de los gases volcánicos de Isla Decepción (Shetland del Sur, Antártida). Revista de la Asociación Geológica Argentina, 57(1).

García A, and DECVOL Working Group (2002) A cross-disciplinary study at Deception Island (South shetland Islands, Antarctica). Evaluation of the recent volcanological status. Documento interno.

Ibáñez J, del Pezzo E, Almendros J, La Rocca M, Alguacil G, Ortiz R, García A (2000) Seismovolcanic signals at Deception Island volcano, Antarctica: Wave Field Analysis and source modelling. J. Gephys. Res., 105, 13905-13931.

Ortiz R, Vila J, Garcia A, Camacho A. G, Diez J. L, Aparicio A, Soto R, Viramonte J. G, Risso C, Menegatti N, Petrinovic I. (1992) Geophysical Features of Deception Island. Recent Progress in Antarctic Earth Science, edited by Y. Yoshida, K. Kaminuma, K, Shiraishi, Terrapub, Tokyo: 443-448.

Ortiz R, Garcia A, Aparicio A, Blanco I, Felpeto A, del Rey R, Villegas M. T, Ibañez J, Morales J, del Pezzo E, Olmedillas J. C, Astiz M, Vila J, Ramos M, Viramonte J. G, del Risso C, Caselli A. (1997) Monitoring of the volcanic activity of Deception Island, South Shetland Islands, Antarctica (1986-1995) The Antarctic Region: geologycal evolution and processes.Ed. By C. A. Rizzi, Terra Antarctica Publ.(Siena): 1071-1076.

Acknowledgements

The realization of this work is being possible thanks to the subproject 'Geodetic Studies in Deception Island: deformation models, geoid determination and SIMAC'- REN 2000-0551-C03-01, included in the project 'Geodynamic Activity in Deception Island (GEODEC)', financed by the Ministry of Science and Technology of Spain through the National Program of Antarctic Research of Natural Resources. We will also thank the Las Palmas ship crew and the members of the Spanish Base Gabriel de Castilla for their collaboration during the last Antarctic campaigns.

Geodetic and Geophysical Repeated Measurements in Geodynamic Monitoring Networks of Estonia

Tarmo Kall
Estonian University of Life Sciences, Kreutzwaldi 5, Tartu 51014, Estonia

Tõnis Oja
Estonian Land Board, Mustamäe tee 51, Tallinn 10602, Estonia

Abstract. Repeated measurements were started in the first Estonian geodynamic network – the Põltsamaa-Lelle geodynamic line in 1961. Three more geodynamic networks were established in the 1970s: in Navesti (Central Estonia), in Viru-Nigula (North–East Estonia) and in Paluküla (West Estonia). The aim of these networks was a detailed study of local crustal movements in the regions where tectonic faults are located. Repeated levellings as well as gravimetric, electrometric and magnetometric observations were made on those networks. On Viru-Nigula network the measurements of baseline lengths for the determination of horizontal movements were performed. Measurements on the networks were stopped at the end of 1980s due to the lack of financial resources. Repeated high precision gravity measurements on Põltsamaa-Lelle geodynamic line are still carried out nowadays.

The present paper gives an overview of the repeated measurements, mainly of repeated levellings and gravity measurements. Based on the repeated levellings, the vertical displacements of benchmarks are calculated and the possible reasons for benchmark displacements, like the change in groundwater level, subsidence of building, fault activity and postglacial rebound, are discussed. In Põltsamaa-Lelle geodynamic line the connection between the vertical movements of benchmarks and observed gravity change is brought out.

Keywords. Levelling, gravity measurements, vertical displacements, gravity change, postglacial rebound.

1. Introduction

In the years 1970–1974, after locating potential tectonic faults, three geodynamic networks were built in Estonia in Navesti, Paluküla and Viru-Nigula (Fig. 1). The aim of these networks was to investigate the local movement of the Earth's crust, based on geodetic and geophysical repeated measurements. For the same reason repeated measurements were started on the Põltsamaa-Lelle geodynamic line in 1961 (Fig. 1).

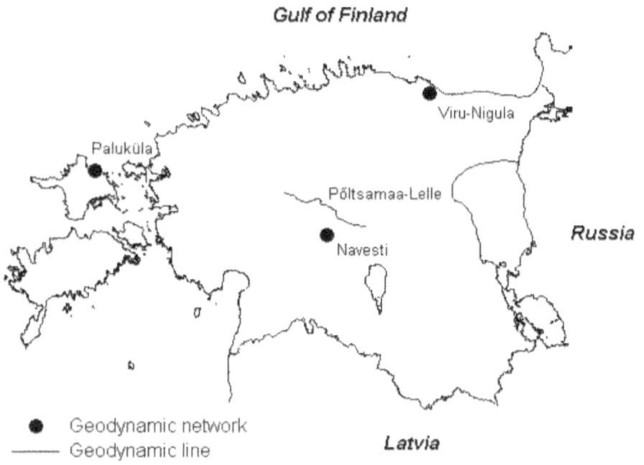

Fig. 1 Geodynamic monitoring networks in Estonia.

2. Geodynamic networks and the geological structure in the network areas

2.1. Navesti geodynamic network

Navesti network (Fig. 2) is situated in Central Estonia in the valley of the Navesti River on the accumulation plain of the Baltic Ice Lake where the thickness of Quaternary sediments fluctuates from a couple of centimetres to six meters.

The network is situated near the Paldiski–Pskov shear zone; permeate through Central Estonia in the NW–SE direction. The magnetic anomaly is mostly negative or slightly positive. On the basis of local anomalies of the gravity field it can be assumed that several faults cross the crystalline basement (depth 365 metres) in the NE–SW direction which appear in the form of 100–200 m wide zones which are at the distance of 1.5–2.0 km from each other.

As the current of the Navesti River has been blocked for a couple of times, the riverbed was deepened at the end of the 1930s and the 1950s (Torim and Sildvee (2002)). The lower reaches of the river often suffer from floods. A tectonic uplift in the SE–NW direction is considered to have been the reason for the blockage (Miidel (1991)).

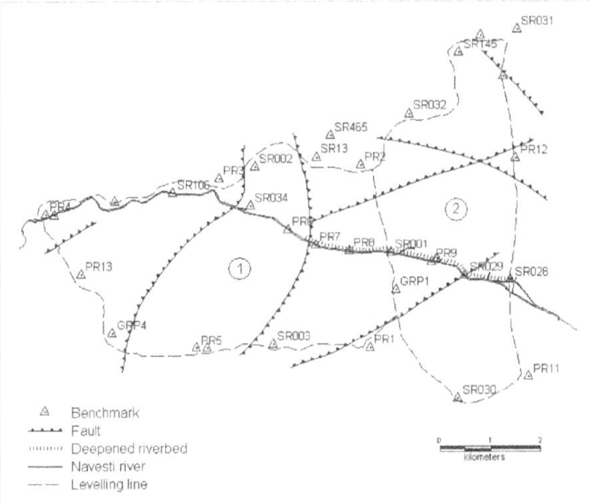

Fig. 2 Navesti geodynamic network. Faults are drawn after Torim and Sildvee (2002).

Navesti network consists of two polygons. Polygon I was established in 1970 when four gravimetric points and five fundamental benchmarks were installed.

In 1981 the network was extended in the eastward direction with Polygon II. In the process of extension four ground and four wall benchmarks were additionally placed. Furthermore, a levelling line was formed along the bank of the Navesti River (Torim and Sildvee (2002)).

2.2. Paluküla geodynamic network

Paluküla network is situated on the North-Eastern slope of the Kärdla meteorite crater, on the island Hiiumaa. In the northern part of Hiiumaa the crystalline basement is at the depth of about 200 m, at Paluküla only of 25–30 m. It was first considered a bank fold that had been formed in the Early Ordovician period and which can also be found elsewhere in Estonia (Sonda, Uljaste). Bedding faults in the Paluküla limestone quarries also indicated an uplift in this region. Repeated measurements in the network had to reveal the possible character and velocity of the local movement.

With the help of geological and geophysical studies the impact structure of the Kärdla crater was revealed. Up to now 160 boreholes have been made and several geophysical research methods have been used. According to the recent data the Kärdla crater (rim-to-rim diameter four km, depth 500 m) was formed in the Late Ordovician period, and later covered by the Late Ordovician sediments (Ainsaar et al. (2002), Suuroja et al. (2002)).

The establishment of the network began in 1971. All in all, eight fundamental and six wall benchmarks were installed.

The levelling lines form three closed loops with the perimeters of 4.9 km, 4.7 km and 5.4 km (Torim and Sildvee (2002)).

2.3. Viru-Nigula geodynamic network

Viru-Nigula network was established for the simultaneous study of horizontal and vertical movements of the Earth's surface. Here the fault separates the basement blocks having the vertical offset of about 52 m. The Aseri fault zone is longer than 75 km and the width in the study area is 200–300 m.

The network was established in 1974. It is of rectangular shape, its two sides intersect the tectonic fault. One side of the network is mounted with four benchmarks, the other with three. To establish horizontal displacement, the distances between benchmarks were measured.

To determine the vertical movements, the first order levelling line was established where additionally four ground benchmarks and two wall benchmarks were installed (Torim (1989)).

2.4. Põltsamaa-Lelle geodynamic line

The aim of establishing the Põltsamaa-Lelle geodynamic line (Fig. 3) was to investigate in detail the recent movements of the Earth's crust in the region of tectonic faults in Central Estonia. The line with the length of 72 km has been supplied with 53 benchmarks and eight gravimetric points. It coincides with the direction of uplift of the postglacial rebound, confirmed by several research outcomes (Zhelnin (1966), Vallner et al. (1988), Ekman and Mäkinen (1996)).

The sedimentary cover is formed by the Ordovician and Silurian carbonate rock, which is covered by the layers of till, sand and clay (the thickness of Quaternary sediments is 0.5–10 m).

The line runs along the Paldiski–Pskov shear zone and cuts the Saaremaa–Mustvee and Pärnu–Rakvere shear zones. The levelling line also crosses the local faults in the crystalline basement or sedimentary cover. The basement is especially divided in the environs of the towns of Türi and Võhma where faults of different directions are intersecting. (Torim and Sildvee (2002)).

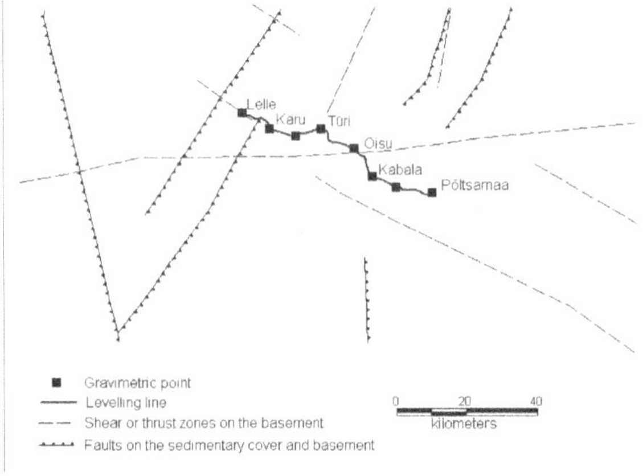

Fig. 3 Põltsamaa-Lelle geodynamic line. Tectonic shear zones and faults are drawn after Soesoo et al (2004).

3. Repeated measurements of geodynamic networks

3.1. Navesti geodynamic network

The initial levelling of the Navesti network took place twice in 1971. Levelling was repeated in 1973 (twice), 1976, 1980, 1983 and 1987. The levelling of the Polygon II was carried out in 1983 and 1987. The line passing along the river Navesti has been levelled three times: in 1983, 1984 and 1987.

The average random error of the levelling of Polygon I was ±0.42 mm/km, the systematic error ±0.06 mm/km and the loop misclosure 2.9 mm. The biggest random error, ±0.59 and ±0.50 mm/km were obtained in 1983 and 1987 respectively. Systematic errors were the biggest in 1983 and 1987, ±0.07 and ±0.09 mm/km respectively. The biggest misclosures characterize the levellings in 1971 and in 1987 when these errors were -4.0 and 4.4 mm respectively.

The average random error of levelling of Polygon II was ±0.50 mm, the systematic error ±0.14 mm/ km and the misclosure 3.6 mm.

3.2. Paluküla geodynamic network.

In Paluküla network all the three loops have been levelled four times: the initial levelling took place in 1972, repeated levellings in 1975, 1977 and 1982.

The average random error of levelling was ±0.50 mm/km, the systematic error ±0.16 mm/km, and the loop misclosure 1.6 mm. The biggest random error was in levelling of Loop I in 1982 – ±0.95mm/km, the biggest systematic error in the levelling of Loop III in 1972 – ±0.45 mm/km, and the biggest loop misclosure 5.0 mm also in levelling of Loop III in 1972.

3.3. Viru-Nigula geodynamic network

The initial levelling of the network was performed in 1975, repeated levellings in 1980 and 1988.

The average random error of levelling was ±0.55 mm/km, the systematic error ±0.18 mm/km and the misclosure 1.5 mm. The lowest precision was obtained in 1988, when the random error was ±0.53 mm/km, the systematic error ±0.26 mm/km and the loop misclosure 2.5 mm. Big random and systematic errors were also obtained during the initial levelling in 1975 – ±0.76 mm/km and ±0.15 mm/km respectively.

3.4. Põltsamaa-Lelle geodynamic line

Põltsamaa-Lelle levelling line belonged to the high precision levelling network, established already in 1933–1943. Its levelling was carried out in 1936. Measurements were repeated in 1961, 1964, 1972, 1980, 1982 and 1987.

The average random error of levellings was ±0.44 mm/km, the systematic error ±0.09 mm/km. The biggest random errors occurred in levelling in 1982 – ±0.58 mm/km, 1987 – ±0.49 mm/km and in 1964 – ±0.48 mm/km. Biggest systematic errors were obtained in levellings in 1961 – ±0.18 mm/km, 1964 – ±0.12 mm/km and 1987 also ±0.12 mm/km.

4. Calculation of benchmark displacements

Prior to vertical displacement calculation, the levelling errors were analyzed in detail. Fore and back levelling discrepancies in sections, allowable discrepancies (2 mm/km, according to instruction of I order levelling (ELB (1994)) and cumulation of discrepancies were calculated.

Vertical displacement of section i between levelling epochs j and k are determined by levelled average height differences h:

$$\Delta h_i = h_k - h_j \,. \tag{1}$$

Uncertainty of displacement between the two benchmarks at the end of the section i is (Hodgkinson et al (1996), Demoulin and Collignon (2000), Camelbeeck et al (2002)):

$$\delta_i = \sqrt{\eta_j^2 + \eta_k^2} \,, \tag{2}$$

where η_j and η_k are random errors of levellings on epochs j and k, that have been determined by loop misclosure f and perimeter P (ELB (1994)):

$$\eta = \sqrt{f^2/P} \,. \tag{3}$$

If closed loop was not formed, levelling random error is calculated by length L and fore and back levelling discrepancy d of n sections (ELB (1994)):

$$\eta^2 = 1/4n \cdot \sum d^2/L \,. \tag{4}$$

Standard deviation of vertical displacement of section i with the length L is calculated from (Hodgkinson et al (1996), Demoulin and Collignon (2000), Camelbeeck et al (2002)):

$$\sigma_i = (\delta_i + \delta_{i+1}) \cdot \sqrt{L} / 2 \,. \tag{5}$$

Significance of vertical displacement is evaluated by calculating signal-to-noise ratio (Demoulin and Collignon (2000), Camelbeeck et al (2002)):

$$S / N = \Delta h_i / \sigma_i \,, \tag{6}$$

which has to be bigger than 2, to confirm that the vertical displacement is statistically significant at the 95% confidence level.

After that, vertical displacement profiles (Fig. 4 and Fig. 5) were compiled by adding the vertical displacement of each consecutive section along the line, where the first benchmark was considered to be fixed. Error bars (1σ, Eq. (5)) are given in relation to the previous benchmark. Profiles are used in order to evaluate the character of displacements. Following aspects were observed (Lenôtre et al (1999), Giménez et al (2000)):

1) Slopes and slope changes of the compiled profiles of vertical displacements;

2) Anomalies of vertical displacements;

3) Information about stability of benchmarks and levelling errors.

5. Vertical displacements in geodynamic networks

5.1. Navesti geodynamic network

In Navesti geodynamic network vertical displacements have been calculated between the consecutive levelling years 1971–1973, 1973–1976, 1976–1980, 1980–1983 1983–1987 and 1971–1987 in Polygon I, and between 1983–1987, 1983–1984 and 1984–1987 in Polygon II and in the levelling line along the river bank. Profiles of vertical displacements are shown in Fig. 4.

On the average, 24% of the benchmarks in Polygon I have a significant vertical displacement. Most of the significant displacements, 62%, were obtained in 1976–1980 and 40% in 1971–1987, whereas, in 1983–1987 none of the displacements were statistically significant.

In order to control the assumption that there exists a tectonic uplift in the direction of the current along the Navesti River (SE–NW) we should look at section 4–12 km on the profiles, where a positive slope should appear, and section 15–20 km, where, on the contrary, a negative slope should appear.

In long–term scale significant positive slope occurs in sections 2–13 km except for one significant slope change in section 8–9 km, where long–term slope is negative. This anomaly is probably related to the subsidence of wall benchmark. An insignificant negative slope occurs in section 12–19 km and an insignificant positive slope occurs again in section 19–21 km. In the long–term scale (1971–1987) displacement values exceed the individual short–term values mostly twice.

Compared to long–term displacements, some significant anomalies of short–term displacements occur:

1) in 1976–1980 in section 7–8 km (significant negative slope, long–term slope is positive); in

section 12–15 km (significant positive slope, long–term slope is negative) and in section 15–19 km (significant positive slope, long–term slope is negative);

2) in 1980–1983 in section 12–19 km (significant negative slope, long–term slope is also negative, but statistically insignificant).

Speaking about vertical displacements in connection with benchmark stability, it is important to mention that most of the significant displacements occur among the wall benchmarks and only a few among the ground benchmarks.

The analysis of gravity measurements in the network did not reveal any significant gravity changes (Torim and Sildvee (2002)).

Altogether we can conclude that the assumption about the tectonic uplift in the SE–NW direction in Navesti network cannot be confirmed based on such a small number of significant displacements. Taking into account that the benchmarks in Navesti network are located on the watertight sediments of the Baltic Ice Lake, preliminary results indicate that the above described significant anomalous displacements are more likely related to changes in the hydrological conditions of this region.

5.2. Põltsamaa-Lelle geodynamic line

Vertical displacements have been calculated between the consecutive levelling epochs in 1936–1961, 1961–1964, 1964–1969, 1969–1972, 1972–1980, 1980–1982, 1982–1987 and also in 1936–1987. The profiles of vertical displacements are presented in Fig. 5.

Põltsamaa-Lelle geodynamic line has more statistically significant vertical displacements than the Navesti geodynamic network, on the average even up to 88% within long–term periods of 1936–1961 and 1936–1987.

For short–term periods 29% of the vertical displacements are statistically significant. The least number of significant displacements (7%) was obtained in 1980–1982.

It is characteristic for the long–term vertical displacements, that sections with different slopes cannot be distinguished. In case of these periods we can talk about a section of the same positive slope from the beginning to the end of the line, with some negative anomalies of vertical displacements.

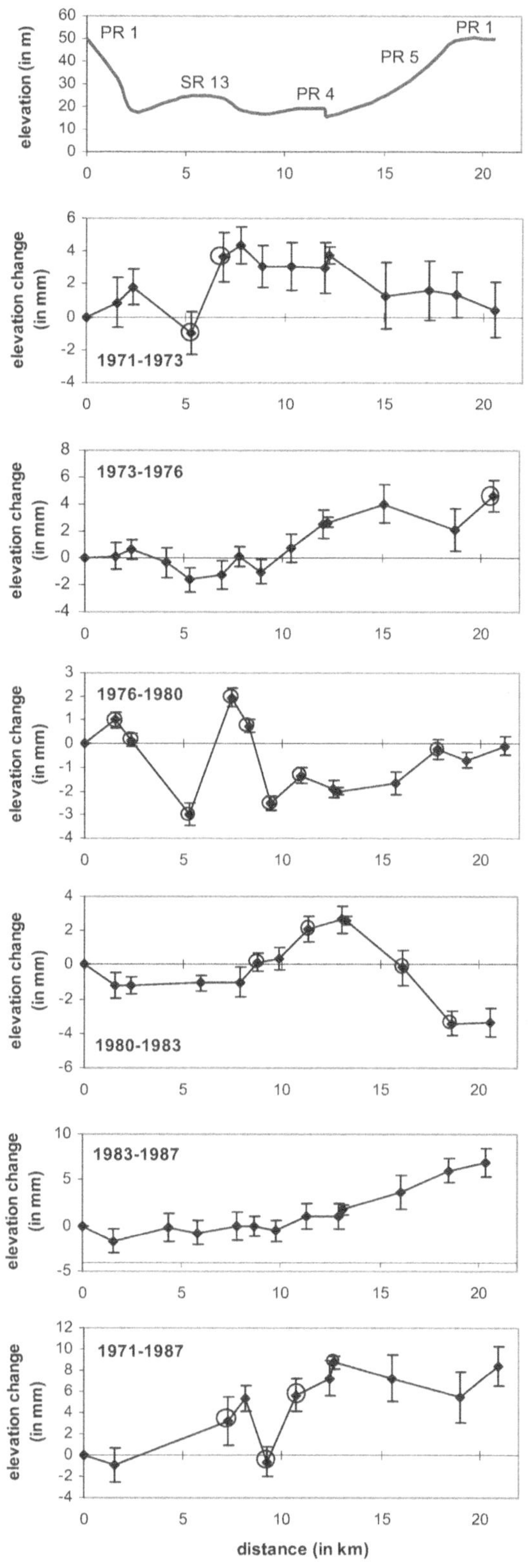

Fig. 4 Vertical displacements in Polygon I of Navesti geodynamic network. On top figure, the topography along the line is shown. Significant displacements are marked with a circle.

Linear regression analysis shows that the Lelle fundamental benchmark has uplifted in reference to the Põltsamaa fundamental benchmark by 15.7±8.8 mm (0.63±0.35 mm/yr) in the period of 1936–1961 and by 20.5±12.2 mm (0.40±0.24 mm/yr) in 1936–1987. These results are in good agreement with the value of postglacial rebound uplift in the given section.

In the background of postglacial uplift (positive slope from the beginning to the end of the line), some slope anomalies (slopes are negative) can be traced in following sections: 1–2.5 km, 12–16 km, 29–32 km, 44–52 km, 55–64 km. In our opinion those anomalies refer to subsidence of benchmark or local tectonic tilt.

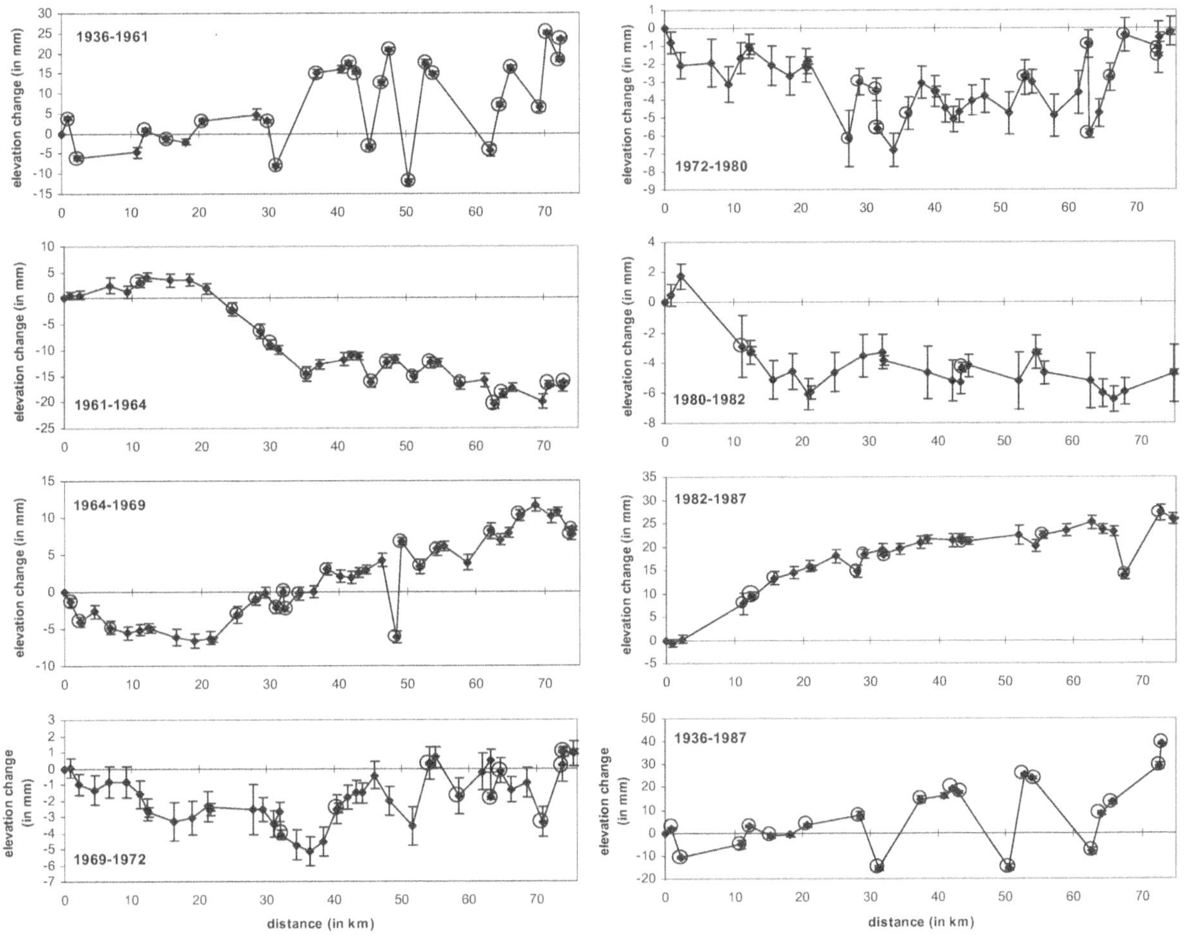

Fig. 5 Vertical displacements in Põltsamaa-Lelle geodynamic line. Significant displacements are marked with a circle.

Compared to long–term displacements, most notable significant anomalies of short–term displacements occur:

1) in 1961–1964 in section 20–28 km (significant negative slope, long–term slope is almost zero);

2) in 1964–1969 in section 66–73 km (significant negative slope, long term slope is positive);

3) in 1980–1982 in section 2–11 km (significant negative slope, probably due to big levelling random error in that section, long–term slope is positive);

4) in 1982–1987 in section 2–11 km (significant positive slope, probably due to big levelling random error in that section, in long–term scale slope is smaller);

5) in 1982–1987 in section 12–16 km (significant positive slope, long–term slope is negative);

6) in 1982–1987 in section 65–68 (significant negative slope, long term slope is positive).

In addition to the levelling errors, these anomalies are most probably caused by local subsidence due to groundwater withdrawal.

Vertical displacements of 1961–1964 and 1982–1987 (also 1980–1987) call for more attention as their values are two–three times bigger than those of other periods and are comparable to the long-term vertical displacements of 1936–1961 and 1936–1987. Levellings in 1961, 1964, 1982 and 1987 can be characterised by a rather big systematic error. If we remove the systematic error from the levelled height differences by linear regression, the vertical displacements decrease by 3 mm on the average, but the displacements still remain several times bigger than the vertical displacements of other periods.

Zhelnin (1966) has explained the big subsidence of surface in 1961–1964 by the drought in 1964, when the level of ground water was extraordinarily low. He also considered that the first 20 km of the line is in the surface where the sedimentary rocks are only 1.3 meters deep and here the sinkage is smaller. In the next parts of the levelling line, the depth of the sedimentary rock increases and as seen from the profile, the subsidence of the surface increases as well. In the period of 1964–1969, the subsidence is almost entirely compensated. The profile shows it as land uplift. As most of subsidence occur in section 20–35 km, then one possible reason could also be systematic errors in levelling of 1964, as cumulation of fore and back discrepancies in that section is -6.6 mm.

At the same time the significant land uplift in 1982–1987 cannot be explained by the rapid change in the ground water level. A remarkable part (7.6 mm) of total uplift results from displacement in section 2–11 km. In that section a big levelling random errors exist in both levelling period.

The possible explanation to anomalous displacements in this period could also be the fault activity, as several earthquakes of magnitude 4–5 were observed nearby to the beginning of the levelling line in 1987–1988. A magnitude 3 earthquake was also registered in 1989, in the vicinity of the end of the levelling line (Nikonov and Sildvee (1991)).

Repeated GPS measurements, started in 2005, should give us additional information on the displacements in the given levelling line.

5.3. Paluküla geodynamic network

Since 1980s, when the impact structure of the Kärdla crater was ascertained (Suuroja et al (2002)), the repeated levellings on the Paluküla network were stopped.

The analysis of vertical displacements does not show any considerable trends. Benchmarks with significant displacements (all in all 36%) show uneven change in heights, most of them subsiding.

5.4. Viru-Nigula geodynamic network

In Viru-Nigula network the benchmark displacements in 1975–1980 remain between -0.7–1.8 mm. When the random errors calculated from the loop misclosures are used for the determination of S/N ratio, most of the displacements (75%) are statistically significant. When random error calculations are based on fore and back discrepancies, only one benchmark (8%) has a statistically significant displacement value.

In 1980–1988, the displacements were between 0–3.4 mm, however only one displacement was statistically significant. Still, both periods show a similar positive slope in section 3–9 km and negative slope in section 9–13 km. The given sections are parallel with the fault. GPS measurements, which were started in 2005, should clarify the character of displacements in the above–mentioned network.

6. Gravity measurements along Põltsamaa-Lelle geodynamic line

Between the years 1970 and 1989, nearly fifteen gravity campaigns were carried out in the Põltsamaa-Lelle geodynamic line with Soviet geodetic gravimeters GAG–2. In all these measurements the standard deviation of a gravity difference on the average ±30 μGal (1 μGal = 10^{-8} m s^{-2}) was achieved (Sildvee (1998)). Since 1998, the line has been re-measured six times with several LCR (LaCoste&Romberg) G–models and Scintrex CG–5 gravimeters. The average standard deviation of those measurements is below 10 μGal. Above–mentioned gravimeters have been repeatedly tested on the calibration lines in Estonia and some of them also in Finland. Moreover, the calibration function of every gravimeter has been updated when required.

Rigorous and uniform methodology has been followed during the campaigns from 1970 up to now. All stations have been measured in the sequence 1–2–1–2–3–2–3–... In addition, air pressure, temperature, instrument height etc. were recorded as well.

Prior to the final data processing several corrections (tidal, air pressure, free air, polar motion) of the readings (after the conversion to the gravity unit mGal) were done. Special attention was paid to drift computation because the data measured with GAG–2 quartz system may contain highly nonlinear and irregular instrument drift (Oja (2004)).

On the basis of corrected readings, the apparent changes of gravity ($\dot{g}$) were computed for every station on the line. Over 800 readings were inserted into a weighted least squares adjustment with simultaneous $\dot{g}$ estimation. Weights were determined in accordance with the average standard deviation of every campaign. Adjusted time rates of changes of gravity with standard deviations (one sigma) in Table 1 are relative to station Põltsamaa, which is fixed in our study.

Table 1. Adjusted time rate of change of gravity and its standard deviation in Põltsamaa-Lelle geodynamic line.

Station	Distance (km)	$\dot{g}$ (μGal/yr)	error (1σ) (μGal/yr)
Põltsamaa	0.0	0.0	±0.0
Pilistvere	12.1	-0.7	±0.2
Kabala	20.4	-1.1	±0.2
Oisu	31.1	-1.6	±0.3
Türi	42.7	-1.2	±0.3
Kolu	52.9	-1.4	±0.4
Käru	62.0	-1.9	±0.4
Lelle	72.1	-2.3	±0.4

To compare the changes of gravity and height in the Põltsamaa-Lelle geodynamic line, the annual change of height ($\dot{h}$) was derived on the basis of vertical displacements within the period of 1936–1987 (see Figure 5).

The time rates of gravity and height (Fig. 6) are negatively correlated which is rather typical for the postglacial rebound areas. However, the average ratio of $\dot{g}/\dot{h} \approx$ -1.5 μGal/mm is far higher than the results of other similar studies, e.g. the ratio of 0.2 μGal/mm obtained on the Fennoscandian 63° land uplift gravity line (Ekman and Mäkinen (1996)).

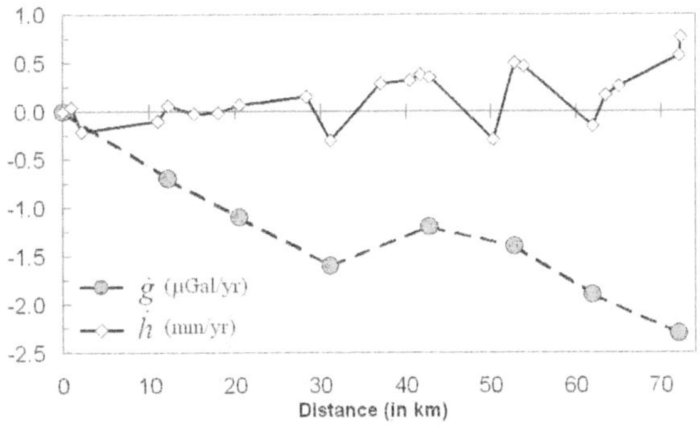

Fig. 6 Annual changes of gravity ($\dot{g}$) and height ($\dot{h}$) in Põltsamaa-Lelle geodynamic line.

Apparently, the high value of the ratio refers to systematic biases in the gravity measurement data which may be caused by the effect of different factors like nonlinear calibration errors of gravimeters, temporal variations in groundwater level, the content of soil moisture and precipitations, local tectonic movements etc. The factors affecting gravity data are currently under study.

7. Conclusions

Vertical displacements in Paluküla and Viru-Nigula geodynamic networks are mostly statistically insignificant. Few significant displacements on those networks are probably caused by the variation of some local factors or levelling errors. Also, fewer repeated measurements have been carried out in the Paluküla and Viru-Nigula networks as compared to Navesti and Põltsamaa-Lelle networks.

Similarly, in Navesti network only a few displacements are statistically significant; therefore it is impossible to draw reliable conclusions about the tectonic uplift in SE–NW direction. On most profiles the positive slope in SE–NW direction is visible, but only in half of the cases it is statistically significant. The fluctuating change of displacements indicates that the cause of displacements is most likely the variation of some local factors, most probably a hydrological one.

In Põltsamaa-Lelle geodynamic line, on a long–term scale, the analysis of vertical displacements and gravity change confirm the postglacial land uplift. The changes of slopes are probably related to subsidence of building or local tectonic tilt. On a short–term scale, anomalies of vertical displacements are determined. Some of displacement anomalies are probably caused by levelling errors or the variation of hydrological factors; others may originate from tectonic movements.

To confirm the presence of tectonic movements on the line, the impact of measurement errors and environmental factors (groundwater, precipitations, air pressure, ocean loading etc) needs ongoing and more comprehensive investigations.

8. References

Ainsaar, L., Suuroja, K., Semidor, M. (2002). Long–term effect of the Kärdla crater (Hiiumaa, Estonia) on Late Ordovician carbonate sedimentation. Deep–Sea Research II 49, pp. 1145–1155

Camelbeeck, T., Van Camp, M., Jongmans, D., Francis, O., Van Dam T. (2002). Comment on „Nature of the recent vertical ground movements inferred from high-precision levelling data in an intraplate setting: NE Ardenne, Belgium" by A. Demoulin and A. Collignon. Journal of Geophysical Research, vol. 107, No. B11, 2281, doi: 10.1029/2001JB000397

Demoulin, A., Collignon, A. (2000). Nature of the recent vertical ground movements inferred from high-precision levelling data in an intraplate setting: NE Ardenne, Belgium. Journal of Geophysical Research, vol. 105, No. B1, pp. 693–705

Ekman, M., Mäkinen, J. (1996). Recent postglacial rebound, gravity change and mantle flow in Fennoscandia. Geophysical Journal International, 126, pp. 229–234

ELB (1994). Instruction of I, II and III order levelling. (in Estonian)

Giménez, J., Suriñach, E., Goula, X. (2000) Quantification of vertical movements in the eastern Betics (Spain) by comparing levelling data. Tectonophysics, 317, pp. 237–258

Hodgkinson, K. M., Stein, R. S., Marshall, G. (1996). Geometry of the 1954 Fairview Peak-Dixie Valley earthquake sequence from a joint inversion of levelling and triangulation data. Journal of Geophysical Research, vol. 101, No. B11, pp. 25,437–25,457

Lenôtre, N., Thierry, P., Blanchin, R., Brochard, G. (1999). Current vertical movement demonstrated by comparative levelling in Brittany (northwestern France). Tectonophysics, 301, pp. 333–344

Miidel, A. (1991). Land uplift blocks rivers: About one idea of A. Tammekand. Eesti Loodus, 9/10, pp. 610–613 (in Estonian)

Nikonov, A. A., Sildvee, H. (1991). Historical earthquakes in Estonia and Their Seismotectonic Position. Geophysica 27, 1-2. Finnish Geophysical Society, Helsinki, 1992, pp. 79-93

Oja, T. (2004). Temporal Variations of Gravity along the Line of Põltsamaa-Lelle, Estonia. Abstract and poster presented on the EGU 1st Gen. Assembly, Nice, France, 25–30 April

Sildvee, H. (1998). Gravity Measurements of Estonia. Reports of the Finnish Geodetic Institute, 98:3

Soesoo, A., Puura, V., Kirs, J., Petersell, V., Niin, M., All, T. (2004). Outlines of Precambrian basement of Estonia. Proc. Estonian Acad. Sci. Geol., 53, 3, pp. 149–164

Suuroja, K., Suuroja, S., All, T., Floden, T. (2002). Kärdla (Hiiumaa Island, Estonia)—the buried and wellpreserved Ordovician marine impact structure. Deep–Sea Research II 49, pp. 1121–1144

Torim A., Sildvee H., (2002). The Geodynamic Polygons of Estonia. Geodeet, 26(50), Tartu, pp. 14–18 (in Estonian)

Torim, A. (1989). About the Movements of the Earth's Crust in Aseri Fault Zone. In Nowadays and Perspectives of Applied Geodesy, vol. 264, pp. 111–116. Latvian Agricultural University, Jelgava (in Russian)

Vallner, L., Sildvee, H., Torim, A. (1988). Recent crustal movements in Estonia. Journal of Geodynamics 9, pp. 215–223

Zhelnin, G. (1966). On the recent movements of the Earth's surface in the Estonian SSR. Proceedings of the Second International Symposium on Recent Crustal Movements. Annales Academiae Scientiarum Fennicae, Series A, III, Geologica–Geographica. Helsinki, pp. 489–493

Session D

Local and Engineering Applications

Chairman: A.J.Gil

Modelling of Strains and Stresses of the Earth-Crust in the Zones of Ecologically Dangerous Objects

C. Aksamitauskas, A. Zakarevicius, A. Stanionis
Dept of Geodesy and Cadastre, VGTU, Sauletekio al. 11, LT-10223 Vilnius-40, Lithuania

Abstract. Up to the last decade of the twentieth century, the main attention to the study of current movements of the Earth-crust was paid in seismically active territories. Such studies performed in platform regions with a little seismic activity were considered to be of scientific value only, and not very important in practice. However, after development of building of complex objects with an increased ecological dangerousness sensitive to deformations and seismic phenomena (nuclear power plants, large chemistry plants, etc.) and upon accumulation of instrumental observations of seismic phenomena, the criteria for estimating the seismic dangerousness in platform territories have changed.

In this work, the modelling of the Earth-crust strains and stress change by applying the results of geodetic surveys is analysed. The tensor analysis is applied to make models. Parameters of models are estimated by the finite element method.

According to the methods suggested, the studies were performed in the Ignalina Nuclear Power Plant (INPP) region.

The conclusion is drawn from the analysis of results that the model of strains and stresses made by the finite element method details the structure of the current tectonic activity in the study district and expands possibilities of the geotectonic interpretation of survey results. The studies performed show the merits of modelling of strains and stresses of the Earth-crust by the finite element method in comparison with other methods and possibilities of applying the program package *Ansys* for geodynamic studies.

Keywords. Finite element method, GPS, geodynamic processes, tectonic stresses, strains.

1 Introduction

By studying the current horizontal movements of the Earth-crust by geodetic methods, the movements of geodetic points that are identified with movements of the Earth-crust are established. The movements of geodetic points are established according to the changes of coordinates within the certain period of time between geodetic measurements. Thus, the reading values of movements of the Earth-crust only in solitary points located sparsely are determined. When making models of deformations of the Earth-crust, the values of deformations in other points are determined in the way of interpolation, by applying the procedures of rectilinear interpolation in most cases. Such method of determining the horizontal deformations of the Earth-crust has many shortcomings. The measured values and directions of geodetic point movements depend on the selection of initial points that are considered stable when calculating movements. When calculating the values of movements of the Earth-crust in non-geodetic points in the way of geometric interpolation, mechanical characteristics of the deformed body, i.e. mechanical model of deformations, are not estimated.

According to the laws of mechanics, movements of the Earth-crust are related with the change of the Earth-crust stresses. Therefore, there is a functional relation between movements and stresses of the Earth-crust. Since the current movements of the Earth-crust are the continuation of tectonic deformations of the Earth-crust occurring for a long period of time, the deformations determined by geodetic methods are the change of deformations of the long period of time within the certain period of time. For this reason, according to the results of geodetic measurements, by using relations between strains and stresses, it is possible to determine only the changes of stresses, but not the absolute values of stresses. However, by applying the direct comparison of coordinates obtained in separate cycles of measuring to determine the movements of the Earth-crust, the results obtained may have more than one meaning. For this reason, the changes of the Earth-crust stresses calculated according to the movements of geodetic points will have also more than one meaning.

Therefore, when modelling the horizontal strains of the Earth-crust and the changes of the Earth-crust stresses according to the results of geodetic measurements, it is necessary to apply the methods of analysis invariant with respect to the systems of coordinates. One of such methods is the tensor analysis of strains and stresses of the Earth-crust (Varadan and Bhaskar (1999); Zadro and Braitenberg (1999)).

The object of this study is to analyze the application of tensor analysis and finite element method in modelling the horizontal strains of the Earth-crust and the change of the Earth-crust stresses according to the data of geodetic measurements and carry out studies by these methods in the INPP region.

Relative linear and shear horizontal strains, changes of normal and shear tectonic stresses and changes of the principal stresses were calculated in the INPP region by using changes of ground point coordinates, taking that the character of plane strains is isotropic. By using the physical relation between horizontal strains and stresses, new results were obtained and the connection between the tectonic structure of the INPP region and current geodynamic processes was estimated basing on these results. It was established that the structure of horizontal deformations of the Earth-crust and tectonic stresses is related with tectonic peculiarities of the territory in the INPP region, i.e. the territory is also active in current period from the geodynamic point of view.

2 Methods of calculating the horizontal deformations

Horizontal deformations of the Earth-crust are determined according to the data of repeated geodetic network measurements. Characteristics of horizontal movements of the Earth-crust that occurred within the certain period of time between repeated measurements may be described according to these data (Calais et al. (2000); Hsu and Li (2004); Ruiz et al. (2003); Sue et al. (2000); Vaníček et al. (2001); Vaníček and Krakiwsky (1986)). The method of determination of the horizontal Earth-crust movements applied widely is the comparison of identical point coordinates calculated according to measurements done at different time (Kuang (1996); Zakarevicius (2003)).

In general, strains ε_{xx}, ε_{yy}, ε_{xy} are linked to shifts u, v by three geometric (Koshi) equations in a horizontal plane at the point of deformed body (Rao (1999); Zakarevicius and Stanionis (2004); Zienkiewicz and Taylor (2000)):

$$\begin{cases} \varepsilon_{xx} = \dfrac{\partial u}{\partial x}, \\ \varepsilon_{yy} = \dfrac{\partial v}{\partial y}, \\ \varepsilon_{xy} = \dfrac{1}{2}\cdot\left(\dfrac{\partial u}{\partial y} + \dfrac{\partial v}{\partial x}\right), \end{cases} \tag{1}$$

where

$$u = \Delta x = u(x,y), \tag{2}$$

$$v = \Delta y = v(x,y), \tag{3}$$

and u, v are the shifts of coordinates in rectilinear functions of coordinates in the Cartesian Coordinate System; ε_{xx}, ε_{yy} are the relative linear strains, ε_{xy} is the relative shear strain.

Strains for plane stress state $\varepsilon_{xz} = 0$, $\varepsilon_{yz} = 0$, $\varepsilon_{zz} \neq 0$ (Varadan and Bhaskar (1999)):

$$\varepsilon_{zz} = -\frac{\nu}{(1-\nu)}\cdot(\varepsilon_{xx} + \varepsilon_{yy}), \tag{4}$$

where ν – the Poisson's ratio (0.25), ε_{xz}, ε_{yz} are the relative shear strains, ε_{zz} is the relative linear strain.

3 Relations between tectonic stresses and horizontal strains

When having horizontal relative linear and shear strains calculated, it is possible to estimate the change of tectonic stresses, i.e. to determine changes of stresses occurred within a certain period of time.

The Hooke's law may be applied to model tectonic stresses in a horizontal plane, by expressing the stresses in strains (stresses for plane stress state: $\sigma_{xz} = 0$, $\sigma_{yz} = 0$, $\sigma_{zz} = 0$) (Varadan and Bhaskar (1999)):

$$\begin{cases} \sigma_{xx} = \dfrac{E}{1-\nu^2} \cdot \left(\varepsilon_{xx} + \nu \cdot \varepsilon_{yy}\right) , \\ \sigma_{yy} = \dfrac{E}{1-\nu^2} \cdot \left(\varepsilon_{yy} + \nu \cdot \varepsilon_{xx}\right) , \\ \sigma_{xy} = G \cdot \varepsilon_{xy} = \dfrac{E}{2 \cdot (1+\nu)} \cdot \varepsilon_{xy} , \end{cases} \tag{5}$$

where G is the shear modulus, E is the Young's modulus $\left(7 \cdot 10^{10} \ \dfrac{\text{N}}{\text{m}^2}\right)$; σ_{xx}, σ_{yy}, σ_{zz} are the normal stresses, σ_{xy}, σ_{xz}, σ_{yz} are the shear stresses.

The principal tectonic stresses are calculated from the quadratic equation:

$$\sigma^2 - I_1 \cdot \sigma + I_2 = 0 , \tag{6}$$

that is obtained by extending the determinant (ANSYS Theory Reference (1998)):

$$\begin{vmatrix} \sigma_{xx} - \sigma & \sigma_{xy} \\ \sigma_{xy} & \sigma_{yy} - \sigma \end{vmatrix} = 0 , \tag{7}$$

$$I_1 = \sigma_{xx} + \sigma_{yy} , \tag{8}$$

$$I_2 = \begin{vmatrix} \sigma_{xx} & \sigma_{xy} \\ \sigma_{xy} & \sigma_{yy} \end{vmatrix} , \tag{9}$$

where σ is the principal stresses, I_1, I_2 are the stress tensor invariants.

After solution of quadratic equation (6), two actual roots σ_1, σ_2 $(\sigma_1 \geq \sigma_2)$ are obtained, i.e. σ_1 is the maximum principal stress, σ_2 is the minimum principal stress.

4 Modelling of horizontal strains and tectonic stresses 2-D of the Earth-crust by the finite element method

A special polygon of 10 ground points deepened up to 2.5 m with forced centering was established for the measurement of horizontal movements of the Earth-crust in the region of INPP (Aksamitauskas et al. (2005); Stanionis (2005); Zakarevicius et al. (2003); Zakarevicius and Stanionis (2004)).

The power plant is located at the crossing of large tectonic breaks. It is located on the intersection of three large structures: Mazurian-Belorussian anteclise, Baltic syneclise and Latvian saddle (Figure 1). According to the topography of the crystalline basement, there are several lower structures distinguished: North Zarasai step, Anisimovichi graben, East Druksiai elevation, Druksiai depression (graben) and South Druksiai elevation.

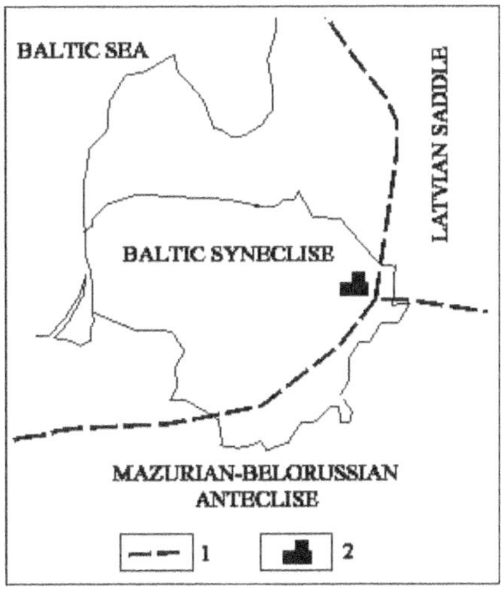

Fig. 1. Scheme of the tectonic macrostructures (1 – tectonic breaks (according to S.Sliaupa), 2 – Ignalina NPP)

Points are arranged so that it would be possible to control the main blocks of the Earth-crust breaks. Measurements in the geodynamic polygon were done in 1998 and September 1999. The measurement programme is presented in Table 1.

Table 1. Measurement programme

Session	Point number									
	1	2	3	4	5	6	7	8	9	10
A	x	x		x	x	x	x		x	x
B	x	x		x	x	x	x		x	x
C	x	x	x	x		x		x	x	x
D	x	x	x	x		x		x	x	x

The measurement was carried out with eight GPS devices of the Firm ASHTECH Z-Surveyor and Z-12. The measurements were carried out by the specialists from VGTU and the Danish company „Nellemann & Bjornkjar". The measurement programme consisted of 4 sessions. The duration of one session was 24 hours. The measurements in points 1, 2, 4, 6, 9, 10 were carried out in all 4 sessions, and the measurements in points 3, 5, 7 and 8 were performed in 2 sessions. The measurement data were processed in the Geodesy Institute of VGTU and the Danish company „Nellemann & Bjornkjar" using the program packages GPPS and FILLNET. The average quadratic errors of spatial coordinates did not exceed 1.2 mm. Changes of point coordinates within the period from September 1998 to September 1999 are presented in Table 2. The scheme of point location is presented in Figure 2.

Table 2. Alterations of point coordinates

Point number	Δx (m)	Δy (m)
1	0.000	0.000
2	0.013	0.004
3	0.008	0.009
4	0.010	0.000
5	0.013	0.004
6	–0.006	0.008
7	0.003	0.002
8	–0.010	–0.002
9	0.004	0.001
10	0.005	0.000

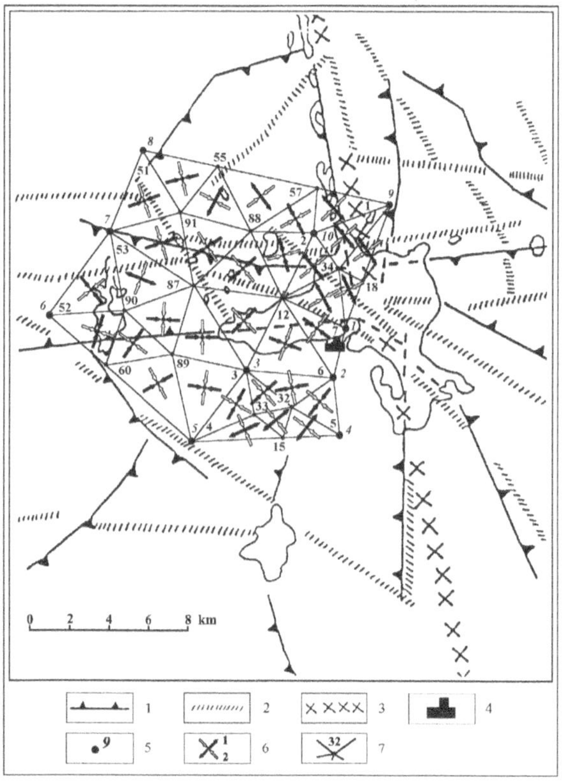

Fig. 2. Directions of the principal tectonic Earth-crust stresses. Local tectonic breaks (according to I. Juknelis, V. Marcinkevicius, I. Sestokas and A. Sliaupa) discovered by: 1 – seismic survey, 2 – aeromagnetic survey, 3 – morphostructural analysis; 4 – Ignalina nuclear power plant, 5 – GPS point, 6 – the principal stresses (1 – maximum change of the principal stress, 2 – minimum change of the principal stress), 7 – corner node

According to the presented method of calculation of horizontal strains and tectonic stresses, the two-dimension (2-D) model of geometric body is made to model the horizontal strains and stresses of the Earth-crust.

Ten ground points were selected to make a two-dimension (2-D) model of horizontal deformations where repeated GPS measurements were carried out. By using these points, the finite element network consisting of 33 triangles was formed (Figure 2). The expansion of territory studied in finite elements, taking into account the positions of initial points and conditions of expansion, was carried out using the program package Ansys (ANSYS Theory Reference (1998)).

The triangle is described by six nodes: I, J, K, L, M and N. The geometry of the finite element, the distribution of nodes and the system of coordinates are presented in Figure 3.

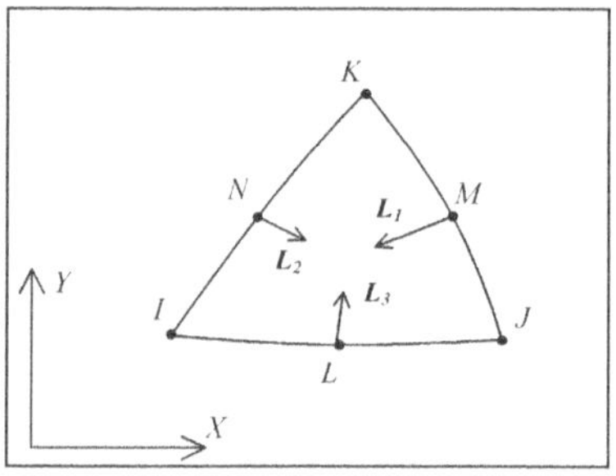

Fig. 3. Geometry of the finite element

Each node of the triangle has two degrees of freedom (north and east shifts). Deformations of the finite element is described by formulae (ANSYS Theory Reference (1998); Zienkiewicz and Taylor (2000)):

$$\begin{aligned} u_i = {} & u_I(2L_1-1)L_1 + u_J(2L_2-1)L_2 + \\ & + u_K(2L_3-1)L_3 + u_L(4L_1L_2) + \\ & + u_M(4L_2L_3) + u_N(4L_3L_1) \ , \end{aligned} \tag{10}$$

$$\begin{aligned} v_i = {} & v_I(2L_1-1)L_1 + v_J(2L_2-1)L_2 + \\ & + v_K(2L_3-1)L_3 + v_L(4L_1L_2) + \\ & + v_M(4L_2L_3) + v_N(4L_3L_1) \ , \end{aligned} \tag{11}$$

where u_I, u_J, u_K, u_L, u_M, u_N, v_I, v_J, v_K, v_L, v_M, v_N are the shifts of node coordinates; L_1, L_2, L_3 are the normalized coordinates (the range from 0 to 1 in the finite element).

The shift of nodes of the finite elements was calculated using *Ansys* code. When calculating shifts of nodes, the obtained mechanical plane model of isotropic body deformations is estimated according to the information of initial points.

It is impossible to determine absolute indices of the Earth-crust deformations from the beginning of deformation development according to the geodetic measuring; it is only possible to determine the change increments of deformations within a certain period of time from initial to repeated measurements. Therefore, basing oneself on geodetic measuring and functional dependency of strains and tectonic stresses, it is possible to calculate changes of tectonic stresses within the period of time between repeated measurements.

Relative linear and shear horizontal strains, changes of normal and shear stresses and changes of the principal stresses were estimated in the finite element nodes. Values of changes of horizontal strains and tectonic stresses are presented in Table 3. Directions of the principal stresses of the finite element are presented in Figure 2. The principal stresses are perpendicular to each other.

When using the results of modelling analysis (Figure 2), some laws of distribution of changes of the principal stresses of the Earth-crust that are related with the tectonic structure of the territory (Marcinkevicius and Laskovas (1996)) and geophysical fields (Petroskevicius (2004)) are noticed in the INPP region. Changes of maximum stresses are almost parallel, and changes of minimum stresses are perpendicular to breaks of the crystal foundation.

The results of horizontal strains and tectonic stresses of the Earth-crust in the INPP region established in modelling by the finite element method do not contradict (Aksamitauskas et al. (2005); Zakarevicius et al. (2003); Zakarevicius and Stanionis (2004)) the results obtained from studies.

Table 3. Strain and stress rates

Corner node	$\dot{\varepsilon}_{xx}\cdot 10^{-6}$	$\dot{\varepsilon}_{yy}\cdot 10^{-6}$	$\dot{\varepsilon}_{zz}\cdot 10^{-6}$	$\dot{\varepsilon}_{xy}\cdot 10^{-6}$	$\dot{\sigma}_{xx}$, MPa	$\dot{\sigma}_{yy}$, MPa	$\dot{\sigma}_{xy}$, MPa	$\dot{\sigma}_1$, MPa	$\dot{\sigma}_2$, MPa
1	–0.060	–0.323	0.128	0.003	–0.010	–0.025	0.000	–0.010	–0.025
2	–0.098	–0.027	0.042	–0.722	–0.008	–0.004	–0.020	0.014	–0.026
3	–0.918	0.146	0.257	–0.690	–0.066	–0.006	–0.019	0.000	–0.072
4	0.107	–1.473	0.455	0.505	–0.019	–0.108	0.014	–0.017	–0.110
5	–2.962	0.522	0.813	2.360	–0.211	–0.016	0.066	0.004	–0.232
6	–0.064	–1.010	0.358	–0.100	–0.024	–0.077	–0.003	–0.024	–0.077
7	–0.884	0.413	0.157	–2.683	–0.058	0.014	–0.075	0.062	–0.105
12	–0.238	–0.270	0.169	0.293	–0.023	–0.025	0.008	–0.015	–0.032
15	0.157	–0.022	–0.045	0.321	0.011	0.001	0.009	0.017	–0.004
18	–0.126	0.438	–0.104	0.123	–0.001	0.030	0.003	0.031	–0.002
32	0.098	–0.702	0.202	1.850	–0.006	–0.051	0.052	0.028	–0.085
33	–0.334	0.139	0.065	0.462	–0.022	0.004	0.013	0.009	–0.028
34	–0.276	0.345	–0.023	–1.511	–0.014	0.021	–0.042	0.049	–0.043
51	0.296	–3.924	1.209	1.663	–0.051	–0.287	0.047	–0.042	–0.296
52	–1.291	0.164	0.375	1.972	–0.093	–0.012	0.055	0.016	–0.121
53	0.415	–0.322	–0.031	–0.927	0.025	–0.016	–0.026	0.038	–0.029
55	0.205	0.364	–0.189	–0.733	0.022	0.031	–0.021	0.048	0.006
57	0.099	0.009	–0.036	–0.232	0.008	0.003	–0.006	0.012	–0.002
60	–0.531	0.020	0.170	0.836	–0.039	–0.008	0.023	0.004	–0.052
87	–0.014	–0.508	0.174	0.403	–0.011	–0.038	0.011	–0.007	–0.042
88	–0.271	–0.038	0.103	0.075	–0.021	–0.008	0.002	–0.008	–0.021
89	0.157	–0.765	0.203	–0.068	–0.003	–0.054	–0.002	–0.003	–0.054
90	–0.265	–0.126	0.130	–0.632	–0.022	–0.014	–0.018	0.000	–0.036
91	–0.282	–0.204	0.162	1.533	–0.025	–0.020	0.043	0.020	–0.066

5 Conclusions

The modelling of horizontal strains and tectonic stresses of the Earth-crust according to the data of geodetic measurements, by using the finite element method, details the structure of current tectonic activity in the research territory and facilitates the geotectonic interpretation of measurement results.

By applying the program package *Ansys* to model strains and stresses of the Earth-crust by the finite element method, the mechanical model of isotropic body deformations is estimated. The model of horizontal strains and stresses of the Earth-crust made in this way corresponds better to the actual physical process occurring in the Earth-crust than models made on the basis of rectilinear geometrical interpolation.

Using the methods of estimation, the plane stresses suggested, it is possible to determine geodynamic laws of current tectonic and local movements occurring in the Earth-crust. The change of the tectonic Earth-crust stresses was estimated and the changes of stress between repeated measurements were determined in the INPP region. It is obvious from the results obtained that the significant change of tectonic stresses occurs in the territory.

The model of the horizontal Earth-crust movements made by the finite element method in the INPP region has clear connections with the tectonic structure of the Earth-crust. The orientation of change directions of the principal maximum stresses is close to the directions of the main breaks of the Earth-crust in this territory and changes from –0.042 MPa to 0.062 MPa, and the orientation of change directions of minimum stresses is perpendicular to the tectonic breaks and changes from –0.296 MPa to 0.006 MPa.

References

Aksamitauskas, C., P. Petroskevicius, A. Zakarevicius, and A. Stanionis (2005). Geodetic measurements for the estimation of eco-geodynamic risks in the area of the Ignalina nuclear power plant. In: *33e Symposium International FESF Strasbourg, Recent Developments in Environmental Protection*, Vol. 40, Peter Lang AG, Bern, pp. 237–251.

ANSYS Theory Reference (1998). Tenth Edition. SAS IP, Inc.

Calais, E., L. Galisson, J.-F. Stéphan, J. Delteil, J. Deverchère, C. Larroque, B. Mercier de Lépinay, M. Popoff, and M. Sosson (2000). Crustal strain in the Southern Alps, France, 1948–1998. *Tectonophysics*, 319, pp. 1–17.

Hsu, R., and S. Li (2004). Decomposition of deformation primitives of horizontal geodetic networks: application to Taiwan's GPS network. *Journal of Geodesy*, 78, pp. 251–262.

Kuang, S. (1996). Geodetic network analysis and optimal design: concepts and applications. Ann Arbor Press, Inc. Chelsea, Michigan, 368 pp.

Marcinkevicius, V., and J. Laskovas (1996). Tectonic conditions of Ignalina Nuclear Power Plant region (in Lithuanian with English abstract). *Geological horizons*, No 1–2, pp. 8–23.

Petroskevicius, P. (2004). Gravitation field effect on geodetic observations (in Lithuanian with English abstract). Vilnius, Technika, 292 pp.

Rao, S. S. (1999). The finite element method in engineering. Third edition. Butterworth–Heinemann, 556 pp.

Ruiz, A. M., G. Ferhat, P. Alfaro, C. Sanz de Galdeano, M.C. de Lacy, G. Rodríguez-Caderot, and A.J. Gil, A (2003). Geodetic measurements of crustal deformation on NW–SE faults of the Betic Cordillera, southern Spain, 1999–2001. *Journal of Geodynamics*, 35, pp. 259–272.

Stanionis, A. (2005). Research of the Earth's crust horizontal movements in the Ignalina nuclear power plant region by geodetic methods. Summary of doctoral dissertation. Vilnius, Technika, 24 pp.

Sue, C., J. Martinod, P. Tricart, F. Thouvenot, J-F. Gamond, J. Fréchet, D. Marinier, J-P. Glot, and J-R. Grasso (2000). Active deformation in the inner western Alps inferred from comparison between 1972-classical and 1996-GPS geodetic surveys. *Tectonophysics*, 320, pp. 17–29.

Vaníček, P., M.R. Craymer, and E.J. Krakiwsky (2001). Robustness analysis of geodetic horizontal networks. *Journal of Geodesy*, 75, pp. 199–209.

Vaníček, P., and E.J. Krakiwsky (1986). Geodesy: The concepts. 2nd rev. ed. North–Holland, Amsterdam, 697 pp.

Varadan, T.K., and K. Bhaskar (1999). Analysis of Plates: theory and problems. Narosa Publishing House, 190 pp.

Zadro, M., and C. Braitenberg (1999). Measurements and interpretations of tilt-strain gauges in seismically active areas. *Earth-Science Reviews*, 47, pp. 151–187.

Zakarevicius, A. (2003). Investigation of the recent geodynamic processes in the territory of Lithuania (in Lithuanian with English abstract). Vilnius, Technika, 195 pp.

Zakarevicius, A., C. Aksamitauskas, and A. Stanionis (2003). Horizontal deformations of the Earth's crust in Ignalina nuclear power plant region (in Russian). *Engineering geodesy*, 49, Київ, pp. 102–110.

Zakarevicius, A., and A. Stanionis (2004). Modelling the horizontal movements and deformations of the earth crust by the finite element method (in Lithuanian with English abstract). *Geodesy and Cartography*, Vol XXX, No 2, Vilnius, Technika, pp. 35–40.

Zienkiewicz, O.C., and R.L. Taylor (2000). The finite element method. Volume 1, Fifth edition. Butterworth-Heinemann, 689 pp.

Accuracy of Displacement Monitoring at Large Dams with GPS

Nuno Lima
Institute of Tropical Sciences Research
Rua da Junqueira, 534, 1300-341 Lisbon, Portugal
E-mail: jnlima@telepac.mail.pt; tel.: ++351 213631862 Fax: ++351 213641947

J. Casaca, M.J. Henriques
Division of Applied Geodesy, National Laboratory for Civil Engineering,
Av. do Brasil 101, 1700-066 Lisbon, Portugal

Abstract. The GPS has dramatically changed the way that surveyors, GIS professionals, engineers and others estimate positions by means coordinates. In fact, since the last years of 20th Century, these scientists and practitioners can determine 3D coordinates of points with centimetre-level accuracy, relative to control points located a few hundred kilometres away. However, GPS positioning isn't still the more used technique in geodetic monitoring of manmade structures, like dams or bridges. There are two main reasons that can explain it: first, GPS techniques have shown some limitations to reach the millimetre-level accuracy; second, the shadow zones generated by the structure itself or by the inclined hillside do not allow good satellite coverage.

In the near future, new developments that are expected for the GPS (L2 civil signal and L5 signal) and the GALILEO implementation will improve the overall system performance accuracy levels. On the other hand, pseudollites that are in a testing phase may improve weak geometry of the GPS satellite constellation. But is static GPS positioning presently competitive with classical geodetic methods?

This paper presents a case study in a Portuguese large dam, where static GPS was compared with classical geodetic measuring methods with submillimetric accuracy. Different observation schemes in static GPS mode were tested to achieve the optimal solution concerning accuracy, cost and reliability.

Keywords. Displacement monitoring, GPS, millimetre-level accuracy.

1 Introduction

After ten years of experimental development followed by ten years of exploration using the complete system, two challenges are still set to the GPS: to reach the millimetre precision and to overcome limitations due to shadow zones. These two factors still limit the use of GPS in geodetic monitoring of structures such as large dams and bridges.

However, new developments and GPS improvements are expected. For example, Block IIR-M satellites will transmit the L2 signal without the script imposed 10 years ago (L2CS, Signal civil L2) and a new opened signal, assigned for L5 frequency of 1176.45 MHz, will be available with the Block IIF satellites. With these new signals it will be obtained, assuredly, an even better level of precision comparing to the one that actually is achieved. On the other hand, in near future, with the implementation of the European GALILEO system, it will be possible to track more satellites, using hybrid GPS-GALILEO receivers, given thus a better satellite configuration. It still has to be pointed out that the pseudollites (pseudo-satellites, terrestrial transmitters of analogous signals to GPS satellites) are now a reality, and permit a significant improvement of the weak geometry of the GPS constellation, benefiting not only the shadow zones, but, also, attenuating the differences between planimetric and altimetric precision.

Nowadays, subcentimetric or even the millimetric precision is not anymore an unattainable platform for the GPS. For instance, recent studies handling real time kinematic positioning (RTK) through permanent stations show that the application of filters allows to improve the precision when eliminating, or at least attenuating significantly, certain errors as, for example, the multipath. One of the used filters is based on the Allan Variance, traditionally used in the study of the stability of atomic clocks that has been applied in the time series of GPS observations for the volcano monitoring of the Papandayan Mount, in Indonesia (Roberts, 2002). Another filter is based on deviation of the absolute minimum (L1 Norm) but requires one or more time series of observations (Kaelber and Jaeger, 2001). Even without these new developments, will the GPS in static mode be able to compete with the classic geodetic methods in the dams and others man-made-structures monitoring? The planned test with the GPS in static mode at Alqueva's dam was an attempt to answer this question.

2 Alqueva's Dam Monitoring with GPS

The Alqueva's dam has a geodetic observation system for the structure displacement monitoring. The system includes one traverse on the crest of the dam that is materialized by several pillars used as stations or as targets. The traverse has fourteen points, as shown in Figure 1. Eight of these points (P1 to P8) are placed on the crest of the dam and are object points; three points (PD, PD1 and PE) are placed in the banks of the river and are the reference points of the traverse. All points are materialized by forced centring devices (Wild type) on the top of concrete pillars. The reference pillars have large concrete foundations on rock and all pillars have thermal isolation (see Figure 2).

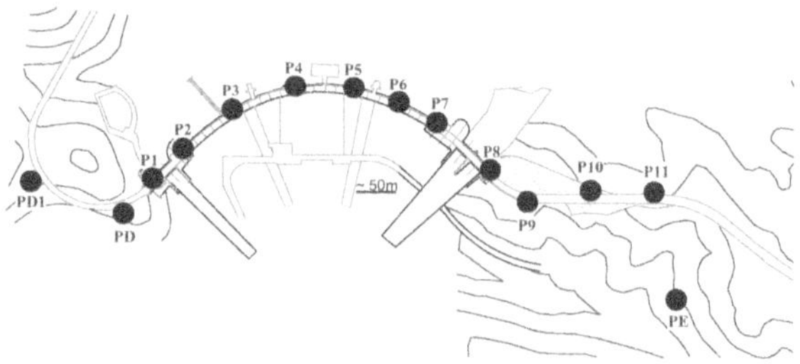

Fig. 1 Alqueva's dam traverse network.

Fig. 2 Reference pillar (PE) with thermal isolation and forced centring device.

For the test, five pillars on the crest were chosen (P2, P3, P4, P5 and P6) and one pillar in the left bank (PE) as a reference station (see Figure 2 and Figure 3).

In January and June of 2004, the National Laboratory for Civil Engineering and the Institute of Tropical Sciences Research had carried out two GPS campaigns. In the same period, two classical geodetic campaigns were also carried out, using a traverse for horizontal displacements monitoring and a geometric levelling for vertical displacements monitoring. For the GPS campaigns three Trimble 4000 SSI receivers were used with Trimble Compact L1/L2 GP antennas. In the first campaign 13 observation sessions of 1 hour duration each had been held out, six in day 20 and seven in day 21 of January. During these 13 sessions, one GPS receiver was on the permanent pillar PE. The observations recorded in this pillar, two series with more than 8 hours each, had allowed the connection of this network to the Portuguese GPS permanent stations, Cascais, Gaia and Mirandela (the stations of Beja and Lagos were not operational in these days). In that way, it was possible to reference the GPS network in ITRF-2000. The 13 sessions allowed to get the network (drawn in Figure 4) characterized by: i) linearly independent vectors between all the points of the network; ii) each pillar was occupied three consecutive sessions, at least. The value of 10º for the cut-of angle (or either, the minimum value from which satellites GPS can be observed) and the value of 15 seconds for the observation rate were chosen.

Fig. 3 Alqueva's dam and the localization of the traverse network pillars.

In the second GPS campaign, at 1 and 2 of June, was followed the observation plan of the first campaign.

The GPS vectors have been processed with two different softwares: TRIMBLE GPSURVEY 2.35 and BERNESE v. 4.2 of the University of Bern. In the case of GPSURVEY, all the linearly independent vectors have been processed, one vector each time (single baseline processing mode) using program WAVE (Trimble, 1999). Afterwards, all these vectors were adjusted with TRIMNET PLUS (Trimble,1999), fixing the PE coordinates, in order to get the coordinates of the reference epoch. In the case of the BERNESE all the linearly independent vectors have been processed, in multistation processing mode, two vectors simultaneously each time. Afterwards, the obtained normal equations of each session were combined using the program ADDNEQ (Hugentobler et al., 2001), constrained the three components of PE to 0.0001 m, and so obtaining the coordinates of the campaign of January 2004.

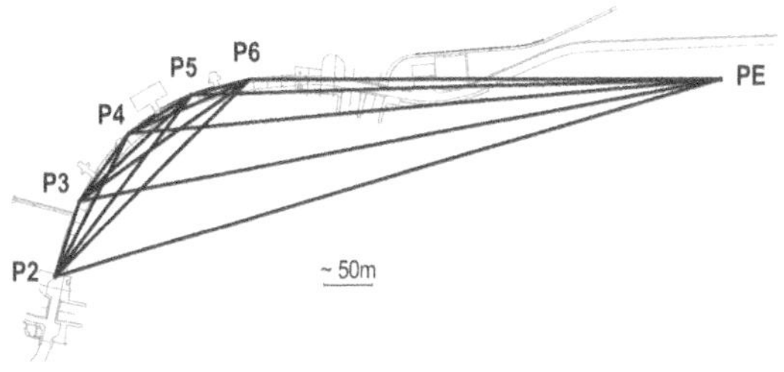

Fig. 4 The GPS network of the Alqueva's dam.

The coordinate differences of the two final solutions, obtained by the two softwares, are shown in Table 1. The agreement between the two solutions is appreciable: in average the values are less than 1 mm in planimetry and less than 5 mm in altimetry.

Table 1. Differences between GPSURVEY and BERNESE, January of 2004.

Station	Latitude difference (m)	Longitude difference (m)	Height difference (m)
P2	0.0003	-0.0003	-0.0020
P3	0.0009	0.0009	0.0074
P4	0.0014	0.0011	0.0062
P5	0.0012	0.0004	-0.0004
P6	-0.0002	0.0005	0.0061
PE	0.0000	0.0000	0.0000

It seems to be very interesting the comparison between the pillars ortometric heights, computed by the geometric levelling, and the pillars ellipsoidal heights, computed by GPSURVEY and BERNESE, Table 2. The agreement between these values (excluding, of course, the geoid undulation that will have to be close to 55.010 m) is very good: in average the values are less than 5 mm.

Table 2. Height differences, January of 2004.

Sta.	H (m)	h1 GPSUR-VEY (m)	h2 BER-NESE (m)	h1 – H (m)	h2 – H (m)
PE	164.16	219.1693	219.1692	55.010	55.010
P2	155.44	210.4499	210.4479	55.010	55.008
P3	155.47	210.4772	210.4846	55.007	55.015
P4	155.41	210.4140	210.4202	55.004	55.010
P5	155.47	210.4861	210.4857	55.016	55.016
P6	155.42	210.4286	210.4347	55.009	55.015

Table 3. Slope distance differences, January of 2004.

Distances S	S1 EDM (m)	S2 GPSURVEY (m)	S3 BERNESE (m)
P2 – P3	83.3948	83.3957	83.3966
P3 – P4	87.8377	87.8381	87.8385
P4 – P5	76.5789	76.5807	76.5800
P4 – P6	136.9478	136.9482	136.9471
P5 – P6	61.8408	61.8390	61.8388
PE – P2	–	719.2137	719.2139
PE – P3	–	675.4068	675.4056
PE – P4	–	614.3332	614.3320
PE – P5	–	546.9728	546.9724
PE – P6	–	487.0457	487.0450
P2 – P6	–	289.1588	289.1590
P2 – P5	–	238.0567	238.0578
P2 – P4	–	169.1099	169.1114
P3 – P5	–	161.3653	161.3651
Differences DS	**S2 – S1 (mm)**	**S3 – S1 (mm)**	**S3 – S2 (mm)**
P2 – P3	0.9	1.8	0.9
P3 – P4	0.4	0.8	0.4
P4 – P5	1.8	1.1	-0.7
P4 – P6	0.4	-0.6	-1.1
P5 – P6	-1.8	-2.0	-0.2
PE – P2	–	–	0.2
PE – P3	–	–	-1.2
PE – P4	–	–	-1.2
PE – P5	–	–	-0.4
PE – P6	–	–	-0.7
P2 – P6	–	–	-1.1
P2 – P5	–	–	1.1
P2 – P4	–	–	1.5
P3 – P5	–	–	-0.2
P3 – P6	–	–	-1.0

In Table 3 the observed distances with the Leica TCA2003 EDM and the obtained ones with GPSURVEY and the BERNESE are compared. Due to the good agreement of the solutions from GPSURVEY and BERNESE, the differences in slope distances are only in the order of 1mm. The differences between these two solutions and the EDM measurements, shown in Table 3, are a little bigger than what already it would be to expect, however less than 2 mm.

The processing of observation of the second GPS campaign was very similar to the one used in the first campaign, with the same softwares and strategies. Table 4 summarizes the planimetric differences estimated by the traverse, the GPSURVEY and the BERNESE.

Table 4 . Planimetric displacement observed between January and June of 2004

Station	(1) EDM DX (mm)	(1) EDM DY (mm)	(2) GPSURVEY DX (mm)	(2) GPSURVEY DY (mm)	(3) BERNESE DX (mm)	(3) BERNESE DY (mm)
PE	0.0	0.0	0.0	0.0	0.0	0.0
P2	0.0	-1.6	4.1	0.8	2.3	0.7
P3	-1.4	0.7	4.3	2.2	3.1	0.6
P4	0.0	-1.5	4.2	-0.5	1.4	-1.4
P5	0.3	-4.2	1.8	0.6	1.2	-0.7
P6	0.7	-3.7	3.0	-2.2	2.8	-2.0
Station	**Difference (2)-(1) DX (mm)**	**Difference (2)-(1) DY (mm)**	**Difference (3)-(1) DX (mm)**	**Difference (3)-(1) DY (mm)**	**Difference (3)-(2) DX (mm)**	**Difference (3)-(2) DY (mm)**
PE	0.0	0.0	0.0	0.0	0.0	0.0
P2	4.1	2.4	2.3	2.3	-1.8	0.1
P3	5.7	1.5	4.5	0.1	-1.2	-1.4
P4	4.2	1.0	1.4	-0.1	-2.8	-0.9
P5	1.5	4.8	0.9	3.5	-0.6	-0.1
P6	2.3	1.5	2.1	1.7	-0.2	-0.2

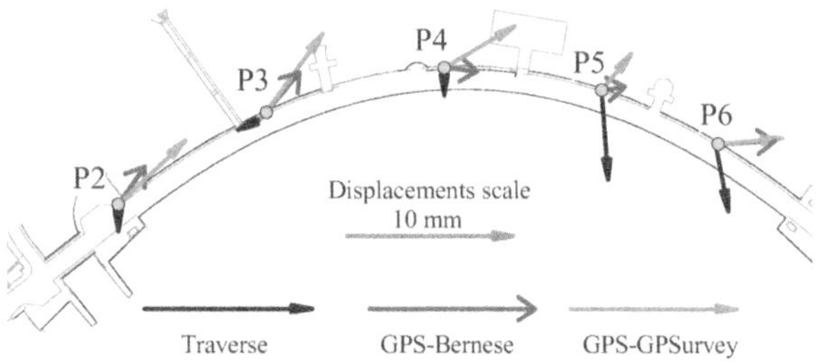

Fig. 5 Planimetric displacements estimated by traverse and GPS, between January and June of 2004 (the black arrows is the traverse solution, the grey arrows is the GPSURVEY solution and the dark grey arrows with large arrowhead is the Bernese solution).

The planimetric displacements observed by the classic geodetic methods between January and June of 2004 are less than 5 mm (only two of these are greater than 2 mm). As these values were less than (or practically less than) the errors of the GPS in static mode, it becomes difficult to estimate them with this last technique, as already mentioned in the Introduction. Table 4 and Figure 5 confirm this difficulty: the "observed" displacements with GPS are not consistent to the observed ones with the classic geodetic methods of submillimetre precision. However, the differences between the observed displacements rarely exceed 5 mm allowing us to foresee the possibility to detect displacements in the order of 1 cm.

The vertical component of the displacements, GPS and geometric levelling, are summarized. The displacements observed with geometric levelling are less than 1 mm (except for P3). Concerning the less precise GPS height component more discrepant results would be expected. However, the differences are always less than 5 mm, if the "observed" displacement of P3 from GPSURVEY adjustment is not considered.

Table 5. Height displacements observed between January and June of 2004

Station	Geometric Levelling Differences (m)	GPSURVEY Height Differences (m)	BERNESE Height Differences (m)
P2	-0.0009	-0.0009	0.0037
P3	-0.0011	0.0127	0.0043
P4	-0.0004	0.0032	0.0048
P5	-0.0006	-0.0050	0.0001
P6	-0.0003	0.0003	0.0014
PE	-	0.0000	0.0000

3 Conclusions

GPS campaigns in the Alqueva's dam were used to test the applicability of this method of positioning for large dam displacement monitoring. The advantage of using scientific software, as the BERNESE, over commercial software, as the GPSURVEY, was not demonstrated. In fact, for short baselines as in the presented network, the results obtained with each one of the two softwares are very similar: the average variation in planimetry was less than 1 mm and the average variation in altimetry was less than 5 mm. The comparison of the results, in the first campaign, between these two softwares with the ones of the classic geodetic methods, seems to guarantee an agreement in the order of 2 mm in planimetry and 5 mm in altimetry. However, the differences between the displacements estimated by these two methods, although being less than 5 mm, are discrepant. Therefore, it is questionable that the value of this agreement is in the order of 2 mm in planimetry, as it could be concluded by the results of the first GPS campaign.

References

Kaelber, Simone e Jaeger, Reiner (2001) – "GPS-Based Online Control and Alarm System", 10th FIG International Symposium on Deformation Measurements, Orange, California, USA, March of 2001.

Hugentobler, U.; Shaer, S. e Fridez, P. (2001), *Bernese GPS Software Version 4.2*. Astronomical Institute, University of Berne.

Roberts, C. (2002) – "A continuous low-cost GPS-based volcano deformation monitoring system in Indonesia", PhD Dissertation, School of Surveying and Spatial Information Systems, University of New South Wales, Sydney, Australia, 287 pp.

Trimble (1999), *GPSurvey V. 2.35 User's Manual*. Trimble Navigation Ltd, Sunnyvale, California, USA.

Analyzing the Deformations of a Bridge Using GPS and Levelling Data

S. Erol, B. Erol, T. Ayan
Department of Geodesy and Photogrammetry Engineering
Istanbul Technical University, Civil Engineering Faculty, Maslak, 34469, Istanbul, Turkey

Abstract. The aim of this study is analysing 1D (in vertical) and 3D movements of a highway viaduct, which crosses over a lake, using GPS and levelling measurement data separately and their combination as well. The data are acquired from the measurement campaigns which include GPS sessions and precise levelling measurements, performed with six–month intervals for two years.

In 1D analysis of the (vertical) deformations, the height differences derived from GPS and levelling data were evaluated. While combining the height–difference sets (GPS derived and levelling derived) in the third stage of 1D analysis, Variance Component Estimation (VCE) techniques according to Helmert's approach and Rao's Minimum Norm Quadratic Unbiased Estimation (MINQUE) approach have been used.

In 3D analysis of the deformations with only GPS data classical S–transformation method was employed.

The theoretical aspects of each method that was used in the data analyses of this study are summarized. The analysis results of the deformation inspections of the highway viaduct are discussed and from the results, an optimal way of combining GPS data and levelling data to provide reliable inputs to deformation investigations is investigated in this study.

Keywords. GPS, Levelling, Deformation Analysis, Variance Component Estimation, S–Transformation

1 Introduction

It has a considerable importance to have the movements of an engineering structure within certain limits for the safety of the community depending on it. To determine whether an engineering structure is safe to use or not, their movements are monitored and possible deformations are detected from the analysis of observations. An appropriate observation technique, which can be geodetic or non–geodetic (geotechnical–structural) according to classification in Chrzanowski and Chrzanowski (1995), is chosen with considering the physical conditions of the observed structure (its shape, size location and so on), environmental conditions (the geologic properties of the based ground, tectonic activities of the region, common atmospheric phenomena around the structure and so on), the type of monitoring (continuous or static) and the required measuring accuracy for being able to recognize the significant movements.

Until the beginning of the 1980's, conventional measurement techniques have been used for detecting the deformations in large engineering structures. After that the advances in space technologies and their geodetic applications provided impetus for their use in deformation measurements (Erol and Ayan (2003)). GPS positioning technique has the biggest benefit of high accuracy 3D positioning; however, the vertical position is the least accurately determined component due to inherent geometric weakness of the system and atmospheric errors (Featherstone et al. (1998)).

Therefore, using GPS measurement technique in deformation measurements at millimeter level accuracy requires some special precautions, such as using forced centering equipment, applying special measuring techniques like the rapid static method for short baselines or designing special equipment for precise antenna height readings (see Erol and Ayan (2003) for their uses in practice). In some cases, even these special precautions remain insufficient and hence, the GPS measurements need to be combined with another measurement technique to improve its accuracy in height component.

In geodetic evaluation of deformations, static observations obtained by terrestrial and/or GPS technique are subject to a *two–epoch analysis*. The two–epoch analysis basically consists of independent Least Squares Estimation (LSE) of the single epochs and geometrical detection of deformations between epochs. Detailed explanations of the methods based on this fundamental idea are found in Niemeier et al. (1982), Chen (1983), Gründig et al. (1985), Fraser and Gründig (1985), Chrzanowski and Chen (1986), Caspary (1987), Cooper (1987), Biacs (1989), Teskey and Biacs (1990), Chrzanowski et al. (1991).

Here, the aim is analysing 1D and 3D deformations of an engineering structure using GPS and levelling measurements data. During the 1D deformation analysis, three different approaches were performed separately. In the first and second approaches, height differences from precise levelling measurements and GPS measurements respectively were input in the analysing algorithm. In the third approach the combination of height differences from both techniques were evaluated for vertical deformation. While combining the two measurement sets, *Helmert Variance Component Estimation* (HVCE) and *Minimum Norm Quadratic Unbiased Estimation* (MINQUE) techniques were used. 3D deformation analysis only with GPS measurements was accomplished using S-transformation technique. The theories behind the used deformation analysis and variance component estimation methods are summarized in the following. Thereafter the optimal solution for combining the GPS and precise levelling data to improve the GPS derived heights and hence to provide reliable inputs via the optimal solution for the deformation investigations are discussed.

The highway viaduct of which deformations were inspected in this study is 2160 meter long and crosses over a lake on 110 piers. It is located in active tectonic region very close to the North Anatolian Fault (NAF). With the aim of monitoring its deformations, four measurement campaigns including GPS sessions and precise levelling measurements were carried out with six–month intervals. The session plans were prepared appropriately for each campaign on a pre–positioned deformation network.

2 Deformation Analysis Using Height Differences

In general, the classical (geometrical) deformation analysis is evaluated *in three steps* in a geodetic network. *In the first step*, the observations, which were recorded at epoch t_1 and epoch t_2, are adjusted separately according to free network adjustment approach. During the computations, all point heights are assumed to be subject to change and the same point heights, which are approximate values, are used in adjustment computation of each epoch. Computations are repeated until all outliers are eliminated.

In the second step, global test procedure is applied to ensure the stability assumptions of network points during the interval $\Delta t = t_2 - t_1$. In the global test, the combined free adjustment is applied to both epoch measurements (t_1 and t_2). In this adjustment computation, the partial–trace minimum solution is applied on the stable points (see Erol and Ayan (2003)).

$$\Omega_1 = \underline{v}_1^T \underline{P_1} \underline{v}_1 ; \quad S_{01}^2 = \frac{\Omega_1}{f_1} \tag{1}$$

$$\Omega_2 = \underline{v}_2^T \underline{P_2} \underline{v}_2 ; \quad S_{02}^2 = \frac{\Omega_2}{f_2} \tag{2}$$

$$\Omega_G = \underline{v}_G^T \underline{P_G} \underline{v}_G ; \quad S_{0G}^2 = \frac{\Omega_G}{f_G} \tag{3}$$

where, f_1 , f_2 , f_G signify the degree of freedom after first, second and third adjustment computations respectively. Equation (1) and equation (2) represent free adjustment computations of the first and second epochs and equation (3) describes combined free adjustment. From the results found in equation (1), (2) and (3), the test value T_G is determined as in the following.

$$\Omega_0 = \Omega_1 + \Omega_2 \; ; \; f_0 = f_1 + f_2$$
$$r = f_G - f_0 \; ; \; T_G = \frac{(\Omega_G - \Omega_0)/r}{(\Omega_0 / f_0)} \tag{4}$$

This test value is independent from the datum and it is in F–distribution. T_G test value is compared with the critical value which is selected from the Fisher–distribution table according to r (rank) and f_0 (degrees of freedom) for S=1-α (0.95) confidence level. If $T_G < F_{r,f_0,1-\alpha}$, the H_0 null hypothesis which implies $H_0 : d = \underline{H}_2 - \underline{H}_1 = 0$ is true for the points of which heights were assumed not to be changed.

On the other hand, if $T_G > F_{r,f_0,1-\alpha}$, in the group of points which had been assumed as stable in the global test procedure, there is/are instable point(s). Then the necessity of localization of the deformations is understood and the combined free adjustment and global test are repeated until only the stable points are left out in the set.

In the last step of the analysis, the following testing procedure is applied to the height changes. Similar to previous steps, T_H test values are calculated for all network points except the stable ones and compared with the critical value of F from the Fisher–distribution table.

$$d = H_2 - H_1 \ ; \ S_0^2 = \frac{\Omega_0}{f_0}$$

$$Q_{dd} = Q_{H_1H_1} + Q_{H_2H_2} \ ; \ T_H = \frac{d^T Q_{dd}^{-1} d}{r S_0^2} \quad (5)$$

If $T_H > F_{r, f_0, 1-\alpha}$, it is concluded that the change in height is significant. Otherwise, it is concluded that the height change d is not significant and it is caused by the random measurement errors.

2.1 Variance Component Estimation

In the method of Least Squares (LS), weights of the observations are the essential prerequisite in order to correctly estimate the unknown parameters. The purpose of variance component estimation (VCE) is basically to find realistic and reliable variances of the measurements for constructing the appropriate a–priori covariance matrix of the observations. Improper stochastic modelling can lead to systematic deviations in the results and these results may appear to include significant deformations. Methods for estimating variance and covariance components, within the context of the LS adjustment, have been intensively investigated in the statistical and geodetic literature. The methods developed so far can be categorized as follows (see Crocetto et al (2000)):

- Functional models
- Stochastic models
- Estimation approaches

When the variance component estimation is concerned, a first solution to the problem was provided by Helmert in 1924, who proposed a method for unbiased variance estimates (Helmert (1924)). In 1970, an independent solution was derived by Rao, who was unaware of Helmert's method, and was called the minimum norm quadratic unbiased estimation (MINQUE) method (Rao (1970)). Under the assumption of normally distributed observations, both Helmert and Rao's MINQUE approaches are equivalent.

2.1.1 Helmert Approach in VCE

A full derivation of the Helmert technique and computational model of variance component estimation is given in Grafarend (1984). A summary of the mathematical model is given below (Kızılsu (1998)). The Helmert equation,

$$H\sigma^2 = c \quad (6)$$

The matrix expression of equation (6) is given in (7),

$$\begin{bmatrix} h_{11} h_{12} \ldots h_{1u} \\ h_{21} h_{22} \ldots h_{2u} \\ \ldots\ldots\ldots\ldots \\ h_{u1} h_{u2} \ldots h_{uu} \end{bmatrix} \begin{bmatrix} \sigma_1^2 \\ \sigma_2^2 \\ \ldots \\ \sigma_u^2 \end{bmatrix} = \begin{bmatrix} c_1 \\ c_2 \\ \ldots \\ c_u \end{bmatrix} \quad (7)$$

where, u is the number of measurement groups

$$c_i = v_i^T P_i v_i \quad (8)$$

$$h_{ii} = n_i - 2\,tr(N^{-1}N_i) + tr(N^{-1}N_i N^{-1}N_i) \quad (9)$$

$$h_{ij} = tr(N^{-1}N_j N^{-1}N_i) \quad (10)$$

where, tr(·) is the trace operator, N is the global normal equation matrix including all measurements, n_i, P_i, N_i, v_i are the number of measurements, assigned weight matrix, normal equation matrix and residuals of the ith group measurements respectively; and σ_i^2 is estimated variance factor.

It can be seen that c_i is a function of P_i. On the other hand P_i is also a function of σ_i^2. Because of this hierarchy, Helmert solution is an iterative computation.

The step by step computation algorithm of Helmert Variance Component Estimation is given as below:

1. Before the adjustment, a unique weight is selected for each of the measurement groups. At the start of iterative procedure, the weights for each measurement groups can be chosen equal to one ($P_1 = P_2 = \ldots\ldots = P_u = 1$)
2. By using the a–priori weights, separate normal equations ($N_1, N_2, \ldots, N_u$) for each measurement group and the general normal equation (N) are composed. Here, the general normal equation is

the summation of normal equations such as: $N = N_1 + N_2 + ... + N_u$.

3. The adjustment process is started, in which the unknowns and residuals are calculated.

$$x = N^{-1}d \ , \ (d = A^T Pb) \tag{11}$$

$$\begin{aligned} v_1 &= A_1 x - b_1 \\ &.................... \\ v_u &= A_u x - b_u \end{aligned} \tag{12}$$

4. Helmert equation is generated (equation (6)).

5. Variance components in Helmert equation $(\sigma_1{}^2, \sigma_2{}^2, ..., \sigma_u{}^2)$ and new weights are calculated.

$$P_{i+1} = P_i / \sigma_i{}^2 \tag{13}$$

6. If the variance component $(\sigma_i{}^2)$ for all groups $(i = 1,2,....,u)$ is equal to one, then the iteration is stopped. If not, the procedure is repeated from the second step using the new weights. The iterations are continued until the variances reach one $(\sigma_i{}^2 = 1)$.

2.1.2 MINQUE Approach in VCE

The general theory and algorithms of the minimum norm quadratic unbiased estimation procedure are described in Rao (1971); Rao and Kleffe (1988). This statistical estimation method has been implemented and proven useful in various applications not only for evaluating the variance–covariance matrix of the observations, but also for modelling the error structure of the observations (Fotopoulos (2003)).

The theory of Minimum Norm Quadratic Estimation is widely regarded as one of the best estimators. Its application to a levelling network has been explained in Chen and Chrzanowski (1985), and it was used for GPS baseline data processing in Wang et al. (1998).

MINQUE is classified as a quadratic–based approach where a quadratic estimator is sought that satisfies the minimum norm optimality criterion. Given the Gauss-Markov functional model v=Ax–b, where b and v are vectors of the observations and residuals, respectively, the selected stochastic model for the data, and the variance covariance matrix are expressed as follows.

$$C_b = \sum_{i=1}^{k} \sigma_i^2 Q_i = \sigma_1^2 Q_1 + \sigma_2^2 Q_2 + ... + \sigma_k^2 Q_k \tag{14}$$

$$C_b = \mathrm{diag}[\sigma_i^2 Q_i] = \begin{bmatrix} \sigma_1^2 Q_1 & 0 & 0 & 0 \\ 0 & \sigma_2^2 Q_2 & 0 & 0 \\ 0 & 0 & .... & 0 \\ 0 & 0 & 0 & \sigma_k^2 Q_k \end{bmatrix} \tag{15}$$

where, only variance components $\sigma_i{}^2$ are to be estimated. Such a model is used extensively for many applications, including Sjöberg (1984), Caspary (1987) and Fotopoulos and Sideris (2003).

The MINQUE problem is reduced to the solution of the following system

$$S\sigma_i{}^2 = q \tag{16}$$

S is a k×k symmetric matrix and each element s_{ij} in the matrix S is computed from the expression

$$s_{ij} = tr(RQ_iRQ_j) \ ; \ i,j = 1,2,....k \tag{17}$$

where,. tr(·) is the trace operator, Q(·) is a positive definite cofactor matrix for each group of observations. R is a symmetric matrix defined by

$$R = C_b^{-1}[I - A(A^T C_b^{-1} A)^{-1} A^T C_b^{-1}] \tag{18}$$

where, I is an identity matrix, A is an appropriate design matrix of full column–rank and C_b is the covariance matrix of the observations. The vector q contains the quadratic forms

$$q = \{q_i\} \ ; \ q_i = v_i{}^T Q_i{}^{-1} v_i = b^T RQ_i Rb \tag{19}$$

where, v_i are the estimated observational residuals for each group of observations b_i. As a result we can generate equation (16) as below.

$$\begin{bmatrix} tr(RQ_1RQ_1) & tr(RQ_1RQ_2) & ...... & tr(RQ_1RQ_k) \\ tr(RQ_2RQ_1) & tr(RQ_2RQ_2) & ...... & tr(RQ_2RQ_k) \\ ...... & ...... & ...... & ...... \\ tr(RQ_kRQ_1) & tr(RQ_kRQ_2) & & tr(RQ_kRQ_k) \end{bmatrix} \begin{bmatrix} \sigma_1^2 \\ \sigma_2^2 \\ .... \\ \sigma_k^2 \end{bmatrix} = \begin{bmatrix} b^T RQ_1 Rb \\ b^T RQ_2 Rb \\ ...... \\ b^T RQ_k Rb \end{bmatrix} \tag{20}$$

The computed values resulting from a first run through the MINQUE algorithm is $(\sigma_i^2)_1$. $(\sigma_i^2)_0$ can be chosen equal to one as a–priori values for each variance factor. The resulting estimates $(\sigma_i^2)_1$ can be used as 'new' a–priori values and the MINQUE procedure is repeated. Performing this process iteratively is referred to as *iterative MINQUE* (IMINQUE). The iteration is repeated until all variance factor estimates approach unity. The final estimated variance component values can be calculated by

$$\sigma_i^2 = \prod_{\alpha=0}^{n} (\sigma_i^2)_\alpha \tag{21}$$

3 Deformation Analysis Using GPS Results

3.1 S–Transformation

The datum consistency between different epochs can be obtained by employing the S–Transformation and also the moving points are determined by applying this transformation (see Baarda (1973); Strang van Hees (1982)). S–transformation is an operation that is used for transition from one datum to another datum without using a new adjustment computation. In other words, S–transformation is a transformation computation of the unknown parameters, which were determined in one datum, from the current datum to the new datum with their cofactor matrix. S–transformation is similar to the free adjustment computations. The equations that give the transition from the datum i to datum k are given below (Demirel (1987), Welsch (1993)).

$$\underline{x}^k = \underline{S}_k \underline{x}^i \tag{22}$$

$$\underline{Q}^k_{xx} = \underline{S}_k \underline{Q}^i_{xx} \underline{S}_k^T \tag{23}$$

$$\underline{S}_k = \underline{I} - \underline{G}\left(\underline{G}^T \underline{E}_k \underline{G}\right)^{-1} \underline{G}^T \underline{E}_k \tag{24}$$

where, $\underline{I}$ is the identity matrix, $\underline{E}_k$ is the datum determining matrix whose diagonal elements are 1 for the datum determining points and 0 for the other points.

$$\underline{E}_k = \begin{bmatrix} \underline{0} & \underline{0} & \underline{0} \\ \underline{0} & \underline{I} & \underline{0} \\ \underline{0} & \underline{0} & \underline{0} \end{bmatrix} \tag{25}$$

$$G^T = \begin{array}{c} \\ \left[\begin{array}{cccccccc} \multicolumn{3}{c}{\text{Point 1}} & \multicolumn{3}{c}{\text{Point 2}} & & \\ 1 & 0 & 0 & 1 & 0 & 0 & \cdots \\ 0 & 1 & 0 & 0 & 1 & 0 & \cdots \\ 0 & 0 & 1 & 0 & 0 & 1 & \cdots \\ 0 & -\bar{z}_{10} & \bar{y}_{10} & 0 & -\bar{z}_{20} & \bar{y}_{20} & \cdots \\ \bar{z}_{10} & 0 & -\bar{x}_{10} & \bar{z}_{20} & 0 & -\bar{x}_{20} & \cdots \\ -\bar{y}_{10} & \bar{x}_{10} & 0 & -\bar{y}_{20} & \bar{x}_{20} & 0 & \cdots \\ \bar{x}_{10} & \bar{y}_{10} & \bar{z}_{10} & \bar{x}_{20} & \bar{y}_{20} & \bar{z}_{20} & \cdots \end{array} \right] \end{array} \begin{array}{l} \\ \\ \text{3 Shifts} \\ \\ \\ \text{3 Rotations} \\ \\ \text{1 Scale} \end{array} \tag{26}$$

where, $\bar{x}_{i0}, \bar{y}_{i0}, \bar{z}_{i0}$ are the shifted approximate coordinates of points to the mass center of control network and i=1,2,3 … p is the number of points.

In the conventional 3D geodetic networks, the number of datum defects due to outer parameters is 7. However, in GPS networks the number of datum defects is 3 as the number of shifts on the three axis directions (Welsch (1993)). On the other hand, the number of datum defects is exactly known in the conventional networks as related to measurements that are performed on the network, where as this number can not be known exactly in the GPS networks because of error sources such as atmospheric effects, using different antenna types, using different satellite ephemerides in very long baseline measurements and directing the antennas to the local north (see Blewitt (1990)).

3.2 Global Test Using S-Transformation

A control network is composed of the datum points and deformation points. With the help of the datum points, the control network, which is measured in t_i and t_j epochs, is transformed to the same datum.

While inspecting the significant movements of the points, continuous datum transformation is necessary. Because of this, first of all, the networks, which are going to be compared to each other, are adjusted in any datum such as using free adjustment technique.

After applying this technique, the coordinates of the control network points, which were measured in t_i epoch, are divided into two groups as f (datum) points and n (deformation) points.

$$\underline{x}^i = \begin{bmatrix} \underline{x}^i_f \\ \underline{x}^i_n \end{bmatrix} \tag{27}$$

$$\underline{Q}^i_{xx} = \begin{bmatrix} \underline{Q}^i_{ff} & \underline{Q}^i_{fn} \\ \underline{Q}^i_{nf} & \underline{Q}^i_{nn} \end{bmatrix} \tag{28}$$

where, $\underline{x}^i$ is the vector of parameters and is $\underline{Q}^i_{xx}$ cofactor matrix in the datum i. The transformation is accomplished from datum i to datum k by the help of $\underline{E}_k$, datum determining matrix. This transformation is as given in equations (29) and (30),

$$\begin{bmatrix} \underline{x}^k_f \\ \underline{x}^k_n \end{bmatrix} = \underline{S}_k \begin{bmatrix} \underline{x}^i_f \\ \underline{x}^i_n \end{bmatrix} \tag{29}$$

$$\underline{Q}^k_{xx} = \begin{bmatrix} \underline{Q}^k_{ff} & \underline{Q}^k_{fn} \\ \underline{Q}^k_{nf} & \underline{Q}^k_{nn} \end{bmatrix} = \underline{S}_k \underline{Q}^i_{xx} \underline{S}_k^T \tag{30}$$

The operations, which are given in equations (27)–(30), are repeated for the transformation from datum j to datum k. In this way, datum i and datum j can be transformed into the same datum k with the help of the datum points. As the result, the vectors of coordinate unknowns $(\underline{x}_f^k)_i$ and $(\underline{x}_f^k)_j$, and also their cofactor matrix $(\underline{Q}_{ff}^k)_i$ and $(\underline{Q}_{ff}^k)_j$ are found for the datum points in the same k datum. With the global (congruency) test, it is determined if there are any significant movements in the datum points or not. For the global test of the datum points, the H_0 null hypothesis and T test value are (Pelzer (1971), Caspary (1987), Fraser and Gründig (1985))

$$H_0 : E(\underline{x}_f^k)_i = E(\underline{x}_f^k)_j \quad \text{(}H_0\text{ null hypothesis)} \tag{31}$$

$$\underline{d}_f = (\underline{x}_f^k)_j - (\underline{x}_f^k)_i \quad \text{(displacement vector)} \tag{32}$$

$$(\underline{Q}_{dd})_f = (\underline{Q}_{ff}^k)_i + (\underline{Q}_{ff}^k)_j \quad \text{(cofactor matrix of } d_f) \tag{33}$$

$$\underline{R}_f = \underline{d}_f^T (\underline{Q}_{dd})_f^+ \underline{d}_f \quad \text{(quadratic form)} \tag{34}$$

$$s_0^2 = \frac{f_i s_{0i}^2 + f_j s_{0j}^2}{f_i + f_j} \quad \text{(pooled variance factor)} \tag{35}$$

$$T = \frac{R_f}{s_0^2 h_f} \quad \text{(test value)} \tag{36}$$

where, the degree of freedom of R_f is $h_f = \text{rank}\{(\underline{Q}_{dd})_f\} = u_f - d$; u_f is the number of unknowns for the datum points, d is datum defect, $(\underline{Q}_{dd})_f^+$ is pseudo inverse of $(\underline{Q}_{dd})_f$ such that $(Q_{dd})_f^+ = ((\underline{Q}_{dd})_f + GG^T)^{-1} - G(G^TGG^TG)^{-1}G^T$ (Caspary, (1987)).

If $T > F_{h_f,f,1-\alpha}$ ($f = f_i + f_j$), it is decided that there is a significant deformation in the part of datum points of the control network

After the result of global test, if it is decided that there is deformation in one part of the datum points, the determination of the significant point movements using S–transformation (localization of the deformations) step is started (Chrzanowski and Chen (1986), Fraser and Gründig (1985)). In this step, it is assumed that each of the datum points might have undergone change in position. For each point, the group of these points is divided into two parts: the first part includes the datum points, which are assumed as stable and the second part includes one point, which is assumed as instable. All the computation steps that are explained above are repeated one by one for each datum points. In this way, all of the points are tested according to whether they are stable or not. In the end, the exact datum points are derived (Caspary (1987), Demirel (1987)).

3.3 Determining the Deformation Values

After determining the significant point movements as in section 3.2., the block of datum points, which do not have any deformations, is to be determined. With the help of these datum points, both epochs are shifted to the same datum and deformation values are computed as explained below (Cooper (1987)).

The deformation vector for point P is:

$$\underline{d} = \begin{bmatrix} x_k^j - x_k^i \\ y_k^j - y_k^i \\ z_k^j - z_k^i \end{bmatrix} = \begin{bmatrix} d_x \\ d_y \\ d_z \end{bmatrix} \tag{37}$$

the magnitude of the vector is:

$$d = \sqrt{\underline{d}^T \underline{d}} \tag{38}$$

To determine the significance of these deformation vectors, which are computed according to above equations, the H_0 null hypothesis is carried out as given below.

$$H_0 : d = 0 \tag{39}$$

and the test value:

$$T = \frac{\underline{d}^T \underline{Q}_{dd}^{-1} \underline{d}}{3s_0^2} \tag{40}$$

This test value is compared with the critical value, $F_{3,f;1-\alpha}$. If $T \rangle F_{3,f;1-\alpha}$, it is concluded that there is a significant 3D deformation in the point P.

4 Numerical Example

In this study, the deformations of a highway viaduct– called as Karasu, were investigated using GPS and precise levelling data. It is 2160 m long and located in the west of Istanbul, Turkey in one part of the European Transit Motorway. The first 1000 meter of this viaduct crosses over the Büyükçekmece Lake, the piers of the structure were constructed in to this lake (see Figure 1).

The viaduct consists of two separate tracks (northern and southern) and was constructed on 110 piers (each track has 55 piers). The distance between two piers is 40 meters and there is one deformation point in every 5 piers.

The deformation measurements of the viaduct involved four measurement campaigns, which include GPS measurements and precise levelling measurements, performed with six–month intervals. Before performing the measurement campaigns, a well designed local geodetic network had been established in order to investigate the deformations of the structure (see Figure 1). It has 6 reference points around and 24 deformation points on the viaduct. The deformation points are established exactly at the top of piers, where are expected as the most stable locations on the body of viaduct. The network was measured using GPS in each of the sessions, which were planned carefully for each campaign, and precise levelling was applied in between network points.

During GPS sessions, static measurement method was applied and dual frequency receivers were used with forced centering equipments. Levelling measurements were carried out using Koni 007 precise level with two invar rods. The relative accuracies of point positions are in millimeter level. The heights derived from precise levelling measurements have the accuracy between 0.2–0.8 millimeters.

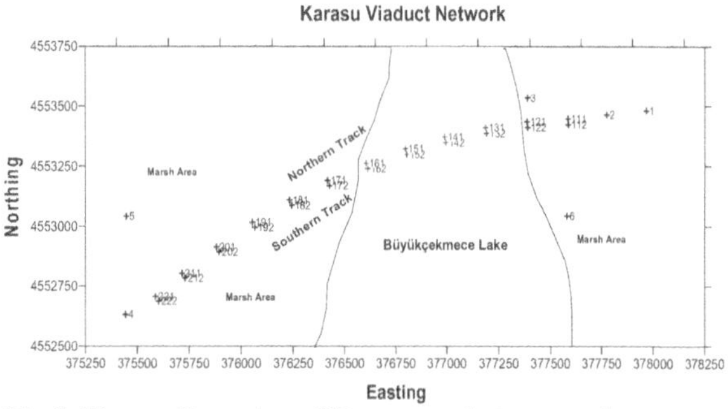

Fig 1. The configuration of Karasu geodetic network

The results of evaluations for the three height difference sets using conventional deformation analysis are seen in the Figures 3, 4 and 5.

In the third evaluation, height differences from both GPS and levelling techniques are used and deformation analysis is applied. Having different accuracies for both height sets derived from both techniques is a very important consideration at this point. Therefore, the stochastic information between these measurement groups (relative to each other) has to be derived. For computing the weights of measurement groups, MINQUE and Helmert Variance Component Estimation (HVCE) techniques have been employed. Figures 2a and 2b show the results of employing the VCE techniques.

In order to both of VCE techniques, although that the similar results were reached after the same number of iteration, the MINQUE results were chosen and applied in combining the data sets, because MINQUE technique provided smoother trend in comparison to Helmert technique.

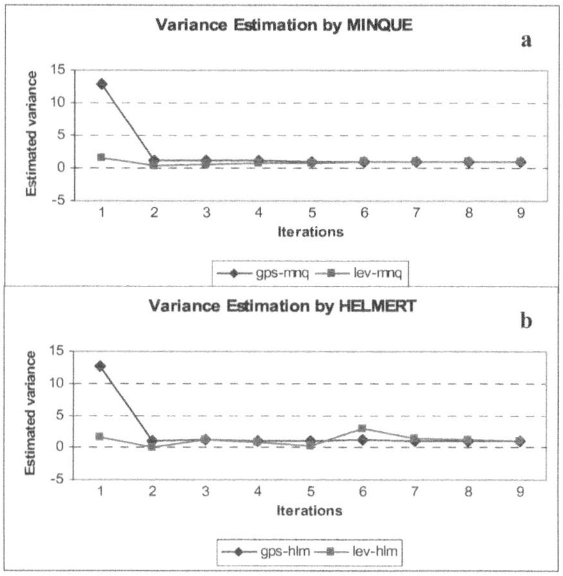

Fig 2. The graph shows the iteration result of VCE techniques; **a** MINQUE, **b** Helmert

It should be noted that the deformations in both tracks represented the same character in analyses results, therefore the only height differences belonging to northern–track points are given in the graphs here.

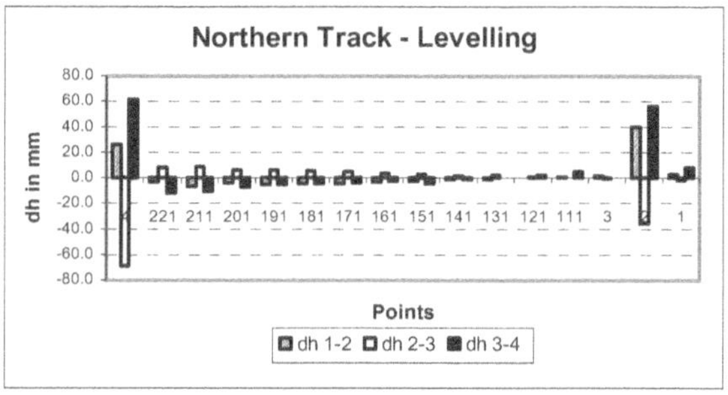

Fig 3. Levelling derived height differences between consecutive epochs for each point in the northern track.

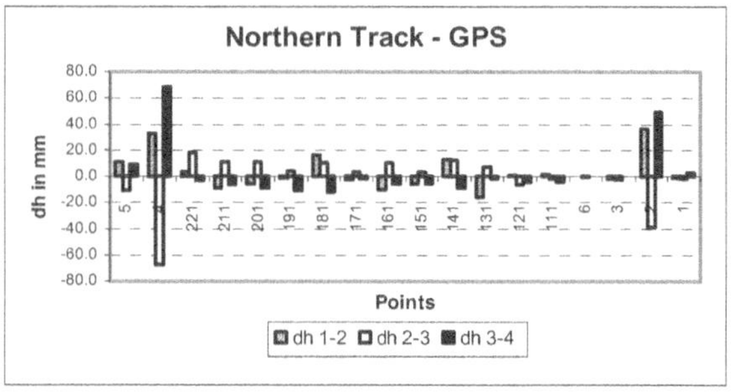

Fig 4. GPS derived height differences between consecutive epochs for each point in the northern track.

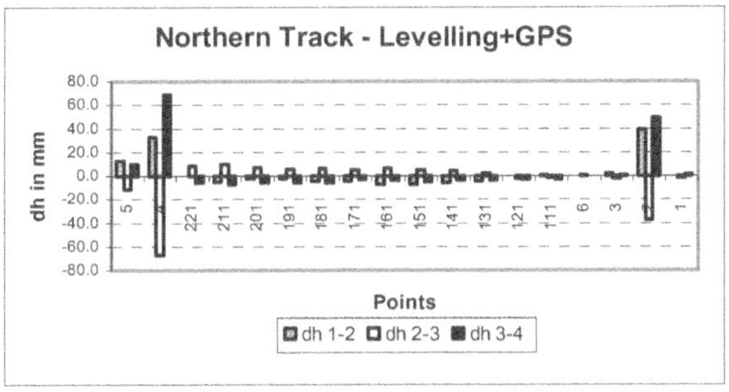

Fig 5. Combined derived height differences between consecutive epochs for each point in the northern track.

The results of evaluations for three height difference sets using conventional deformation analysis are confirmed each other. According to figures 3, 4 and 5, all reference and deformation points have similar characteristic of movements between consecutive epochs. Maximum movements were realised in point 2 and point 4 and these movements were interpreted as deformation.

In figure 6, geoid undulation changes at some of the viaduct points are seen. However, when the figures 3 and 4 were investigated, it is understood that these changes were caused by errors stem from GPS observations and does not give any idea related to deformations of the points. Because, levelling derived height differences between consecutive epochs don't show any height change for these points but GPS derived height differences do.

As the result, it was seen that levelling measurements provide check for GPS derived heights and possible antenna height problems occurred during GPS sessions, and thus benefit to GPS measurements. This is considerable contribution of levelling measurements to GPS measurements in deformation monitoring and analysing.

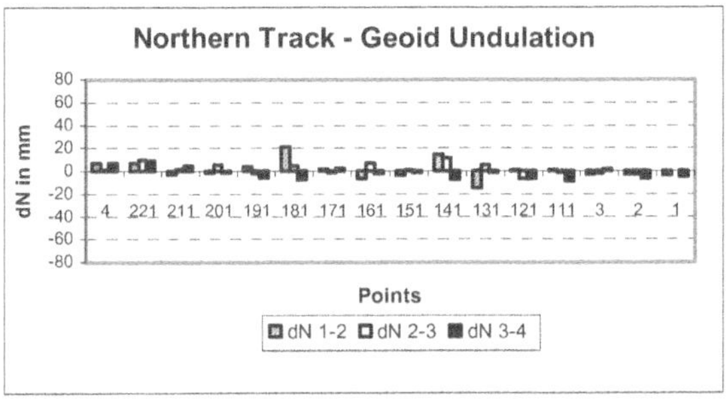

Fig 6. GPS and Levelling derived Geoid Undulation differences between consecutive epochs for each point in the northern track.

After the 1D deformation analysis, 3D deformation analysis was accomplished as mentioned in previous sections. The horizontal displacements, which were found in the 3D analysis results, can be seen in figure 7.

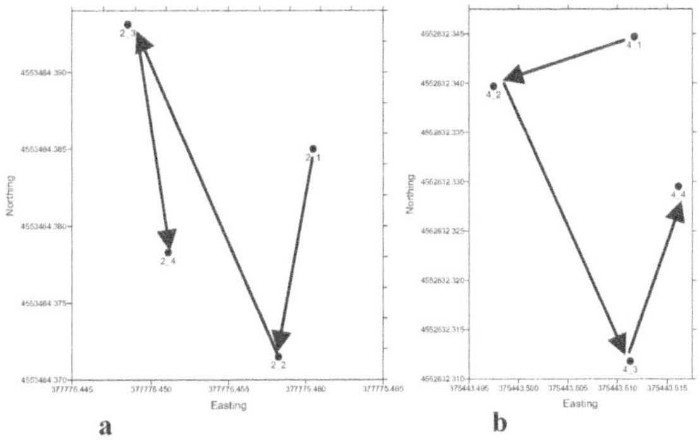

Fig 7. Horizontal displacements; **a** in point 2 (maximum 2 centimeter), **b** in point 4 (maximum 3 centimeter)

5 Results and Conclusion

It is well known from numerous scientific researches that the weakest component in GPS derived position is the height component, mainly because of the geometric structure of GPS. Therefore, in determining vertical deformations, GPS derived heights need to be supported by precise levelling measurements in order to improve their accuracies.

Herein, the 1D and 3D deformations of a large engineering structure has been investigated and analysed using GPS and levelling data separately and also, their combinations. On the other hand an optimal algorithm for combining GPS derived and levelling derived heights in order to improve GPS quality in deformation investigations were analysed using the case study results.

When the figures 3, 4 and 5 were investigated, surprisingly the maximum height changes were seen in point 2 and point 4 even though they are pillars and they had been assumed to be stable at the beginning of the study. According to analysis results of the GPS observations, height changes for some deformation points on the viaduct were recognized. However, while the evaluations with the levelling and combined data were considered, it is understood that these changes, seemed to be deformations on the deformation points, are not significant and caused by the error sources in GPS measurements.

In the second stage of the study, 1D analysis results have supplied input into the 3D analysis of the deformations, while determining whether the network points are stable or not. Horizontal displacements were detected in points 2 and 4 in the result of 3D analysis, (see figure 7) whereas points 1, 3 and 5 were stable. At a first glance, these displacements in 2 and 4 were unexpected. However after geological and geophysical investigations, the origin of these results was understood. The area is a marsh area and this characteristic might widen also underneath of these two reference points. The uppermost soil layer in the region does not seem to be stable and the foundations of the constructions of the reference points 2 and 4 are not founded as deep as the piers of the viaduct. So, they are affected by the environmental conditions easily. Also Points 1, 3 and 5 are not goes as deep as the piers, but their foundations are not similar with point 2 and 4 and are steel marks on the 3×3×3 meters concrete block. The variety of soil layer in the region, known according to geological investigations, also might have role in this result.

On the other hand, it is possible to mention about the correlation between vertical movements in two reference points, 2 and 4, and wet/ dry seasons, because the uplift and sinking movements of these reference points seems to be very synchronous when compared to seasonal changes in the amount of water.

The results of this study, experienced with measurements of the viaduct, are thought to be important remarks for deformation analysis studies using GPS measurements. As the first remark, GPS measurement technique can be used for determining deformations with some special precautions like using forced centering mechanisms to avoid centering errors, using special equipments for precision antenna height readings, using special antenna types to avoid multipath effects etc. However, even though these precautions are taken to provide better results in 1D and 3D deformation analysis, GPS measurements have to be supported with Precise Levelling measurements.

Acknowledgement. The authors acknowledge Volkswagen–Stiftung, Germany, who sponsored the project, the 17th Division of General Directorate of Highway Organization, who provided security on highway during measurement campaigns and also the Institute of Physical Geodesy, Darmstadt University of Technology, who were the partners of the project. The reviewers are kindly acknowledged for their contributions in the improvement of the article.

References

Baarda, W. (1973). S-transformation and criterion matrices, *Netherlands Geodetic Commission Publications on Geodesy*, New Series, 5(1)

Biacs, Z. F. (1989). Estimation and hypothesis testing for deformation analysis in special purpose networks, *UCSE Report 20032,* Department of Surveying Engineering, University of Calgary.

Blewitt, G. (1990). GPS Techniques for Monitoring Geodynamics at Regional Scales. *In Proceeding of Second International Symposium on GPS*, Canada, pp. 495-507.

Caspary, W. F. (1987). Concept of network and deformation analysis. *Wales, Monograph 11*, School of Surveying, University of New South Wales, pp. 1–183.

Chen, Y. G. and A. Chrzanowski (1985). Assessment of levelling measurements using the theory of MINQUE. *Proceeding of 3rd International Symposium North American Vertical Datum*, Rockville, pp. 389-400.

Chen, Y. Q. (1983). Analysis of deformation surveys – a generalized method. *Technical Report No: 94*, Department of Surveying Engineering, University of New Brunswick, pp. 1–262.

Chrzanowski, A. and Y. Q. Chen (1986). Report of the ad hoc committee on the analysis of deformation surveys. *In Proc. of the FIG XVIII International Congress. Paper No: 608.1,* Toronto 6, pp. 167-185.

Chrzanowski A., Y. Q. Chen, J. M. Secord, A. S. Chrzanowski (1991). Problems and solutions in the integrated monitoring and analysis of dam deformations, *CISM Journal*, 45(4). pp. 547-560.

Chrzanowski, A., A. S. Chrzanowski (1995). Identification of dam deformation mechanism. *In Proceedings of the MWA International Conference on Dam Engineering*, Kuala Lumpur, pp. 179-187.

Cooper, M. A. R. (1987). Control surveys in civil engineering. *Blackwell Scientific*, Oxford, pp. 1–381.

Crocetto, N., M. Gatti, P. Russo (2000). Simplified Formulae for The BIQUE Estimation of Variance Components in Disjunctive Observation Groups. *Journal of Geodesy*, 74 pp. 447-457.

Demirel, H. (1987). S-Transformation and Deformation Analysis. *In Proc. of First Turkey Technical Surveying Congress,* Chamber of Surveying Engineers of Turkey, Ankara, Turkey, February 23-27, (in Turkish)

Erol, S., T. Ayan (2003). An Investigation on Deformation Measurements of Engineering Structures With GPS And Levelling Data Case Study. *In: Proc. International Symposium on Modern Technologies, Education and*

Professional Practice in the Globalizing World, Sofia, Bulgaria, November 6-7.

Featherstone, W. E., M. C. Denith, J. F. Kirby (1998). Strategies for the Accurate Determination of Orthometric Heights from GPS. *Survey Review* 34, 267, pp 278-296.

Fotopoulos, G. (2003). An Analysis on the Optimal Combination of Geoid, Orthometric and Ellipsoidal Height Data. PhD. Thesis, *UCGE Report No. 20185*, Department of Geomatics Engineering, The University of Calgary.

Fotopoulos, G., and M. G. Sideris (2003) On the estimation of variance components using GPS, geoid and levelling data, *Canadian Geophysical Union Annual meeting: Challenges and opportunities for geophysics in Canada*, Banff, Canada, May 10-14.

Fraser, C. S., and L. Gründig (1985). The analysis of photogrammetric deformation measurements on Turtle Mountain, *Photogrammetric Engineering and Remote Sensing*, 51(2), pp. 207-216.

Grafarend, E. W. (1984). Variance-Covariance Components Estimation, Theoretical Results and Geodetic Applications, *16th European Meeting of Statisticians*, Marburg.

Gründig, L., M. Neureither, J. Bahndorf (1985). Detection and localization of geometrical movements. *ASCE Journal of Surveying Engineering*, 111(2), pp. 118-132.

Helmert, F. R. (1924). Die Ausgleichungsrechnung nach der Methode der Kleinsten Quadratic, *3. AUFL.*, Leipzig/Berlin.

Kızılsu, G. (1998) Precision Analysis for Lageos I and Lageos II, PhD Thesis, Istanbul Technical University, Institute of Science and Technology, Istanbul, Turkey (in Turkish).

Niemeier, W., W. F. Teskey, R. G. Lyall (1982). Precision, reliability and sensitivity aspects of an openpit monitoring network. *Australian Journal of Geodesy*, Photogrammetry and Surveying, 37, pp. 1-27.

Pelzer, H. (1971). Zur analyse geodatischer deformationsmessungen, *Deutsche Geodatische Kommission, Munich,* Reihe C, Heft 164, pp. 1–86.

Rao, C. R. (1970). Estimation of Heterogeneous Variances in Linear Models, *Journal of American Statistical Association*, vol. 65, pp. 161 – 172.

Rao, C. R. (1971). Estimation of Variance Components – MINQUE Theory. *Journal of Multivariate Statistics*, vol.1, pp. 257-275.

Rao, C. R., and J. Kleffe (1988). Estimation of Variance Components and Applications, *North-Holland Series in Statistics and Probability*, 3.

Sjöberg, L. (1984). Non-negative Variance Component Estimation in the Gauss-Helmert Adjustment Model. *Manuscripta Geodaetica*, vol. 9, pp. 247-280.

Strang van Hees, G. L. (1982). Variance-covariance transformations of geodetic networks, *Manuscripta Geodaetica*, 7, pp. 1–20.

Teskey, W. F., and Z. F. Biacs (1990). A PC-based program system for adjustment and deformation analysis of precise engineering and monitoring networks. *Australian Journal of Geodesy, Photogrammetry and Surveying*, 52, pp. 37-55.

Wang, J., M. Stewart, M. Tsakiri (1998). Stochastic Modelling for Static GPS Baseline Data Processing. *Journal of Surveying Engineering*, 124(4), pp. 171-181.

Welsch, W. M. (1993). A General 7-parameter Transformation for the Combination, Comparison and Accuracy Control of Terrestrial and Satellite Network Observations, *Manuscripta Geodaetica*, 18, pp. 295-305.

GPS for Subsidence Detection, the Case Study of Aguascalientes

R. Esquivel, A. Hernández
Dirección General de Geografía, Instituto Nacional de Estadística Geografía e Informática (INEGI)
Av. Héroe de Nacozari Sur 2301, C.P. 20270, Aguascalientes, Ags., México

M. E. Zermeño
Universidad Autónoma de Aguascalientes, Av. Universidad S/N, C.P. 20100 Aguascalientes, Ags., México

Abstract. The paper describes the land subsidence monitoring project for the city of Aguascalientes, Mexico, and includes a brief analysis of the results obtained during the first year of work. The monitoring is carried out by means of geodetic surveys with GPS equipment over 60 points established on selected sites as the most representative of the ground faults detected within the urban spot.

The objective is to describe the horizontal and vertical behaviour, in order to later infer, by using mathematical models, the faulting behaviour of the land in the years to come. The results show subsidence of 18 centimetres in some points of the city and a close relationship between groundwater extraction and land subsidence.

The works performed are part of an interdisciplinary project in which geotechnics, geophysics, geohydrology and civil engineering studies are carried out, where the National Institute of Statistics, Geography and Informatics (INEGI) is participating as a member of the "Interinstitutional Committee for Geologic Faults and Ground Failures" mainly conformed by some state and municipal governmental institutions, decentralized public offices and some organizations representative of civil society.

Keywords. Subsidence, GPS, ground failures, water extraction

1 Introduction

The city of Aguascalientes is located on the southern part of the valley of Aguascalientes (Figure 1), near the geographic centre of Mexico. Weather in the valley is semidesertic, with a mean annual rainfall of 526 mm. Subsidence in the city and related ground failures were detected 25 years ago and have been directly associated with aquifer depletion due to groundwater overexploitation.

Yearly groundwater extraction is almost 100% bigger than aquifer recharge. Some consequences of ground failures are: aquifer contamination because of damage of sewage lines; structural damages in housing, public buildings, and roads. According to the experts (Arroyo (2003)), aquifer depletion has also become a factor for induced seismicity.

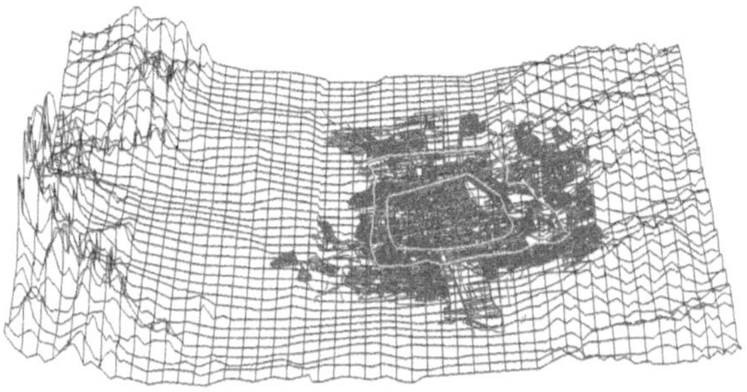

Fig. 1. The city of Aguascalientes on the south of the valley.

2 Previous Studies

Some studies were done during the 80's and 90's, mainly concluding that the main cause of the problem is aquifer overexploitation (Lermo (1994)) and presenting a global description of the geological structures that determine the ground failure behaviour. Nevertheless it is required a further study of the urban zone.

2.1 Topographic Surveys

In March 2003 the Autonomous University of Aguascalientes (UAA) performed relevelling surveys over seven groups of four benchmarks, each placed on some ground failures (Figure 2). These marks were established and measured with topographic equipment in 1985 (Aranda and Aranda (1985)) and remeasurements were made in 1998 and 1993. Unfortunately, in the references there are no details about equipment employed.

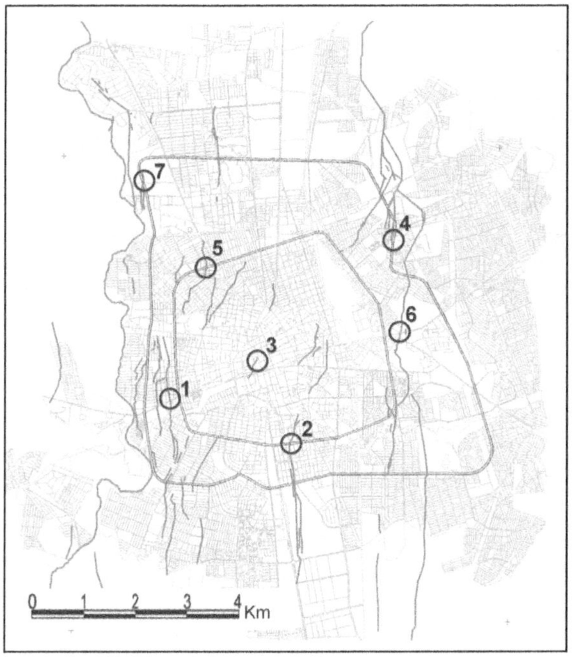

Fig. 2. Location of the sites where topographic measurements were made.

Initially, absolute and relative surveys were planned; nevertheless remeasurements over the complete network were not performed. For relative measurements over each group of four benchmarks a height of 100 metres was given to the highest one, named BM1 (Figure 3).

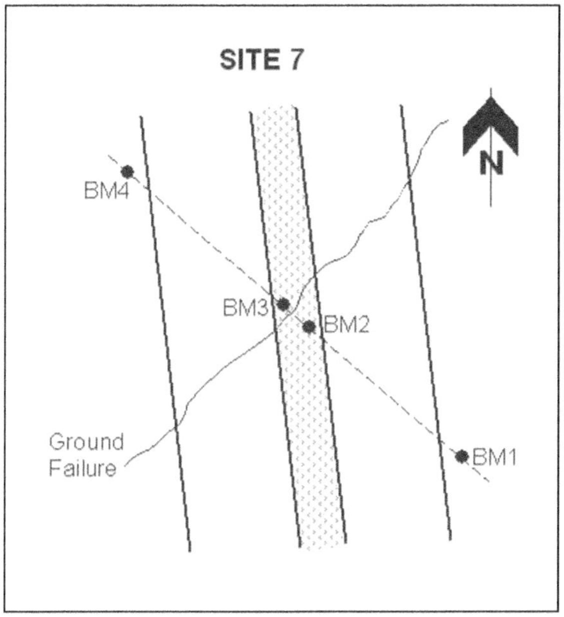

Fig. 3. Example of the distribution of the four benchmarks for topographic measurements.

Table 1 shows the four observations for each BM4 at the seven sites with uncertainties of ±1 centimetre, the derived relative drop (vertical displacement between both sides of the failure) from the first to the last observation, and the relative velocity in the 18 years of the study.

Table 1. Results from previous topographic surveys.

Site	Observations (metres)				Drop	Velocity
	1985	1993	1998	2003	(m)	(m/yr)
1	99.35	99.12	99.11	98.80	0.55	0.031
2	98.91	98.82	98.79	98.76	0.15	0.008
3	99.71	99.62	99.61	99.58	0.13	0.007
4	97.90	97.78	97.64	97.56	0.34	0.019
5	99.19	99.03	98.85	98.81	0.38	0.021
6	98.36	98.10	97.80	97.60	0.76	0.042
7	98.83	98.82	98.79	98.76	0.07	0.004

Largest vertical displacements at sites 1 and 6 of 55 and 76 (±2 centimetres) in 18 years respectively, give an idea of the magnitude of the subsidence problem in the city.

2.2 Data Analysis of the National Geodetic Active Network

In 2003, Márquez and DeMets published a regional crustal deformation research based on the National Geodetic Active Network GPS data provided by INEGI. In that work the special case of the station INEG, located on the INEGI building in the city of Aguascalientes, was detected. The station presented an unusual vertical displacement with respect to ITRF00. The study was made with data between 1993 and 2002 (Figure 4). The estimated vertical displacement velocity is 11.18 ±0.96 cm/year, nevertheless a diminishment in the last years was observed.

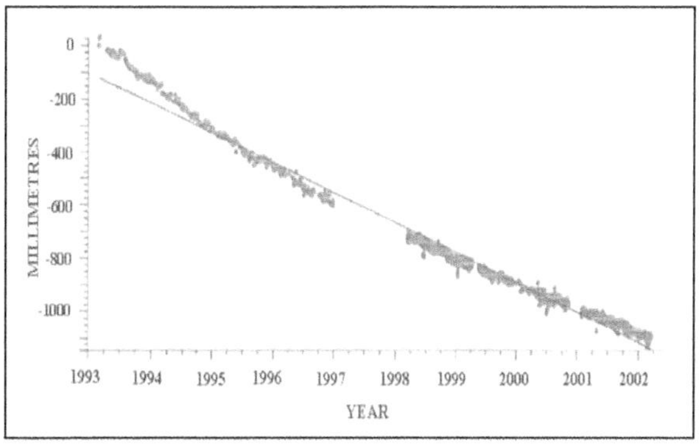

Fig. 4. Behaviour of INEG station height between 1993 and 2002. Márquez and DeMets (2003).

3 GPS Surveys for Subsidence Analysis

The Project consists of three hour bimonthly static GPS measurements on the 60 marks established by pairs (Figure 5), each one 15 metres approximately from the ground failure. GPS surveys at these monitoring points started in the second week of April 2003. The equipment used were two Trimble 5700 receivers with Zephir Geodetic antenna, and two Trimble 4700 receivers with Compact Geodetic antenna. Also, in the first and second campaigns two Ashtech Z12 and two Leica SR530 receivers were used.

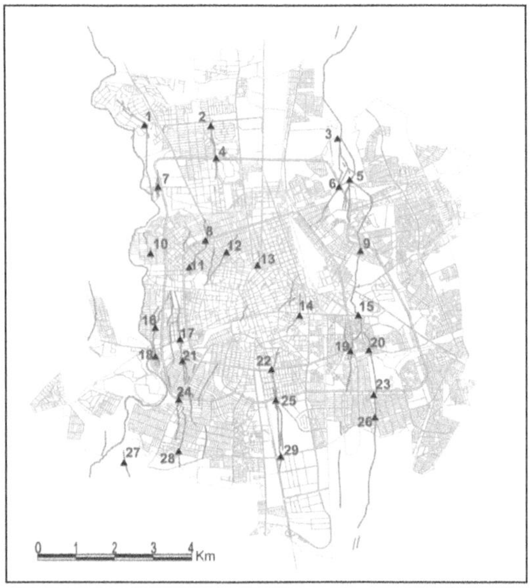

Fig. 5. Location of the pairs of GPS monitoring points.

3.1 Post-processing Method

Station INEG is located in the North American plate, so the continuously operating GPS stations TAMP, CULI and TOL2 in the same plate, were used in a differential process to get weekly sets of coordinates for INEG referenced to ITRF2000 epoch 2003.3. These stations presented similar motions (-1±1 millimetres in Latitude, 3±1 millimetres in Longitude and -2±2 millimetres in Height) with respect to ITRF2000 in recent neotectonics studies (Márquez and DeMets (2003)).

At least 3 days of continuous GPS data and precise ephemeris were used in data processing to obtain the weekly time series with commercial software. Elevation mask was set to 10° and a maximum of 13° for problematic vectors, using troposphere modelling for a better estimation of heights. With this method, effects of diurnal and semi-diurnal frequencies of earth tides and ocean loading were averaged in the resulting positions. The weekly updated coordinates were used to process the static GPS data of the monitoring points resulted from the repeated measurements. Monitoring points are at a maximum distance of 9 kilometres from station INEG.

4 Results

4.1 Displacements at Station INEG

The recent weekly behaviour, from the second week of April 2003 to the second week of April 2004, of the station INEG in the three components with its related uncertainties is shown in Figures 6, 7 and 8. The vertical displacement of the station is 5.6 ±0.3 centimetres for the year of study.

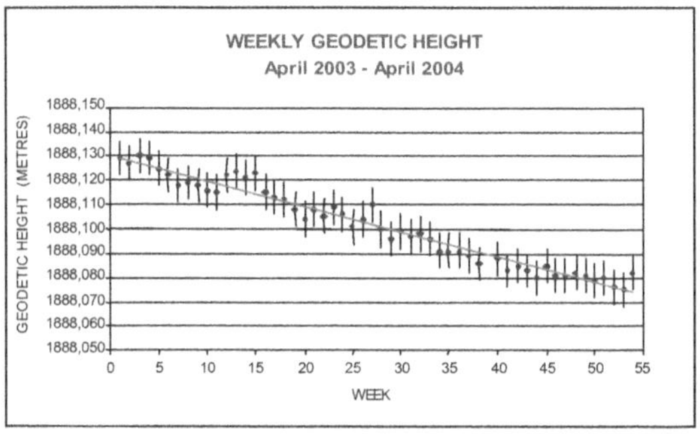

Fig. 6. Weekly vertical behaviour of station INEG in the first year.

Standard deviations of 1, 3 and 6 millimetres in average were achieved for latitude, longitude and geodetic height respectively. Results show remaining effects of long term frequencies of earth tides and ocean loading phenomena.

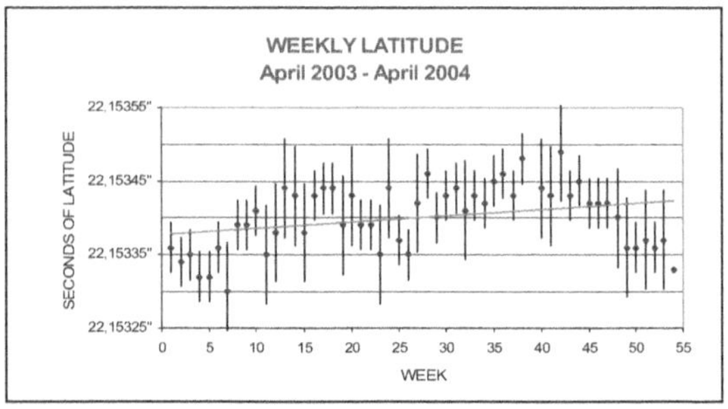

Fig. 7. Weekly behaviour of the latitude component of the station INEG.

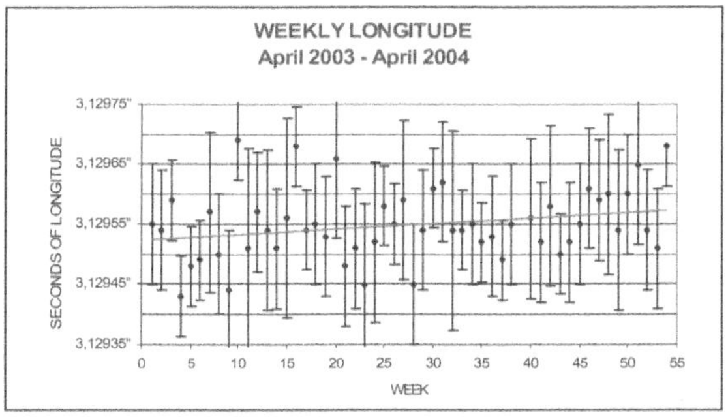

Fig. 8. Weekly behaviour of the longitude component of the station INEG.

4.2 Displacements at Monitoring Points

Although results of differential post-process of monitoring points show mean standard deviations of 1.5 millimetres in horizontal components and 8 millimetres in vertical ones, the quality of the positions resulted from the three-hour static measurements are significant lower than results at station INEG due to the less amount of data and conditions for the measurements in urban areas. Nevertheless, repeated measurements allow spatial and temporal interpolation that, in this case, gives a good description of the deformations at each point. Figure 9 shows an example of the vertical displacement at a pair of monitoring points. The absolute subsidence for the period of study at each point was obtained with the differences between the adjusted values for initial and final observations obtained by linear regression, using the least squares form (Kenney and Keeping, (1962)):

$$\hat{y} = a + bx_i$$

where b is the vertical coordinate of the best-fit line.

Regression coefficient b for x_i (epoch) and y_i (geodetic height) given values is calculated by:

$$b = \frac{\left(\sum_{i=1}^{n} x_i y_i\right) - n\,\bar{x}\,\bar{y}}{\sum_{i=1}^{n} x_i^2 - n\,\bar{x}^2}$$

and

$$a = \bar{y} - b\bar{x}$$

whereas the uncertainties were calculated with the b standard error estimation from Student's t distribution with its respective degrees of freedom (n-2) for a 95% confidence (Acton (1966)):

$$SE(b) = t_{e/2}^{n-2}\,\frac{\sqrt{\sum_{i=1}^{n}\dfrac{(y_i - \hat{y}_i)}{n-2}}}{\sqrt{\sum_{i=1}^{n}(x_i - \bar{x})^2}}$$

Absolute subsidence at monitoring points varies from zero at a point in the southeast extreme of the city to 19 centimetres (R square of 0.8538) at a point in the northwest. 33 of the 60 monitoring points presented vertical displacements largest than 5 centimetres, 97% of these points presented P-values smaller than 0.05 for the t-test of significance (level of confidence of 95%); uncertainties for absolute subsidence are of ±4 centimetres.

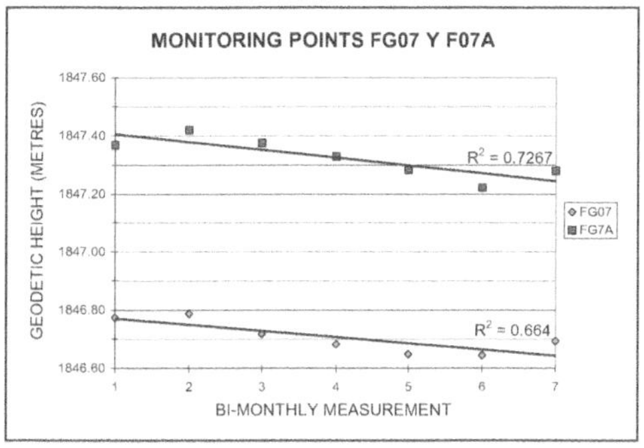

Fig. 9. Subsidence at monitoring points FG07 and F07A.

The subsidence map in Figure 10 was made by means of graphic interpolation of absolute subsidence at monitoring points.

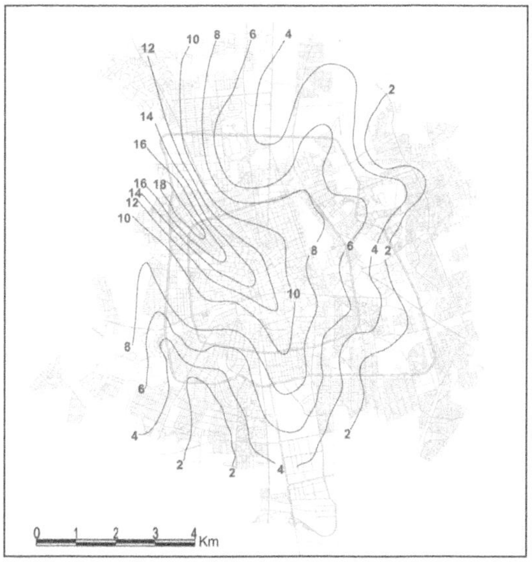

Fig. 10. Subsidence map for the period of study. Contour lines in centimetres.

The relative displacements between pairs of points (both sides of the fault) were derived from absolute subsidence. The biggest relative vertical displacement is of six centimetres and the observed mean is of 3 centimetres. To date, results at monitoring points do not allow an estimation of horizontal displacements.

5 Subsidence and Water Extraction

The Institute in charge of managing potable water in the state made maps of the ground water level based on measurements on the wells that supply water to the city. Map in Figure 11 shows contour lines of ground water level from registers of 2000 to 2002. The area with lowest level of 1710 metres above mean sea level matches with biggest subsidence of the subsidence map in Figure 10.

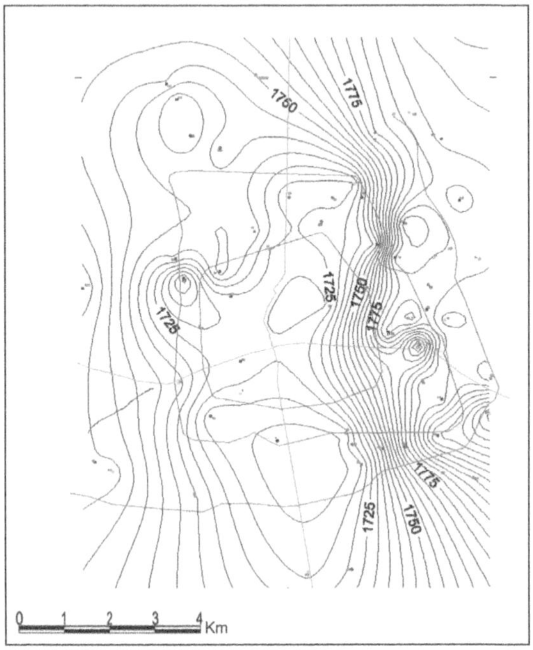

Fig. 11. Ground water level map 2002. Contour lines in metres above mean sea level. Source CCAPAMA.

6 Conclusions

The results of the study confirm that GPS is an efficient tool for subsidence measurement. In the case of Aguascalientes, continuous monitoring at station INEG and repeated measurements at monitoring points allow to obtain reliable information about ground deformations in order to detect the magnitude of the problem. With the monitoring continuation, we expect to get more information about the behavior of subsidence and ground failures and their relationship with respect to water extraction and rainfall seasons.

Once more information is obtained, we will be able to determine if horizontal displacements are experienced at monitoring points. Volumetric loss theory about mechanical behavior of the soil due to groundwater level decline considers small horizontal displacements, especially at the shoulders of the graben (Rojas et al. (2004)).

Commercial software have provided good results, although the future goal is to process current data with a scientific software so to reduce some remaining sources of error.

References

Acton, F. S. (1966). Analysis of Straight-Line Data. New York: Dover.

Arroyo M., et al (2003). Causa y Efectos de las Grietas y Fallas en el Valle de Aguascalientes. Reporte Técnico. Proyecto SIHGO 1998. 02. 04.012. Consejo Nacional de Ciencia y Tecnología.

Aranda, J. M. and Aranda, J. J. (1985). Estudio sobre Agrietamientos en la Ciudad de Aguascalientes, Ags. Reporte Técnico. Universidad Autónoma de Aguascalientes.

Kenney, J. F. and Keeping, E. S. (1962). Linear Regression and Correlation. Princeton, NJ: Van Nostrand, pp. 252-285.

Lermo J. (1994). Monitoreo Sísmico y Análisis Estructural de las Grietas en la Ciudad de Aguascalientes. Reporte Técnico. Instituto de Ingeniería, UNAM.

Márquez, B. and DeMets, Ch. (2003). Estudio de las deformaciones de la Corteza Terrestre a través de los datos de la Red Geodésica Nacional Activa (RGNA) del Instituto Nacional de Estadística, Geografía e Informática. Reporte Técnico. Universidad de Guadalajara.

Rojas, E., Arzate, A., and Arroyo, M. (2004). Predicción de las Zonas de Agrietamiento por Pérdida Volumétrica. III. Simposio de Ingeniería Civil, Sistemas Acuíferos y Fallamiento de Suelos. Aguascalientes, Ags., México. March 18-20

Aerial Digital Photogrammetric Monitoring for Landslide Susceptibility Mapping

J. Cardenal, J. Delgado, E. Mata
Dpto. Ingeniería Cartográfica, Geodésica y Fotogrametría. Universidad de Jaén
Campus de las Lagunillas, Edif. A-3. 23071 Jaén (Spain)

A. González, J. Remondo, R. Díaz de Terán, E. Francés, L. Salas, J. Bonaechea, I. Olague
Dpto. de Ciencias de la Tierra y Física de la Materia Condensada. Universidad de Cantabria
Avda. de los Castros s/n, 39005 Santander (Spain)

A.M. Felicisimo
Escuela Politécnica. Universidad de Extremadura. 10071 Cáceres (Spain)

Ch. J. Chung
ESS/Geological Survey of Canada. Natural Resources Canada. 601 Booth St, Ottawa, ON. Canada K1A 0E8

A. Fabbri
SPINlab. Vrije Universiteit Amsterdam. De Boelelaan 1087, 1081 HV Amsterdam (Netherlands)

A.Soares
Centro de Modelizaçao de Reservorios Petroliferos -Grupo do Ambiente-. Instituto Técnico Superior. Av Rovisco Pais. 1096 Lisboa (Portugal)

Abstract.
Landslides are one of the most frequent deformations in the shallow layers of the Earth and have very important social and economic consequences. It is one of the more hazardous natural risks. At present landslide knowledge and modelling are an interesting research line in order to establish the susceptibility of a certain zone to be affected by one of these processes. In this kind of studies, it is also very important to establish the conditioning factors that produce these phenomena and the foresight (prediction) of when they are going to act and its effects.

An optimum susceptibility model must be based in both the quality of the model and the quality of the input information. In this way the reliability of the entrance information in the model of susceptibility is guaranteed. This is the basic objective of this project that has given rise to this work (the FODISPIL project): the capture of very high quality information for application in landslide susceptibility maps generation using aerial digital photogrammetric techniques.

In this work, the preliminaries results derived from the aforementioned project are presented. These results are basically an input information quality improvement thanks to the use of ad-hoc large scale photogrammetric colour and infrared flights (1/5000 and larger), a best humid areas localization using colour infrared films (that reveal the hydric stress of the vegetation in a determinate area), as well as the use of historical photographs in order to increment the historical record of processes that allow a more rigorous temporal analysis of the conditioning factors (for example, human effects), premonitory features and processes.

Keywords. Deformation monitoring, Landslides, Digital Photogrammetry, Aerial images

1 Introduction

Mass movements are an important geomorphologic process and their development causes potential threat for human lives and assets. This type of natural processes constitutes one of the natural processes with a bigger associated risk (number of victims and economic risk). Brabb and Harrods (1989) analyse the socioeconomic influences of the landslides in reference to direct costs (human lives and direct economic losses –indemnifications, reconstructions and maintenance–) and indirect costs (spare expenses, dismissed profit, etc.). Ayala-Carcedo (1994) analyzes the 1000-1994 period and concludes that landslides have supposed human losses above 280,000 people, being the third natural hazard (after earthquakes and floods) in number of victims. It is important to bear in mind that losses have grown upon the last decades due to the demographic rising and to the increasing human environment impact.

The economic costs of phenomena are important. Thus 115 million of US$/year are used for the maintenance of the federal highways in USA (Schuster, 1994). In Spain, the Spanish Geological Survey (IGME) estimated in 36 million of euros/year the costs associated with the landslides for the five year period 1990-1994.

So there is an actual interest in establishing the mechanisms of landslide movements in order to obtain susceptibility maps that allows an adequate land planning. Also a proper knowledge of conditioning factors can allow correlate the risk to any phenomenon (for example, rainy period). One of the most useful tools to mitigate the development of this type of processes is the elaboration of susceptibility, hazard and risk maps. However, the elaboration of such maps is not easy due to the complexity of these phenomena and the difficulty to properly assess their future behaviour.

The process of prediction is not a simple task and although there has been successful cases such as the prediction of the breaking of one of slopes in the copper mine of Chuquicamata (Chile) (Kennedy and Niermeyer, 1971) also substantial errors have been reported. Perhaps the most severe error was the Vaiont's catastrophe, where prediction suggested a breaking the November 14, 1963 but it was produced at October 10, 1963. The slide of the Toc Mountain (Adriatics Alps) into the Vaiont reservoir produced a wave that killed 2000 people. The associated costs relating this catastrophe was estimated in 425 million $ (dam), 100 million $ (properties) and 68 million of $ (personnel) (Kiersh, 1964).

2 Proposed methodology

The FODISPIL project aims with an improvement of susceptibility maps using high quality input information which means high spatial accuracy. This information is mainly cartographical and it has been generated by means of digital photogrammetric techniques.

Digital photogrammetry has allowed the generation of digital elevation models (DEM) and spatial feature collection (in three dimensions) of conditioning factors –vegetation, lithology, fractures, landslide crowns, etc.–. In order to correlate to rainy period and storms, weather information was available from continuous record automatized weather stations.

The new methodological approach aims to improve the prediction capacity of the landslide hazard maps, clearing some of the uncertainties contained in the landslide susceptibility maps, relative to: where and when the new instability processes will be developed. These uncertainties remain because lack of reliability and resolution in the DEM`s usually used in such maps.

So the new methodology is based on the combined use of information related to landslide conditioning factors (lithology, landcover, soils, slopes, etc.; which is obtained by means of field works and analysed with Geographic Information Systems –GIS–) and aerial photographs processed with digital photogrammetric techniques (figure 1).

Reliable high-resolution DEM will be used to the susceptibility analysis. These DEM are obtained from aerial photographs ("ad hoc" photogrammetric flights at detailed scale with both colour and colour infrared films) and Digital Photogrammetric Workstation (DPW).

The reliability of a DEM has been analysed by means of estimation and simulation geostatistical techniques. Moreover, by means of digital photogrammetric stereo-plotting a new type of information (the future rupture areas) is introduced into the analysis. These rupture areas are defined from the "premonitory slide features" (open cracks, bulkiness, internal depressions, etc.) and the humid areas (areas in which the surface waters are accumulated).

The premonitory features were mapped from stereomodels generated from colour photographs whereas the humid areas were mapped from infrared photographs. Using digital photogrammetric systems it is possible to avoid the positional errors related to traditional photointerpretation and further transfer of the extracted features to the maps.

All the conditioning factor layers considered into the analysis are being introduced in the susceptibility analysis as a continuous or discrete variable. Diverse statistics (Favourability functions, Bayesian probability models, etc.) are being applied. Every result of susceptibility will be validated from the dates of landslide temporal occurrence found.

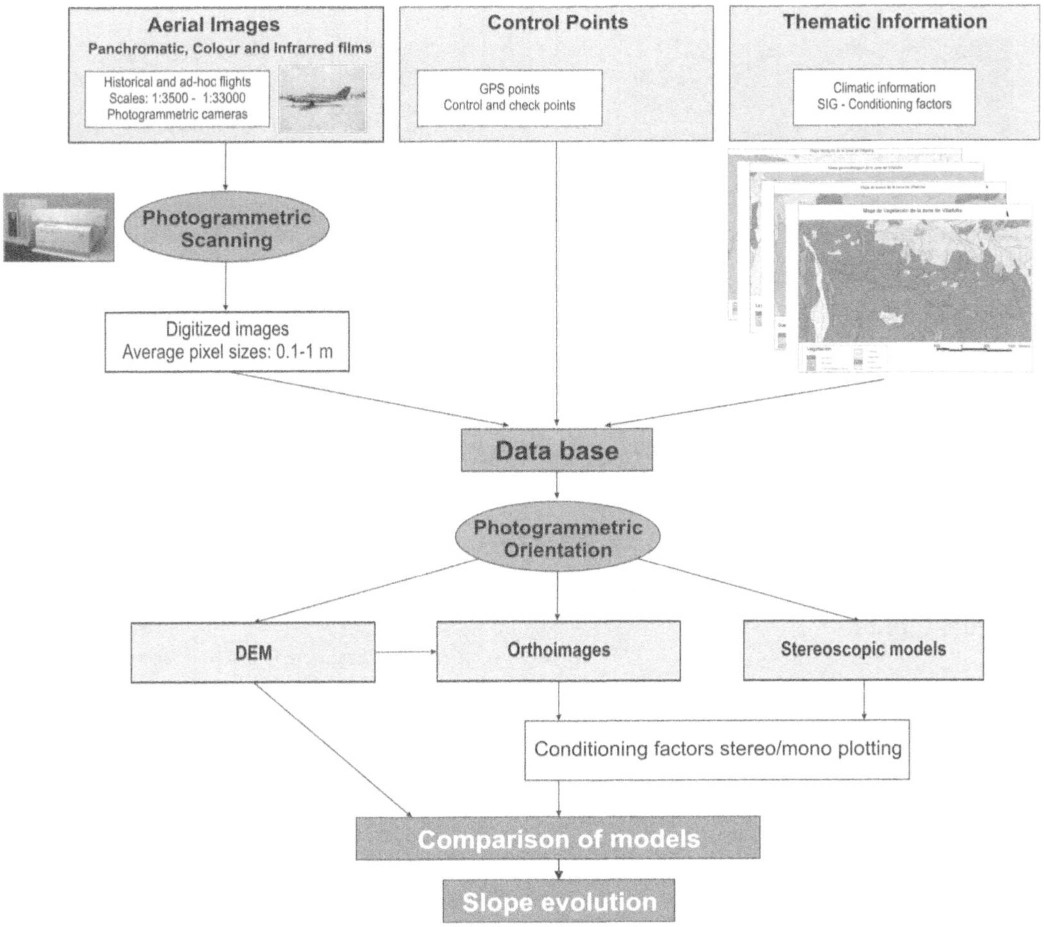

Figure 1. FODISPIL Project general schema

3 Study areas

This methodology has been applied in two study areas localized in the internal valleys of the Cantabrian Range (Northern Spain): Miera and Villafufre areas (figure 2).

Miera area (2km^2) is a deep valley with a highly rough relief with elevations between 600m (bottom of the valley) to 1400m (summit line). The materials are sedimentary carbonate materials with sandstone interbedded folded in a synclinal structure (Aptiense) and covered by till deposits. This is a very interesting area due to its topography that produces several problems in the different phases of the photogrammetric process (flight mission, ground control, digital elevation model generation, etc.).

Villafufre area (3km^2) has also rough relief with elevation range between 100m to 450m. The materials are basically Triassic (Keuper clays, gypsum and ophyttes) and Jurassic materials on top. Drainage preferential pathways that communicate dolines with the valley lower parts are present.

4 Works

4.1 Database

The first objective of FODISPIL project consisted in the database design and creation. This database must include all of the useful information of the survey areas for the modelling and analysis of the processes (climatology, lithology, vegetation, premonitory features, etc.). All information was georeferenced and included in a geographical information system (ArcGIS) for further data processing.

Fig.2. FODISPIL study areas.

4.2 Historical flights

An interesting aspect of the project has been the data processing of available historical flights. This has permitted increase the historical record of available information.

A total of 5 flights on the study areas were available from 1958 to 2001, with scales among 1/33.000 and 1/15.000 and with colour and panchromatic films. The final quality of available originals (obtaining negatives in good condition for scanning has been difficult) and the available information (in some cases, there is no information about the camera used in the flight) have taken to select two flights: a 1/20.000 scale panchromatic flown at 1970 and a 1/15.000 scale colour flown at 1988.

For image orientation, considering that the information usually was incomplete, self-calibration processes have been used. This procedure has allowed the inner camera parameters adjustment using the ORIMA® software of the LPS® (Leica Photogrammetric Suite) system. In Table 1 the final residuals of the transformation (expressed as terrain coordinates) and the DEM accuracy (considering control points) are presented.

4.3 FODISPIL ad-hoc flights

Within the FODISPIL project a total of 8 flights have been made. The flights are usually 1/5000 scale (also 1/3500 and 1/10000 scales have been used in order to check optimal scales) with colour and colour infrared films.

Table 1. Characteristics and accuracy of the historical flights used in FODISPIL project

Year	Scale	GSD (m)	Block residuals (m)	DEM accuracy (m)
1970	1:20000	0.23	X=0.74 Y=0.97 Z=1.98	XY=1.21 Z=1.76
1988	1:15000	0.24	X=0.42 Y=0.49 Z=1.06	XY=0.63 Z=1.58

These flights have been oriented using a control point network using ORIMA® program. This network has been designed and observed for this project and check points have been included (all the points were localized in stable areas). The final block adjustments obtained show that it is possible to obtain errors lower than 0.1m (terrain coordinates) on the image stereoscopic measurements. Because the high quality images (terrain pixel size around 0.075m) it has been possible to improve the premonitory features mapping (figure 3).

In addition, it is also possible to obtain temporal evolutions from processes. Figure 4 shows an example of the evolution in a premonitory feature detected in 1988 image. In the 2003 flight the features have been developed and a small house was destroyed.

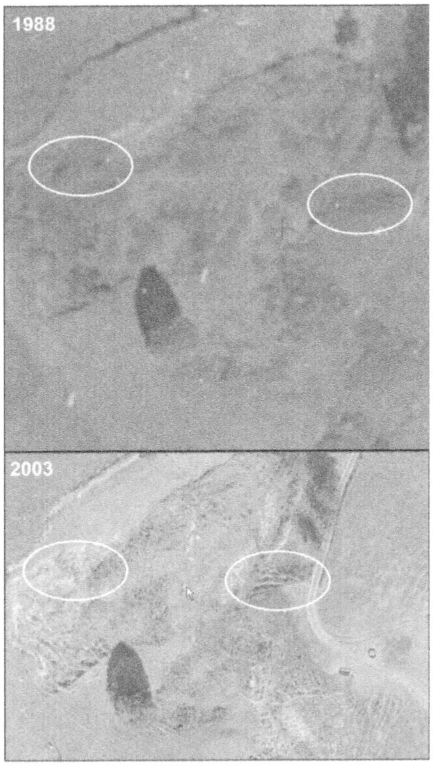

Fig.3. Premonitory features mapping improvement using high-quality photogrammetric images. Top: 1988 pancromatic (1/15000), Bottom: 2003 color (1/5000).

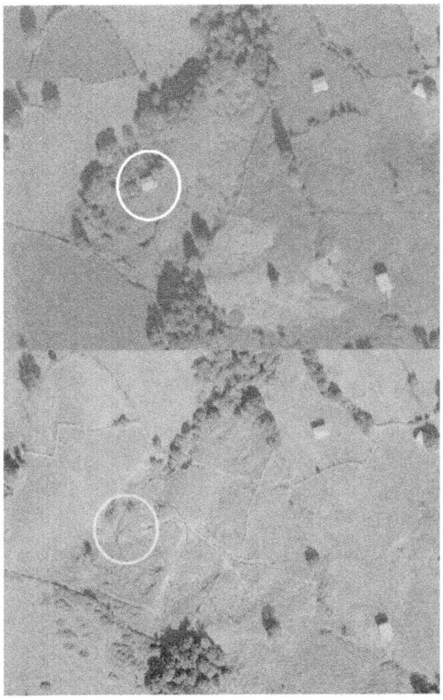

Fig.4. Landslide evolution in 1970 (top) - 2003 (bottom) period. The circle marks the position of a house that has been destroyed by the landslide. It is possible to observe the loss of the trees due to the instability of the terrain.

Figure 5 shows an example of landslide evolution. The temporal evolution has been established using stereoplottings from the 1970, 1988 and 2003 flights. Table 2 shows the length variations and displacement annual values.

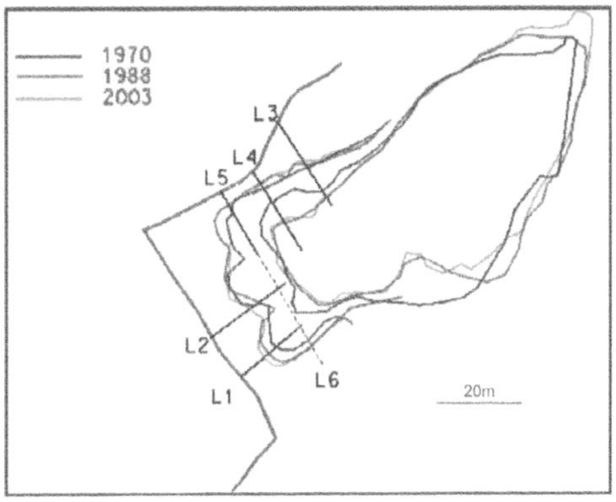

	L1	L2	L3	L4	L5	L6
1970	10.89	14.70	13.16	9.09	5.91	40.15
1988	8.30	14.57	12.16	9.09	3.63	41.18
2003	7.00	11.46	11.51	7.83		41.18

Fig.5. Temporal evolution of landslide crowns (m).

Table 2. Temporal evolution of landslide crowns (annual rates –m/year– obtained from the values of the figure 5).

	1970-2003	1970-1988	1988-2003
L1	0.12	0.14	0.09
L2	0.10	0.01	0.21
L3	0.05	0.06	0.04
Mean	0.09	0.07	0.11

Also colour infrared imagery has been used for vegetation index calculation using the expression proposed by Murtha (1989). The main problem for this application is derived from the film characteristics, the photographic process and scanning. These problems produce films density variations and the vegetation index only can be considered from a qualitative point of view, considering only one image. To avoid the above mentioned problems, further studies with comparison of new flights using infrared images provided by a digital aerial photogrammetric camera will be explored.

But infrared imagery has been very interesting in order to define humid areas. In figure 6, an image from of Sept'03 flight is shown. In this image, it is evident the presence of domestic wastewater discharges. This is an additional water supply that can influence the terrain instability.

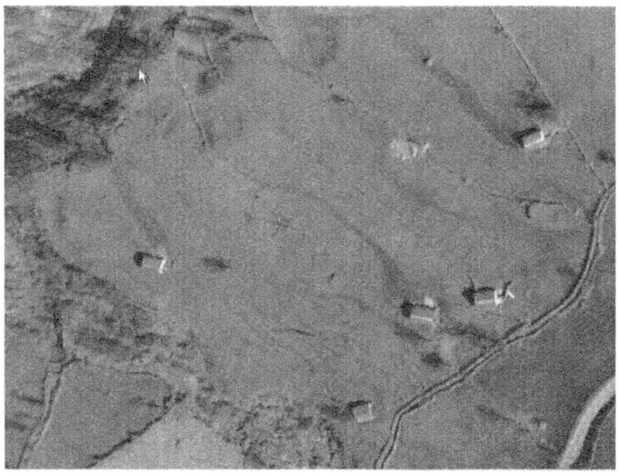

Fig. 6. Infrared aerial image. It is possible to observe the domestic wastewater discharges emitted into the environment from the rural houses.

5 Conclusions and Future works

The application of the digital photogrammetric methodology in order to obtain very high quality information for landslide susceptibility maps generation and process modelling has several advantages:

a) The photogrammetric flights allow the coverage of relatively extent areas (10-100 km2) with an adequate accuracy level (0.1 to 1m) being a low cost technique, at least compared with other remote sensing techniques (high resolution satellite images).

b) Photogrammetry is a well known and well established technique that has been applied for mapping in the last decades. Photogrammetry has experienced many changes that have simplified the process (especially with the use of digital imagery). These changes have allowed the use of these techniques for non photogrammetric specialists. Photogrammetric techniques provide the important advantage to provide high-quality imagery that is very useful for the establishment of landslide susceptibilities (photointerpretation). Recent techniques for monitoring deformation (for example inSAR) do not provide easily photo-interpretable images.

c) Quantity and quality of the information that can be extracted from the imagery is favoured by the stereoscopic visualization. Thus it is important to bear in mind that the obtained information is three-dimensional georeferenced information.

d) Use of historical flights can increment the historical record of the processes. These images can provide information with metric quality (it will depend on their characteristics and the available information of the flight). This information is very useful in the analysis of the evolution of the conditioning factors and processes.

e) Use of non-visible films (or sensors) allows obtaining additional information that can not be reached directly by the human eye. Thus it is possible to obtain the vegetation hydric stress (that indicates water presencc in the ground). This information can be useful in order to establish any relation between the movements and the presence of water in the vadose zone.

Present work lines in the FODISPIL project are the use of last generation aerial sensors (LIDAR systems for DEM generation and digital cameras for high resolution imagery -0.1m GSD-) as well as the use of terrestrial systems (using a schema similar to the aerial one, with laser-scanner for DEM and terrestrial cameras for imagery) in order to obtain detailed surveying of landslide that are active at present.

Acknowledges

The present study by grant REN 2002/0073/RIES from the R+D+I program of the spanish Ministerio de Ciencia y Tecnología, partially financed by FEDER funds of the European Union.

References

Ayala-Carcedo, F.J., (1994). Socioeconomic impacts and vulnerability resulting from slope movements. In: *Proc. of the US-Spain Workshop on Natural Hazards.* Corominas & Georgakakos (eds), Iowa University, pp 235-254.

Brabb, E.E. and B. Harrods, (1989). Landslides: Extent and Economic Significance. Balkema, Rotterdam.

Kennedy, B. A. and K. E. Niermeyer (1971). Slope monitoring systems used in the prediction of a major slope failure at the Chuquicamata Mine, Chile. *Planning Open Pit Mines Symp.*, Johannesburg, Balkema, Rotterdam, pp. 215-225.

Kiersch, G.A. (1964). Vaiont Reservoir Disaster. *ACSE Civil Engineering*, March, pp. 32-37.

Murtha, P.A. and Wiart, R.J. (1989). Cluster analysis of pine crown foliage patterns air identification of mountain pine beetle current attack. *Photogrammetric Engineering & Remote Sensing*, 55, pp. 83-86.

Schuster, R.L. (1996). Socioeconomic Significance of Landslides. In: *Landslides. Investigation and Mitigation.* Turner & Schuster (eds), Transportation Research Board, Special Report 247, National Academy Press, Washington DC.

Deformation Analysis to Study Erosion in Sloped Olive Orchards.

M.I. Ramos, A.J. Gil.
Dept.Ingeniería Cartográfica, Geodésica y Fotogrametría.
Universidad de Jaén. Campus Las Lagunilla Edf. A3. s/n. 23071. Jaén.

F.R. Feito.
Dept. Informática.
Universidad de Jaén. Campus Las Lagunilla Edf. A3. s/n. 23071. Jaén.

Abstract. Erosion is a problem that produces an important impact on the landscape and especially in agricultural areas. This process is accentuated by the effects of meteorological factors, agricultural machinery and the slope of the land. Different methods exist for evaluating the soil loss, some of these are qualitative and others are quantitative. However, despite the methodology or parameters used, it is evident that it is necessary to manage spatial data in order to quantify the changes that take place. In this study we analyse the effects of erosion on a property of olive orchards located on variable sloping land. This property is located in the province of Jaén (Andalusia, Spain), as it is one of the main producers of olive oil in the world and therefore is the most representative place. This study examines erosion process causes important problems of drag materials until it produces displacements of olive trees. The study of the land displacement has its methodological base in the revision of the position of the object points throughout several campaigns. In this area, the most suitable instrumentation for taking measurements is GPS, due to the lack of visibility through the olive trees. Until now three campaigns have been carried out. We have attempted to make each campaign coincide with the hardest ploughing periods, as this is when more erosion takes place.

Keywords. Deformation analysis, erosion, GPS, olive orchards.

1 Introduction

Nowadays agricultural areas and the landscape in general are facing a serious problem of degradation and subsequent soil loss. This situation is caused by the erosion process, whose effects are accentuated by many factors such as the climate, agricultural machinery and the slope of the land. This latter effect is of greater importance because it leads to surface runoff which in turn causes the loss of vital topsoil. The fact that the soil does not absorb the rain water causes it to dry up and blow away (Martínez A., 2003).

In the Mediterranean olive orchards, millions of tons of soil are annually swept away by runoff water, having a disastrous effect on the olive harvest and creating a gradual loss of farmland productivity. In orchards where mechanized farming practices are used the erosion phenomenon is accelerated because the agricultural machinery removes important surface vegetation which would otherwise reduce soil degradation caused by the rain. Nevertheless this is not the only cause. The intermittent rain, characteristic in Mediterranean regions, increases the runoff process in sloping orchards. The main consequence of this situation is the deformation of the land. Sometimes the erosion is so aggressive that it causes displacement of the olive trees themselves.

This process requires analysis; in particular it is necessary to study the variation of the topography. Many scientists have been involved in soil erosion research for years, and many models for the erosion process have been developed. For this kind of research it is necessary to work with spatial data to better quantify the changes that take place.

This work studies how the erosion process causes important problems of drag materials until it produces a considerable modification of the topography. In our study we attempt to quantify the way in which continuous topsoil erosion produces eventual land movement and the subsequent displacement of olive trees.

Our experimental plot is a property of olive orchards situated on variable sloping land. This property is located in the province of Jaén (Andalusia, Spain), as it is one of the main producers of olive oil in the world and therefore is the best representative sample. The parameters we use to analyze these effects are the trees' displacements and their connection with the elevation and slope models. The study of land displacement has its methodological base in the revision of the position of the object points throughout several campaigns.

The most suitable instrumentation for taking measurements in this zone is GPS because of the lack of visibility through the olive trees. Until now three campaigns have been carried out. We have attempted to make each campaign coincide with the hardest ploughing periods, as this is when more erosion takes place.

The first part of the field measurement process includes the monitoring of the control networks. There are two networks: one inside the study area and the other outside. The object points are positioned in the base of the trunks of olive trees in order to analyse their movements. This displacement data, combined with erosion parameters, gives us valuable information which improves the study of topsoil loss and the subsequent loss of farmland productivity.

2 Data and method

In the initial phase of the study it is very important to select the experimental plot carefully. Here we deal with a study of a particular plot of olive orchards located on variable sloping land. Our objective is to calculate displacement values accurately and establish the connection between such movements and slope.

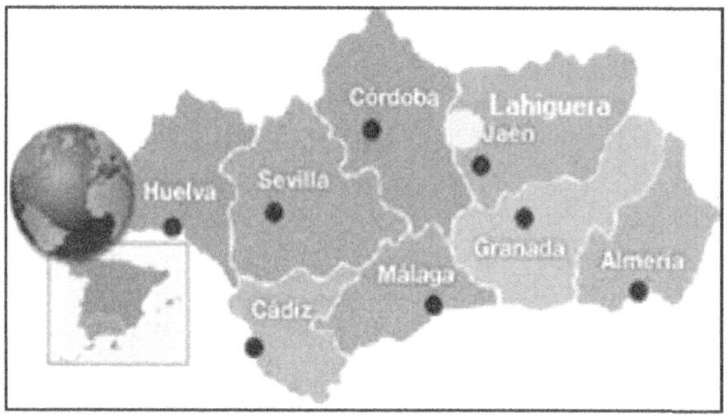

Fig. 1 Location of study area in Andalusia (Spain).

In this article our study is carried out in an unirrigated olive orchard where the slope gradient values are between 2% and 20%. The property selected is located in a village called Lahiguera in the province of Jaén, Fig 1. This is one of the main olive oil-producing regions in the country. In the province of Jaén olive orchards cover 589532Ha. which represents 25% of the spanish and 42% of the andalusian surface. We have chosen a property located in a place called Loma del Madero. This is a property of 1.190Ha and 120 olive trees, Fig. 3, of variable gradient obtaining values over 20%. All the olive trees here are thirty years of age, and their mean fruit harvest is 60 kg per year.

Fig. 2 Maximum sloped values in the olive orchard property.

The Good Agricultural Practices Code (Dirección General de la Producción Agraria. Consejería de Agricultura y Pesca. Junta de Andalucía, 2003) determines land classification as the following: slope gradient < 3% is considered even land on which agricultural machinery can be used, on land of slope gradient > 20% it can not, but if the gradient is between 3 and 20% it is allowed, although there are important problems of run-off and erosion. The latter case is our plot, Fig. 2.

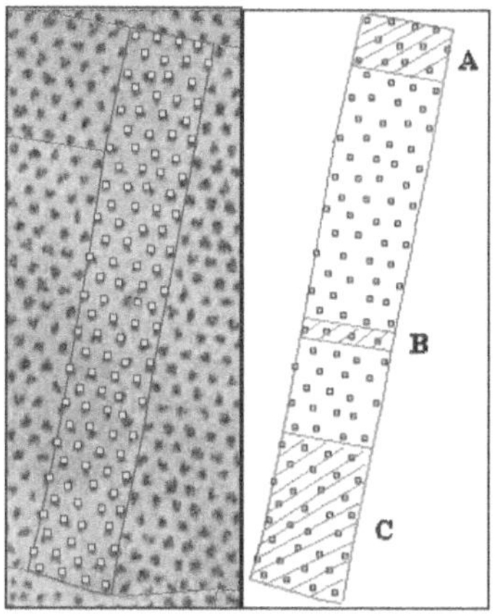

Fig 3. Aerial photography of the olive orchard property

The property is divided into three zones, Fig. 3, following the criteria of the gradient of the slope of the land. Zones A and C have higher gradients than B, a plain zone. The classification is flat as follows: zone A: medium gradient, zone B: minimum gradient and zone C: highest gradient. We are aware of the usefulness of this classification for analysing zones with similar properties and isolating them from the whole property (Zhang, 2002).

2. 1 Monitoring GPS network

The basis of the methodology of land deformation analysis is the monitoring of the position of several points in each survey. We occupy these points in order to take several measurements at the same time. This research can be carried out using two different methods (Caspary, 1988): In the first, we analyse relative movements between the object points, Fig, 4; in the second, we analyse the relative movements of object points with respect to the control network, Fig, 5.

Fig 4. Object Points.

The control network is a non-permanent GPS network which is observed periodically. Field measurement campaigns are spaced over time in order to monitor its fixed position. This methodology has already been developed by other authors (Betti, 1999).

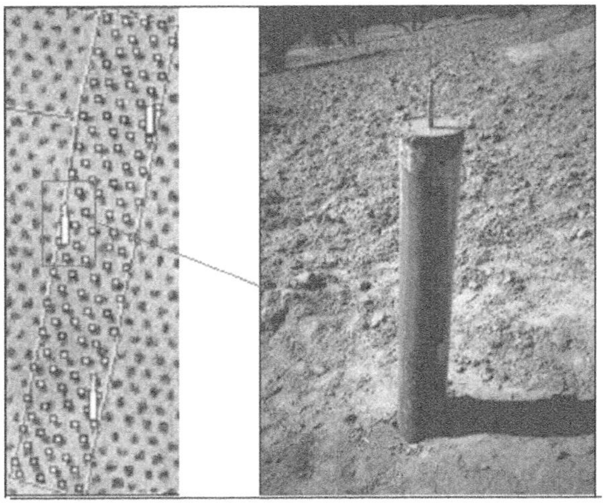

Fig 5. Local control network.

The concrete pillars which make up the local control network ensure that the antennae will be placed exactly at the same position in different campaigns. These antennae fit exactly into the pillars by the use of an embedded forced centring system. Similar systems used in studies were of deformation monitoring (Gil et al. 2002).

We assume this local control network to be vulnerable to external factors. They are closed to the olive trees. The pillars could be displaced due to the use of farming machinery, occasional deep land movements or even human vandalism.

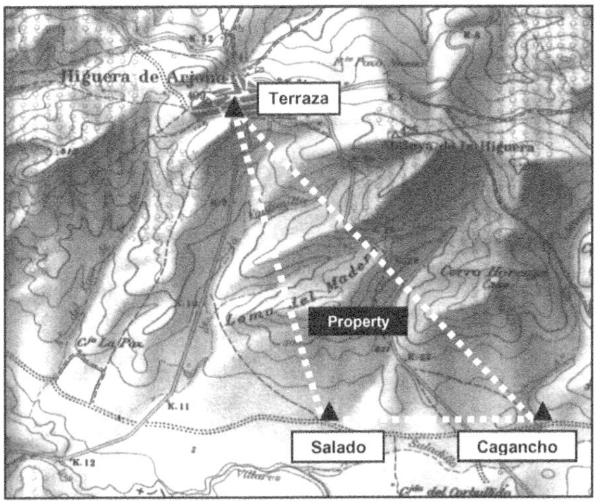

Fig 6. External control network.

Therefore another control nertwork has been designed (the external control network) to monitor the fixed position of the local one, Fig. 6.

This latter network is placed far from the study area in order to keep it away from the local movements. The network consists of three pillars which meet the same construction requirements as the ones implemented to build the local network.

2. 2 Object Points

Once the monitoring networks have been established the next stage is to mark the olive trees to be studied. These marks, nails, must be embedded in the base of the trunk in order to avoid the movements of the branches coming from their growth.

The first survey was carried out in November 2003, the second one in June 2004 and the third in October 2004. The equipment used was three Leica SR399, two Leica SR9500 and one Leica S500 dual frequency carrier phase GPS receivers belonging to the University of Jaén.

The networks have been tied to the European Terrestrial Reference System 1989 (ETRS89) from which the coordinates of the Terraza point were computed. GPS observations data processing was performed using Leica Ski-Pro 2.5 software. In each campign we computed the baselines and their least squares adjustments separately. The network coordinates which were obtained in November 2003 have been considered as the reference frame. This means that in comparisons between the campaigns the first one fixed.

3 Results

To identify an olive tree whose coordinates have changed we analyse two types of data. Firstly, the variation on planimetric position, and secondly, the altimetric. The data used to study planimetry is the displacement vector defined by its module and geodetic acimut. The study of altimetric movements is based on the comparison of altitude values of a point on each campaign. These differences in altimetry are represented by vertical vectors whose starting point belongs to the first campaign, November 2003. We use the confidence regions from the adjustment of each point in the three campaigns. The planimetric error ellipses and altimetric error ratio are shown in figures 7 and 8.

We assume the altimetric movements to be significant, at 1% significance level, if vector displacements are out of the error interval. In addition we are aware of planimetric displacements, at 99% reliability, if the error ellipses do not intersect.

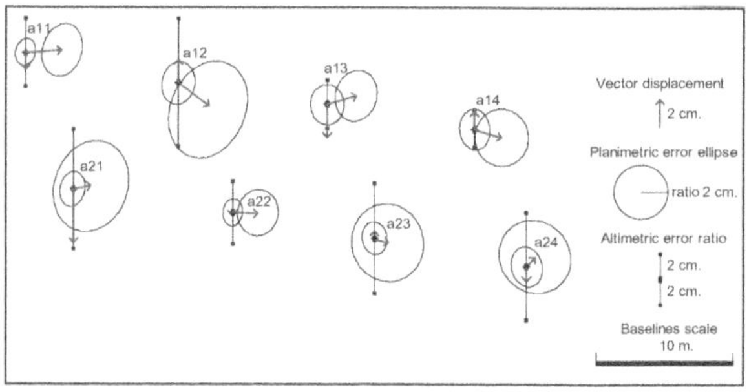

Fig 7. Analysis of displacements from November 2003 to June 2004. Zone A

Figures 7 and 8 show the analysis results from November 2003 to June 2004 and October 2004 respectively. They both include some olive trees of zone A which have been displaced.

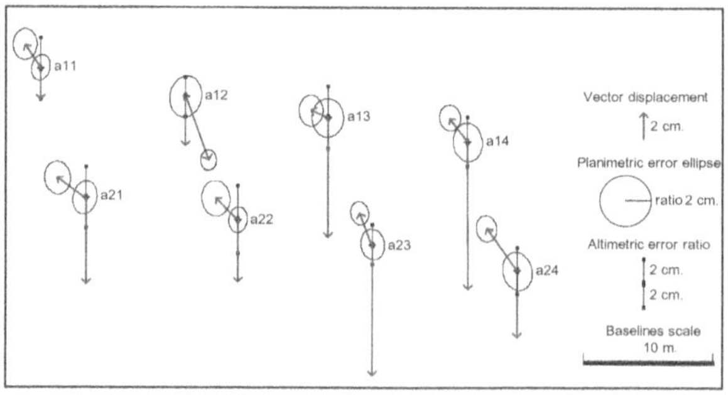

Fig 8. Analysis of displacements from November 2003 to October 2004. Zone A

4 Conclusions

In Figs. 7 and 8 we can observe that many points do not have significant planimetric movements at 99% reliability. Nevertheless, those points which have moved, from one campaign to another, are those which are placed in the greatest sloping area. In addition, their movements are downhill. We can say this because the direction of the vector displacements is very similar to that of the slope.

The greatest differences are detected in the altimetric values. In fact the area studied is sensitive to variations in altitude. However, an aspect that must be taken into account is that GPS precision on altimetry is not as accurate as on planimetry

There are some trees which have displaced in a different direction. This event confirms the idea that movements are not always cause by slope and climatic agents. Agricultural machinery causes substantial deformation of the terrain and subsequent slope variation. The machinery used in the olive harvest campaigns (from November to February) moves the trunk to make the olives fall down. This movement is random, so this could be the reason why some movements change the vector displacement direction from one campaign to another.

References

Betti B, Biagi L, Crespi M, Riguzzi F (1999) GPS sensitivity analysis applied to non-permanent deformation control networks. Journal of Geodesy 73: 158-167.

Caspary WF (1988) Concepts of network and deformation analysis. 2nd corr. imp. School of Surveying. Kensington.

Dirección General de la Producción Agraria. Consejería de Agricultura y Pesca. Viceconsejería. (2003) El Olivar Andaluz. Servicio de Publicaciones y Divulgación. Junta de Andalucía. España.

Gil AJ, Rodríguez-Caderot G, Lacy MC, Sanz de Galdeano C, Alfaro P (2002) Establishment a non-permanent GPS Network to monitor the deformation in the Granada Basin (Betic Cordillera, South Spain). StudiaGeo s.r.o., Prague. Stud. Geophys. Geod., 46: 395-410.

Leica Geosystems AG, (2000) CH-9435 Heerbrugg, Switzerland. 722021en-VI.00-RDV.

Neményi M, Mesterházi PÁ, Pecze Zs, Stépán (2003) The role of GIS and GPS in precision farming. Computers and Electronics in Agriculture 40: 45-55.

Martínez Raya A (2003) La erosión hídrica y sus efectos en el olivar. Vida Rural. nº 162 1 de febrero de 2003: 60-63.

Stafford John V (2000) Implementing Precision Agriculture in the 21st Century. J. Agric. Eng. Rest. 76 (3): 267-275.

Zhang N, Wangb M, Wanga N (2002) Precision agriculture - a worldwide overview. Computers and Electronics in Agriculture 36 : 113-132.

Geodetic Inspections Using Distance Measurements: Applications to "La Cohilla" Dam (Cantabria, Spain)

Ferrer Torío, R., Piña Patón, B., De Luis Ruiz; J. M., Ruiz Bedia, Mª L. & Castillo López, E.

Cartographic Engineering, Geodesy and Photogrammetry R + D + I Group

Civil Engineering School, University of Cantabria (Spain)

Abstract

Large dam construction with the so-called double curve arch structure was started around 1940. This structural type achieved a considerable saving in concrete but at the same time made these thin high structures sensitive to the effects of hydrostatic pressure and temperature. A dam with these characteristics located in a high narrow gorge of a given river undergoes the greatest stress and, therefore, the greatest displacement downstream under high hydrostatic pressure (full reservoir) and low temperatures. The greatest displacement upstream occurs under low hydrostatic pressure and high temperatures. These two limiting situations usually coincide with winter and summer, respectively. Because of the state of the technology in those early years, the only method available to monitor these displacements and to maintain an acceptable safety level was the so-called "geodetic inspection" method, which allowed measurement of movements with a standard deviation of 1 mm. These measurements, the most accurate that could be achieved in the topographic field, were termed "geodetic" in allusion to the difficult long-range observations that always required the best instruments, the best methods and the most highly trained observers.

This article details the change from inspection with the initial angle observation to inspection with distance observation at the La Cohilla dam (Cantabria, Spain), and justifies the peculiarities of the method, the appropriateness of the change and the expected benefits.

Key Words: microgeodesy, geodetic inspections, angle observation, distance observation, deformation status, co-ordinate variation

1. Introduction

When it came into service, the La Cohilla dam was the highest in Spain, and the company responsible "Saltos del Nansa" considered it wise to investigate the best way to monitor its movements by sending their engineers to Switzerland, the most advanced country in that type of technology. The result of the analysis advised evaluation of the downstream face monitored from external fixed points, and positioning of a pendulum anchored to the crown of the dam at its axis. Topographic measurements would be made with a Wild T-3 theodolite No. 18,602, which was purchased for this purpose, and based on the following set-up: four inspection pillars, two remote fixed reference points to establish the initial angles, and an array of seven rows and five columns of target points. Eight safety points were also included to guarantee immobility of the pillars and, if these should be displaced, to measure the displacement. The first microgeodetic measurements were made on the 15th of May, 1950 and measurements have continued almost uninterrupted over a period of 55 years. The photograph (figure 1) was taken downstream of the 116-metre high dam and shows the four shelters protecting the four inspection pillars.

Fig. 1: La Cohilla dam. Downstream face

The pendulum was installed in October, 1951, with its upper end situated close to target point 7-c. The lower end of the pendulum has clearly defined X,Y measurement axes. At first, measurements were taken weekly, but are now made every three days at the same time as seepage is assessed, and the deformation status of the dam is determined.

Until now, the microgeodetic measurements have been made by observing the azimuth angle with the above-mentioned optical theodolite. The Wild T-3 theodolites were characterised by an azimuth circle resolution of 0.1" (0.03 mgon) and a telescope magnification of 40X. To make an accurate reading in the azimuth circle was difficult and required the expertise of experienced observers. Later, in the setting of total topographic stations, electronic theodolites proved more convenient, and provided greater cost-effectiveness because of the automated data capture, but with a slightly lower accuracy.

In spite of the improvement in instrumentation, it is still necessary to make a large number of serial angle measurements of the target points from the inspection pillars with a forced centred device, and of the safety points from the initial angle-setting points (polar points). These respective direct and inverse intersections are then treated by conventional graphic and analytical methods.

At the present time, angle data are collected with the Leica TC-2003 topographic station (Wild, see figure 2), which allows the captured data to be recorded, thereby making the operation much quicker.

T-3 TC-2003

Fig 2 Former and current instruments used for data capture

Over the last three years, parallel research carried out by our R + D group has established a change in the microgeodetic methodology. The usual components employed for this type of work (inspection pillars, target points and safety points) are still relevant, but the type of data capture has changed with the introduction of the more accurate distance meters of topographic stations, and with the use of new techniques to correct the delays due to meteorological considerations. The target points are reflection prisms and the topographic instruments used are stations that have a distance-meter with an accuracy of 1 mm + 1 ppm. This allows measurement of movements in soil and structures with a standard deviation close to 1 mm, and constitutes a much simpler and, therefore, a much quicker and more economic inspection method. The use of polar points is no longer necessary, the safety points can disappear because the pillars can be placed further from the target points (this is impossible with angle observation), since distance observation is more stable than angle observation, and the number of series can be reduced in data capture. In angle observation, however, since the pillars must remain, so must the safety points.

In 2006, the conversion of La Cohilla dam will be completed, with the positioning of the remaining prisms (50%). The prisms, which comprise a network similar to the current one, are offset one metre downwards from the present target points, with the columns in the same positions. The range of movements is characterised by vectors with components (dx, dy) which establish the origin-destination of the position for each target point in two consecutive campaigns (n-1) and (n).

2. General Reference Systems

The two control methods (geodetic inspection and the pendulum) have two theoretically independent reference systems, although they share an approximately common node, which is the axis of the dam. Since the different structural elements of the dam are circumference arcs, there is a dam axis that can be considered the fundamental element that characterises the movement.

2.1. Reference System for Geodetic Inspection

Arbitrary local co-ordinates (x,y) are assigned to the set of fixed elements (two polar points, the inspection pillars, the target points and the safety points), which constitute a cartesian system. Inspection is made from the four pillars (figure 3), which consist of four concrete supports, each perfectly embedded in the floor of the protective shelter. They are denominated Pillar I, Pillar II, Pillar III and Pillar IV and on them rests the topographic instrument for observation. The protective shelters have windows from which it is possible to see not only the signals and the reference points but also the other pillars. The local co-ordinates of the pillars that comprise the fundamental vertices (and bases) of the control system are as follows:

	X	Y
Pilar I	1000,000	1000,000
Pilar II	1092,055	961,153
Pilar III	1033,480	1014,112
Pilar IV	1086,299,	992,664

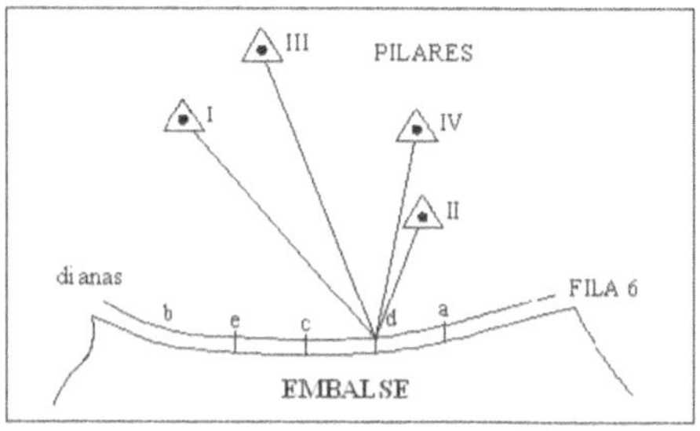

Fig 3 Dam and Pillars. Objective of geodetic inspection

The target points are mobile points that define monitoring of the face of the dam. There are 35 target points, or reference points, located in the downstream face of the dam. They are arranged in seven rows and five columns. The other points are reference points of the dam abutments. The points are identified by a number and a letter, corresponding to the row and the column for the target points, with a Greek letter being used to identify the abutment reference points.

Figure 4 shows the distribution of the target points in the face of the dam.

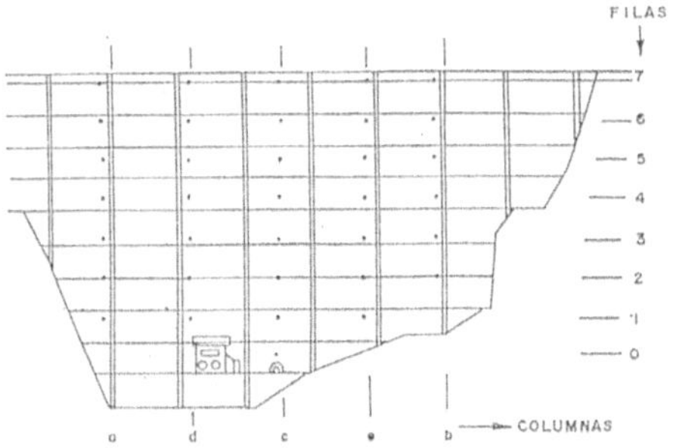

7-a	7-d	7-c	7-e	7-b
6-a	6-d	6-c	6-e	6-b
5-a	5-d	5-c	5-e	5-b
4-a	4-d	4-c	4-e	4-b
3-a	3-d	3-c	3-e	3-b
2-a	2-d	2-c	2-e	2-b
1-a	1-d	1-c	1-e	1-b
		0-c		

Fig 4 Distribution of the target points in the dam

2.2. Pendulum Reference System

The pendulum (figure 5) is a direct pendulum whose anchor point is situated on the crown of the dam close to its axis and not far from target point (7-c).

Fig 5 Pipe – vertical shaft encasing the direct pendulum wire

The weight on the lower end rests in an oil tank, and in the service gallery there is a device that assesses two movements, as shown in the following diagram

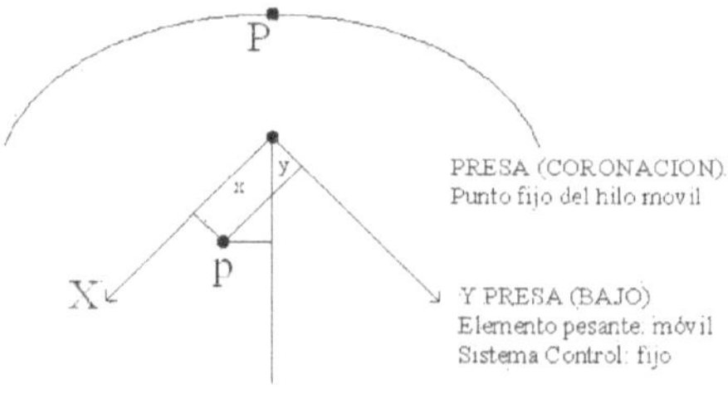

Fig 6: Displacement of the pendulum in a given campaign

The control carried out corresponds to a point close to target point (7-c) which is assessed in terms of a fixed (XY) system, thus providing a series of (x y) data for subsequent analysis. The bisector of this reference system need not coincide with the Y axis of the geodetic inspection reference system, but will approximate very closely to it, since this is the desired result. Achieving the superposition of the axes when the two systems were set up proved extremely difficult.

In the pendulum study from 1959 through 2004, the first thing was to determine the control referential axes (pendulum reference system). Once these imaginary and a posteriori elected axes had been established, the radial and tangential components were decomposed. The main axis was selected as the regression line of the point cloud and represents the ordinate axis. Its perpendicular is the abscissa axis, as in the geodetic inspection system. The component in the direction of the "ordinate" axis (regression line) is the radial component, and represents movement downstream if this value is positive. The tangential component perpendicular to the regression line represents movement towards the left abutment if the value is positive.

As can be seen, the pendulum reference system depends on whatever axes are considered appropriate, and when the study was undertaken, this was the first decision that had to be made. Data were taken from those available for the period 3-1-1959 to 24-12-2004, comprising a total of over four thousand recordings. In addition to the above-mentioned X,Y data, external data were also included: height of reservoir water and temperature at the crown of the dam. The level refers to a height-measuring system based on scales located close to the dam that have the following interpretation: the dam top level, 777.85 m, corresponds to the mark 61.25 m and the lower level, 718.70 m, to the 2.10 m mark. Temperature in degrees centigrade corresponds to the mean temperature on the day of measurement.

Taking the data for 1990-1999 as representative and using robust regression according to Huber's method (figure 7) gives

$$Y = 1{,}029\ x + 3{,}428$$

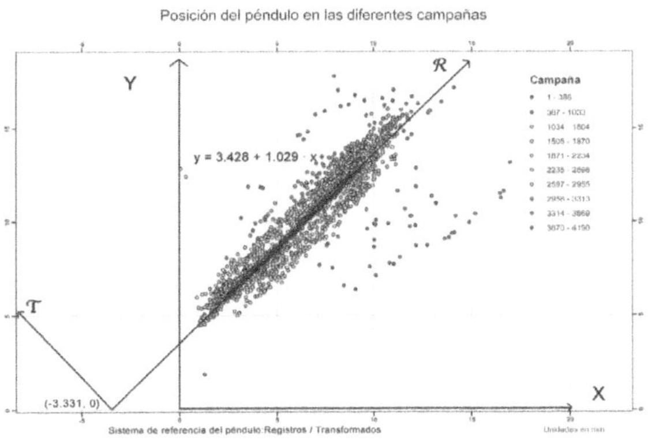

Fig 7 Pendulum axis system

3. Relationship between the Pendulum Movement and Target Points

Although an exact relation cannot be established between pendulum movement and that of target point (7-c) in the geodetic inspection method, it is possible to achieve an approximation close enough to draw some conclusions (figure 8). This target point is situated on the downstream face, with the pendulum anchor point on the crown of the dam and, therefore, the mobile control point for the dam situated at some distance towards the left abutment. The relation between the two must be approximate, but they should conserve the deformation trend in the direction both of the movements and of the absolute value of the deformation. A real study involves consideration of uncertainties in the movements, which can be evaluated by the co-ordinate variation method used in the microgeodetic system, which is completely specified. In the case of the pendulum, determination of the error ellipses is more difficult.

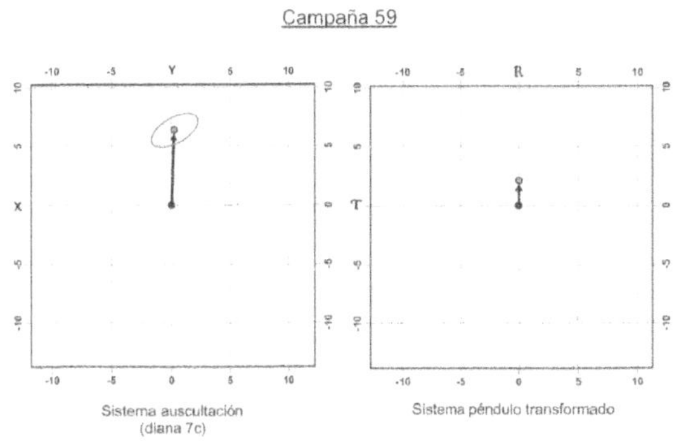

Fig 8 Error ellipses. Geodetic inspection system target point (7-c)

With these considerations on uncertainties, the movements of target point (7-c) and of the anchor point of the pendulum can be related between individual campaigns. Figure 9 shows the relation for the '56 campaign.

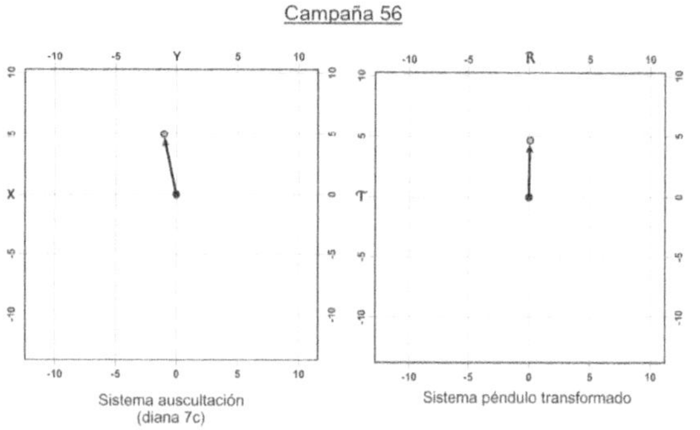

Fig 9 Deformation relation: target point (7-c) and anchor point of the pendulum

4. Integrated Inspection with Distance Measurement

4.1. Foreseen Plan of Action

For the next campaign (February 2006) measurements will be taken not only by observing the azimuth angle but also with distance observation, since all the prisms have now been inserted in the face of the dam. For the most recent campaigns, comparisons of behaviour have been made with a sample of prisms, and the movement obtained with distance observation coincided with that obtained with angle observation. For the current campaign, a comparison will be made with the full set of prisms. The displacement data evaluated by means of the pendulum will also be compared with the new series of measurements. This, together with maintenance of the existing external safety point network, will guarantee control of planimetric deformation.

4.2. Co-ordinate Variation with Distance as the Observable

Distance had been used as the observable before the introduction of high precision infrared distance meters. The ME-5000 Mekometer, which emitted a visible signal (λ = 0.632 µm), was a geometric distance meter in the submillimetre range, but its exorbitant cost meant that it was used only sporadically. By 1992, Leica had launched their TC-2000 topographic station with an angle measuring accuracy of 0.5" (0.15 mgon) and a distance measuring accuracy of 3mm + 3 ppm, and five years later produced their TC-2003 topographic station with the same angle measuring accuracy but with an enhanced distance measuring accuracy of 1 mm + 1 ppm. After that, the method became widely used and new applications were planned.

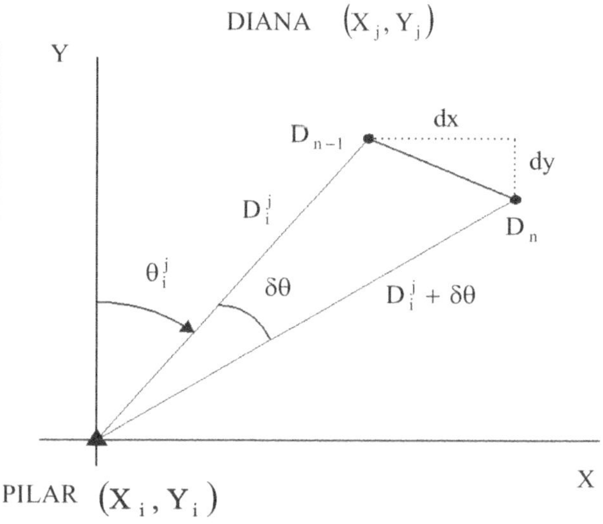

Campaña (n-1)
Angulo -------- $[\theta_i^j]_n$ Distancia ----- $[D_i^j]_{n-1}$
Campaña (n)
Angulo -------- $[\theta_i^j]_n$ Distancia ----- $[D_i^j]_n$
Base de control planimétrico
Angulo -------- $(\delta\theta_i^j)_{n-1}^n$ Distancia ----- $(\delta D_i^j)_{n-1}^n$

Fig 10 Difference between angle and distance observation

With all the elements planned and constructed, the series of observations is established and initially the co-ordinates of all the targets and pillars involved are evaluated and form the initial geometry. After this evaluation, appropriate methods other than the classical methods used in topography are applied in order to obtain the movements of the target points (figure 10).

In a system of fundamental pillar/target elements, the following equality can be considered

$$D_i^j = \left[(x_j - x_i)^2 + (y_j - y_i)^2\right]^{1/2}$$

where D_i^j is the reduced topographic distance corrected for meteorological effects and expressed in terms of the co-ordinate components of the target points and pillars. Differentiating gives the equation:

$$dD_i^j = \frac{x_j - x_i}{D_i^j} dx + \frac{y_j - y_i}{D_i^j} dy$$

where:

j: is the target involved in the movement

i: is the pillar from which the distance is measured

dx,dy: are the movements between campaigns of target j

The movement of the different target points between consecutive campaigns (n-1) and (n) after data collection is determined from the difference by obtaining the value {dDij}. If p pillars are considered, the movement of target q is obtained as follows (for a more detailed development see refs I and III):

$$\begin{Bmatrix} dD_1^q \\ dD_2^q \\ \cdot \\ \cdot \\ dD_p^q \end{Bmatrix} = \begin{bmatrix} \frac{x_q - x_1}{D_1^q} & \frac{y_q - y_1}{D_1^q} \\ \dots & \dots \\ \frac{x_q - x_p}{D_p^q} & \frac{y_q - y_p}{D_p^q} \end{bmatrix} \begin{Bmatrix} dx \\ dy \end{Bmatrix}$$

or by expressing this equation in matrix form

$$D = A \cdot X$$

It should be noted that the A matrix, denominated the design or configuration matrix, depends on the initial geometry of the set of pillars and target points and is constant for each pillar-target pair.

Premultiplying by the transpose of the design matrix

$$A^t \cdot D = A^t \cdot AX$$

and solving for X gives:

$$X = \left[A^t \cdot A\right]^{-1} \cdot A^t \cdot D$$

denominating:

$$A^t \cdot A = N$$

and the inverse of N … N-1:

$$A^t \cdot D = P$$

yields:

$$X = N^{-1} \cdot P$$

In the case of La Cohilla dam, as there are four inspection pillars, the system is composed of four equations with two unknowns [m = 4; n = 2].

It can be seen that in the system of equations a crucial role is played by data acquisition from the different pillar-target distances. The inaccuracy of the distance meter, which is of the order of 2 mm per kilometre (1 mm in the present case) is less than that derived from the incorrect application of the correction factor for meteorological effects, which in this method is of utmost importance. It can be said that the success of geodetic inspection by observing distance depends on the proper use of this correction factor. The most extreme case is associated with monitoring systems at long range (over 1000 metres) and with large height differences (300 metres) when fixed reflexion prisms are used in the target points, i.e., targets that are not readily accessible, where it is difficult to obtain meteorological data and these must be estimated with a suitable model.

Once the movement vector has been obtained $[dx, dy]_{n-1}^{n}$, the residual matrix is also obtained, since there is a redundancy of 2 (m-n=2)

$$R = X - \left[A^t A\right]^{-1} \cdot A^t D$$

When the column matrix of the residuals has been defined, the unit weight variance is given by: (For La Cohilla: : $R^t = \left[r_I^j, r_{II}^j; r_{III}^j, r_{IV}^j\right]$)

$$s^2 = \frac{R^t Q R}{m - n}$$

where:

R: is the residual matrix

R^t :is the transpose of the residual matrix

Q: is the weight matrix (unit matrix in the present case)

m: is the number of equations

n: is the number of unknowns

thus giving the unit weight variance, which will allow determination of the covariance matrix:

$$c = s^2 N^{-1} = \begin{bmatrix} \sigma_x^2 & \sigma_{xy} \\ \sigma_{xy} & \sigma_y^2 \end{bmatrix}$$

thereby obtaining σ_x, σ_y, σ_{xy}, the eigenvalues and corresponding eigenvectors, and allowing determination of σ_{max} and σ_{min} (semi major axis and semi minor axis of the error ellipse) in the following way:

$$\sigma^2 = \frac{1}{2}\left[\sigma_x^2 + \sigma_y^2 \pm \sqrt{\left(\sigma_x^2 - \sigma_y^2\right)^2 + 4\sigma_{xy}^2}\right]$$

These values are reflected in figure 11, where the slope (azimuth) of the main axis of the error ellipse is also determined.

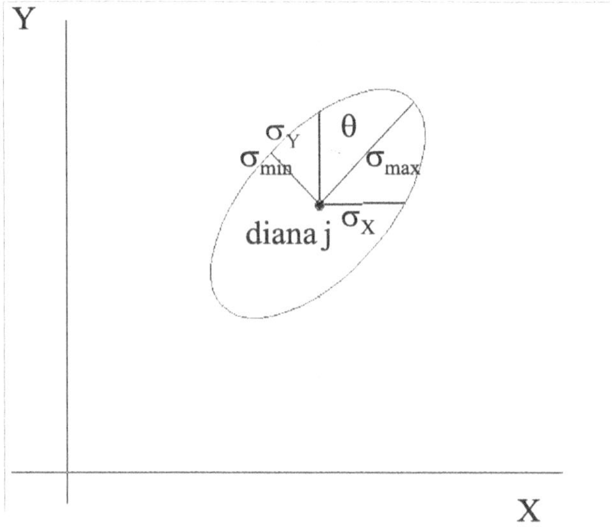

Fig 11 Error ellipse

This is obtained by assessing the value of θ:

$$tg2\theta = 2\frac{\sigma_{xy}}{\sigma_y^2 - \sigma_x^2}$$

5. References

[I] Ferrer, R.; Piña, B.: *Topografía de proyectos y obras.* Servicio de Publicaciones de la E.T.S. de Ingenieros de Caminos. Santander, 1991.

[II] Herbert, W.: "Development and application of refractivity correction: formula for optical and infrared EDM observations". *Surveying and Land Information Systems.* Vol. 53, Nº2. 1993.

[III] Ferrer, R.; Piña, B.; Castillo, C.: "Microgeodesia en la Ingeniería Civil. Aplicación a la auscultación de presas". *Revista de Obras Públicas,* nº3338, 1994. 7-21.

[IV] Ferrer, R.: "Redes topográfico-geodésicas de auscultación". *Cartografía, Geodesia y Fotogrametría.* XIV Cursos de Verano de la Universidad de Cantabria. 1998.

[V] Ferrer, R.: "Topografía clásica. Estaciones topográficas totales y niveles digitales. Influencia de los parámetros atmosféricos en la determinación de distancias". *Topografía Aplicada. XVI Cursos de Verano de la Universidad de Cantabria.* 2000.

[VI] García, A.: *Métodos avanzados de estadística aplicada. Métodos robustos y de remuestreo.* UNED. Madrid. 2001.

[VII] Ferrer, R.; Del Jesús, M. et al.: "Auscultación geodésica utilizando observable distancia. Incorporación de los resultados obtenidos en el puerto de montaña de Las Estacas de Trueba (Cantabria. España)". *4ª Asamblea Hispano Portuguesa de Geodesia y Geofísica.* 2004.

Investigation of a Building's Response in Tunnel Excavation

George D. Georgopoulos
Lecturer, Rural & Surveying Engineering School, National Technical University of Athens,
9 Heroon Polytechniou St., GR - 157 80 Zografos, Athens, Greece.

Abstract. Tunnelling in urban regions might cause ground surface deformations, which, in their turn, are responsible for the damages observed on surface constructions.

In this paper the response of a five – floor building, situated just above one of the Athens METRO tunnels is investigated. It must be pointed that out in the above area significant ground surface settlements ranging up to 96 mm were detected during the tunnel's excavation.

For this purpose an horizontal control network was established in the vicinity and on the construction. The network's control points were established on the construction in especially selected positions in order to describe the construction's bearing frame behavior. After the adjustment, the slabs' deviations from the horizontal plane (ranging up to 12cm) as well as the absolute and relative horizontal displacement vectors of the control points (ranging up to 10cm) were estimated using the control points' coordinates.

Finally the conclusions withdrawn from the study of the displacement elements concerning the response of the building are presented.

Keywords. Structures' kinematic behavior, constructions' deformations, buildings' displacements

1 Introduction

Tunnel excavations are usually the cause of ground's movements on the surface. Particularly, in densely populated urban areas, ground displacements cause deformations on the nearby constructions. These deformations are responsible for buildings' damages, which occur during underground works (R. J. Mair, R. N. Taylor, J. B. Burland, 1996).

The monitoring of the structures' displacements is indispensable for the establishment of an early warning system against collapses during the tunnel's excavation. On the other hand, the estimated displacements of the constructions are essential for the study of the buildings' reconstruction.

The present paper is focusing upon the ground movements and buildings' displacements due to tunneling. More particularly in this article the Geodetic Methodology is described, which is a very effective procedure for the monitoring of the constructions' deformations. The response of a five – floor building, situated just above one of the Athens METRO tunnels (station Ag.Ioannis D.Latta Street and Ag.Ioannis Square), is also investigated. Finally conclusions and suggestions withdrawn from this study are presented.

2 Ground Displacements

During the excavation of a tunnel and after its completion movements appear in a wide region of the ground mass under the tunnel. Particular on the ground surface a settlements trough develops above and ahead the tunnel's axis due to underground works.

It has been found out, both theoretically and from an amount of data from field measurements, that the shape of the settlement trough is well described by a Gaussian distribution (equation 1.), as it is shown in perspective figure 1. (Peck 1969, O'Reilly and New 1982).

$$S_V = S_{\max} \exp(-y^2 / 2 \cdot i^2) \qquad (1)$$

Where: S_V is settlement, S_{max} is the maximum settlement (above the tunnel's center line), y is the horizontal distance from the center line, i is the horizontal distance from the tunnel center line to the point of inflexion on the settlement trough.

The geometrical features of the Gaussian curve depend on the depth z_0 of the axis' tunnel and the diameter of the excavation; another basic parameter is the type of the soil (cohesive or granular). The width of the settlement trough is about $6 \cdot i$, where $i = \mathrm{K} \cdot z_0$, K is a parameter depending on the type of the ground. (O'Reilly and New 1982).

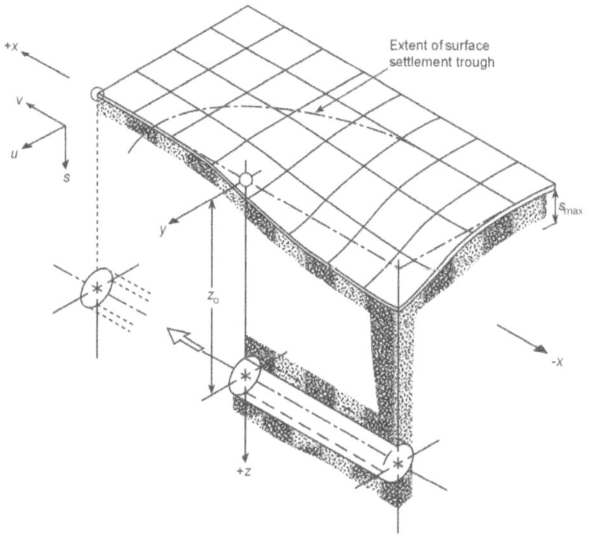

Fig. 1 Settlement above tunnel
(Attewell, P.B. et al 1982)

Apart from ground settlements, mainly above the tunnel's axis, horizontal movements appear during the tunnel excavations. The norm of the horizontal ground displacement vector S_h, is related to that of the settlements S_V, through the equation 2.

$$S_h = \frac{y}{z_0} \cdot S_V \qquad (2)$$

The maximum values of the horizontal displacements' vectors appear at the inflection points of the Gaussian distribution of the settlements having a direction towards the tunnel's axis. Both the distributions of the horizontal movements and of the vertical displacements are illustrated in figure 2.

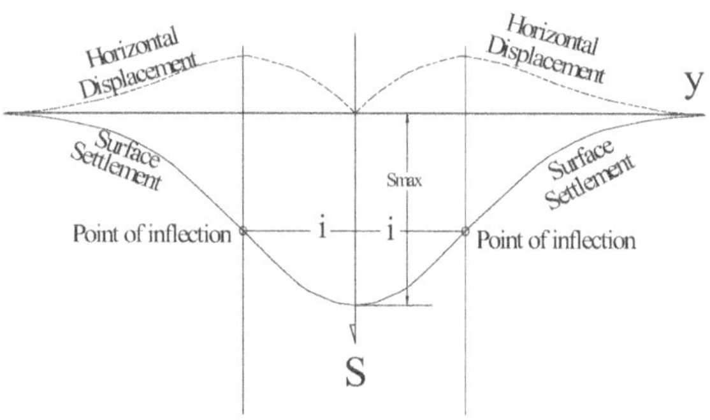

Fig. 2 Ground movements (vertical, horizontal)
(Mair et al 1996)

3 Buildings Deformation

As a consequence of the ground displacements, deformations occur on the constructions, situated inside of the settlements' trough. As the underground excavations go on, damages on brick walls and concrete elements (mainly columns and beams) appear because of the buildings' deformations.

A typical case of a building (with height H and length L) affected by the tunnel's trough settlement is depicted on figure 3. The middle point of the construction is considered to be located, as presented on figure 3, on the inflection point of the ground settlements distribution.

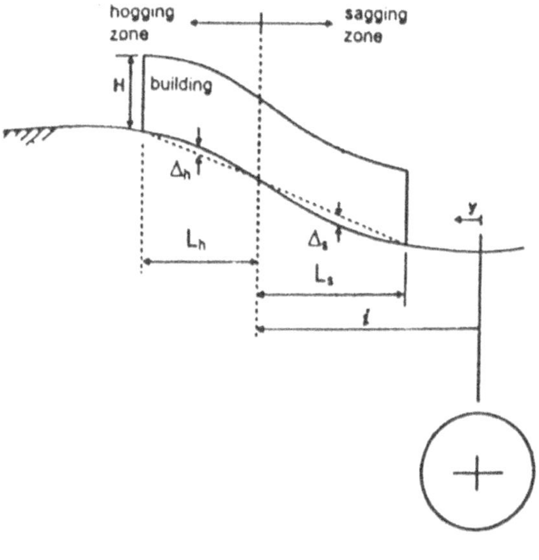

Fig. 3 Building deformation
(Mair et al 1996)

In this case, two zones are developing on the building from both sides of the inflection point. The construction is thus situated in a hogging and a sagging zone (figure 3) and of this its bearing frame undergoes tensile and shear strain.

Based on the building's dimensions (L, H), the relative settlement's amount Δ, Young's and shear modulus of the building, Burland et al (1977) have derived expressions which relate the deflection ratio Δ/L with the maximum strain of the building, for the case of $L/H = 1$ and for hogging mode, Burland (1995) has presented a diagram which determines the damage category of the building according to the deflection ratio and the horizontal strain.

4 Monitoring Structures

Because of the buildings deformations, displacements of the constructions appear before the bearing frame's cracking. These displacements are usually differential subsidences among the columns' foundations (vertical movements) and displacements of the constructions' slabs and beams (horizontal movements).

The aforementioned movements may induce damages on the brick walls or the bearing frame of the building. Moreover, collapse of a part or even the complete building is probable if the norm (the magnitude) of the displacements' vector is significant.

For this purpose the monitoring of the structures, during the underground excavations, protects them from unpredictable damages and collapses. In this way an early warning system is established against unfavorable conditions. On the other hand, the estimated displacements are indispensable for the study of the buildings' reinforcement.

The structures' monitoring can be accurately perform by instruments installed on the buildings (inclinometer, extensometer, et al). Nevertheless, nowadays the most reliable tool for the detection of the buildings' displacements, for both vertical and horizontal direction, is the geodetic procedure.

5 Geodetic Procedure

The Geodetic Procedure uses geodetic control networks (vertical, horizontal or three-dimensional). The coordinates of the networks' nodes are usually determined in a local reference system, in more than one epochs. The displacements' vectors are then determined by comparing the coordinates between every two successive epochs. This procedure includes the instrumentation for the field works as well as the steps, which are the components of the Geodetic Methodology.

The instrumentation used for the measurements, are usually total stations of high precision (reflectorless or not), for the detection of the horizontal and the vertical displacements of the structures, and digital levels for the buildings' settlements. The instruments must be checked before and after the measurements, for the detection of gross errors'.

As it is well known (E. Graffarent 1981, E. Krakiwsky 1990, W. Niemeier 1982), the successive steps of the Geodetic Methodology are the following:

- Design of the Geodetic Control Network (E. Graffarent 1981):
  - Zero Order Design. Use of a local geodetic reference system (minimum constraints 1 point and 1 azimuth fixed)
  - First Order Design. Determination of the number and the position of the reference and the control points.
  - Second Order Design. Determination of the accuracy of the observations.
- Optimization of the Geodetic Control Network (Kuang 1991). Determination of the following qualitative features:
  - Precision (high)
  - Reliability (high)
  - Sensitivity (high)
  - Cost (low)
- Measurement of the Network's elements (angles, lengths and height differences).
- Adjustment of the observations – Estimation of the points' coordinates
- Post adjustment tests (Van Mierlo 1981).
  - Global test for the acceptance of the null hypothesis.
  - Data snooping for the detection of (possible) systematic and gross errors in the observations' vector.
- Estimation of the displacements' magnitude. The statistically significant displacements are estimated.

6 A Building above Athens Metro

6.1 Geodetic Control Network

A geodetic control network was established in order to detect the building's displacements. The network consists of the reference points in the vicinity of the building, and the control points established on the building, in especially selected positions. Distances, horizontal and vertical angles and height differences between the reference points are measured. Distances and angles (horizontal and vertical) are also measured from the reference to the control points, since the latter are inaccessible. The measurements are adjusted by Least Squares and the coordinates of the network's points together with their standard deviations are estimated in a local, arbitrarily chosen reference coordinate system. Five reference points (S_1, S_2, …, S_5) were established, in the vicinity of the building, and were marked with special markers. Four control points on every floor were established. Thus 16 control points exist on the building, as a whole.

In order to determine the declinations of the slabs from the horizontal plane, the control points are situated 20cm under the cantilever of the 1st, 2nd, 3rd and 4th floor roof slabs. Therefore, the control points of each floor belong on the plane, which is parallel to the roof's slab plane and lies underneath it, at a distance of 20cm. The declination of each slab is determined through the comparison of the height differences of the control points belonging on the same floor.
In order to detect the relative horizontal displacements of the slabs, the control points were established in selected distances from the three edges of the building, in such a way that they form sets of four (one point per floor), each set belonging to the same plumb line.
In order to determine the floors' displacements relatively to the ground floor, five more points were established on the building, at the ground floor level.

6.2 Observations

The observations of the horizontal and vertical angles together with the side length measurements between the reference points were carried out with the total station TC1600 WILD, having an accuracy of ±(3mm ± 2ppm) in distance measurements and 0.3mgon in angle measurements. Both the horizontal and vertical angles were measured in two full sets.
The total station TCR303 WILD, which has the ability of measuring distances without reflector prisms and has the same accuracy with TC1600, was used for the side length measurements between the reference and control points. Both of the instruments were tested in laboratory before the field campaign in order to eliminate the possibility of the presence of systematic errors in observations, due to incorrect instrument function.
The heights of the reference points were determined by double geometric leveling, using a level of high accuracy. The network was connected with the benchmark Y.A. 10 which is one of the benchmarks that the Attiko Metro had established in the area (code nr. ID3). The benchmark is situated on the building, on the corner of D. Latta Street and Ag.Ioannis Square. Its height was assumed to be the same with the one determined during the campaign for the determination of the subsidences in the wider area of the building, on March 2003 ($H_{Y.A.10}$= 81.045m).

6.3 Adjustment of observations – Estimation of the points' coordinates

The geodetic control network consists of the 5 reference points and the 21 control points (16 points at the four floors and 5 on the ground floor level). Thus, the network as a whole consists of 26 points. 66 distances and 136 angular measurements formed the observational scheme.
The coordinates of the network points were estimated through least squares adjustment of the observations, in a local reference system. The reference system has the positive axis Ox+ coinciding with the building's façade on D. Latta street, the positive axis Oy+ perpendicular to the Ox+, while the coordinates (100, 100) were assigned at the corner of the building, at the crossing of D. Latta Street and Ag.Ioannis Square. Thus, the points' relative displacements ΔX, ΔY on the Ox, Oy axis can be directly related with the building's facades.
The adjustment of the observations was performed with the minimum external constraints: the coordinates of reference point S_1 together with the azimuth of side S_1 S_4 were kept fixed (X_{S1} = +89.250 m, Y_{S1} = 110.415 m, H_{S1} = +80.915 m, a_{S1-S4} = 159.7122^g). From the adjustment of observations the estimates of the coordinates of the network's points, together with their covariance matrix were determined. The null hypothesis was accepted from the global test (χ^2 - test) applied on the a posteriori standard deviation of the unit weight $\hat{\sigma}_0\,(=\pm 0.935)$, while no observation was rejected when Baarda's data snooping was applied for the detection of outliers. The coordinates (X, Y) and the heights of the network's reference points (RP) and of the five control points at the ground floor level, together with their standard deviations, are depicted in Table 1.

Table 1. Coordinates of the reference and the ground floor points

(RP)	X (m)	σ_X (mm)	Y (m)	σ_Y (mm)	Z (m)	σ_Z (mm)
S1	89.250	± 0.0	110.415	± 0.0	80.915	± 2.0
S2	83.940	± 0.8	86.686	± 1.4	80.834	± 2.0
S3	97.043	± 0.8	89.591	± 1.2	80.877	± 2.0
S4	103.228	± 0.8	91.357	± 1.1	80.889	± 2.0
S5	124.624	± 1.7	90.567	± 2.0	81.128	± 2.0
GROUND FLOOR POINTS						
A	100.400	± 1.6	110.634	± 0.9	82.371	± 3.0
B	99.998	± 0.9	100.014	± 1.3	82.049	± 2.0
C	100.015	± 1.5	100.000	± 3.3	82.527	± 3.0
D	121.883	± 1.6	99.977	± 1.8	82.768	± 3.0
E	125.838	± 1.7	100.000	± 1.9	83.082	± 3.0

The coordinates of four sets of three new points (three points on every floor: points A1, A2, A3, B1, B2, B3, C1, C2, C3, D1, D2, D3) were determined, using the distances of the control points from the three edges of the building and the estimates of the control points' coordinates. These sets of points determine the three edges of the building and are lying on the plane of the façade at the ground floor level, as it extends up to the fourth floor of the building. The new points were used for the determination of the magnitude of the slabs' displacements. The coordinates of three corresponding points at the ground floor level were also determined from the five control points of the ground floor (points I1, I2, I3). These points were used for the determination of the slabs' displacements with respect to the ground floor level. The coordinates and heights of these 15 points are depicted in Table 2.

The gap of the vertical joint at the building's party wall has a magnitude of 2cm, at the ground floor level. Since it was not possible to establish control points at the foundation level, it was assumed that at this level no horizontal movement had taken place. Therefore, the assumption was made that a parallel movement of 2cm of the ground floor with respect to the foundation level had occurred, having a direction towards D. Latta Str., i.e. towards the negative axis Oy. Under this assumption, the coordinates of three points (G1, G2, G3) at the foundation level were determined; these points define the edges of the building at this level (Table 2), and were used for the determination of the slabs' displacements with respect to the foundation level.

Table 2. Coordinates and heights of the control points at the building's edges.

EDGES' POINTS	X (m)	Y (m)	Z (m)
FOUNDATION LEVEL			
G1	100.400	110.657	-
G2	100.000	100.020	-
G3	125.841	100.020	-
GROUND FLOOR LEVEL			
I1	100.400	110.637	81.146
I2	100.000	100.000	81.012
I3	125.841	100.000	81.112
1st FLOOR LEVEL			
A 1	100.414	110.594	89.133
A 2	100.008	99.968	89.123
A 3	125.888	99.963	89.022
EDGES' POINTS	X (m)	Y (m)	Z (m)
2nd FLOOR LEVEL			
B 1	100.415	110.603	92.326
B 2	100.029	99.985	92.316
B 3	125.892	99.948	92.210
3rd FLOOR LEVEL			
C 1	100.409	110.595	95.530
C 2	100.021	99.961	95.519
C 3	125.904	99.938	95.397
4th FLOOR LEVEL			
D 1	100.407	110.591	98.728
D 2	99.999	99.983	98.707
D 3	125.903	99.919	98.616

6.4 Estimation of displacements

The horizontal displacements of the successive floors along the axis Ox, Oy were determined from the comparison of the coordinates of the corresponding control points at the edges of the building. The relative horizontal displacement vectors were also computed and their statistical significance was tested for a selected significance level. The slope (‰) of each vector as well as its direction from the positive axis Ox was determined too (Table 3). The relative horizontal displacements are drawn and presented (figure 4.) at the end of this paper.

Table 3. Relative horizontal displacements of the control points (C.P.) between the successive floors of the construction.

C. P.	ΔX (cm)	ΔY (cm)	DISPL. VECTOR S (cm)	SLOPE ‰	DIRECTION FROM THE AXIS X+
FOUNDATION LEVEL – GROUND FLOOR LEVEL					
G 1 – I 1	0.0	- 2.0	2.0	0	90°
G 2 – I 2	0.0	- 2.0	2.0	0	90°
G 3 – I 3	0.0	- 2.0	2.0	0	90°
GROUND FLOOR LEVEL – 1st FLOOR LEVEL					
I 1 – A 1	+ 1.4	-4.3	4.5	5.6	72.0
I 2 – A 2	+ 0.8	-3.2	3.3	4.1	76.0
I 3 – A 3	+ 4.7	-3.7	6.0	7.5	38.2
1st FLOOR LEVEL – 2nd FLOOR LEVEL					
A 1 - B 1	+ 0.1	+0.9	0.9	2.8	276.3°
A 2 – B 2	+2.1	+1.7	2.7	8.4	321.0°
A 3 – B 3	+0.4	-1.5	1.6	4.9	75.1°
2nd FLOOR LEVEL – 3rd FLOOR LEVEL					
B 1 – C 1	-0.6	-0.8	1.0	3.1	126.9°
B 2 – C 2	-0.8	-2.4	2.5	7.9	108.4°
B 3 – C 3	+1.2	-1.0	1.6	4.9	39.8°
3rd FLOOR LEVEL – 4th FLOOR LEVEL					
C 1 – D 1	-0.2	-0.4	0.4	1.4	116.6°
C 2 – D 2	-2.2	+2.2	3.1	9.7	225.0°
C 3 - D 3	-0.1	-1.9	1.9	5.9	93.0°

The slabs' declinations from the horizontal plane were also determined. These slabs' deformations are represented by the vertical differences between the successive control points of each floor on Table 4.

The slabs' vertical movements are perspective drawn and presented (figure 5.) at the end of this paper.

Table 4. Relative vertical displacements of the control points (C.P.) (declinations from the horizontal plane) of the building's four floors.

C. P.	ΔZ (cm)	DISTANCE (m)	SLOPE (‰)	DIFFER. SUBSIDENCE (mm)
1st FLOOR				
A 1 – A 2	- 1.0	10.65	- 1.0	- 13
A 2 – A 3	- 10.1	25.85	- 3.9	-70
2nd FLOOR				
B 1 - B 2	- 1.0	10.65	- 1.0	- 13
B 2 – B3	- 10.6	25.85	- 4.1	-70
3rd FLOOR				
C 1 – C 2	- 1.0	10.65	- 1.0	- 13
C 2 – C 3	- 12.2	25.85	- 4.7	-70
4th FLOOR				
D 1 – D 2	- 2.1	9.25	- 1.9	- 13
D 2 – D 3	- 9.1	25.85	- 3.5	-70

7 Conclusions - Suggestions

From the discussion presented in the above sections the following conclusions-suggestions are withdrawn regarding:

- The monitoring ability of Geodetic Procedure. Displacements of various magnitudes can be determined precisely and accurately by this procedure. Therefore, Geodetic Methodology reliably monitors the response of a construction during tunnel excavation.
- The above-mentioned building's response. Underground works are the cause of both horizontal and vertical displacements on the buildings situated above tunnel's axis. More specifically the settlements of the columns' foundations are greater than the slabs' deformations.
- The Geodetic Procedure must be used in urban regions' tunnelling in order to detect as soon as possible the building's displacements. Thus an early alarm can be given and the risk of damages or collapses is minimized.

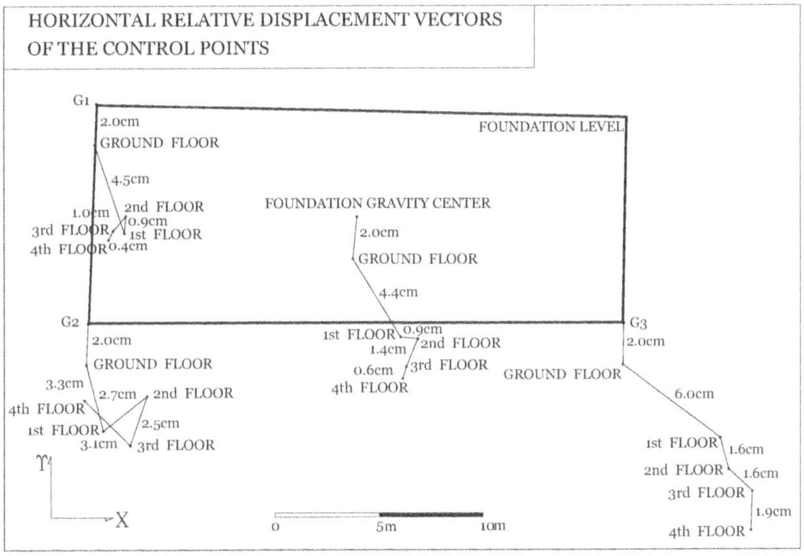

Fig. 4 The relative horizontal displacements

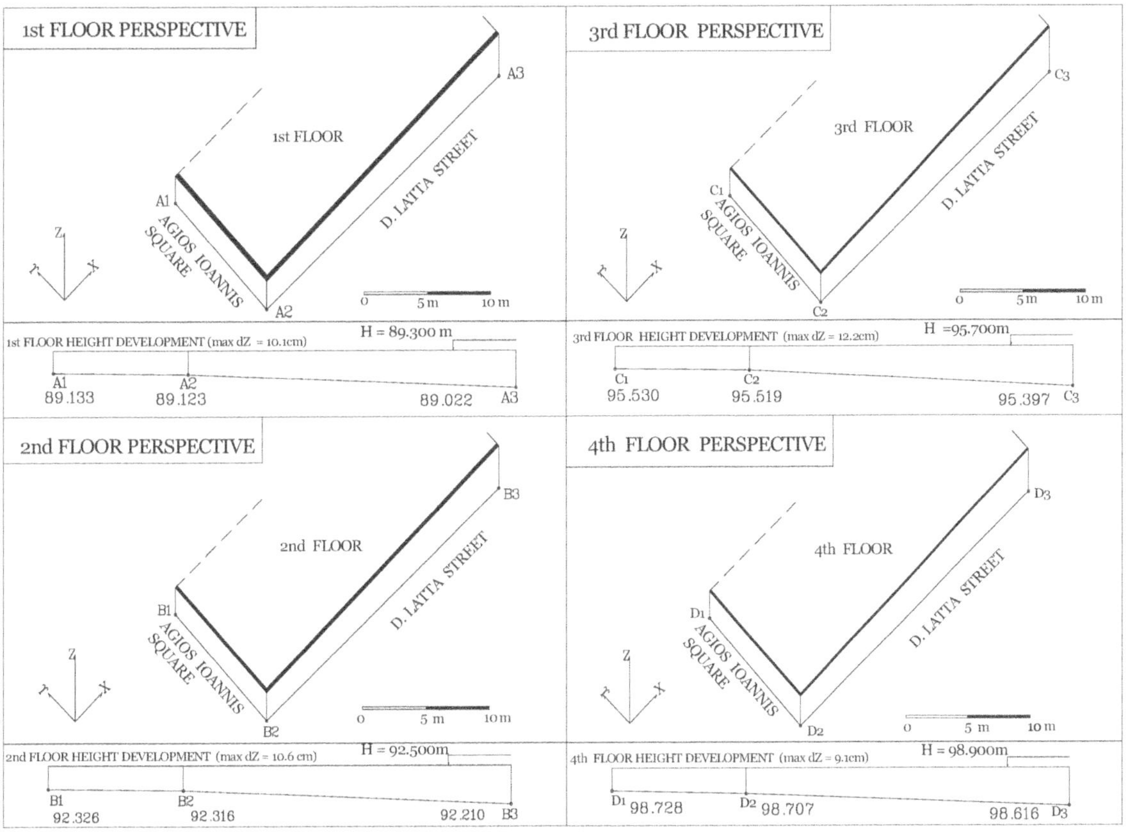

Fig. 5 The slabs' vertical movements

Acknowledgments

The author would like to thank Dr. E. C. Telioni, Lecturer in National Technical University of Athens, for her comments to this paper.

References

Attewell, P.B,. and J.P. Woodman (1982). Predicting the dynamics of ground settlements and its derivatives caused by tunnelling in soil. In: *Ground Engineering,* Vol. 15, No 8, pp. 13-22, 36.

Burland, J.B., B. Broms and V.F.B. DeMello (1977). Behavior of foundations and structures. In: *SOA Report, Session 2, Proc. 9th Int. Conf. SMFE,* Tokyo, 2; 495-546.

Burland, J.B. (1995). Assessment of risk of damage to buildings due to tunnelling and excavations. *Invited Special Lecture to IS-Tokyo 95*: 1st Int. Conf. On Earthquake Geotechnical Engineering.

Georgopoulos, G.D. (2004). Detection of the Displacements of the construction on 10, D. Latta Street, in the area of Dafni, Attica, Greece. *Study has been submitted to insurance Company Generali Hellas.*

Graffarent, E. (1981). Optimization of Geodetic Networks. In: *Boletino di Geodesia a science affini,* Vol. 33, No 4.

Kavvadas, M (2003). Deformation Measurements: Applications in Tunnelling. In: *11th FIG International Symposium on Deformation Measurements Santorini Greece*, 25-28 May 2003.

Krakiwsky, E. (1990). The method of Least Squares, Synthesis of advances, *Dept. of Surveying Engineering University of Calgary.*

Kuang Shan-Long, (1991). Optimization and Design of Deformation Monitoring Schemes. *Dep. Surveying Engineering University of New Brunswick.*

Mair, R.J., R.N. Taylor and J.B. Burland (1996). Prediction of ground movements and assessment of risk of building damage due to bored tunneling. In: *Geotechnical Aspects of Underground Construction in Soft Ground*, Balkema, Rotterdam. ISBN 9054108568.

Niemeier, W. (1982). Principal components analysis and geodetic networks, some basic considerations. *Proc. of the meeting of FIG study group 5B.*

O'Reilly, M.P., and B.M New (1982). Settlements above tunnels in the United Kingdom-their magnitude and prediction. In: *Tunnelling '82, London,* IMM 173-181.

Peck, R.B. (1969). Deep excavations and tunneling in soft ground. In: *SOA Report, 7th Int. Conf. SMFE* Mexico City, State of the Art vol. Pp 225-290.

Telioni, E. C. (2005). Ground Surface Settlements due to Underground Works. (Under print) In: *Proc. International Symposium on Geodetic Monitoring: from Geophysical to Geodetic Roles,* 17-19 March 2005 in Jaén, Spain.

Van Mierlo, J. (1981). A Review of Model Checks and Reability. In: *International Symposium On Geodetic Networks and computations* Munich.

Ground Surface Settlements due to Underground Works

Dr. Elisavet Telioni
Lecturer, Rural & Surveying Engineering School, National Technical University of Athens
9, Heroon Polytechniou str., GR – 157 80, Zografos, Athens, Greece.

Abstract. Ground surface deformations due to tunneling are crucial in urban regions where damages to surface structures and utilities should be prevented. In this paper the detection and monitoring of the vertical displacements of an area situated above the axis of one of the Athens METRO tunnels, where damages on constructions were observed during the excavations, is presented. For this purpose a geodetic vertical control network is established (March 2003). The features of the network, especially those regarding its sensitivity and reliability are presented. Significant displacements ranging up to 96mm were detected from the comparison of the adjusted heights of the network's points against those of the reference epoch t_0 before the beginning of the underground works. Finally the conclusions withdrawn are presented.

Keywords. Tunneling in urban regions, surface settlements, ground deformation monitoring, geodetic procedure.

1 Introduction

The construction of tunnels has as result the appearance of settlements on the ground surface. In many cases underground works proceed in unfavorable conditions such as intense tectonization, heterogeneous geology, adverse groundwater regimes. Moreover tunnels may be constructed in densely populated areas and, thus, they might affect structures and utilities situated above them. In order to ensure safety, not only of the working personnel and equipment but also of the surface structures, ground deformations must be measured throughout the construction, and, in many cases, even after the completion of the underground works. Especially in urban areas ground deformation monitoring can be used as a warning system against damages on structures and utilities at the surface.

Since the measured deformations of the ground reflect the in-situ conditions, they are also used as a feed back for the design verification and the adaptation to local conditions.

2 Ground settlements

Ground settlements appear as a result of underground excavations. Their amount depends on the soil characteristics, the groundwater conditions, the size and depth of the underground works, the method of excavation and support and the speed of advance (Szechy, 1973).

As it has been mentioned above, the presence of settlements, although inevitable, is of particular importance in urban environment because of their influence on buildings and utilities. It is therefore important to estimate their magnitude, and take it into consideration during the consecutive stages of underground works (planning, design and construction) in order to minimize their effect on the superimposed buildings.

A variety of methods is used for the modeling and prediction of ground movements during tunneling. These methods are categorized as follows:

- Empirical methods,
- Closed form solutions, and
- Finite Element analysis.

The most commonly used method is the empirical one, which is used for the estimation of vertical displacements and more especially of surface settlements.

According to this method the construction of a tunnel, at depth z_0 below ground surface level, has as result ground settlements, which take the shape of a trough that develops above and ahead of the tunnel's axis (Fig. 1). Martos (1958), Peck (1969), Schmidt (1969), Clough and Schmidt (1981), Attewell and Yeates (1984) assume that the transverse settlement profile can be approximated by an inverse error (Gaussian) curve, whose mathematical expression is given by the following equation:

$$S(y) = S_{\max} \cdot e^{\frac{-y_i^2}{2i^2}} \tag{1}$$

where : $S(y)$: the settlement at a distance y from the center line,

$S_{\max}$: the maximum (center line) settlement,

i : the standard deviation of the settlement curve.

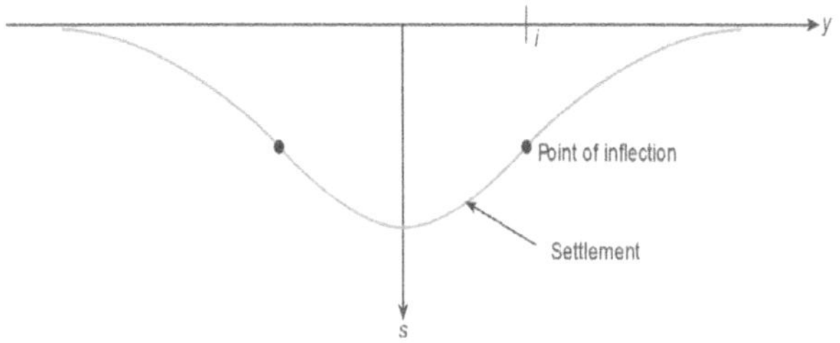

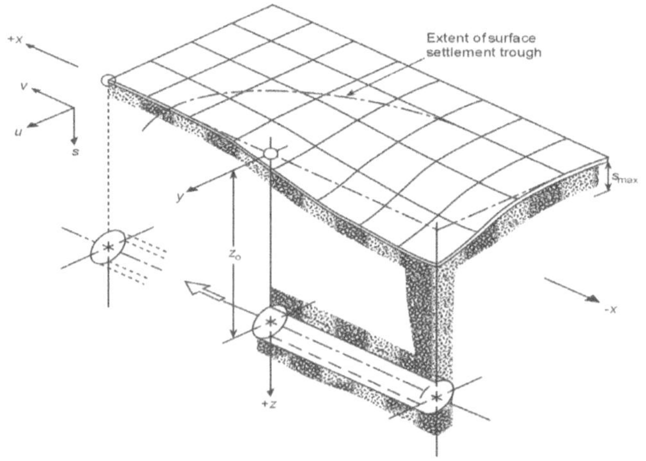

Figure 1. Settlements above advancing tunnel (Attewell et al. 1986)

The value of the quantity i practically determines the width of the settlement trough, and corresponds to the inflection point of the curve (point with maximum gradient). The settlement at this inflection point is $0.61 \cdot S_{\max}$, while the width of the settlement trough is practically taken equal to $6 \cdot i$.

The equation (1) was derived through analysis of a considerable number of case studies concerning tunneling in variable soil conditions.

Through consecutive derivations of equation (1) the following relations concerning the gradient and curvature, respectively, of the settlements curve are derived:

$$\frac{dS}{dy} = -S_{\max} \cdot e^{-\frac{y^2}{2i^2}} \quad \text{(Gradient)} \tag{2}$$

$$\frac{d^2S}{dy^2} = \frac{S_{\max}}{i^2} \cdot \left(\frac{y^2}{i^2} - 1 \right) \cdot e^{-\frac{y^2}{2i^2}} \quad \text{(Curvature)} \tag{3}$$

The volume V_s of ground settlements per meter length of the tunnel, when the deformation takes place in undrained conditions, is equal to the absolute value of the ground loss i.e. of the volume of soil between the final (after the deformation) and the initial position of the tunnel wall. It is therefore evident that, if the quantities V_s and i are known, then the settlement magnitude as well as the gradient and curvature of every point of the trough can be estimated through equations (1), (2) and (3) respectively.

The ground loss can be estimated through nomograms (Clough and Schmidt), or, by integrating equation (1), thus, having:

$$V_s = \sqrt{2\pi} \cdot i \cdot S_{\max} \tag{4}$$

According to Peck (1969) the width of the settlement trough depends primarily on the depth of the tunnel's axis. Nowadays, it has been found, from studies of a great amount of tunneling in various kinds of soil, that the value of the quantity i, and therefore, the magnitude of the settlement trough width, is a linear function of the depth z_0 of the tunnel.

Horizontal ground movements also occur during tunneling, causing damages to structures from horizontal tensile strain. The vector of ground movements has a direction towards the tunnel axis (O' Reilly and New). If S is the vertical and S_h the horizontal component of the ground movement vector, then S_h can be estimated through the following equation:

$$S_h = \frac{y}{z_0} \cdot S \qquad (5)$$

As it can be seen in Figure 1 a longitudinal settlement trough is also generated ahead of the tunnel face. It is assumed that this trough has the shape of a cumulative probability curve (Attewell & Woodman, New & O' Reilly, New & Bowers). According this assumption, the settlement directly above the tunnel face equals $0.5S$. It must be pointed out here that, there have been reported cases where buildings, situated close to or directly above the tunnel face, have experienced more damage from this longitudinal trough than from the transverse settlement curve, generated after the tunnel face had passed beneath the building.

3 Monitoring ground deformation

From the predicted ground deformations the appropriate method of tunnel excavation and support is selected. However, due to inherent ground variability, the real conditions may significantly differ from those adopted at the stage of the tunnel design. It is therefore important to measure ground deformation during tunnel construction. Through these measurements construction practices are adapted to the in – situ conditions, while, by comparing predicted against measured deformations, the appropriate typical section for the following excavation step is selected. Moreover, ground deformation monitoring usually covers a wider zone around the tunnel, thus, the ground quality is assessed as a whole.

Ground deformation measurements have proved to be indispensable as an early warning of incipient failures, as well as for the control of damages of sensitive structures located at the ground surface.

Finally, by combining the observed ground quality, the height of overburden and the observed (measured) ground deformations, the design of the final lining of the tunnel is selected.

Ground deformation can be monitored with geodetic and/or geotechnical methods. Geotechnical methods use geotechnical instruments (extensometers, micrometers, inclinometers, deflector meters, electrolytic tilt sensors), installed at selected points of the ground, in order to measure ground deformation. The main advantage of the geotechnical methods is the high accuracy of the measurements (less than 1mm), but they provide relative displacements of the points with respect to an initial time epoch, when the first measurement took place.

On the contrary, geodetic methods determine absolute displacements of the points of a geodetic control network, with respect to a reference coordinate system. Thus, the displacement vectors, although discrete in time, can be correlated and an overall view of the deformation pattern is obtained. Moreover, especially in ground surface settlements monitoring, kinematic models are applied in order to describe the ground behavior and early predict any change of it.

It must be also pointed out that, if the position of the geotechnical instruments is determined by geodetic means at the reference system of the geodetic control network, geodetic and geotechnical measurements can be combined and a more detailed pattern of the ground deformations can be obtained.

4 Geodetic monitoring of ground surface settlements

For the monitoring of ground surface settlements due to tunneling, a vertical control network is established. The shape of the network is imposed from the settlement trough, predicted at the stage of the tunnel design. It is evident, that the distribution of the control points depends on the width of the zone (magnitude of the quantity i), determined from the inflection points from both sides of the tunnel axis. Moreover, control points must be established above the tunnel axis, in order to detect movements directly above the tunnel face (longitudinal settlement trough), as well as on sensitive structures existing within the zone of influence of the tunnel. The density of the control points' distribution depends on the ground quality, determined through geological mapping and geotechnical and geophysical investigations. At least one reference point must also be established in a geologically stable area outside of the zone of influence. It is, therefore, obvious that the cooperation of geodesists and geotechnical engineers is indispensable for the design of the vertical control network.

The optimization primarily concerns the sensitivity of the network, i.e. the capability of the network to detect, accurately and reliably, vertical displacements of a predesignated magnitude. Since ground deformations induced by tunneling are usually less than 10mm or even 5mm, the measurement precision is of utmost importance.

The networks elements (height differences) are determined by double geometric leveling using digital levels. The use of digital levels has augmented the speed of measurements, thus the time needed for the observations can be considered as instant.

The network's elements are observed in more than one measuring epochs; the first of which usually takes place before the beginning of the excavations (reference epoch, t_0). The time span between the successive measuring epochs depends on the progress of the underground works, the magnitude of the expected (predicted) settlements and, primarily, on their development.

Each measuring epoch is adjusted separately, with the minimum external constraints and the same initial conditions. Post adjustment tests are applied for the detection of gross and systematic errors. Finally, the vector of vertical displacements is determined, by comparing the control points' heights for every two successive epochs and the statistical significance of the settlements is tested.

From the displacements vector of the successive epochs, velocities and accelerations are, following, determined and a kinematic model can be applied in order to describe the ground behavior and early predict any change of it.

5 Detection of surface settlements above a tunnel in urban area

The case presented here refers to the ground settlements detected during tunneling of Line 2 of Athens Metro. Tunneling for the Athens Metro was performed mostly in the Athens Schist formation, a thick sequence of flysch-type sediments consisting of calcareous sandstones, siltstones, phylites, shales, and (occasionally) limestones and marls. The tunnels were excavated by open face Tunnel Boring Machines (TBM); the underground stations were excavated using either the New Austrian Tunneling Method (NATM), either the cut and cover technique. During tunneling of Line 2, significant damages were observed on buildings above the tunnel, in the area of Dafni, in the vicinity of Agios Ioannis station. These damages are due to the ground settlements that occurred because of the excavation. On March 2003, the elements of a vertical control network were remeasured in order to detect if the development of the ground movements still continued. The network consists of 35 benchmarks installed at street level on the road surface, pavements and building columns (Fig. 2). The network was established by *Attiko Metro S.A.*, owner of the project, in order to monitor the ground settlements, during the progress of underground works.

44 height differences between the control points were determined by double leveling. The digital level Wild NA3003, having an accuracy of $1.2mm/\sqrt{km}$ was used for the measurements in combination with bar-coded staves.

The observations were adjusted with the minimum external constraints and initial conditions the same as those used by *Attico Metro S.A.*: The height of benchmark R20 was kept fixed ($H_{R20} = +84.079m$). The standard errors of the estimated heights after the adjustment were of the range of $\pm 1mm$.

The null hypothesis was accepted from the global test of the network ($\hat{\sigma}_0 = \pm 0.82mm/\sqrt{km}$) while no observation was rejected when Baarda's data snooping was applied.

The vertical displacements δH_i of the control points were, following, determined from the comparison of the adjusted heights of the measuring epoch of March 2003 and the initial one (t_0):

$$\delta H_i = \hat{H}_i^{2003} - \hat{H}_i^{t_0} \tag{6}$$

The statistical significance of the determined displacements was tested for confidence level 95%, by applying the following one – sided statistical test

$$\left|\frac{\delta H_i}{\sigma_{\delta H_i}}\right| \leq z_{0.95} \tag{7}$$

where:

$$\sigma_{\delta H_i} = \sqrt{\sigma^2_{H_{i,2003}} + \sigma^2_{H_{i,t0}}} \tag{8}$$

the standard deviation of the displacement of i control point, and

$z_{0.95}$: the corresponding value of normal distribution for confidence level 95%. ($z_{0.95} = 1.96$).

Since no information was given about the standard errors of the adjusted heights of the previous measuring epochs it was assumed that they were of the same range ($\pm 1mm$).

All vertical displacements for which the following inequality was valid :

$$|\delta H_i| \geq 1.96 \cdot \sqrt{\sigma^2_{i,t_0} + \sigma^2_{i,2003}} \tag{8}$$

i.e. all vertical displacements having magnitude greater ±3*mm* were statistically significant, their magnitude ranged up to –96*mm* (Table 1).
The vertical displacements of the control points, for the time interval between the end of underground works (when the last measuring campaign *of Attico Metro S.A.* took place) till March 2003, were also computed and their statistical significance was tested using formula (8). From the application of the above mentioned formula, it was found out that no significant settlement had occurred during the above time interval. Therefore the development of the surface settlements has come to an end.

Table 1. Statistically significant vertical displacements of the control points.

Control Point	δH_i (mm)
R 4	- 4
R 6	- 20
R 7	- 23
R 8	0
R 9	- 4
R 10	- 26
R 12	- 7
R 13	- 38
R 16	- 96
R 17	- 24
R 18	- 11
R 19	- 3
R 20	fixed
R 22	0
R 23	0

Control Point	δH_i (mm)
R 24	0
R 25	0
R 26	- 13
R 27	- 19
R 29	- 75
R 30	- 6
R 31	- 5
R 32	- 3
R 33	- 4
R 34	- 50
R 35	- 78
R 36	- 21
R 37	- 17
R 38	- 6
R 39	0

Using the determined vertical displacements of the control points of the network, the settlement isolines were drawn and are depicted in Fig.3. The axis of the tunnel, together with the predicted, during the design of the tunneling, settlement trough, as it was given by *Attico Metro S.A.* are also plotted in the same figure. It can be seen that the settlements with the greater magnitude have occurred just above the tunnel axis, where the structures with the most serious damages are situated. Moreover the width of the settlement trough, as determined through the settlement isolines, coincides well with the predicted one.

6 Conclusions

Geodetic methodology is an indispensable tool for the monitoring of the ground settlements. The establishment of a control network and the measurement of its elements in different measuring epochs, covering the total time span needed for the underground works, can not only determine the magnitude of vertical displacements but also monitors the ground behavior. Thus, an alert signal can be given as soon as possible whenever an unpredicted change takes place.
Geodetic and geotechnical measurements should be combined in the same reference frame in order to obtain a detailed pattern of the ground deformations.
Moreover, based on the ground behavior monitoring, the model of the ground response, adopted at the stage of tunnel design, can be adequately updated, and used for back analysis for the support of the front of the tunnel, during the underground works.
Since the volume loss can be detected in real time conditions with the afore-mentioned methodology, reinforcement of the tunnel lining can be decided, if necessary, immediately.

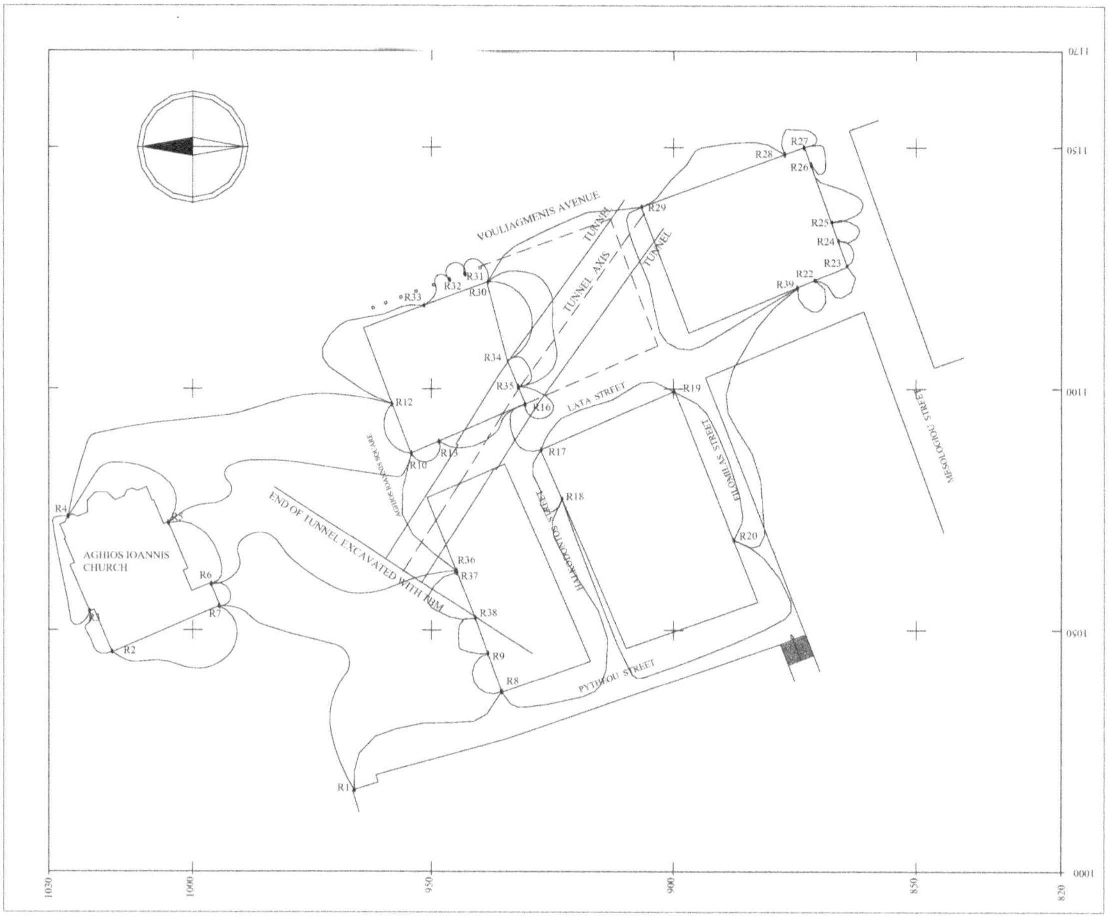

Figure 2. The vertical control network

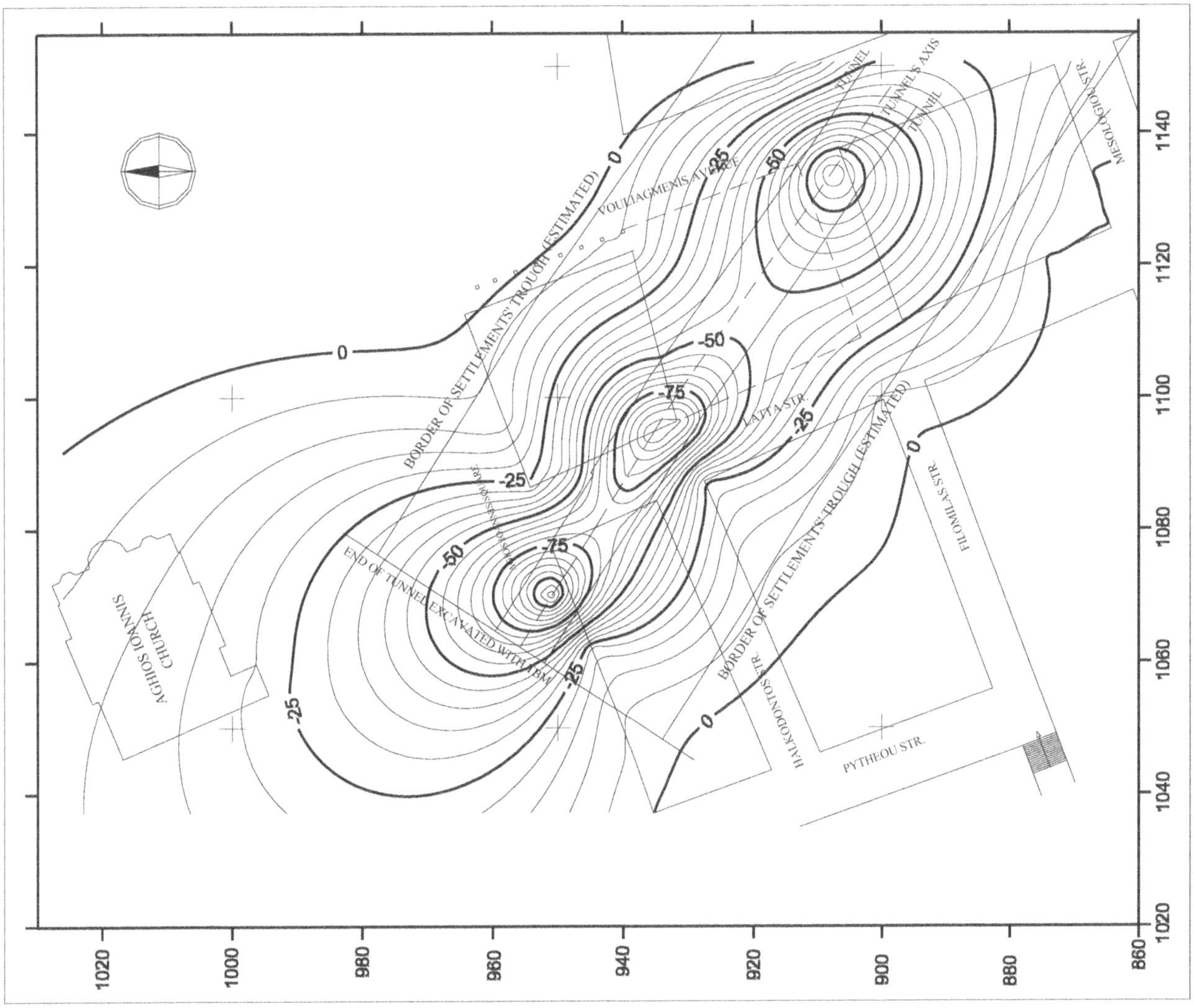

Figure 3 Surface settlements' trough – Subsidence isolines in the area under consideration.

References

Attico Metro S.A. Personal Comunication (2003).

Attewell, P.B. and Woodman J.P. (1982). Predicting the dynamics of ground settlements and its derivatives caused by tunnelling in soil. In: *Ground Engineering*, Vol. 15, No 8, pp. 13-22, 36.

Burland, J.B., B. Bioms and V.F.B. DeMello (1977). Behavior of foundations and structures. In: *SOA Report, Session 2, Proc. 9th Int. Conf. SMFE*, Tokyo, 2; 495-546.

Burland, J.B. (1995). Assessment of risk of damage to buildings due to tunnelling and excavations. *Invited Special Lecture to IS-Tokyo 95*: 1st Int. Conf. On Earthquake Geotechnical Engineering.

Georgopoulos, G.D., (2004). Detection of the Displacements of the construction on 10, D. Latta Street, in the area of Dafni, Attica, Greece. *Study has been submitted to insurance Company Generali Hellas.*

Graffarent E. (1981). Optimization of Geodetic Networks. In: *Boletino di Geodesia a science affini*, Vol. 33, No 4.

Kavvadas, M (2003). Deformation Measurements: Applications in Tunnelling. In: *11th FIG International Symposium on Deformation Measurements Santorini Greece*, 25-28 May 2003.

Krakiwsky E. (1990). The method of Least Squares, Synthesis of advances, *Dept. of Surveying Engineering University of Calgary.*

Kuang Shan-Long, (1991). Optimization and Design of Deformation Monitoring Schemes. *Dep. Surveying Engineering University of New Brunswick.*

Mair, R.J., R.N. Taylor, J.B. Burland, (1996). Prediction of ground movements and assessment of risk of building damage due to bored tunneling. In: *Geotechnical Aspects of Underground Construction in Soft Ground*, Balkema, Rotterdam. ISBN 9054108568.

Niemeier W. (1982). Principal components analysis and geodetic networks, some basic considerations. *Proc. of the meeting of FIG study group 5B.*

O'Reilly, M.P., B.M New, (1982). Settlements above tunnels in the United Kingdom-their magnitude and prediction. In: *Tunnelling '82, London*, IMM 173-181.

Peck, R.B., (1969). Deep excavations and tunneling in soft ground. In: *SOA Report, 7th Int. Conf. SMFE* Mexico City, State of the Art vol. Pp 225-290.

Telioni, E. C., (2003). Kinematic Modeling of Subsidences In: *Proc. 11th FIG International Symposium on Deformation Measurements Santorini Greece*, 25-28 May 2003.

Van Mierlo J. (1981). A Review of Model Checks and Reability. In: *International Symposium On Geodetic Networks and computations* Munich.

GNSS/LPS Based Online Control and Alarm System (GOCA) - Mathematical Models and Technical Realization of a System for Natural and Geotechnical Deformation Monitoring and Hazard Prevention

R. Jäger and F.González,

Institute of Applied Research (IAF), Hochschule Karlsruhe für Technik und Wirtschaft - University of Applied Sciences. Moltkestrasse 30, D-76133 Karlsruhe. URL: www.goca.info

Abstract. The research and development project GOCA (GNSS/LPS based Online Control and Alarm System) applies GNSS as well as classical local positioning sensors (LPS), such as e.g. total stations, for a real-time deformation monitoring. The GOCA system may be set up as an early-warning-system for natural hazards (landslides, volcanos) and also for the monitoring of geotechnical installation and buildings (mining, tunneling activities, bridges). The GOCA system consists of GNSS and LPS sensors, which are set up in the monitoring area as a permanent array or as a mobile monitoring system. The GOCA hardware-control and communication software modules collect, in different kind of communication modes, the GNSS/LPS data in a user defined sampling rate. The GOCA deformation-analysis software is responsible for a further processing of that data in a three steps sequential adjustment procedure. Both least squares and robust estimation techniques (L1-norm and other bounded influence M-estimators) are applied. A first focus is set on the robust online displacement estimation, statistical testing and alarm setting. Further the algorithmic scheme of a L2- and a robust L1-norm Kalman filter is treated, which is applied in the GOCA system for the estimation of the object-point state vector of displacements, velocities and accelerations. The further development of the deformation analysis concern the integration of further parameters (e.g. material parameters and damage models) as well as of additional geotechnical sensors (e.g. strain-meters) and lead to system analysis based approaches. Here FEM-based approaches for static and dynamics processes are proposed as appropriate models. Typical project applications are shown as examples for the GOCA-system.

Keywords: GNSS and LPS real-time monitoring system, geodetic deformation analysis, robust estimation, Kalman-filtering, natural hazard and disaster prevention.

1 Introduction

The GOCA system consists of a set of GPS sensors and communication units set up in the monitoring area and two software packages, namely the GOCA sensor communication software and the GOCA deformation analysis software. The GOCA computation unit, called the "GOCA-Center", consists accordingly of a computer, where these two software components are running. Based on the GKA data interface, the GOCA communication software packages of the GOCA cooperation partners, namely MONITOR of the GeoNav_Trimble and S+H Systemtechnik companies (www.geonav.de; www.sh-systemtechnik.de) and GOCA_DC3 of the company Dr. Bertges (www.drbertges.de) are able to control any array of GNSS and LPS sensors (fig. 1). They provide the GNSS and LPS data for the GOCA deformation analysis software (briefly GOCA software). The structure of the GNSS and LPS data interface for the GOCA deformation analysis software, the so-called GKA files, are adapted to the GNSS baseline standard output of respective coordinates and covariance matrices and the LPS data standard (zenith angles, distances, directions, height differences). Any local GOCA-center may be linked over a long distance to another PC, which serves as a remote control station, e.g. by internet or a telephone link. The further processing of the GKA baseline and LPS data and respective online modeling of a three-dimensional displacement, velocity and acceleration field and related deformation functions are performed by the GOCA software, which has been developed by the GOCA team Karlsruhe starting in 1998 (www.goca.info). The GOCA software sounds alarm, if a specified critical state becomes significant during an online monitoring. The complete deformation analysis functionality is provided in a online, a near online and in post-processing mode respectively. So the GOCA system may be set up in an object area as a permanent array or as a mobile "task force system" in areas, where danger becomes imminent (Borchers and Heer, 2002; Bulowski, 2001; Korittke and Palte 2001; Lauterbach and Krauter 2002; Schäfer, 2004; Kälber and Jäger 1999-2005).

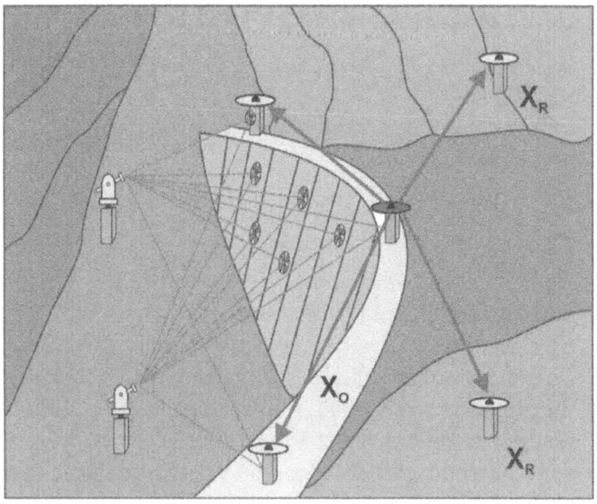

Fig. 1 Design of a classical absolute deformation network realized by the GOCA system.

The GOCA GNSS and LPS sensor array (fig. 1) sets up a classical geodetic deformation network online, which is parted into a stable reference point frame $\mathbf{x}_R$ and the domain of moving object points $\mathbf{x}_O$ (Pelzer 1974). The sensor observations $\mathbf{l}$, in the dam example fig. 1 GNSS baseline observations and total stations observations (LPS) enable the permanent estimation of the deformation state vector $\mathbf{y}(t)$ depending on time t.

2 GOCA Deformation Analysis Software – Basic Concepts

The deformation analysis concept implemented in the GOCA software is due to a classical deformation analysis (Pelzer 1974; Kälber and Jäger 2001; Feldmeth et. al. 2004; Jäger and Bertges 2004). It is based on a strict network adjustment and is realized in three subsequent (1st, 2nd and 3rd) adjustment steps. The monitoring network is physically realized by an array of GPS/GNSS and LPS sensors, while the respective deformation network design has to be specified in the initialization step (1st step).

The 1st step provides the initialization of the so-called reference or stable point frame $\mathbf{x}_R$ of the monitoring network. Here, as well as in the following monitoring, every GNSS receiver, LPS sensor and target, has to be specified to be either a (stable) reference point $\mathbf{x}_R$ or a (moving) object point $\mathbf{x}_O$. So GNSS base stations may be situated in the object part $\mathbf{x}_O$, as well as rovers may be used to represent the stable part $\mathbf{x}_R$ (fig. 1). In that way an optimum design and sensitivity (Jäger 1997) of a monitoring array - e.g. with respect to short GNSS baselines - is enabled. In the context with an adjustment concept behind all steps of the GOCA deformation analysis, the deformation network design may be set up as a redundant (e.g. by using two GNSS reference stations or additionally LPS sensors) or as a non-redundant configuration.

The initialization, namely the 1st adjustment step, is based on a least squares (L2-norm) free network adjustment of the GNSS baseline and LPS data with respect to a user-defined starting time. It is robustified with respect to gross errors by an automatic iterative data snooping including a stepwise variance component estimation. By the aim of realizing a classical deformation analysis in a permanent online mode, this 1st step has to precede the deformation monitoring, as it provides the network datum $\mathbf{x}_R$ (fig. 1) for the permanent georeferencing of the object points $\mathbf{x}_O$ (fig. 1) and the deformation analysis, which run parallel in the following 2nd and 3rd adjustment step.

The 2nd step performed by the GOCA software is again based on the above mentioned GKA GNSS baseline and LPS data and is running completely automatically during the online monitoring. Any online monitoring project can however also be processed in a post-processing mode. The 2nd step comprises the permanent L2-norm adjustment of the GNSS and LPS data (distances, zenith angles, directions, height differences), and the mathematical model provides the georeferencing of the 3D object-point position time series $\mathbf{x}_O(t_i)$. The reference frame $\mathbf{x}_R$ (fig. 1) is constituted by the stable reference points, while both the coordinates and the covariance matrix of the reference points - as the result of the 1st step - are considered in the mathematical model. Based on a L2-norm adjustment and including automatic data-snooping, the 2nd step works automatically, both for redundant or for non-redundant GNSS and LPS array configurations. With respect to a necessary discretization in time, the expression "epochs" at times t_i and t_j are to be understood in the following as extended time intervals with a duration ΔT. The subsequent adjustment intervals ΔT have to be equal or larger than the sensor data sampling-interval Δt, which is needed for the data collecting in the GOCA-array. The time stamp t_i of an "epoch" accordingly refers then to the centre of the ith interval $\Delta T(t_i)$ (2a, 2b).

The 3rd step, the deformation analysis itself, implemented in the GOCA software, deals with the estimation of the parameters of different so-called deformation functions and runs parallel online to the 2nd step. The parameter estimation is related to the results of the 2nd step, namely the 3D object-point position time series $\mathbf{x}_O(t_i)$ and their stochastic model. The 2nd and the 3rd step are handled online as seamless adjustment processes. The estimation of respective deformation functions and parameters can be performed in the 3rd step online either as a

- L2-norm estimation, (5b) or a
- Robust L1-norm estimation (5c), or a
- Robust M-Estimation (5d).

The estimation principle can be chosen by the GOCA software user. The following object-point related deformation functions and respective parameter estimation algorithms are available in the online monitoring mode of the GOCA software:

- Moving average including the detection of critical displacements
- Automatic displacement estimation ((8a), (8b), fig. 4, fig. 5) between different extended epochs $\Delta T(t_i)$ (e.g. 3 hours)
  - 1st epoch = Initialization (1st step) and 2nd epoch is repeated periodically (e.g. every 12 hours).
  - 1st epoch = Fixed by user definition and 2nd epoch repeated periodically.
  - 1st epoch = Repeated also periodically such as the 2nd epoch.
- Kalman-Filtering (9a), (9b) with the state vector of three-dimensional
  - displacements $\mathbf{u}(t)$,
  - velocities $\dot{\mathbf{u}}(t)$ and
  - accelerations $\ddot{\mathbf{u}}(t)$.

The deformation functions and the respective parameters described above can be referred either to critical values or to significant changes, e.g. in the displacements **u**(t) ((7a), (7b); (8a), (8b), (9a), (9b); fig. 4; fig. 5). So an automatic alarm can be sounded according to the user-defined priorities (fig. 4) and alarm modes (email, SMS, fax, etc). The above online deformation functions can also be used in a near-online or a post-processing mode (e.g. in a daily processing of the data). In the near-online and post-processing of the 3rd step the complete spectrum of the above mentioned online deformation functions is again available, and additionally the deformation functions of a

- Polynomial based trend-estimation, a
- Leap detection and the estimation of
- Displacements between two epochs defined by an individual interval length (e.g. one day, fig. 5).

Besides L2-norm the robust estimation principles (Huber 1973; Bickel 1975; Holland 1977; Jäger et al. 2005) of a L1-norm and of the Huber-estimator can be chosen (fig. 4).

Fig. 2 Example of a GPS receiver and telemetric equipment within a GOCA sensor array in case a slope monitoring in open cast mining at RWE Power AG, Garzweiler using the GOCA system.

3 Mathematical Models implemented in the GOCA software

3.1 Adjustment steps 1 and 2 – Reference Point Initialisation and Georeferencing of the Object Points

In the 1st and in the 2nd step all GNSS and LPS observations **l**, which may according to the sensor design (fig. 1) either take place between the reference points $\mathbf{x}_R$ or between the object points $\mathbf{x}_O$ or between reference points $\mathbf{x}_R$ and object points $\mathbf{x}_O$ are included in an online geodetic network adjustment. The observations $\mathbf{l}$ - like shown in fig. 1 at the example of the GNSS baseline vectors and/or the LPS observations of total-stations (distances, zenith angles and distances) - are collected and provided by the GKA data interface together with their respective covariance matrices $\mathbf{C}_{li}$. In general the relation

$$\mathbf{l} = \mathbf{l}(\mathbf{y}) \quad (1)$$

between the sensor observations $\mathbf{l}$ and the state vector $\mathbf{y}$ of the deformation monitoring network is nonlinear, and the so-called functional model (1) has to be linearised by introducing approximate parameters $\mathbf{y}^0$. This is done automatically by the GOCA software.

The sensor observations $\mathbf{l}$ (1) are collected at a tracking rate Δt_{sens} (e.g. $\Delta t_{sens} = 1\,\text{sec}$ for GNSS (fig. 2)) and Δt is the time, which is need to collect all sensor data in a GOCA array. The definition of discrete adjustment "epochs" and thereby the discretization of the object point movements is referring to an interval ΔT with

$$\Delta t \leq \Delta T \quad . \quad (2a)$$

The lower and upper border of the that interval- $\Delta T(t_i)$, which refers to the epoch t_i and has a duration of ΔT, reads:

$$\Delta T(t_i) = [t_i - \tfrac{\Delta T}{2}, t_i + \tfrac{\Delta T}{2}] \quad (2b)$$

According to (2b) the time dicretisation of the object point's displacements is done in subsequent intervals (2b) of duration ΔT. So all observations $\mathbf{l}_i$ within the borders of the interval $\Delta T(t_i)$ (2b) imply the estimation of a constant state vector $\mathbf{x}_O(t_i) = \mathbf{x}_{Oi}$ of the object point coordinates referring to the discretization epoch t_i. Observation sets $\mathbf{l}_i$ and $\mathbf{l}_j$ belonging to two general different epochs t_i and t_j accordingly lead to two different sets of object-point coordinates $\mathbf{x}_{Oi}$ and $\mathbf{x}_{Oj}$, which are be set up in the time invariant coordinate frame $\mathbf{x}_R$ of the reference points (fig. 1). For two general epochs t_i and t_j we get, after the linearization of (1), the following system of an in general over-determined system of observation equations:

$$\begin{aligned}(\mathbf{l}_i - \mathbf{l}_i(\mathbf{y}^0)) + \mathbf{v}_i &= \\ = \mathbf{A}_{Ri} \cdot \mathbf{dx}_{Ri} + \mathbf{A}_{Oi} \cdot \mathbf{dx}_{Oi} + \mathbf{0} \cdot \mathbf{dx}_{Oj} & \quad (3a)\end{aligned}$$

and $\mathbf{C}_{li}$

$$\begin{aligned}(\mathbf{l}_j - \mathbf{l}_j(\mathbf{y}^0)) + \mathbf{v}_j &= \\ = \mathbf{A}_{Rj} \cdot \mathbf{dx}_{Rj} + \mathbf{0} \cdot \mathbf{dx}_{Oi} + \mathbf{A}_{Oj} \cdot \mathbf{dx}_{Oj} & \quad (3b)\end{aligned}$$

and $\mathbf{C}_{lj}$

With $\mathbf{A}(\mathbf{y}^0)$ we introduce the different design matrices of the adjustment, which are set up by the first derivatives of the observations $\mathbf{l}$ with respect to the unknown parameters $\mathbf{y}$ and approximations $\mathbf{y}^0$. With $\mathbf{v}_i$ and $\mathbf{v}_j$ we describe the vectors of the corrections for the observation vectors $\mathbf{l}_i$ and $\mathbf{l}_j$. Upon demanding the stability constraint $\mathbf{x}_{Ri} = \mathbf{x}_{Rj} = \mathbf{x}_R$ for the reference points (fig. 1), we obtain as the basic state vector of the deformation process modeling $\mathbf{y}$ in the 1st and 2nd step for two general epochs t_i and t_j reading

$$\mathbf{dy} = (\mathbf{dx}_R, \mathbf{dx}_O(t_i), \mathbf{dx}_O(t_j))^T \quad \text{, with} \quad (4a)$$

$$\mathbf{y} = \mathbf{y}^0 + \mathbf{dy} \quad . \quad (4b)$$

According to (3a), (3b) and (4a), (4b) the mathematical model and the basic deformation process modeling of the 2nd GOCA adjustment step, and the related state vector $\mathbf{y}$ for m=2 epochs, can easily be extended to any number of m monitoring epochs.

3.2 Least Squares Adjustment and Robust M-Estimation

The so-called M-estimation (Huber 1973; Bickel 1975; Holland 1975; Jäger et al., 2005) applied to n observations l_k (k=1,n) in all m monitoring epochs reads:

$$\begin{aligned}&\sum_{k=1}^{n} \rho(\bar{v}_k) = \\ &= \sum_{k=1}^{n} \rho((\mathbf{C}_l^{-\frac{1}{2}} \cdot \mathbf{A})_k \cdot d\hat{\mathbf{y}} - (\mathbf{C}_l^{-\frac{1}{2}} \cdot (\mathbf{l} - \mathbf{l}(\mathbf{y}^0)))_k) \\ &= \sum_{k=1}^{n} \rho(\overline{\mathbf{A}}_k \cdot d\hat{\mathbf{y}} - (\bar{\mathbf{l}} - \bar{\mathbf{l}}(\mathbf{y}^0))_k) = \text{Min}\big|_{d\hat{\mathbf{y}}} \quad . \quad (5a)\end{aligned}$$

It leads to the estimated state vector **y** (4a), (4b) by minimizing the total sum of the so-called loss function $\rho(\overline{v}_k)$ of the standardized residuals $\overline{v}_k$. With $\overline{\mathbf{A}}$ and $\overline{\mathbf{l}}$ we introduce the so-called homogenized design matrix and homogenized observations, respectively. The algorithmic solution of (5a) and the determination of the estimation $\hat{y}$ and the covariance $\mathbf{C}_{\hat{y}}$ are described e.g. in Hollan (1997) and in Jäger et. al. (2005).

Depending on the type of the loss-function $\rho(\overline{v}_k)$ the estimation of the parameters **y** in the 1st, 2nd and 3rd GOCA-adjustment step is either due to a least squares (L2-Norm), a L1-Norm or a Huber-estimation. The respective loss functions (Jäger et al. 2005, Holland 1975, Bickel 1975) are given by the equations (5b,c,d) and read:

- Least squares estimation (L2-norm)

$$\rho(\overline{v}_i) = \frac{1}{2}\overline{v}_i^{\,2} \tag{5b}$$

- Robust L1-norm estimation

$$\rho(\overline{v}_i) = \frac{1}{2}\,|\,\overline{v}_i\,| \tag{5c}$$

- Weak robust Huber-Estimation

$$\rho(\overline{v}_i) = \begin{cases} \frac{1}{2}\overline{v}_i^2 & \forall\ |\overline{v}_i| \le k \\ |\overline{v}_i| & \forall\ |\overline{v}_i| > k \end{cases} \tag{5d}$$

The loss function $\rho(\overline{v}_k)$ (5b) is optimal for normal distributed observation errors, but it is not robust against gross observation errors ∇l_k, and has to be robustified in that case by an explicit datasnooping. The loss-functions (5c) and (5d) imply however implicitely a certain kind of robustness and bounded influence (Huber 1973; Bickel 1973; Holland 1977) concerning gross errors. Especially (5c) shows a high robustness - meaning a breaking-point (Jäger et. al 2005) up to 50% - in case of the object point time-series $\mathbf{x}_0(t_i)$ ((6a,b), fig. 3, fig. 5). This is due to the fact, that the $\mathbf{x}_0(t_i)$ are - with respect to the GNSS and LPS observations **l** in the epochs -, mathematically uncorrelated direct observations in the deformation-analysis (3rd step), with regard to the three step adjustment model of the GOCA deformation-analysis software.

The 1st adjustment step (initialization) is based on a L2-norm (5b) and is robustified by the procedure of iterative datasnooping. The 1st step provides the reference point frame information for all the subsequent monitoring steps. The steps 2 and 3 are running parallel. The essential result of the initialization is the information about the reference point frame $\mathbf{x}_R$ (fig. 1), which is completely represented by adjusted coordinates $\mathbf{x}_R$ and the covariance matrix $\mathbf{C}_{\mathbf{x}_R}$. Both are stored in the GOCA project database.

The mathematical model of the 2nd adjustment step is also based on a L2-norm estimation (5c) and iterative datasnooping and provides the time series $\mathbf{x}_O(t_i)$ of the object points and the covariance matrices $\mathbf{C}_{\mathbf{x}_O(t_i)}$ as

$$\mathbf{x}_O(t_i) \quad \text{and} \quad \mathbf{C}_{\mathbf{x}_O}(t_i) \tag{6a,b}$$

According to (3a), (3b) the time series $\mathbf{x}_O(t_i)$ (6a) (see fig. 3) are georeferenced in the reference point frame $\mathbf{x}_R$. In order to increase essentially the real-time performance of the 2nd step adjustment the coordinates $\mathbf{x}_R$ of the reference points resulting of the initialization are held fix on setting $\mathbf{dx}_R = \mathbf{0}$. The covariance matrix $\mathbf{C}_{\mathbf{x}_R}$ is however taken into account for the computation of the covariance matrices $\mathbf{C}_{\mathbf{x}_O}(t_i)$ (6b) of the object point time series $\mathbf{x}_O(t_i)$. The raw object point displacements t_0 are resulting by defining a discrete reference epoch t_0, and we have

$$\mathbf{u}_O(t_i) = \mathbf{x}_O(t_i) - \mathbf{x}_O(t_0) \quad \text{and} \quad \mathbf{C}_{\mathbf{u}_O}(t_i) \tag{7a,b}$$

The covariance matrices $\mathbf{C}_{\mathbf{u}_O}(t_i)$ (7b) are resulting by applying the so-called law of error propagation to (7a).

3.3 Online Displacement Estimations

The mathematical model of the object-point related deformation functions (chap. 2.1) and the respective parameter estimation algorithms in the 3rd step of the GOCA deformation analysis are referring to the object points $\mathbf{x}_O(t_i)$ (6a) and $\mathbf{u}_O(t_i)$ (7a) respectively. The object points $\mathbf{x}_O(t_i)$ (6a) and $\mathbf{u}_O(t_i)$ (7a) are used as observations $\mathbf{l}(t_i)$ together with the stochastical models $\mathbf{C}_{\mathbf{x}_O}(t_i)$ (6b) and $\mathbf{C}_{\mathbf{u}_O}(t_i)$ (7b), respectively. Like in the 2nd adjustment step, the "epochs" in the 3rd deformation analysis step are also referring to extended intervals, which are centred around t_i and named as $\Delta t(t_i)$.

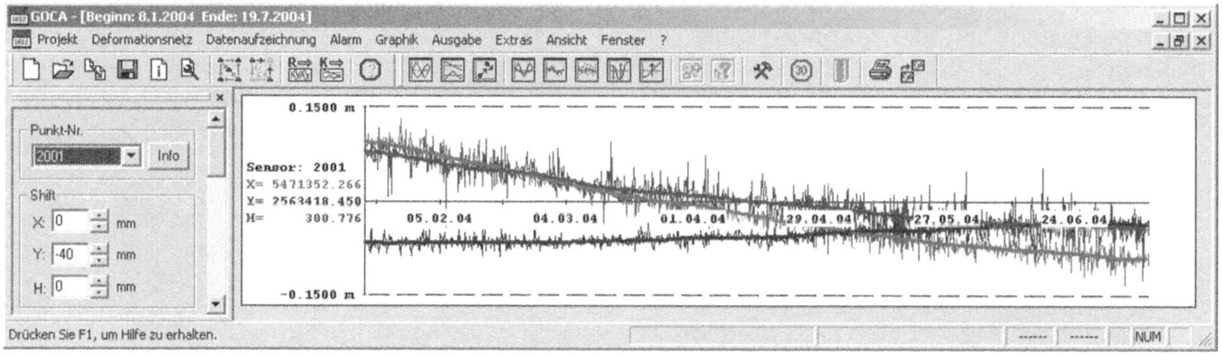

Fig. 3 GOCA object-point time series $\mathbf{x}_0(t_i)$ as result of the 2nd adjustment step of the GOCA-software. The thick lines show the smoothing by a robust moving average estimation applying the M-estimation (5c).

The different M-estimation principles (5b), (5c) and (5d) can be selected arbitrarily be the user in the GOCA settings dialogs for the deformation function estimations of the 3rd step (see e.g. fig 4).

The functional model of the GOCA object point displacement estimation $\mathbf{u}(t)$ (Jäger und Bertges, 2004) at a time t with respect to the reference time t_0 and state vector $\mathbf{y}(t)$ reads:

$$\begin{bmatrix} \mathbf{l}_{t_0} \\ \mathbf{l}_t \end{bmatrix} + \begin{bmatrix} \mathbf{v}_{t_0} \\ \mathbf{v}_t \end{bmatrix} = \begin{bmatrix} \mathbf{E}_1 & \mathbf{0} \\ \mathbf{E}_2 & \mathbf{E}_2 \end{bmatrix} \cdot \begin{bmatrix} \mathbf{x}_0 \\ \mathbf{u}(t) \end{bmatrix}$$
(8a)

$$\mathbf{y}(t) = (\mathbf{x}_0, \mathbf{u}(t))^T \quad (8b)$$

The reference time t_0 is again defined with respect to an extended time interval $\Delta t(t_0)$ (fig. 4, left) for the observations taken from the time series $\mathbf{x}_O(t_i)$ (6a). The same holds for the epoch t_i referring also to an extended interval $\Delta t(i)$ (fig. 5, right). Accordingly the two observations groups $\mathbf{l}_{t_0}$ and $\mathbf{l}_{t_i}$ (8a) taken from the time series vector $\mathbf{x}_O$ (6a) provide in general redundancy with respect to the state vector **y** (8b) of the displacement model (8a) with only 6 parameters for each object point $\mathbf{x}_O$. The six parameters y(t) are the 3-dimensional position $\mathbf{x}_0$ at the reference time t_0 and the 3-dimensional displacement $\mathbf{u}(t)$ at the estimation epoch time t_i. The matrices $\mathbf{E}_1$ and $\mathbf{E}_2$ are column matrices composed of (3x3)-unit matrices for each three-dimensional observation **l** in the respective group.

Fig. 4 shows the GOCA-software settings dialog for the online displacement estimation according to the mathematical model (8a) (8b) in the deformation analysis (3rd step). The different settings concern the choice of the object points, the epoch definition for the displacement estimation $\mathbf{u}(t)$, the settings for adjustment and statistical testing, and the settings for an automatic alert.

3.4 Kalman-Filtering as General M-Estimation

The GOCA Kalman-Filtering (Kälber and Jäger 1999, Feldmeth et al. 2004, Kälber et al. 2000) as a second example for the deformation parameter estimation in the 3rd step is related to the following so-called transition equation (9a,b), and to the state vector **y** (9c) reading:

$$\mathbf{y}(t) = \mathbf{T} \cdot \mathbf{y}(t - dt) \quad , \quad (9a)$$

$$\mathbf{y}(t) = \begin{bmatrix} \mathbf{u}(t) \\ \dot{\mathbf{u}}(t) \\ \ddot{\mathbf{u}}(t) \end{bmatrix} = \begin{bmatrix} \mathbf{I} & [dt] & \left[\frac{1}{2} dt^2\right] \\ \mathbf{0} & \mathbf{I} & [dt] \\ \mathbf{0} & \mathbf{0} & \mathbf{I} \end{bmatrix} \cdot \begin{bmatrix} \mathbf{u}(t-dt) \\ \dot{\mathbf{u}}(t-dt) \\ \ddot{\mathbf{u}}(t-dt) \end{bmatrix} \quad (9b)$$

$$\text{with} \quad \mathbf{y}(t) = (\mathbf{u}, \dot{\mathbf{u}}, \ddot{\mathbf{u}})^T \quad . \quad (9c)$$

The state vector **y**(t) of the GOCA-Kalman-Filtering comprises the individual three-dimensional displacement $\mathbf{u}(t)$, velocity $\dot{\mathbf{u}}(t)$ and acceleration $\ddot{\mathbf{u}}(t)$ of the object points between subsequent small time intervals dt. The state transition model (9b) is widely used in geodetic monitoring of slow and medium fast object points movements, e.g. land-slides, mining etc. (Kahmen und Suhre 1983; Jäger und Kälber 1999). In case of fast kinematic processes - e.g. for monitoring and modeling the vibration of structures - the assumption of a constant acceleration $\ddot{\mathbf{u}}(t)$ may be replaced by a process--related parametrization (Jäger and Kälber 1999; Eichhorn 2005).

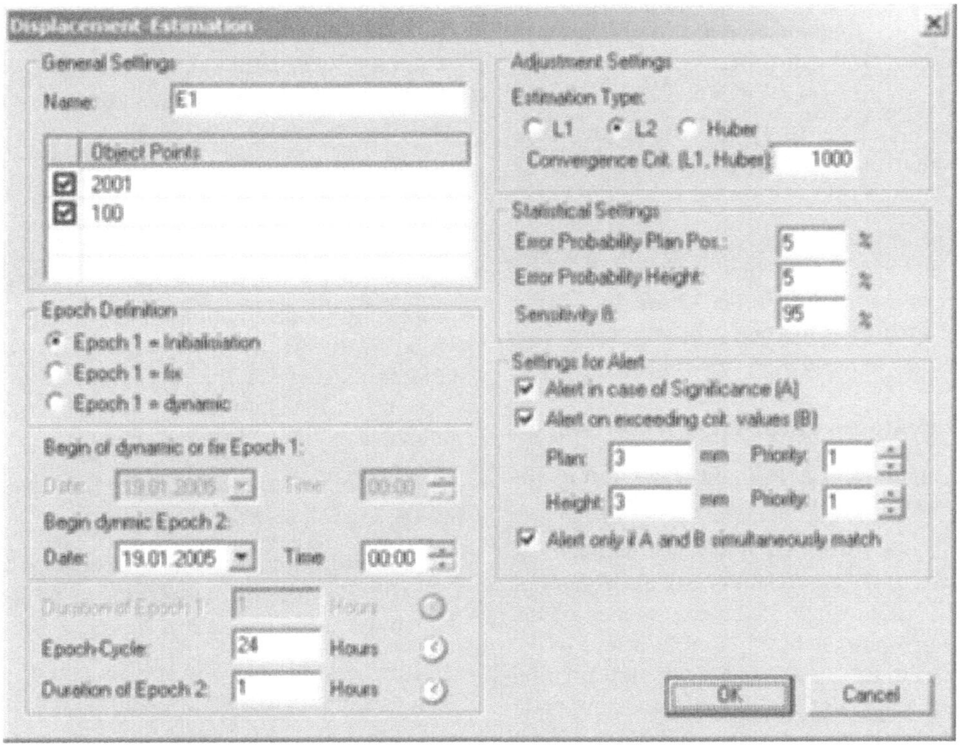

Fig. 4. GOCA online deformation (3rd step). Settings for the online displacement estimation and alerting.

A Kalman-Filtering a time t is equivalent to the common adjustment of the prediction of the state vector $\mathbf{y(t)}$ (9a,b,c) and the state vector related observations $\mathbf{l(y(t))}$ at time t. So the system of observation equations of a Kalman-Filtering generally reads:

$$\mathbf{y}(t) + \mathbf{v}_y = \hat{\mathbf{y}}(t) \text{ and } \mathbf{C}_y(t) \tag{10a}$$

$$\mathbf{l}(t) + \mathbf{v}_l = \mathbf{l}(\hat{\mathbf{y}}(t)) = \mathbf{A}\cdot d\hat{\mathbf{y}} + \mathbf{l}(\mathbf{y}(t)) \text{ and } \mathbf{C}_l\,. \tag{10b}$$

The covariance matrix $\mathbf{C}_y(t)$ (10a) of the prediction is received by applying the law of error propagation to the system transition equation (9a,b). In case of GOCA the observations $\mathbf{l}(t)$ for the Kalman-Filtering are taken from the object point time series (2nd adjustment step) and relate only to the object point displacements $\mathbf{u}_O(t_i)$ (7a). So the design matrix $\mathbf{A}$ (10b) of the GOCA observations' component is filled with zero elements at the position of the velocity and acceleration parameters within $\hat{\mathbf{y}}(t)$. The linearization in (10b) is done with respect to the prediction $\mathbf{y}(t)$ (9a,b). The generalized M-estimation of the parameters $\hat{\mathbf{y}}(t)^{(j)}$ of the Kalman-Filter model (10a,b) can be computed iteratively as:

$$\hat{\mathbf{y}}(t)^{(j)} = \mathbf{y}(t) + \mathbf{K}^{(j)}\cdot(\bar{\mathbf{l}}(t) - \bar{\mathbf{l}}(\mathbf{y}(t)))\,, \text{ with} \tag{11a}$$

$$\mathbf{K}^{(j)} = (\mathbf{C}_y^{-\frac{1}{2}}\mathbf{W}_y^{(j)}\mathbf{C}_y^{-\frac{1}{2}} + \bar{\mathbf{A}}\mathbf{W}_l^{(j)}\bar{\mathbf{A}})^{-1}\cdot\bar{\mathbf{A}}\mathbf{W}_l^{(j)}) \tag{11b}$$

(González 2005). Again the above homogenizition (5a) is used in (11a,b). The iterative Kalman matrix $\mathbf{K}^{(j)}$ has to be recomputed stepwise (j=j-th step). In the j-th step the diagonal weigth-matrices $\mathbf{W}_y^{(j)}$ and $\mathbf{W}_l^{(j)}$ within $\mathbf{K}^{(j)}$ are set up a weighting function related to the argument of the homogenized corrections $\bar{v}_{y,i}$ and $\bar{v}_{l,i}$ of the parameter and the observation component (10b,c). The specific weighting function $w(\bar{v}_i)$ of any M-estimator is strictly based its loss function $\rho(\bar{v}_i)$ (5b,c,d). It reads (Huber 1973; Bickel 1975; Holland 1977; Jäger et al. 2005):

$$w_i = w(\bar{v}_i) = \frac{\partial\rho(\bar{v}_i)/\partial\bar{v}_i}{\bar{v}_i} = \frac{\partial\psi(\bar{v}_i)}{\bar{v}_i}\,. \tag{11c}$$

The first derivative $\psi(\overline{v}_i)$ of the loss function is the so-called influence function. In case of an L2-norm (least squares) Kalman-Filter (5b) it holds that $w_i = 1.0$. So the general iterative Kalman filter M-estimation (11a,b) is (only) in case of the classical least squares (L2 norm) Kalman filter already finished in one step $j-1$.

Like the online displacement estimation (8a,b), with the settings shown in fig. 4, right, the above GOCA-Kalman-Filtering can also be performed by setting either a least squares estimation (5b) or a robust L1-norm estimation (5c) principle by the user.

4 Further Features of the GOCA-Software

4.1 Congruency Testing and Detection of Instable References Points

A special feature of the GOCA deformation analysis software consists in the automatic procedure of a statistically strict testing of the congruency of the plan and height component of a GNSS-array. In the context of setting up a classical deformation network, the stable point test procedure is applied to detect instable reference points (Kälber and Jäger 1999, 2001). In this way distortions of the object-points or better "pseudo deformations", which would in case of undetected instable reference points occur of the object space, are excluded of the deformation process modelling in the 3rd step. With respect to a maximum sensitivity the detection of instable reference points between different epochs is performed as a "1:(n-m) process". This means, that - in analogy to the classical observation-related iterative data-snooping - a significant 3D displacement of each reference point is tested in the mth step relative to (n-m) stable points, starting with m=1. The iterative stable point testing is again accompanied and robustified, respectively, by an iterative data snooping concerning the GPS/GNSS baseline observations (GKA files) and by a variance-component estimation.

4.2 Deformation Integrity Monitoring of Reference Station Networks

The GOCA software and the implemented stable point testing algorithm (chap. 4.1) can be applied for the statistically strict detection of possible movements of stations in GNSS positioning service such as ***SAPOS*** (www.sapos.de) and ***ascos***. (www.ascos.de). This feature of the GOCA system is called deformation integrity monitoring of reference station network, and is applied e.g. in the German ***SAPOS*** network.

5 Present Developments of the GOCA-Software

The present development of the GOCA software is dealing with the implementation of different so-called

- GNSS-baseline processing engines.

This kind of GNSS-processing packages will enable the GOCA software to work in a near-online mode with respect to the integration of a GNSS raw data processing (code and phase measurements), which is based on the world-wide standard of RINEX data. One application of the GNSS raw data processing capacity is the deformation integrity monitoring of reference station networks (chap. 4.2).

A third topic is concerning the further development of the GOCA software for a

- Monitoring in the higher frequency domain (e.g. 100 Hz for the monitoring a structural vibrations).

Further developments in the domain of deformation analysis theory are dealing with

- System analysis related deformation process modeling and a respective sensor and system parameter integrating modeling (Jäger and Bertges, 2004).

Appropriate approaches both for static and dynamic process modeling and system parameter estimation are provided by finite element models (FEM) of structures (Teskey 1988; Jäger and Kälber 1999; Welsch and Heunecke 1999; Jäger and Bertges 2004). Concerning the FEM theory of mechanical and geotechnical structures it is referred to Zienkiwicz (1988) and concerning the helpful analogies between FEM and geodetic network adjustment to Jäger (1988), Jäger and Kälber (1999) and Jäger and Bertges (2004). In that way, the displacement-, velocity- and acceleration-field, which are resulting from the GOCA software in the 2nd and 3rd step form GNSS and LPS data, as well as data from other local sensors (e.g. strain- or tilt-meters), can be used commonly as additional observation sources in an FEM approach (Kälber et al. 1999; Kälber und Jäger 2001; Jäger und Bertges 2004).

6 Examples GOCA-Software and Projects

The GOCA system is applied at different enterprises all over the world at more than 20 different GOCA installation locations and purposes. There is also an increasing trend for the use of the

GOCA system in different areas and disciplines of teaching and research, presently at five universities.

Reports of the GOCA system users and downloads can be found at the GOCA web-site www.goca.info (Kälber and Jäger 1999-2005). The fig. 3 shows the visualisation of the object point time series, resulting in the above mentioned 2nd adjustment step, which implies the georeferencing of the object points $\mathbf{x}_0(t_i)$ in the datum of the stable points $\mathbf{x}_R$ together with the parallel running moving average estimator (3rd adjustment step). The example fig. 3 shows the vertical and the horizontal displacements of a GOCA monitoring of an underground coal mining in a depth of 800 m.

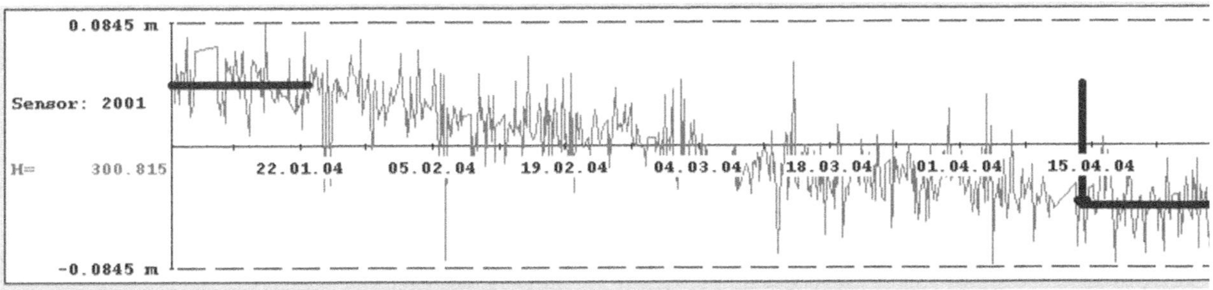

Fig.5 Visualization of the GOCA displacement estimation (3rd step). Thick horizontal line left shows the estimation of $\mathbf{x}_0$ (8a), (8b) by a number of observations $\mathbf{x}_0(t_i)$ belonging to the reference time t_0 and the estimation interval $\Delta t(t_0)$. The thick horizontal line right shows the estimated position $\mathbf{x}(t) = \mathbf{x}_0 + \mathbf{u}(t)$. The thick arrow shows the estimation vertical displacement $\mathbf{u}(t)$.

The fig. 5 shows the visualization of the result of a displacement estimation **u**(t) between a reference epoch t_0 (left) with a finite observation interval, and the estimation epoch t (right). The fig. 5 clearly reveals the benefits of an online monitoring, where - compared to a classical discrete monitoring – a large number of single positions $\mathbf{x}_0(t_i)$ contribute to the estimation of the state vector $\mathbf{y}(t) = (\mathbf{x}_0, \mathbf{u}(t))^T$ (8b), while at the same time a robust estimation (5c), (5d) is able to prevent the influence of gross errors in the observations **l**(t) given by the time series $\mathbf{x}_0(t_i)$ on the deformation parameters **y**(t).

Fig. 6 shows the location of the two object point GNSS sensors during the Kops dam (Austria) monitoring as an example of a GOCA-installation aiming at the monitoring and deformation process modeling of buildings and geotechnical structures, respectively.

The Fig. 7 above shows one of the 5 GPS sensors of the Gotthard tunnel monitoring GPS array in the high Alps of Switzerland. Starting in 2002, the GOCA-system will be installed there for at least 12 years, namely during the complete duration of the construction of the 57 km long Gotthard railway tunnel.

Fig. 6 GOCA installation at Kops dam, Austria.

Conclusions

The contribution treats the GNSS/LPS-based Online Control and Alarm system (GOCA). The GOCA-software components set up an array of GNSS- and/or LPS-sensors, which are equiped with a telemetric, GSM, LAN or internet communication. GOCA can be used for a continuous monitoring of natural hazards (landslides, volcanos, water level), and for the monitoring of geotechnical structures (mining, tunneling) and buildings (constructions, dams). The deformation network points may be occupied both in a permanently mode as well as intermittently. All online computation steps can be performed also in a post-processing or a near-online mode.

Fig. 7 GPS-receiver on pillar with solar panel energy supply as part of the GOCA array for the monitoring of the Gotthard railway tunnel construction (2002-2014).

The GOCA deformation-analysis software is responsible for a further processing of the GPS/LPS data in a three steps sequential adjustment procedure, which models in total an online classical three-dimensional deformation network. The network adjustment concept behind the GOCA-deformation-analysis software provides a unique georeferencing of the GNSS/LPS-occupied object point positions in the coordinate frame of the reference points. The online deformation analysis of the object-point can be set up as flexible user-defined displacement estimation or by a Kalman filter (displacements, velocities and accelerations).Both least squares and robust estimation techniques are applied, so that a reliable setting of an alarm is enabled, in case that a critical state is reached and proved by statistical testing.

The worldwide use of the GOCA system in different domains of environmental monitoring and research is documented and shown by examples. The evaluation of continuous time series of the objects displacement field, provided by GOCA, opens new perspectives in deformation analysis and model calibration. This concerns the transition from the classical geometric deformation analysis to so-called system analysis based approaches, as required by the interests of geodesists and of other disciplines such as geodynamics, geotechnics and civil engineering.

References

Bickel, P. (1975): One-step Huber estimates in the linear model. Journal of the American Statistical Association (70). 428-434.

Borchers, S. und R. Heer (2002): Bauwerksüberwachung mit GOKA (GPS-basiertes Online Kontroll- und Alarmsystem) an der Schleuse Uelzen I. Schriftenreihe des DVW, Band 44, Konrad Wittwer Verlag. ISBN 3-87919-281-2, ISSN 0940-4260, S. 65-84.

Bulowski, T. (2001): Kontinuierliche Überwachung von Tagebauböschungen mit dem System GOCA. 43. Wissenschaftliche Tagung der DMV (Deutsche Markscheider Vereinigung), Sept. 01, Trier. DMV Mitteilungen, 2001.

Eichhorn, A. (2005): Ein Beitrag zur Identifikation von dynamischen Strukturmodellen mit Methoden der adaptiven Kalman-Filterung. DGK-C Nr. 585. Deutsche Geodäische Kommisssion, München.

Feldmeth, I., Jäger, R. Und R. Zischinsky (2004): GPS-based Online Control and Alarmsystem GOCA – Leistungsstandards am Einsatzbeispiel Staumauer Kops (Vorarlberger Illwerke AG, Österreich). Wasserwirtschaft - Zeitschrift für Wasser und Umwelt, Heft 1-2, 2004. Vieweg Verlag, Stuttgart, ISSN 0043-0978.

Holland, P.W. (1977): Robust Regression using iteratively reweighted Least Squares. Commun. Statist.-Theor. Meth. A6(9), S. 813-827.

Huber, P (1973): Robust Regression: Asymptotics, Conjectures and Monte Carlo. Ann. Statist. (1). S. 799-821.

González, F. (2005): Improvement, Redesign, C++-based Implementation and Testing of the GOCA-Kalmanfiltering Modules. Masterthesis im Masterstudiengang Geomatics der Fachhochschule Karlsruhe.

Jäger, R. (1988): Analyse und Optimierung geodätischer Netze nach spektralen Kriterien und mechanische Analogien. Deutsche Geodätische Kommission, Reihe C, Nr. 342, München.

Jäger, R., Weber, A. und R. Haas (1997): Ein ISO 9000 Handbuch für Überwachungsmessungen, DVW-Schriftenreihe, Heft Nr. 27, Wichmann Verlag, Karlsruhe.

Jäger, R.; Müller, T.; Saler, H. and R. Schwäble (2005): Klassische und robuste Ausgleichungsverfahren - Ein Leitfaden für Ausbildung und Praxis von Geodäten und Geoinformatikern. ISBN 3-87907-370. Wichmann Ver-

lag
Jäger, R. und M. Bertges (2004): Integrierte Modellbildung zum permanenten Monitoring von Bauwerken und geotechnischen Anlagen. Beitrag zum 61. DVW-Fortbildungsseminar, 27./28. September 2004, Bauhaus-Universität Weimar. DVW-Schriftenreihe, Band 46/2004. ISBN 3-89639-451-7. S. 101-140.

Kälber, S. and R. Jäger (1999-2005): GOCA-Homepage. www.goca.info.

Kälber, S. and R. Jäger (1999): Realization of a GPS-based Online Control and Alarm System (GOCA) and Preview on Appropriate System Analysis Models for an Online Monitoring. Proceedings of the 9th FIG-Symposium on Deformation Measurement and Analysis. Sept. 1999, Olsztyn, Poland. p. 98 -117.

Kälber, S.; Jäger, R; Schwäble, R.: Heimberg, F. and K. Kast (2000): GPS-based Online Control and Alarm System (GOCA). Report presented to the 20. ICOLD-Congress, 2000, Peking. Int. Commission on Large Dams, Paris.

Kälber, S. und R. Jäger (2001): GPS-based Online Control and Alarm System (GOCA). Proceedings 10th FIG Internat. Symposium on Deformation Measurements. Orange, Californien, USA, 19.-22. March 01. CDROM.

Kahmen, H. und H. Suhre (1983): Ein lernfähiges tachymetrisches Vermessungsystem zur Überwachung kinematischer Vorgänge ohne Beobachter. ZfV 108 (1983). S. 345-351.

Korittke, N. und G. Palte (2001): Real-Time-Monitoring von bergbaubedingten Gebäudebewegungen mit GPS. Schriftenreihe des DVW, Band 43, Konrad Wittwer Verlag. ISBN 3-87919-279-0, ISSN 0940-4260, S. 183-195.

Lauterbach, M. und E. Krauter (2002): Satellitengestütztes Monitoring einer Großrutschung im Bereich eines Autobahndammes bei Landstuhl/Pfalz. Messtechnik – Geotechnik, Nr. 25, 2002.

Pelzer, H. (1974): Zur Analyse geodätischer Deformationsmessungen. DGK, Reihe C, Nr. 164, München.

Schäfer, W. (2004): GPS-gestützte 3D-Permanent-Überwachung bewegungsempfindlicher Bergschadensobjekte. Zeitschrift Markscheidewesen.

Teskey, W. (1988): Integrierte Analyse geodätischer und geotechnischer Daten sowie physikalischer Modelldaten zur Beschreibung des Deformationsverhaltens großer Erddämme unter statischer Belastung. Deutsche Geodätische Kommission, Reihe C, München

Welsch, W. and O. Heunecke (1999): Terminology and classification of deformation models - final report of FIG ad-hoc-Committee of WG 6.1. Proceeding of the 9th International FIG Symposium on Deformation Measurements, Olsztyn, 27-30 September, 1999.

Zienkiewicz, O.C. (1984): Methode der finiten Elemente. Carl Hanser Verlag, München.

Author Index

GPSR Compliance

The European Union's (EU) General Product Safety Regulation (GPSR) is a set of rules that requires consumer products to be safe and our obligations to ensure this.

If you have any concerns about our products, you can contact us on ProductSafety@springernature.com

In case Publisher is established outside the EU, the EU authorized representative is:

Springer Nature Customer Service Center GmbH
Europaplatz 3
69115 Heidelberg, Germany

Batch number: 10755939

Printed by Printforce, the Netherlands